Well-being, Sustainability and Social Development

Harry Lintsen • Frank Veraart • Jan-Pieter Smits
John Grin

Well-being, Sustainability and Social Development

The Netherlands 1850–2050

With Contributions from
Fred Lambert, Ben Gales, Rick Hölsgens, Lilianne Laan
and Frank Notten

With the Assistance of
Martijn Anthonissen, Önder Nomaler and Truus Lintsen

Harry Lintsen
Eindhoven University of Technology
Eindhoven, The Netherlands

Frank Veraart
Eindhoven University of Technology
Eindhoven, The Netherlands

Jan-Pieter Smits
Statistics Netherlands
The Hague, The Netherlands

John Grin
University of Amsterdam
Amsterdam, The Netherlands

Additional material to this book can be downloaded from http://extras.springer.com.

ISBN 978-3-030-09556-7   ISBN 978-3-319-76696-6   (eBook)
https://doi.org/10.1007/978-3-319-76696-6

Cover Photo In the 1970s, concerns about nuclear waste and the impossibility of managing
nuclear power not only stimulated a Broad Societal Debate in the Netherlands but also gave rise
to massive protests ranging from refusals to pay a surcharge on the electricity bill – the so-called
Kalkar levy – to demonstrations in Kalkar, the town where a joint German-Dutch-Belgian
experimental reactor was being built. The picture was taken on September 28th, 1974.

Mieremet, Rob, National Archives of the Netherlands /collection: Anefo, Public Domain,
927-4808 http://www.gahetna.nl

Printed on acid-free paper

This Springer imprint is published by the registered company Springer International Publishing AG part
of Springer Nature.
The registered company address is: Gewerbestrasse 11, 6330 Cham, Switzerland

Colophon

This work is part of the research program *Historical Roots of the Dutch Sustainability Challenge: The Impact of the Utilization of Material Resources on the Modernization of Dutch Society, 1850–2010* with project number 360-69-010, which is (partly) financed by the Netherlands Organization for Scientific Research (NWO). Additional financial support was granted by the Knowledge Network for System Innovations and Transitions (KSI).

Research was conducted at Eindhoven University of Technology, University of Groningen, University of Amsterdam and Statistics Netherlands (CBS).

This publication is made possible with financial support of Netherlands Organization for Scientific Research (NWO) and Simac Techniek NV.

 Netherlands Organisation
for Scientific Research

Acknowledgements

This book has long antecedents. In 2004, a large international research programme into sustainable development and transitions was launched. More than 80 researchers from different disciplinary backgrounds and located at twelve universities and research institutes had organised themselves in the 'Dutch Knowledge Network on System Innovation and Transitions' (KSI). The aim was to acquire insights into transitions based on research into long-term developments, system innovations and current policy. The programme, financed by the Dutch government, generated policy recommendations, courses, scientific publications and six books in the series 'Routledge Studies in Sustainability Transitions'.

In 2009, KSI approached the Technical University of Eindhoven (TU/e) and the Foundation for the History of Technology (SHT) to undertake preliminary research into sustainable development in the Netherlands from the nineteenth century on. Harry Lintsen, Fred Lambert, Frank Veraart, Giel van Hooff and Hans Schippers carried out the research in cooperation with John Grin of the University of Amsterdam (UvA). In this phase, the researchers also co-operated with the Rathenau Institute in a study into the sustainability promises of a bio based economy. The preliminary research programme led to a research proposal: 'Historical Roots of the Dutch Sustainability Challenge: The Impact of the Utilization of Material Resources on the Modernization of Dutch Society, 1850–2010'. The Dutch Organisation for Scientific Research (NWO) funded the proposal with the TU/e as main recipient, and Statistics Netherlands (CBS) and the University of Groningen (RUG) as co-recipients. The research team was composed of Harry Lintsen (TU/e, project leader), Ben Gales (RUG), John Grin (UvA), Rick Hölsgens (RUG), Lilianne Laan (TU/e), Fred Lambert (TU/e), Frank Notten (Statistics Netherlands), Jan Pieter Smits (Statistics Netherlands) and Frank Veraart (TU/e). Deliverables consisted of a dissertation, scientific publications, conference papers and research reports. In addition, provisions were made to produce a synthesis of the different research results.

Harry Lintsen, Frank Veraart, Jan Pieter Smits and John Grin wrote the two publications that synthesized the research programme. The first, entitled *Well-being, Sustainability and Social Development: The Netherlands 1850–2050*, was an open access publication published by Springer in 2018. The second, entitled *De kwetsbare welvaart van Nederland 1850-2050. Naar een circulaire economie*, was published as a book by Prometheus, also in 2018. Contributions to various chapters were made by Fred Lambert (Chaps. 2, 4, 5, 9, 14, 19 and 24), Rick Hölsgens (Chaps. 2, 6, 10, 17 and 22), Ben Gales (Chaps. 2, 3, 6, 10, 17 and 22), Lilianne Laan (Chap. 3) and Frank Notten. Research support was additionally provided by Martijn Anthonissen (TU/e, Chaps. 3 and 9), Truus Lintsen (Chap. 5) and Önder Nomaler (TU/e, Chaps. 3 and 5). André Burger (Cement and Concrete Centre), Petra van Dam (Free University of Amsterdam), Rutger Hoekstra (CBS), Hans Renes (University of Utrecht) and Ab Stevels (Technical University Delft) were all solicited for advice on various aspects of the programme. Giel van Hooff researched the illustrations and wrote the accompanying captions. Brigitte Möller, Nena van As and Marissa Damink helped to prepare the texts for the two publishers.

The manuscript, *Well-being and Sustainability in the Netherlands 1850–2010*, that served as the basis for both the above publications, was discussed at length in the course of an international workshop held on March 3 and 4, 2017, in Eindhoven. Participants were Vanesa Castan Broto (University College London), Nil Disco (University of Twente), Rutger Hoekstra (Statistics Netherlands), Elena Kochetkova (National Research University, Higher School of Economics of St. Petersburg), Fridolin Krausman (Alpen Adia University), Magnus Lindmark (Umea University), Peter Scholiers (Free University Brussels), Johan Schot (SPRU, University of Sussex), Geert Verbong (TU/e) and Jan Luiten van Zanden (University of Utrecht). The members of the research team also participated in the workshop. The English-language manuscript discussed at the workshop as well as the English-language publication with Springer was based on a prior Dutch text. Nil Disco was responsible for the translation. The authors revised the publications for Springer and Prometheus in line with the workshop commentary.

The chair in history of technology at the TU/e (in particular Mila Davids and Ruth Oldenziel) and the Foundation for the History of Technology, both under the leadership of Erik van der Vleuten, were sources of inspiration for the research programme. Simac Techniek NV and the TU/e were crucial in preparing the publications with Springer and Prometheus.

We are extremely grateful to all participants for the contributions they made to the research programme.

Eindhoven, The Netherlands   Harry Lintsen
Eindhoven, The Netherlands   Frank Veraart
The Hague, The Netherlands   Jan-Pieter Smits
Amsterdam, The Netherlands   John Grin
January 2018

Contents

Part III: The Great Turnabout 1970–2010

**17   The Point of Departure Around 1970: Overabundance
and Discontent**.. 375
Frank Veraart

**18   Agriculture and Foods: Overproduction
and Overconsumption**...................................... 397
Frank Veraart

**19   Building Materials and Construction: Sustainability,
Dependency and Foreign Suppliers**........................... 417
Frank Veraart

Abbreviations

ABP	*Algemeen Burgerlijk Pensioenfonds* – General Civil Pension Fund
AKU	*Algemene Kunstzijde Unie* – General Rayon Union (now AKZO-Nobel)
ANWB	*Algemene Nederlandsche Wielrijders-Bond* – General Netherlands Bicycle Association (now Dutch touring association)
AOW	*Algemene Ouderdomswet* – General Old Age Law
BMI	Body Mass Index
BBC	British Broadcasting Cooperation
BSE	Bovine Spongiform Encephalopathy (mad cow disease)
CAO	*Collectieve Arbeidsvoorwaarden* – Collective Labour Agreements
CB	*Centraal Bureau van het Nederlands Landbouw Comité* – Central Bureau of the Netherlands Agricultural Committee
CBS	*Centraal Bureau voor de Statistiek* – Statistics Netherlands
CHV	*Brabantse Coöperatieve Handelsvereniging* – Co-operative Trade Association Brabant
CNA	*Compagnie Neérlandaise d´Azote* – Dutch Nitrogen Company
CO_2	carbon dioxide
CCC	*Centrale Cultuurtechnische commissie* – Central Commission for Cultivation Technology
CDA	*Christen-Democratisch Appèl* – Christian Democratic Party
CHU	*Christelijk-Historische Unie* – Christian Historical Union
CSV	*Centraal Stikstof Verkoopbureau* – Central Nitrogen Sales Office
CVO	*Centraal Insituut voor Voedingsonderzoek* – Central Institute for Nutritional Research
D' 66	*Democraten '66* – Democrats '66 (Liberal Party)
DALY	Disability Adjusted Life Years
DDT	dichloro-diphenyl-trichloroethane
DSM	Dutch State Mines
ECN	*Energieonderzoek Centrum Nederland* – Energy research Centre Netherland
ECSC	European Coal and Steel Community
EEC	European Economic Community

EHS	*Ecologische Hoofdstructuur* – Ecological Main Structure
ENCI	*Eerste Nederlandse Cement Industrie* – First Dutch Cement Industries
EPZ	*Elektriciteits Produktiemaatschappij Zuid-Nederland* – Electricity Production Company South Netherlands
EU	European Union
FODI	*Federatie de Oppervlaktedelfstoffenwinnende Industrieën* – Federation of Shallow Subsoil Resource Extraction Industries
FWS	*Nederlandse Vereniging Frisdraken Water Sappen* – Netherlands' Association of Soft drinks, Water and Juices
GATT	General Agreement on Tariffs and Trade
GDP	Gross domestic product
GJ	Gigajoule (1000 MJ)
GFT	*Groente, fruit en tuinafval* – vegetable, fruit and garden waste
IPCC	International Panel on Climate Change
IWC	International Whaling Commission
KIvI	*Koninklijk Instituut van Ingenieurs* – Royal Netherlands Society of Engineers
kcal	kilo calories (= 4,184 kJ)
KIDV	*Kennisinstituut Duurzame Verpakkingen* – Knowledge Institute for Sustainable Packaging
kJ	Kilojoule
KLM	*Koninklijke Luchtvaart Maatschappij* – Royal Dutch Airlines
KNMI	*Koninklijk Nederlands Meteorologisch Instituut* – Royal Dutch Meteorological Institute
KNWV	*Koninklijke Nederlandse Watersport Vereninging* – Royal Dutch Yachting Association
KVP	*Katholieke Volkspartij* – Catholic People's Party
LCCO	*Landelijke Commissie voor de Coördinatie van het Ontgrondingsbeleid* – National Commission for the Coordination of Excavation Policy
MAO	*Mest Afzet Overeenkomst* – Manure Disposal Agreement
MEKOG	*Maatschappij tot Exploitatie van Kookovengassen* – Company for the Exploitation of Cokes Oven Gas
MINAS	*Mineralen Aangiftesysteem* – mineral denotation system
MJ	Megajoule (1000 kJ)
MSA	Mean Species Abundance
NCB	*Noord-Brabantse Christelijke boerenbond* – North-Brabant Christian Farmers' Union
NAM	*Nederlandse Aardolie Maatschappij* – Netherlands Oil Company
NCL	*Nederlands Landbouw Comité* – Netherlands Agricultural Committee
NHM	Nederlandse Handel-Maatschappij
NH_3	ammonia
NIPO	Netherlands Institute for Public Opinion
NIVE	*Nederlands Instituut voor Efficiency* – Netherlands Institute for Efficiency
NMP	*Nationaal Milieubeleidsplan* – Environmental Policy Plan

NO$_x$	Nitrogen oxides
NS	Nederlandse Spoorwegen
NRK	*Federatie Nederlandse Rubber- en Kunststofindustrie* – Federation Dutch Rubber and Plastics Industries
NVVH	*Nederlandsche Vereniging van Huisvrouwen* – Dutch Association of Housewives
OECD	Organization for Economic Cooperation and Development
OPEC	Organization of the Petroleum Exporting Countries
PGM	platina group metals
PJ	Petajoule (1000 TJ)
PV	Photo Voltaic
PVC	Polyvinyl chloride
PvdA	*Partij van de Arbeid* – Labor Party
PVV	*Partij voor de Vrijheid* – Party for Freedom (nationalist)
R&D	Research and Development
RIV	*Rijksinstituut voor Volksgezondheid* – National Institute for Public Health
RIVM	*Rijksinstituut voor Volksgezondheid en Milieu* – National Institute for Public Health and the Environment
RIZA	*Rijksinstituut voor de Zuivering van Afvalwater* – Government Institute for the Purification of Waste-water
SBB	*Stikstofbindingsbedijf* – Nitrogen Binding Company
SDAP	Sociaal Democratische Arbeiderspartij – Social Democratic Labour Party (Labour Party)
SDG	Sustainable Development Goals
SEP	*Samenwerkende Elektriciteits-Productiebedrijven* – Cooperating Electricity Production Companies
SER	*Sociaal-Economische Raad* – Social and Economic Council of the Netherlands
SNM	*Stichting Natuur & Milieu* – Foundation for Nature and Environment protection
SO$_2$	Sulfur dioxide
TNO	*Nederlandse Organisatie voor Toegepast-Natuurwetenschappelijk Onderzoek* – Netherlands Organization for Applied Scientific Research
TJ	Terajoule (1 million MJ)
UAC	United Africa Company
UN	United Nations
VINEX	*Vierde Nota Ruimtelijke Ordeningen Extra* – Fourth Spatial Memorandum Extra
VVD	*Volkspartij voor Vrijheid en Democratie* – People's Party of Freedom and Democracy (Liberal Party)
VROM	*Ministerie van Volksgezondheid Ruimtelijkeordening en Milieu* – Ministry of Public Housing, Spatial Planning and Environmental Policy
WCED	World Commission on Environment and Development (also known as Brundtland Commission)

List of Figures

List of Graphs

List of Tables

Prologue: Well-being and Sustainability in a Long-Term Perspective

Chapter 1
Well-being and Sustainability: Measurement System and Institutional Framework

Jan-Pieter Smits

Contents

Abstract This study aims to analyse the development of human well-being and sustainable development from a long-term perspective. Though often used, these concepts are generally ill-defined. This chapter proposes to clear the decks by proposing unambiguous definitions of well-being and sustainable development. This emphasis on definitions and measurements echoes international efforts to develop standardised measurement protocols that may help society (and policy makers in particular) to assess the extent to which a society is moving in a more sustainable direction. The framework adopted in this chapter (the so-called *CES Recommendations on Measuring Sustainable Development*) charts the well-being of a country at a specific point in time, and also reveals the extent to which the precise way such well-being is generated hampers the well-being of later generations or people elsewhere on the planet. Natural capital is a particular focus, inasmuch as this is a vital resource on which, in the end, all life depends. In this analysis of two centuries of well-being and sustainable development, the institutional evolution of society over time also plays a prominent role.

Keywords Well-being · Sustainable development · Economic growth · Natural capital

This chapter is written by Jan-Pieter Smits with contributions by Harry Lintsen.

© The Author(s) 2018
H. Lintsen et al., *Well-being, Sustainability and Social Development*,
https://doi.org/10.1007/978-3-319-76696-6_1

1.1 Toward a Better Understanding of Well-being

> The gross domestic product is an almighty figure in public and political debate: how well is
> it going with the economy, for that matter with you as a person?... [But] the gdp does not
> measure how the country is doing, not even how the economy is doing. It only measures the
> magnitude of market production. In politics there is a need for figures that say more, namely
> figures that say something about well-being. This is more than only gdp (gross domestic
> product).[1]

These critical words were spoken by Agnes Mulder, parliamentarian for the Christian
Democratic Alliance (CDA) in a parliamentary debate on well-being, in June 2016.
Parliament was debating recommendations by the Temporary Parliamentary
Commission on Well-Being, a commission charged with investigating how the gross
domestic product can eventually be replaced by a system of measurement that does
justice to well-being in the broadest sense of the term.[2] Verhoeven of the Democrats
66 party explained the shortcomings of the gdp in yet another way:

> If a contractor builds a house this month, the Dutch economy grows. This everyone under-
> stands. If that house is demolished next month, the Dutch economy grows even faster. If the
> house is subsequently rebuilt, there is even more growth. This is the strange effect you get
> if the gross domestic product is a synonym for growth. The gdp also increases if more gas
> is pumped up in Groningen. Is that good for our well-being? No, think the people in
> Groningen, and rightly so...

Marianne Thieme, leader of the Party for Animals, was explicitly negative: gdp
includes economic activities ...

> ... that in and of themselves are anything but a positive contribution to well-being. Think of
> traffic accidents, air pollution, animal misery, animal disease epidemics and the incineration
> of whole forests in power plants. ... Why do we trap ourselves in economic indicators that
> say nothing of the power of our sense of community, care for one another, care for the vul-
> nerable, or the opportunities to develop your talents? Why do we want to measure and
> express everything in money, except that which makes life really worth living?

However, these critical remarks did not mean that the Second Chamber was about
to reject the gdp as indicator. Parliamentarian Veen of the People's Party for Freedom
and Democracy (VVD) emphasized that the gdp was still relevant:

> I share the conclusion that the gross domestic product is inadequate as an indicator of well-
> being, broadly conceived. It only says something quantitatively about the magnitude of our
> economy. But I also share the conclusion that this is the best available alternative for deter-
> mining well-being and international comparisons ... tax incomes and employment levels ...

---

[1] All quotes from this debate are from the official edited minutes. See https://www.tweedekamer.nl/
kamerstukken/plenaire_verslagen/detail?vj=2015-2016&nr=99&version=2#idd1e668761

[2] For the report of this temporary parliamentary commission, see: https://www.tweedekamer.nl/
sites/default/files/atoms/files/34298-3.pdf

Klever, of the Party for Freedom (PVV) struck a different note. She failed to see the need to develop new measures of well-being. Citizens know full well what the concept well-being means for them, namely if at the end of the month they have enough money left in their pockets:

> Of course health and safety are of importance, but if you want to achieve these you first have to ask yourself how we are going to pay for them. In order to pay for this you need a gross domestic product and you have to look and see what part of the gross domestic product you want to spend on our safety, our care system, our education and so forth. That is ultimately a political decision. For the average nurse or police officer who works 40 hours a week and who sees ever more of their tax money lost, the concept of 'well-being' doesn't change in the least, because someone like that just looks at what is left over in their pockets month by month as spending money.

In the end Dutch Parliament agreed to put well-being and sustainable development high on the political agenda and to devote an annual debate to the topic. This is not an isolated development. Globally there is a growing conviction that society is on the wrong road. This conviction is not limited to activist environmental groups, but is also starting to enter into the discourse of mainstream politics. In 2008 the former French president Nicolas Sarkozy invited two Nobel Prize winners, Joseph Stiglitz and Amartya Sen, to prepare a report aimed at investigating how new benchmarks for well-being and sustainability could be developed.[3]

This Stiglitz Report marked the start of a search for a new 'compass' that would enable politicians and policy makers to determine their course. Up to then the concept of economic growth had been dominant – as measured with the aid of the key indicator gdp. Now for the first time this standard vision was being critically assessed. To what exactly does the concept of well-being refer? And what determines this well-being and its sustainability?

These questions have also moved to the forefront within international organizations like the OESO, in the *Progress of Societies*[4] projects and the *Better Life Initiative.*[5] Both projects aim at finding better indicators for measuring well-being and sustainability. In Europe the European Commission has taken the lead with its *GDP and Beyond*[6] initiative, in which various societal organizations, policy makers and academics are asked to reflect on better ways to measure and shape societal progress.

---

[3] J. E. Stiglitz, A. Sen and J.P. Fitoussi, *Report by the commission on the measurement of economic performance and social progress*: http://www.insee.fr/fr/publications-et-services/dossiers_web/stiglitz/doc-commission/RAPPORT_anglais.pdf

[4] http://www.oecd.org/statistics/measuring-well-being-and-progress.htm

[5] http://www.oecdbetterlifeindex.org/

[6] http://ec.europa.eu/environment/beyond_gdp/index_en.html

The crisis of 2008 added fuel to these fires. According to the organizations involved, this crisis demonstrated that the structure of our economic system only creates ever more problems.[7] The sustainability of our well-being over the long term is ever more uncertain. On the economic front there is the chronic instability of the financial sector with its associated 'bubbles' that have produced serious economic disruptions. In addition to this financial and economic crisis there is also talk of an ecological crisis. The realisation is dawning that problems like climate change and global decline in biodiversity are endangering life on earth. Moreover, tensions among different populations, both within and among nations, are also increasing to the extent that one can also speak of a social crisis.

The most prominent example of the striving for a more sustainable world is doubtless the *Post-2015 Agenda* of the United Nations.[8] This ambitious policy initiative, kicked off by world leaders at a UN top level conference in September 2015, contains a list of policy goals aimed at making the world more sustainable by 2030. These *Global Goals for Sustainable Development* will have a major effect on political and policy processes, both on global and national levels. The challenge will be to build a world in which all people have an acceptable quality of life, in which well-being is more equally distributed, and where the interests of future generations are also taken into account.

The debate in the Dutch Second Chamber on gdp reflects this development. Cabinet and parliament have a clear need for robust knowledge on the basis of which decisions regarding well-being and sustainability can be made. At present cost-benefit analyses to support policy decisions are based on a rather limited economic perspective. In addition, social and political debates are generally dominated by short-term visions. Long-term consequences of the behavior of producers and consumers or of government policy are not at present an integral feature of political discourse. The Temporary Second Chamber Commission on Well-Being has suggested a framework for the annual debate on questions of well-being and sustainability. The commission recommends monitoring well-being and sustainability on the basis of the measuring system that also forms the basis of this study.

1.2 Why Study Well-being and Sustainability in a Historical Perspective?

Where political debate in general has eyes only for the most recent developments, this study will examine long-term societal developments throughout a period of more than a century and a half in terms of changes in well-being and sustainability. In contrast with most traditional economic-historical studies we will not focus on economic growth, but rather on the growth of well-being. To what extent does the

---

[7]This is beautifully expressed in the OECD's NAEC project (New Approaches to Economic Challenges): http://www.oecd.org/naec/

[8]https://sustainabledevelopment.un.org/?menu=1300

growth of well-being depend on depleting vital resources, not only in the Netherlands but also elsewhere in the world?

This is obviously crucial for assessing the extent to which a certain level of well-being can be sustained over longer periods of time. Has well-being been generated on the basis of sustainable development, i.e. are vital resources -whether domestically or abroad- not being depleted over time? Significantly, well-being and sustainable development can have identical meanings, but only on condition that the current generation in a given country also concerns itself with the aspirations to well-being of people in other parts of the world and of future generations. Only when these aspects are incorporated into present-day preferences can well-being and sustainability be seen as identical concepts.

An historical analysis can provide deeper insights if only because sustainability is in essence an intergenerational issue. Choices we make about well-being in the past influence the well-being of later generations due to the possible depletion of vital resources like nature and the environment. Thus, present-day sustainability problems such as climate change and loss of biodiversity have their roots in choices made by prior generations. An historical analysis may reveal which aspects of producer and consumer behavior have wreaked such havoc and what kinds of technologies and institutions have facilitated the present sustainability problem. In this connection it is especially important to describe how societal developments can sometimes lead to outcomes that are both undesired and intolerable. This requires a sensitivity to emergent institutional rigidities that make it difficult to bend developments in a more positive direction.

In addition, the theme of sustainability is not just a present-day phenomenon. Sustainability, as defined by Brundtland et al. in the report *Our Common Future*, is an issue that also cropped up in the past, accompanying attempts to achieve progress through modernisation.[9] Then too, efforts were made to solve the sustainability problems of the moment. This study questions the paths along which and the methods by which, between 1850 and 2010, historical actors succeeded in solving their problems; and the ways in which, more often than not, the solutions chosen also generated new problems. We maintain a focus on how the processes of economic growth and sustainable development were entangled. Which technologies and institutional frameworks made it possible for economic development to be associated – or not – with sustainable development? This volume will provide a deeper insight into the complex relationship between economic growth and sustainability.

1.3   The Dutch Case

There are some long-run features of Dutch society that make it an interesting case for a historical analysis of changes in well-being and its sustainability. Compared to other countries, the agricultural productivity, population density and energy

[9] J.A. Du Pisani, 'Sustainable development - historical roots of the concept', *Environmental Sciences* 3(2006) nr. 2, 83–96.

intensity of the economy are high. Besides, even at an early stage the Dutch economy was quite open and imported large quantities of natural resources. The resulting strong impact on the natural environment has deep historical roots, going back to the Golden Age of the Dutch Republic (the period 1580–1640) or even earlier. The economic success of the Dutch Republic is well documented, especially by Anglo-Saxon scholars.[10] So far, the sustainability aspects of economic development have not been systematically analysed from a long-run perspective (even though the phenomenon was signalled by Anglo-Saxon contemporaries such as William Petty and Matthew Hale).

The roots of the modernisation of the Dutch economy can be traced as far back as the late Middle Ages. In this period the rising levels of the North Sea –the so-called late medieval transgression- had a far-reaching impact on arable production in the coastal provinces. As a result of the rising levels of seawater and of soil subsidence, agricultural land became less suitable for arable production.[11] Hence, farmers reoriented their output toward livestock farming. As livestock farming is much more capital-intensive than arable production, a substantial labour surplus was generated in the coastal provinces. This surplus could be relatively easily absorbed by the urban sector, especially because from the last quarter of the sixteenth century onwards, Amsterdam had become one of the main staple markets in world trade.[12] Not only trade, transport and international financial services flourished in the economy of Holland, but industrial activities also witnessed strong growth, above all the 'trafieken', industries which processed imported raw materials. Most of these 'refined' products were subsequently exported.

This quite specific form of economic specialisation had far-reaching environmental consequences. Even at an early stage the Netherlands was a strongly urbanised area characterised by a high population density. In *European urbanization 1500–1800*, Jan de Vries presents figures which show that in Europe around 1700 9% of the population lived in cities, as against 34% in the Netherlands. This was much higher even than Great Britain and Wales with their 13% urban population.[13] Population pressure continued to increase throughout modern times.. During the twentieth century population growth in the Netherlands was quite high in comparison with other Northwest European countries.[14] The high and increasing population pressure had a substantial effect on the natural environment, in terms of the pollution of soil and water and the subsequent loss of biodiversity.

The high urbanisation was made possible by a highly productive agricultural sector. Jan de Vries has analysed the roots of the high productivity of the rural Dutch

---

[10] J. de Vries, 'On the modernity of the Dutch Republic', *Journal of Economic* History, 32 (1973) 191–202; J. Israël, *The Dutch Republic: its rise, greatness and fall, 1477–1806* (Oxford 1995).

[11] J.L. van Zanden, 'Op zoek naar de 'missing link' ', *Tijdschrift voor Sociale Geschiedenis* 14 (1988), pp. 359–386.

[12] T.P. van der Kooij, *Hollands stapelmarkt en haar verval* (Amsterdam 1931).

[13] J. de Vries, *European urbanization 1500–1800* (London 1984).

[14] J.L. van Zanden, *Een klein land in de 20e eeuw. Economische geschiedenis van Nederland 1914– 1995* (Utrecht 1997), pp.29–31.

economy in the Golden Age in great detail.[15] In the nineteenth and twentieth centuries Dutch agriculture continued to operate at the technological frontier. However, productivity growth in agriculture resulted in quite severe environmental problems, especially due to the quantities of manure that were produced and used.

During the twentieth century the industrial sector also began to burden the natural environment. In the second half of the twentieth century the energy intensity in the Netherlands increased faster than anywhere else in the Western world.[16] Nowadays energy consumption per capita is among the highest in the European Union, whereas the share of sustainable energy is quite low by international standards.[17]

Lastly, compared to the rest of the European Union the Netherlands imports large quantities of natural capital from the least developed countries. Quantitative analysis shows that former colonial powers like the Netherlands still stake huge claims on the natural capital of the poorest countries. Dutch trade data indeed reveal a strong path dependency in terms of trade specialisation.[18]

The specific dynamics of economic development and structural change makes the Netherlands an interesting case with which to study human well-being and sustainable development from a long-run perspective.

1.4 New Perspectives on Growth and Measurement

This study evaluates more than 150 years of societal development from the perspective of human well-being and its sustainability. It tries to identify some of the main determinants of well-being and sustainability and explores how these findings may help in shaping a path toward a low-carbon society which may foster well-being in the broadest sense of the word. It also describes which institutional arrangements governed societal changes in the past and asks what light this sheds on the institutional prerequisites for the present transformation toward a green, sustainable economy.

The historical analysis presented in this book has yielded two important findings. First, a well-being paradox is identified. Nowadays, compared to other countries, the Netherlands ranks high in terms of the life satisfaction of its citizens. Nevertheless, a substantial segment of society does not share this positive outlook on life. This phenomenon is not unique for the Netherlands, as indicated by the substantial rise of populism throughout the western world. This study shows that also in earlier periods, a high ranking on various 'lists' went hand in hand with feelings of strong discontent in society. This study may help to better understand this paradox of well-being.

---

[15] J. de Vries, *The Dutch rural economy in the Golden Age, 1500–1700* (London 1974).

[16] A. Maddison, *Dynamic forces in capitalist development. A long-run comparative view* (New York 1991).

[17] CBS, *Monitor Duurzaam Nederland 2014: indicatorenrapport* (Den Haag 2014).

[18] H. Langenberg and J.P. Smits, 'Invoer van grondstoffen uit LDCs: Geworteld in koloniale tijden?', *Internationaliseringsmonitor* 2015 (4), pp. 42–57.

A second finding of this study concerns the significant impact of the Dutch economy on scarce natural resources. Until the 1960s the Netherlands were on a more or less sustainable growth trajectory. Of course, economic development laid a claim on natural resources, but within the limits of sustainability norms as defined by the present generation. In the course of the 1960s, on the one hand the coupling between GDP and improvements in quality of life became much looser (it even inverted on some indicators), while, on the other hand growth became decisively non-sustainable. From this moment on energy consumption and the resulting emissions of CO_2 skyrocketed. The depletion of resources was not restricted to the Netherlands. Natural capital in the least developed countries was also depleted due to the relatively high level of Dutch imports.

This trade pattern has deep historical roots. In order to arrive at a more sustainable way of life, it is important to move toward a more circular economic system in which vital product chains are closed. However, given the open nature of the Dutch economy and in the light of the outsourcing of economic activities by firms and the increasing complexity of global value chains, the closing of many of these chains has become quite difficult. Due to the continuous efforts of countries and firms to specialise on the basis of their revealed comparative advantages, today's economy has become strongly globalised. However, the downside of these economic dynamics is a world with long and complex product chains which put a considerable burden on natural resources.

These two main problems – the increasing distance between state and society as well as the sustainability challenge – cannot be dealt with in isolation from each other. After all, in order to make a fundamental transition to a more sustainable, low-carbon, society, drastic measures need to be taken. In the best case, this will be supported by the active involvement and passive consent of citizens, who recognise it as an opportunity to co-shape well-being. In the worst case, this transition will be experienced by particular groups in society as just another episode in the never-ending story of the neglect of their needs and wants, reinforcing the first problem and undermining the legitimacy of strategies to address the second one.

Such great transformations also occurred at earlier periods in history. In fact, this book deals with the great transformation that took place in the period 1850–1960. The greatest challenge facing Dutch society were the extremely low levels of human well-being. Around the mid-nineteenth century a substantial part of the population was barely able to survive. These problems were alleviated partly thanks to the intervention of physicians and other professionals who got the poverty problem and especially the poor state of local hygiene on to the political agenda. The poor quality of life of the masses was also improved by decreasing income inequality, especially as a result of the tax reforms of the 1860s. Lastly, economic growth ultimately produced higher incomes and an increase in material well-being.

However, this economic growth was in itself the cause of new problems. New technologies proved to be more energy and material intensive. From the 1960s onwards this resulted in increasing claims on natural resources. The 'great transformation' which occurred in the nineteenth and early twentieth centuries was analysed in depth by Karl Polanyi.[19] He argues that some of the negative effects of economic development – such as the poverty problem – were overcome because the state became more active from the nineteenth century onwards, compensating for some of the social problems created by the functioning of liberal markets.

Following Polanyi, some authors now suggest that a new 'great transformation' is needed, one in which mechanisms will be developed to fight the present sustainability crisis. It is suggested that two types of (re) embedding of economic development are needed.[20] First of all, an embedding in society is necessary. Increasingly, groups of citizens feel that their voices are not heard by politicians. New types of social structures – or the re-discovery or restoration of old organisational forms – are needed, which would enable the state and different segments of society to be in closer contact with each other. This may also yield more, and more widespread, legitimacy for policies promoting the transition to a more sustainable society. But this will require a second re-embedding of the economy, nowin nature. More than ever, ecological limits need to be observed and respected. Economic policy as well as producer and consumer behaviour should be adapted to these limits.

This of course leads to the question whether the coming great transformation, possibly characterised by these two re-embeddings, can be realised within the paradigm in which modernisation is dominated and evaluated by gdp growth. There is a rich body of literature which claims that environmental burdens can be eased within the present-day growth paradigm.[21] Through the introduction of green technologies we could have the best of both worlds: economic growth and the preservation of nature and the environment. Others claim that this is not an option. According to them the only answer to the increasing environmental problems is to abandon the desire for economic growth. In the post-growth paradigm, sustaining the present-day well-being of western society is judged sufficient.[22] Not only would more growth result in unacceptable environmental damage, from a well-being perspective there is also no longer a need for further growth.

Recently, the environmental economist Jeroen van den Bergh has suggested an alternative position.[23] In his proposal he makes a careful distinction between the issue ofwhether economic growth is still possible in the light of ecological constraints, and the critique of gdp as the leading indicator in policy making. Van den Bergh is quite clear about the value of gdp as a leading indicator. It is a wrong com-

[19] K. Polanyi, *The great transformation: The political and economic origins of our time* (Boston 1944).

[20] R. Kemp et al., 'The humanization of the economy through social innovation', Paper for SPRU 50th anniversary conference (version august 2016).

[21] Nowadays especially the green growth paradigm maintains that economic growth and an acceptable pressure on the natural environment can go hand in hand, see for example: http://www.oecd.org/greengrowth/

[22] See for example: F. Schneider, G. Kallis and J. Martinez-Alier, Crisis or opportunity? Economic degrowth for social equity and ecological sustainability. *Journal of Cleaner Production* (2010) 18(6): 511–518.

[23] J. van den Bergh, 'Green agrowth as a third option: Removing the GDP-growth constraint on human progress', *EU Policy paper (WelfareWealthWork)*, no 19 (2015).

pass and should therefore be abandoned as soon as possible. He is especially critical of what he calls 'gdp growth fetishism'. In that sense his position reflects the gdp critique as voiced by a majority in the parliamentary debate quoted at the beginning of this chapter.

Van den Bergh coins his perspective as 'a-growth.' Rejecting gdp as a guiding principle, he is 'agnostic' when it comes to the possibility of its growth. He claims that society should first define the environmental damage it deems acceptable in relation to sustaining an acceptable level of well-being. Perhaps, he says, gdp growth is possible, perhaps not. He makes no a priori statements about the possibility of growth in a context that respects environmental limits and that focuses on well-being in the broadest sense. Systematic comparison between gdp series and indicators of well-being and sustainability will answer that question.

The next section will briefly indicate the methodological framework used to analyse these developments. Those not interested in a more detailed description of that framework can make do with reading this paragraph and then move on to Chap. 2.

1.5   Research Methodology in Brief

The following points of departure have guided the selection of themes, the use of concepts, the empirical descriptions and the analysis:

The Measurement Framework for Well-being and Sustainability Researchers at Statistics Netherlands (CBS) have developed a measurement framework that enables them to map well-being, broadly conceived, and its sustainability. This framework measures economic, ecological and societal aspects of existence. The core question is whether the way in which well-being is achieved in the 'here and now' is not achieved at the cost of the depletion of vital resources, whether in the Netherlands or in the rest of the world. But whereas the CBS measurement framework relies on about 100 indicators, this study, in accord with international usage, employs a so-called 'small indicator set.'

Well-being in its original meaning includes whatever contributes to quality of life, insofar as scarce resources are mobilised to constitute this quality of life. But in politics and societal debates the concept of well-being has to a remarkable extent been equated with the gross domestic product (gdp). In response a much broader concept of well-being, broadly conceived, has been proposed and gained wide political support: In April, 2016, a parliamentary committee decided that discussing the state of well-being, broadly conceived should become a yearly routine. This concept of 'well-being, broadly conceived' truly denotes quality of life in the broadest sense. That said, in the remainder of this study, with few exceptions, we will simply   ad sense, to denote our quality of life or the extent to which people have been able meet their preferences. This well-being can be measured in objective as well as subjective terms. The latter relates to the ways in which people perceive their quality of life.

Natural Capital Natural capital occupies a special place within the debate on sustainability. That is because it is a critical resource, essential for the continued existence of life on earth. Natural capital forms, as it were, the physical basis on which well-being is built. The development of this natural capital is mapped through time by means of material flows. The analysis includes the entire chain from nature via producers to consumers.

Multilevel Dynamics, Including Institutional Evolution Transitions are coherent changes in a society's (or sector's, city's) set of interrelated and interacted dominant practices and the structures in which these are embedded, informed by a new set of problems and normative orientations.[24] Regarding structure, in the analyses special attention is devoted to the evolution of the material dimension of structure (physical infrastructures) as well as to its institutional dimension: the so-called institutional rectangle, i.e. the specific nature of and alignment between state, market, civil society and knowledge infrastructure.[25] Also, given the nature of our unit of analysis (a country), our focus on the relations between social practices, nature and environment and our interest in capital flows, we will also pay attention to place and connections.

1.6 The Measurement Framework for Well-being and Sustainability

1.6.1 Brundtland Definition

The analyses of well-being and sustainability in this book are based on a measurement framework for sustainable development that was constructed by employees at the Dutch Central Bureau of Statistics (CBS).[26] Statistics Netherlands developed this framework as an alternative for the usual indicators that are derived from the system of national accounts (including the gdp) and that illuminate only the economic aspects of well-being. The measurement framework developed by CBS has been embraced by an important part of the statistical community. Meanwhile 60

---

[24] J. Grin, J. Rotmans, J. Schot, *Transitions to sustainable development: New directions in the study of long term structural change* (New York 2010), 1–3.

[25] J. Grin, 'Understanding transitions from a governance perspective', in: J. Grin, J. Rotmans, J. Schot, *Transitions to sustainable development: New directions in the study of long term structural change* (New York 2010): 237–248.

[26] Under chairmanship of Statistics Netherlands an international working group developed the so-called *Conference of European Statisticians Recommendations on Measuring Sustainable Development*, United Nations Economic Commission for Europe prepared in cooperation with the Organisation for Economic Co-operation and Development and the Statistical Office of the European Union (Eurostat) (New York and Geneva 2014). In short, *CES Recommendations on Sustainable Development* (2014): https://www.unece.org/fileadmin/DAM/stats/publications/2013/CES_SD_web.pdf. The CES measurement framework is followed almost seamlessly in the indicator report of the Monitor Sustainable Netherlands: http://www.monitorduurzaamnederland.nl/

countries have indicated that this framework is the best way to map well-being and sustainability. That of course does not make it an internationally accepted standard. Universal acceptance will be a complex and time-consuming process, as indeed that of the gdp once was.

The measurement framework elaborates on the definition of sustainable development given in the Brundtland Report,[27] written by the UN Commission for Environment and Development (WCED):

> Sustainable development is development that meets the needs of the present, without compromising the ability of future generations to meet their own needs, here or in other parts of the world.

In other words, well-being and sustainability are viewed in the light of a just distribution of well-being and of the resources necessary to generate well-being. This pertains not only to a just distribution *between* generations, but also to the justice of the distribution *within* the present generation. The latter stipulation is above all a question of a just distribution *among* prosperous countries and the (poorest) developing countries. The Brundtland Report has canonized a strand of thought about sustainability that conceives sustainability as not solely an ecological problem. Economic and societal phenomena were explicitly included in the analysis. In addition the Brundtland Commission emphatically addressed international differences in well-being, in particular the growing gap in well-being between North and South.

While well-being and sustainability are inherently concerned with the 'here and now', they also speak to the effects of maintaining our present quality of life, both on future generations ('later' dimension) and on people elsewhere in the world ('elsewhere' dimension).

1.6.2   Well-being 'Here and Now'

Well-being is a much broader concept than simply material well-being, a condition that is often measured in terms of consumption of goods and services. Well-being also includes our health, level of education and the quality of the natural and social habitat. The resources or capitals necessary to build up well-being are also broadly defined. These include not only economic capital like machines and tools, but also human capital (health, knowledge and skills), social capital (the quality of social networks and institutions) and of course, above all, natural capital.

Natural capital occupies a special position, because it can be seen as the foundation of well-being. The usual economic view of natural capital is anthropocentric, seeing nature only as 'useful and productive' and as a purveyor of 'ecological services' to humans. This ignores the fact that that certain forms of natural capital, like biodiversity, have an intrinsic value quite apart from the specific 'utility' it might have for humans.

---

[27] World Commission on Environment and Development, *Our Common Future* (London 1987).

1.6.3 'Here and Now' Versus 'Later'

From an intergenerational perspective a development is sustainable only if it makes it possible for future generations to achieve at least the same level of well-being as we now enjoy. Of course we have no inkling of the preferences of future generations; we do not know what and to what extent they will want to consume. The only way we can determine whether the present generation is on the road to sustainability, is to check whether we are leaving enough resources for coming generations. A necessary condition for sustainability is therefore also that the per capita quantity of resources (of all the distinct forms of capital) may not decline. Future generations must have access to the same (per capita) quantity of economic, human, natural and social capital in order to meet their needs for well-being.

Tensions may exist between population growth and economy on the one hand, and demands made on natural capital on the other. This is also the reason why in this study population growth is an important contextual variable. Societal developments are often related to population growth, in order to assess whether enough per capita resources are handed over to coming generations. This should not be taken to suggest that economic growth and demands on the environment are necessarily indissolubly linked. Within a more circular economy the demands made on nature and the environment can be significantly reduced even in times of economic expansion.

It should be noted that the above figure only shows the *potential* for sustainable development. On the one hand there are no guarantees that future generations will also mobilize the resources they have inherited in a sustainable way. On the other hand it is entirely possible that in consequence of technological developments, resources may be more efficiently exploited in the future. In that case society could generate the same well-being with *fewer* resources.

1.6.4 'Here and Now' Versus 'Elsewhere'

The capital-based approach sketched above addresses a fundamental aspect of the Brundtland definition, namely the intergenerational aspect of sustainability. This approach is also useful for studying the transborder effects of sustainable development, that is to say the extent to which countries can influence one another with respect to the construction of their well-being.

A country can exploit its own domestic resources for the construction of a certain level of well-being, but it can also choose to import them. In an analysis of well-being and sustainability it is therefore also important to consider international flows of different forms of capital, and in particular to evaluate how the production and consumption activities of one country affect the natural capital of other countries. The other forms of capital are of course also important. In this way, the quality of life in a country can be influenced by the import and export of economic capital (machines and tools), as well as human capital (namely the flow of knowledge associated with migration). This makes the 'elsewhere' dimension (that indicates the effects on the rest of the world of the striving for well-being in a given country) the final element in the definition of sustainability. In this study we shall look especially at the extent to which natural capital elsewhere in the world was utilized by the Netherlands.

Core Concepts in the CBS Measurement Framework for Well-being and Sustainable Development

Sustainable development: Development that meets the needs of the present generation without compromising the ability of future generations to meet their own needs, here and elsewhere in the world.

Well-being (broadly conceived) or quality of life: A broad concept that is not limited to the 'utility' that people derive from the consumption of goods and services, but that is also related to the freedom and the possibilities people have for satisfying their needs for well-being.

Resources (capitals): A supply or resource from which a yield can be derived. Initially capital was regarded strictly as an economic, material concept (like machines, tools, buildings and infrastructure). The concept of capital has gradually been broadened, to the point that it now also includes natural, human and social capital.

Ecological quality of life: This concept focuses on the intrinsic value of nature and its ecosystems, quite apart from the question whether these have a direct economic value for humans.

1.6.5   The Measurement Framework of Well-being and Sustainability (the CES Recommendations)

CBS regularly publishes a Monitor Sustainable Netherlands, in which well-being 'here and now,' 'later' and 'elsewhere' are described. These Monitors map trends since 2000. In this study we shall for the first time attempt to apply this measurement framework to a long term analysis. Of course not all the indicators used in the CBS framework will be available for the period since 1850.[28]

We have therefore opted to work with a 'small set' of indicators. The so-called *CES Recommendations for Sustainable Development*[29] are taken as a point of

---

[28] In certain cases indicators had to be defined somewhat differently than in the present-day measuring system, due to the lack of availability of sources.

[29] See the report of the *Conference of European Statisticians Recommendations on Measuring Sustainable Development* (New York and Geneva 2014).

Table 1.1 Dashboard well-being 'Here and Now'

Theme		Indicator
Well-being		
	Consumption, Income	Consumer expenditure
		General income inequality
		Gender income inequality
	Subjective well-being	Life satisfaction
Personal characteristics		
	Health	Life expectancy
	Nutrition	Height
	Housing	Housing quality
		Public water supply
	Physical safety	Murder victims
	Work	Unemployment
	Education	Level of education
	Leisure time	Leisure time
Natural environment		
	Biodiversity	MSA
	Air quality	SO_2 in air
		Greenhouse gasses
	Water quality	Public water supply
Institutional environment		
	Trust	Generalised trust
	Political institutions	Democracy

departure. These international guidelines indicate how, for countries with limited databases, the large set of indicators can be reduced to a set of 24 indicators. The strategy was to choose one indicator for each discrete theme, one that could act as a kind of proxy guiding indicator that in essence covers a broader terrain. The theme health, for example, can be charted by means of a large number of indicators, but life expectancy can be seen as an overarching indicator that provides a good summary of general trends in changing health patterns. In certain cases where themes had a rather heterogenous character it was found necessary to adopt multiple indicators in order to cover the theme adequately.

The themes and indicators are presented in the CBS Monitor in three 'dashboards', as indicated in Tables 1.1, 1.2 and 1.3. This measurement system covers the full scope of societal development. But the debate on well-being and sustainability emphasises natural capital. That is why we will delve into natural capital more deeply.

Table 1.2 Dashboard well-being 'Later'

Theme		Indicator
Natural capital		
	Energy	Energy consumption
	Non-energetic resources	Gross domestic consumption
	Biodiversity	MSA
	Air quality	SO$_2$ emissies
		Greenhouse gas emissions
	Water quality	Public water supply
Economic capital		
	Physical capital	Economic capital stock
	Financial capital	Gross national debt
	Knowledge	Stock knowledge capital
Human capital		
	Health	Life expectancy
	Work	Unemployment
	Level of eduction	Schooling
Social capital		
	Trust	Generalised trust
	Institutions	Democracy

Table 1.3 Dashboard well-being 'Elsewhere'

Theme		Indicator
Well-being		
	Support	Development aid
Natural capital		
	Natural resources	Import of raw materials

1.7 Natural Capital and the Three Material Flows

The Monitor of Well-Being utilises a broad concept of well-being. Concerning the question of the depletion of vital resources, the Monitor distinguishes four capitals: natural, economic, human and social capital. This book, however, devotes particular attention to natural capital, and that for two reasons. First, natural capital can be seen as the foundation on which the entire economic and social system is built. Humans are able to construct (material) well-being by utilising natural resources. In the second place we can view a portion of available natural capital as *critical* capital. Were economic, human or social capital to be seriously threatened, well-being would doubtless be severely compromised. But if nature and the environment are damaged beyond repair, ecosystems can collapse and with them the very foundations of economy and society.

Natural capital is rooted in ecosystems that a society has at its disposal 'by nature' in order to produce useful goods and services. Humans have had little to no

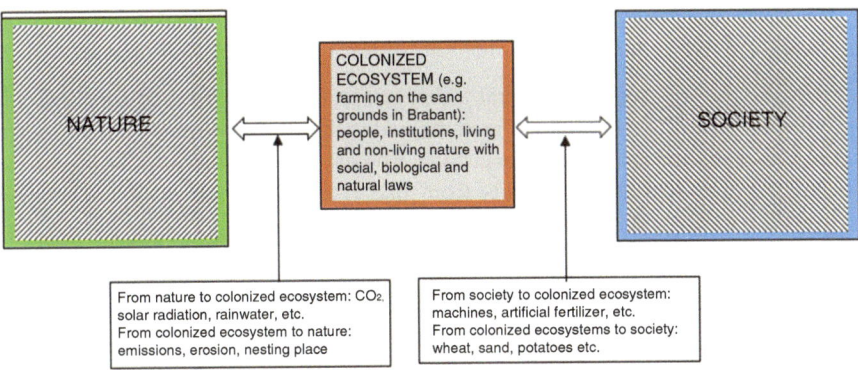

Fig. 1.1 The colonised ecosystem

part in creating these ecosystems, hence the term 'by nature.' Natural capital comprises non-living nature – the ground, the air, water – as well as living nature, the trees, the insects, the birds and other organisms. Living and non-living nature have their own dynamics and exchange matter and energy. Every organisms lives by virtue of metabolic processes or metabolic relationships. A set of close interactions between living and non-living entities is called an ecosystem.[30]

A natural ecosystem in its pure form can hardly be found in the Netherlands because its territory is the product of human intervention. The territory of the Netherlands has been worked to its remotest reaches and in part wrested from the water. In this case we speak of colonised ecosystems, ecosystems influenced by human practices. The Dutch live in such systems also thanks to metabolic relations, but these relations provide even more, namely a stream of products with which the Dutch produce their existence, material culture and society.

This is why we define natural capital as the ability of colonised ecosystems to provide so-called ecosystem services, in other words the ability to generate yields of food, energy, potable water etc.[31] According to this definition, a colonised ecosystem is a separate complex of internal interactions among people, water, air and soil with plants, animals and micro-organisms. In addition to this internal dynamic, a colonised ecosystem has an exchange of matter and energy with nature (for example solar radiation and rainwater) and society (for example, machines, artificial fertiliser, and wheat) (see Fig. 1.1).

---

[30] R.C. Hoffmann, *An environmental history of medieval Europe* (Cambridge 2014), 5–7 and his inspiring introduction, 1–20.

[31] See for the distinction between natural and colonised ecosystems: Hoffmann, *An environmental history*, 7–11. There is a plethora of literature on ecosystems and ecosystem services. See among others: TEEB, *The economics of ecosystems and biodiversity: An interim report* (Cambridge 2008), 12; and J. Dirkx et al., *Balans van de leefomgeving 2014 deel 7: Natuurlijk kapitaal als nieuw beleidsconcept* (Den Haag 2014), 7.

As point of departure for understanding the yields of colonised ecosystems in this study we will examine three categories of raw materials and sub-soil assets:

- Bio-raw materials: all agricultural yields (crops, cattle, fishing, and forestry)
- Mineral sub-soil assets: all yields from underground extraction of an inorganic nature, including sand, clay, marl, stone, iron ore and other ores.
- Fossil sub-soil assets: all yields from underground extraction of an organic nature, namely turf, coal, oil and gas.

With these categories as point of departure, an all but complete image of material flows in the economy can be constructed. We shall follow the development of raw materials and underground resources through time in great detail, including not only domestic, but also foreign sources. We will also investigate the economic processes in which these raw and underground materials are used. In this way we can reconstruct quite precisely the way in which the Netherlands exploited its natural capital from 1850 on. Special attention will be devoted to the supply chains along which these raw and underground materials find their way to the end user. In this way we can analyse how deeply these material flows reach into the capillaries of the social order, and which role they play in processes of economic growth and social change.

1.8   Institutions and Dynamics

Up to this point the promotion of well-being and sustainability has been sketched as a structural and rather abstract process. In this process, institutional and material structures and their evolution are key factors, as is spatiality (place and connections). But, in the end, structuration theory tells us, in John Law's (1992) famous shorthand, that 'structure is a verb': structure matters to the extent that, and through the ways in which, actors reproduce or transform them in their practices (of production, consumption, policy making, innovation etc.). This study therefore pays due attention to agency and practices. Most specifically, we will discuss the (contesting) ways in which, in a specific period, different groups of actors defined the problem of well-being as well as the normative orientations guiding practices (like 'intensification of production').

An illuminating framework for understanding such long term processes of transformative change is offered by transition studies. A basic theorem from that field is that transitions may occur if fundamental changes in (or the emergence of novel) practices, structural change and long term trends (such as individualisation or Europeanization) coincide and dynamically reinforce each other over time. Absent such interconnected dynamics, the coherence of the system of practices and structures reproducs itself and resists change. This theorem is rooted in a range of historical studies[32] as well as a review of literature on innovation.[33]

As noted above, in discussing structure and structural change, we will pay special attention to physical infrastructure (above) and to institutional context. More specifically, we will focus on institutions in the domain of (1) politics and government, (2) the economy, (3) a societal 'midfield,' i.e. civil society and (4) science, technology and innovation.

Institutions may be defined as rules maintained by humans that structure their political, economic and social behavior.[34] There is a distinction between formal and informal institutions. Formal institutions are associated with the legal system (in particular the Constitution), rules for decision-making, property rights, agreements and contracts. Next to these there exist informal institutions like rules of behavior, customs and conventions. Such rules generally have deep socio-cultural roots and are implied in e.g. families, schools and neighbourhoods.

The state, the market and the societal mid-field (civil society) can be seen as crucial institutions and principles of social order.[35] It should not be assumed that there is some ideal combination of these various institutions. Every generation will have to provide its own answers to the challenges of its time, answers that consist of repeated restructurings of the institutional system.

The State, or Political and Governmental Institutions These institutions determine the legal framework by means of which personal freedoms, property rights, and the legal security of citizens are specified. They fulfill functions in the anchoring of stability, safety, security and the prevention of corruption and preferential treatment. The state also co-determines the rules of the game in the market, e.g. as a safety net for market failure and as an inhibitor of monopolistic entrepreneurial practices like cartel-forming. The state can also intervene at moments when market forces generate intolerable levels of social inequality.

The Market or the Economic Institutions These institutions determine the economic rules of the game. They specify the scope of action for entrepreneurs, capitalists, workers, consumers and other groups. They set out the playing field for

---

[32] J. Schot, 'The usefulness of evolutionary models for explaining innovation: The case of the Netherlands in the nineteenth century', *History and Technology* 14 (1998) nr.3, 173–200; F.W. Geels, *Technological transitions and system innovations: A co-evolutionary and socio-technical analysis* (Cheltenham 2005); J. Schot and F. Geels, 'Typology of sociotechnical transition pathways', *Research Policy* 36(2007), nr.3, 399–417; F.W. Geels et al., 'The enactment of sociotechnical transition pathways: A reformulated typology and a comparative multi-level analysis of the German and UK low-carbon electricity transitions (1990–2014)', *Research Policy* 45 (2016), nr.4, 896–913.

[33] A. Rip and R. Kemp, 'Techological change', in: S. Rayner and E.L. Malone (Eds.), *Human choice and climate change. Vol II, Resources and technology* (Columbus1998) 327–399

[34] D.C. North, *Institutions, institutional change and economic performance* (Cambridge-New York 1990).

[35] Zie: Y. Hayami, *Development economics: From the poverty to the wealth of nations (second edition)* (Oxford 2001), chapters 8 en 9.

economic initiatives and competition. They provide the security necessary for market actors to engage in transactions. Countless economic regulatory systems ensure that supply and demand are coordinated as well as possible, so that citizens have access to goods and services to improve their quality of life.

Civil Society, or the Societal Mid-Field The institutions in this quadrant are crucial for the political process. Politics is here considered as the game through which the rules prevailing in economy and society are continually adapted to new situations. Groups coalescing around specific interests attempt to influence that game. In this way it is possible for specific social groups to gain the advantage over others and for vested interests to be protected to the extent that modernisation processes can take off only with the greatest difficulty.[36] In such cases the question is how the societal mid-field is organised and which initiatives from within this institutional quadrant can lead to change. *The science, technology and innovation* system is a fourth crucial institution. While this may possibly not immediately be seen as a principle of social ordering, it is certainly of essential importance for the analysis of well-being and sustainability. After all, technology and innovation are among the important motors of improvement in the quality of life. Not only can technology and science be seen as engines of the economy (and thereby of the increase of material prosperity),[37] but the development of new technologies can make the employment of scarce resources – like natural capital – more efficient. At the same time the role of science and technology in relation to sustainability is suspect because an intensive exploitation of natural capital has also led to negative effects.

In this chapter the methodological framework has been presented by means of which we propose to analyse 150 years of well-being and sustainability.. The CBS measurement framework enables us to trace the main lines along which well-being 'here and now' has developed and the extent to which that has depleted resources in the Netherlands or elsewhere in the world. We pay special attention to natural capital and will chart its historical development with the aid of detailed descriptions of material flows. These data will enable us to assess the extent to which economic developments in the period 1850 to the present have culminated in an increase of well-being and sustainability, or perhaps not. These descriptions will be consistently framed within the 'institutional quadrants,' a strategy that reveals how the structural contextual factors with which actors saw themselves confronted did or did not lead social development in the direction of greater well-being and sustainability.

After this methodological exposition the following chapter wil sketch the historical situation prevailing at the outset of this study. What was the state of Dutch soci-

---

[36] Jan de Vries remarks how in the course of the eighteenth century the institutions became rigid because the political elite forbade economic innovations that would compromise their position: J. de Vries, 'Barges and Capitalism', *AAG Bijdragen* 21 (1978). Until the 1860s social change was inhibited by institutional ossification, see: J.L. van Zanden and A. van Riel, *Nederland 1780–1814. Staat, instituties en economische ontwikkeling* (Amsterdam 2000), 48–56 and 96–101.

[37] Maddison, *Dynamic forces in capitalist development: A long-run comparative view.*

ety around 1850? How can we compare the Netherlands in 1850 with its present-day condition? How can we characterise the big changes in the domain of well-being and sustainability?

Literature

Bergh, J. van den (2015). 'Green agrowth as a third option: Removing the GDP-growth constraint on human progress'. *EU Policy paper (WelfareWealthWork)* no. 19.

Centraal Bureau voor de Statistiek (2014). *Monitor Duurzaam Nederland 2014: Indicatorenrapport.* Den Haag: CBS.

Conference of European Statisticians Recommendations on Measuring Sustainable Development. (2014). New York and Geneva: United Nations Economic Commission for Europe prepared in cooperation with the Organisation for Economic Co-operation and Development and the Statistical Office of the European Union (Eurostat).

Dirkx, J. and B. de Knegt (2014). *Balans van de leefomgeving 2014 deel 7: Natuurlijk kapitaal als nieuw beleidsconcept.* Den Haag: Planbureau voor de Leefomgeving.

Du Pisani, J.A. (2006). 'Sustainable development – historical roots of the concept'. *Environmental Sciences*, 3(2), 83–96.

Geels, F.W. (2005). *Technological transitions and system innovations: A co-evolutionary and socio-technical analysis.* Cheltenham: Edward Elgar Publishing.

Grin, J., J. Rotmans, J. Schot (2010), *Transitions to sustainable development: New directions in the study of long term structural change.* New York: Routledge Press.

Grin, J. (2010) 'Understanding transitions from a governance perspective'. In J. Grin, J. Rotmans, J. Schot (Eds.), *Transitions to sustainable development: New directions in the study of long term structural change.* New York: Routledge Press.

Geels, F.W. et al. (2016). 'The enactment of socio-technical transition pathways: A reformulated typology and a comparative multi-level analysis of the German and UK low-carbon electricity transitions (1990–2014)'. *Research Policy*, 45(4), 896–913.

Hayami, Y. (2001). *Development economics: From the poverty to the wealth of nations (second edition).* Oxford: Oxford University Press.

Hoffmann R.C. (2014). *An environmental history of medieval Europe.* Cambridge: Cambridge University Press.

Israël, J. (1995). *The Dutch Republic: Its rise, greatness and fall, 1477–1806.* Oxford: Oxford University Press.

Kooij, T.P. van der (1931). *Hollands stapelmarkt en haar verval.* Amsterdam: H.J. Paris.

Langenberg, H. and J.P. Smits (2015). 'Invoer van grondstoffen uit LDCs: Geworteld in koloniale tijden?' *Internationaliseringsmonitor*, 4, 42–57.

Maddison A, (1991). *Dynamic forces in capitalist development: A long-run comparative view.* New York: Oxford University Press.

North, D.C. (1990). *Institutions, institutional change and economic performance.* Cambridge: Cambridge University Press.

Polanyi, K. (1944). *The great transformation: The political and economic origins of our time.* Boston: Beacon Press.

Rip, A. and R. Kemp (1998). 'Techological change'. In S. Rayner and E.L. Malone (Eds.), *Human choice and climate change. Vol. II, Resources and technology.* Columbus: Battelle Press.

Schneider, F., G. Kallis and J. Martinez-Alier (2010). Crisis or opportunity? Economic degrowth for social equity and ecological sustainability. *Journal of Cleaner Production*, 18(6), 511–518.

Schot, J. (1998). 'The usefulness of evolutionary models for explaining innovation: The case of the Netherlands in the nineteenth century'. *History and Technology*, 14(3), 173–200.

Schot, J. and F. Geels (2007). 'Typology of sociotechnical transition pathways'. *Research Policy*, 36(3), 399–417.

Stiglitz, J.E., A. Sen and J.P. Fitoussi, *Report by the commission on the measurement of economic performance and social progress*: http://www.insee.fr/fr/publications-et-services/dossiers_web/stiglitz/doc-commission/RAPPORT_anglais.pdf.

TEEB (2008). *The economics of ecosystems and biodiversity: An interim report*. Cambridge: European Communities.

Vries, J. de (1973). 'On the modernity of the Dutch Republic'. *Journal of Economic History* 32, 191–202.

Vries, J. de (1974). The Dutch rural economy in the Golden Age, 1500–1700. London: Yale University Press.

Vries, J. de (1978). 'Barges and capitalism: Passenger transportation in the Dutch economy 1632–1839'. *AAG Bijdragen* 21: 33–398.

Vries, J. de (1984). *European urbanization, 1500–1800*. London: Routledge Press.

World Commission on Environment and Development (1987). *Our common future*. London: Oxford University Press.

Zanden, J.L. van (1988). 'Op zoek naar de 'missing link''. *Tijdschrift voor Sociale Geschiedenis*, 14, 359–386.

Zanden, J.L. van (1997). *Een klein land in de twintigste eeuw: Economische geschiedenis van Nederland*. Utrecht: Spectrum.

Zanden, J.L. van and A. van Riel (2000). *Nederland 1780–1814: Staat, instituties en economische ontwikkeling*. Amsterdam: Balans.

Chapter 2
The Great Transformation and the Questions

Jan-Pieter Smits and Harry Lintsen

Contents

Abstract This chapter opens with an overview of Dutch society midway through the nineteenth century, showing that at the time poverty was the main social problem. This is followed by a comparison between the benchmark years 1850 and 2010. The comparison shows that over the course of the intervening 160 years, Dutch society realised a vast increase in well-being. Extreme poverty was gradually eliminated, though in so doing sustainability issues emerged. Economic modernisation made inordinate demands on natural capital. After 1960, in particular, economic growth had an increasingly negative impact on human well-being and on sustainability.

Keywords Poverty · Well-being · Sustainability · Natural capital

---

This chapter is written by Jan-Pieter Smits and Harry Lintsen with contributions by Fred Lambert.

© The Author(s) 2018
H. Lintsen et al., *Well-being, Sustainability and Social Development*,
https://doi.org/10.1007/978-3-319-76696-6_2

2.1 A Landscape of Horrors[1]

In 1845, for the first time in history, the potato harvest failed nearly everywhere in Europe. The Netherlands were not spared.[2] In the hot and moist summer of that year the mold *Phytophthora Infestans* had spread unusually quickly. The leaves of the potato plant were infested first, after which the underground tuber quickly began to rot. Farmers looked on, astonished. This they had never seen before. The consequences were catastrophic. Since the eighteenth century the potato had been a staple of the popular diet. Potato prices skyrocketed in response to the extreme scarcity and prices of wheat and bread soared with them. There was widespread hunger. The population grew restless and Leiden, Delft, The Hague and Haarlem were the scene of food riots.[3] The police were powerless against mobs, rowdiness, vandalism and sometimes plunder. Grocers who were suspected of profiteering were especially targeted. The army had to be called in to restore order. The situation deteriorated with the failure of the rye harvest in 1846.

This was only the beginning of a war of attrition. The Netherlands was being lashed by a variety of scourges. Many lost the gamble with death. In the summer of 1846 typhus and typhoid fever, among others, reared their ugly heads. By the end of August a malaria epidemic had developed. The sick barely had time to recover due to the cold winter that followed. Bronchitis and flu became epidemic, in many cases ending in death. In subsequent months malaria returned and, abetted by dysentery, measles and associated ills, decimated the population. '... The sick are extremely weakened by the repeated fevers,' as Amsterdam's Medical Commission noted:

> at every illness, even without accompanying fevers, all the sufferers experienced an almost instantaneous loss of vitality and even the healthy complained of a certain degree of sluggishness and apathy. The apathy and exhaustion...were...the precursors and harbingers of the tremendous storm that was about to break... namely the widely feared *Cholera Asiatica*.[4]

The cholera epidemic broke out halfway through 1848 and reached its peak in 1849. After that the Netherlands was able to lick its wounds, but not for long. New dramas followed. 1853 witnessed the start of the Crimean War, a struggle between the Russian Empire and an alliance of among others Turkey, France and England. Its effects were felt, even though the war, that lasted two years, took place at a distance of more than 2000 km. It drove international wheat prices up and seriously depressed

---

[1] The introduction is based on: H. Lintsen et al., *Made in Holland. Een techniekgeschiedenis van Nederland [1800–2000]* (Zutphen 2005), 23.

[2] J. Bieleman, *Geschiedenis van de landbouw in Nederland 1500–1950* (Amsterdam 1992), 132–133.

[3] R. van der Wal, *Of geweld zal worden gebruikt. Militaire bijstand bij de handhaving en het herstel van de openbare orde 1840–1920* (Hilversum 2003), 57–66.

[4] Quoted in: J.M.M. de Meere, *Economische ontwikkeling en levensstandaard in Nederland gedurende de eerste helft van de negentiende eeuw* ('s-Gravenhage 1982), 112.

Dutch wheat imports. Access to food in those years was at least as problematic as during the potato crisis.[5]

In March 1855 the Grebbe Dike was breached over a length of 150 meters. The Land of Maas and Waal and the Gelderland Valley were flooded. Other parts of Gelderland and North Brabant followed. Water and ice floes caused great damage.

> Severe floods and dikes have covered parts of our fatherland with disasters and have cast thousands of our countrymen, robbed of all property and goods, into frightful misery. Every single source of life has ceased to flow, there where water covers the otherwise so fertile fields...[6]

The king visited the affected region. The Dutch opened their purses to the victims. Fund-raising events were organized.[7]

Around 1850 the Netherlands experienced bad times. There was hunger. Illness reigned everywhere. The death rate increased and life expectancy declined. The social order was under duress and criminality increased. In the region of the big rivers and along the coast floodwaters had to be kept at bay.

Is this the background against which we have to paint a picture of well-being and sustainability in the Netherlands since 1850? Further analysis demands the specification of an initial situation. This consists in the first place of a brief characterization of Dutch society.

In the second place, we want to confront the situation at that time with the situation now. We tackle this with the Monitor from the previous chapter. The three 'dashboards' with indicators serve as points of departure. (1) well-being in the Netherlands 'here and now,' (2) resources available for future generations, 'later' and (3) the Netherlands in the world, 'elsewhere'. We fill in the 'dashboards' with a series of numbers for the various indicators pertaining to diverse themes. The starting period of the investigation (the middle of the nineteenth century) can thereby be contrasted with the present day. This exercise should be seen as a preliminary exploration. What do the numbers say?

In the third place we provide an impression of the development of natural capital between 1850 and 2010.

As might be expected, the conclusion is that we can speak of a Great Transformation. It is almost impossible to imagine how fundamentally the Netherlands has changed. This chapter provides a characterization and poses fundamental questions raised by the transformation.

---

[5] M.T. Knibbe, 'De hoofdelijke beschikbaarheid van voedsel en de levensstandaard in Nederland, 1807–1913', *Tijdschrift voor Sociaal-Economische Geschiedenis* 4(2007), nr. 4, 98.

[6] 'Watersnood', *Nieuwe Rotterdamsche Courant*, 11-03-1855.

[7] See the numerous advertisements in the newspapers in that year. Source: Delpher Kranten, keyword 'watersnood' (flooding).

2.2   The Netherlands Around 1850

The historical literature provides a mixed image of Dutch society halfway through the nineteenth century. On the one hand it is evident that in this period the Netherlands is highly developed. Per capita income is among the highest in the world. Moneywise, the Netherlands remained the world's richest country until about 1800. After that, Great Britain, where the industrial revolution raged in the first half of the nineteenth century, took the lead from the Netherlands.[8]

The high level of economic development is also reflected in a number of structural features of Dutch society in this period. The degree of urbanization, amounting to 21% in 1850, was one of the highest in the world.[9] European countries with levels of income similar to the Netherlands hovered around only 6%.[10] This high level of urbanization was made possible to a large degree by the high productivity of Dutch cattle husbandry in particular. This freed an important part of the agricultural labor force to find work in trade, transport and services in the strongly commercialized economies of the coastal provinces.[11] Finally we can point to the high level of literacy, often ascribed to the Protestant church that of course attached great importance to the ability of believers to read the bible themselves.

Nonetheless, in the first half of the nineteenth century this high level of economic development also had its dark sides.[12] The province of North Holland for example, traditionally one of the most prosperous regions of the country, suffered from de-urbanization. Especially in the cities around the Zuiderzee, employment declined to the extent of encouraging a sizeable migration to the countryside. Data on consumpiton also show that material well-being was threatened. Consumption of meat exhibited a significant drop, which is taken in the literature to be a sign of strong social-economic decline. Finally, we can point to the state of government finances. At the outset of the 1840s this was in a deplorable condition. In 1844 'bankruptcy' of the Dutch state could barely be prevented.[13]

This is a diffuse image. The Netherlands exhibits aspects of a highly developed society, while at the same time there are indications of stagnating social development. How can this be understood? Richard Griffiths asked this questions years ago in a publication entitled *Achterlijk, achter of anders?* (*Backward, behind or*

---

[8] A. Maddison, *Monitoring the world economy* (Parijs 1995), 23. For updates of this database see: http://www.ggdc.net/maddison/maddison-project/data.htm

[9] J. de Vries, *European urbanization 1500–1800* (London 1984).

[10] A. Burger, 'Dutch patterns of development: economic growth and structural change in the Netherlands', *Economic and social history in the Netherlands* 7 (1996), 109–132.

[11] For a elaborate analysis of the commercialization of the agrarian sector during the Dutch Republic see: J. de Vries, *The rural Dutch economy in the golden age, 1500–1700* (New Haven 1974).

[12] De Meere, *Economische ontwikkeling en levensstandaard.*

[13] J.T. Buys, *De Nederlandsche staatsschuld* (Haarlem 1857), 149–158; I.J. Brugmans, *Paardenkracht en mensenmacht* (Den Haag 1961), 187.

different?).[14] In certain respects the nineteenth century Dutch economy certainly lagged behind. The process of industrialization that was so successful in Great Britain and Belgium barely took off in the Netherlands. But 'backward' is certainly not a justifiable qualification in view of a number of 'modern' characteristics of Dutch society sketched above.

The key to the paradox can be found in the 'different' nature of Dutch society. Throughout the first half of the nineteenth century the Netherlands found itself in the middle of a far-reaching structural transformation, framed in the literature as the transition from a commercial capitalistic to an industrial capitalistic system.[15] During the Dutch so-called Golden Age (1580–1672) the Netherlands witnessed an economic dynamism unknown in those days. Amsterdam's staple market became one of the most prominent centers of world trade, around which countless processing industries and a flourishing financial and transport sector developed.[16] Around 1650 material well-being had climbed to such heights that per capita income was the highest in the world. This exceeded that of Great Britain by no less than 30%.[17]

The tide turned in the eighteenth century. The Republic's economy declined into a 'stationary state' and all but ceased growing.[18] This can in part be attributed to external factors. The currents of world trade shifted, threatening the functioning of Amsterdam's central staple market.[19] But this is not the whole story. Economic dynamism gradually disappeared and the obsolete institutional system hindered the emergence of new entrepreneurial initiatives. In the eighteenth century economic and institutional rigor mortis assumed such proportions that the literature speaks of an 'obsolete economy.'[20]

The greatest obstacle to innovation was the political and economic elite of the Republic, that clung to its privileges and sources of income and that consciously resisted new and potentially more profitable economic activities as a threat to their vested interests. Halfway through the nineteenth century the Netherlands was forced to eat the bitter fruits of this shortsighted attitude.[21] Certainly, in a number of respects one could always point to a positive inheritance from the period of the Republic (high income levels, robust level of commercialization) but the institutional

---

[14] R.T. Griffiths, *Achterlijk, achter of anders? Aspecten van de economische ontwikkeling van Nederland in de 19ᵉ eeuw* (Amsterdam 1980).

[15] J.L. van Zanden, *Arbeid tijdens het handelskapitalisme. Opkomst en neergang van de Hollandse economie 1350–1850* (Bergen 1991).

[16] T.P. van der Kooij, *Hollands stapelmarkt en haar verval* (Amsterdam 1931).

[17] Zie hiervoor: http://www.ggdc.net/maddison/maddison-project/data.htm

[18] J.L. van Zanden, 'The Dutch economy in the very long run. Growth in production, energy consumption and capital in Holland (1500–1805) and the Netherlands (1805–1910)', in: A. Szirmai, B. van Ark en D. Pilat, red., *Explaining economic growth. Essays in honour of Angus Maddison* (Amsterdam 1993), 267–283.

[19] Joh. De Vries, *De economische achteruitgang van de Republiek in de achttiende eeuw* (Leiden 1959).

[20] J. de Vries, 'Barges and capitalism', *AAG Bijdragen* 21 (1978).

[21] J.L. van Zanden and A. van Riel, *Nederland 1780–1914. Staat, instituties en economische ontwikkeling* (Amsterdam 2000), chapters I.3 en 2.4.

system functioned as a brake on social renewal. On the one hand the old commercial capitalist system no longer generated sufficient growth and employment, and on the other hand the institutions hindered the modernization of social life.

There was also a large income differential. While the per capita income might have been very impressive by international standards, the bulk of this income was pocketed by the political and economic elite, especially the merchants of the province of Holland. A large part of the population lived in poverty. This poverty was exacerbated by the tax system. Tax burdens were distributed quite unevenly across the population and lay especially heavy on workers' families.[22]

In the course of the second half of the nineteenth century the obstacles to economic renewal were gradually eliminated. From the 1860s on the Netherlands experienced a successful transformation from a pre-modern commercial capitalistic to a modern industrial capitalistic structure. What did this transformation bring to the Netherlands in terms of well-being and sustainability? This is the central issue of this book. For a first assessment we compare the situation in the middle of the nineteenth century with the present-day situation and take the CBS monitor as our point of departure..[23]

---

[22] E. Horlings and J.P. Smits, 'Private consumer expenditure in the Netherlands, 1800–1913', *Economic and social history in the Netherlands* 7 (1996), 15–40.

[23] The CBS measurement framework relies on about 100 indicators, this study, in accord with international usage, employs a so-called 'small indicator set.'The choice of the themes is justified in part II of the *CES Recommendations on Sustainable Development* (New York/Geneva 2014). We have of course tried when choosing indicators to follow these recommendations as closely as possible. Sometimes however, laking sufficient data, we had to make use of 'proxies' that could be considered an adequate measure of the given theme. The following differences with the *CES Recommendations on Sustainable Development* (New York and Geneva 2014) should be noted:

- Obesitas as indicator for nutrition is replaced by height.
- The bird index as indicator for biodiversity is replaced by 'Mean Species Abundance' (MSA)
- Fine dust as indicator for air quality is replace by SO_2.

In a number of cases, due to the availability of sources or because of the representation of the historical theme, indicators had to be defined slightly differently than in the measurement system. For example, 'attendance at elections' as an indicator for political institutions has been replaced by the 'democracy index'.In the recommendations population size is used as a general indicator for the context. That is how it is used in this study.

The values of the indicators for the reference dates 1850, 1910, 1970 and 2010 come from the following sources:

- J.L. van Zanden et al. (red.), *How was life? Global well-being since* 1820 (OECD Publishing 2014): general income inequality, life expectancy, height, murder victims, level of education, MSA, democract-index.
- F. Lambert, *Massastromen in Nederland. In de jaren 1850, 1913, 1970, 2010* (researchreport Eindhoven University of Technology, oktober 2016): gross domestic consumption non-fossil raw materials, import of raw materials.
- H. Hölsgens, *Energy transitions in the Netherlands: Sustainability challenges in a historical and comparative perspective* (Groningen 2016), appendix VIII: CO_2.
- K. Breedveld, M. Cloïn en A. van den Broek, *Ruimte voor tijd. Op weg naar een monitor tijdsordening* (Sociaal en Cultureel Planbureau, Den Haag 2002): free time
- J.P. Smits, E. Horlings en J.L. van Zanden, *Dutch GNP and its components, 1800–1913*

2.3 Well-being 'Here and Now': 1850 Versus 2010

Well-being 'here and now' is the monitor's first dashboard and includes a broad palette of themes, among which well-being, labor, air quality and trust in institutions. Quite in line with expectations, material well-being – measured as per capita consumption – increased in the period 1850–2010 (Table 2.1). In fact it grew by a factor of almost six, while the Dutch population increased from 3.1 million to about 16.6 million inhabitants.

Well-being also increased in a broader sense. The average life expectancy, an indicator of health and the physical condition of the Dutch population, was 37 years in 1850. That is less than half the present life expectancy. The nutritional and housing situations are also much improved. The number of years of schooling, at present an important factor in personal development and economic life-chances has increased by eight years on average. An indication of political participation is the democracy index, for which 100% denotes the theoretical maximum.[24] In the 1840s this was 0.31% and at present 39%. By today's standards the quality of life in 1850 as measured by these indicators must be judged to be low or very low. Well-being in terms of satisfaction with life cannot be measured for 1850. It is, however, hardly a daring supposition that the well-being of the Dutch has increased.[25]

In several respects the years around 1850 distinguished themselves in a positive sense. The number of murders per 100,000 inhabitants, a measure of personal safety, was low (0.8), even lower than at present in the Netherlands (1.1). Also, the quality of the natural environment was high. Biodiversity, for example – expressed in 'Mean Species Abundance' (MSA) – where 100% represents the original pre-human

---

(Groningen 2000): unemployment in 1850 and 1910
- Centraal Bureau voor de Statistiek (CBS), among others Statlines: the values of the other indicators.
- For statistical substantiation of the values of the indicators, see the relevant sources.

[24] The democracy index is defined as the product of the participation index (the percentage of the adult population that votes in elections) and the competition index (one minus the proportion of the vote going to the winning party in national elections). J.L. van Zanden et al. (eds.), *How was life? Global well-being since 1820* (OECD Publishing 2014), 163.

[25] An indication could be a comparison between the Netherlands at that time and present-day countries with the same level of material well-being. Then it cannot be otherwise than that the Dutch must have been unsatisfied or dissatisfied with their lives. In 1850 the Netherlands had a gdp/capita of 2330$ (1990 PPP$). From the *Human Development Report* (2010, table 1, 143–145) we select the first country with a gdp/capita below 2350$ (2008 PPP$). That turns out to be Kyrgyzia with a satisfaction score of 6. Subsequently the lowest country with a gdp/capita above 2300$. That turns out to be Yemen with a satisfaction score of 4. In this table the Netherlands occupies seventh place with a score of 9. All scores are rounded off.

One could also imagine that the Dutch may have felt themselves relatively happier than the inhabitants of other countries. After all, the Netherlands was among the richest countries in the world.

The problem was that means of communication were limited and most Dutchmen were not acquainted with the situation elsewhere. They were familiar with their own situation where there was a huge gap between rich and poor. Poverty and inequality have a big impact on well-being.

Table 2.1 Dashboard well-being 'here and now,' 1850 versus 2010

Theme	Indicator	Unit	±1850	±2010	Evaluation corresponding CBS methods
Population		Million inhabitants	3.1	16.6	
Material well-being and well-being					
Consumption, income	Consumer expenditures per capita, constant prices	annual expenses per capita. Index: 1850=100	100	581	⬆
	Income inequality, general	gini coefficient: 0–1	0.48	0.32	⬆
	Gendered income inequality	% difference in hourly wage M/F	?	19	⬆
Subjective well-being	Satisfaction with life	score 0–10	?	8	⬆
Personal characteristics					
Health	Life expectancy	years	37	81	⬆
Nutrition	Height	cm	165	(183)	⬆
Housing	Housing quality	% slums and hovels	30 á 50	<1	⬆
	Public water supply	m³/capita	0	120	⬆
Physical safety	Murder victims	number per 100.000 inhabitants	0.8	(1.1)	⬇
Labour	Unemployment	% workforce	6.4	5.0	⬆
Education	Level of education	years	3	(11)	⬆
Free time	Free time	hours/week	?	44.7	⬆
Natural environment					
Biodiversity	MSA	% of original biodiversity	73	(63)	⬇
Air quality	SO₂	kg SO₂/capita	1.3	4	⬇
	Greenhouse gas emissions	ton CO₂/capita	1.2	10.6	⬇
Water quality	Public Water supply	m³/capita	0	120	⬆
Institutional environment					
Trust	Generalised trust	% population with adequate trust	?	66.7	⬆
Political institutions	Democracy	democracy-index 0–100	0.31	(39)	⬆

Legend

	Positive development
	Negative development

Note: The numbers in brackets are from J.L. van Zanden et al. (ed.), *How was life? Global well-being since 1820* (OECD Publishing 2014) and relate to the year 2000. Numbers for these indicators – measured according to the same methodology – are not available for 2010
Source: See note 23 of this chapter

biodiversity – was higher than at present (73% compared to 63%).[26] However, there was in that time also a major environmental problem, namely the organic pollution of surface water and hence the drinking water. That was the cause of serious public health problems.

These figures are not so much the end, but rather the beginning of stories and analyses about the environment, safety, democracy, health etc. In those stories the historical context has to be prominent. In the tables, we have compared the figures for 1850 with those for today. But how should we judge these figures in the context of the middle of the nineteenth century? What did Dutch people at the time think about the different themes?

Take for example the democracy index, indicating that at that time democratic quality was utterly lacking. This indicator has to be seen in relation to the inception of parliamentary democracy. The 1840s was a turbulent decade in politics. King William I, who for many years had ruled as an autocrat, abdicated his throne. This precipitated a discussion on the organization of the state. This was resolved in 1848, at least for the time being, with Thorbecke's new constitution stipulating direct parliamentary elections and introducing ministerial responsibility. This constitutional revolution did not, however, imply the simultaneous emergence of a corresponding political culture. Political instability persisted in subsequent years and there cannot have been much trust in the new political institutions.[27]

Political instability must certainly be seen as an issue and as an important well-being problem in terms of the monitor. In relation to quality of life another important theme must also be considered. This has to do with two indicators, namely the level of consumer expenditures (150 guilders/year per capita) in combination with income inequality (Gini coefficient of 0.48, where 1 is maximum and 0 no inequality). These reveal – from today's perspective – a disturbing fact, namely that an estimated 21% of the population must have lived *under* the poverty line, in other words must have been 'extremely' poor (for the calculation and the norms see Chap. 4).

On the basis of a first orientation to the quality of life around 1850 two important well-being themes have cropped up: political instability and poverty. There may be other themes of equivalent importance. We will come to these in the course of our investigation. Of the two themes mentioned, we will devote most attention to poverty. As we noted earlier, we have chosen to focus on physical existence and natural capital in the Netherlands. The political situation will be considered primarily in relation to this material aspect.

Poverty as a grave sustainability problem in terms of minimal quality of life demands closer study. Which categories of the poor can be discerned and what was the nature of their poverty? Was the Netherlands in a position to solve poverty with the resources then available and in particular with the available natural capital?

---

[26] For the rest, the emission of CO_2 – one of the greenhouse gasses – was low: 1.2 tons of CO_2 per gdp/capita over against 10.1 nowadays. That of SO_2 – an indicator of air pollution – 0 as compared to 80 kton SO_2 per gdp/capita.

[27] P. de Rooy, *Ons stipje op de wereldkaart. De politieke cultuur van Nederland in de negentiende en twintigste eeuw* (Amsterdam 2014), 75–79.

The problem of poverty was not unique for the Netherlands around 1850, but an issue indissolubly connected with the history of humanity. This raises an important question. From today's perspective poverty is a big problem, but did it have the same importance in 1850? Contemporaries may have seen poverty as a destiny about which little could be done or as an issue that was under control thanks to *charitas* and poor relief. But if that was not the case, the question becomes: what initiatives did contemporaries undertake to solve the problem and what consequences did these have for the available resources?

It is also important to appreciate how standards for poverty and quality of life have changed over the course of the last century and a half. People demanded an ever higher quality of life. Many inhabitants of the Netherlands would have perceived a normal worker's home of 1850 as a hovel only 50 years later. This study is alert to how norms change in the course of time and how increases in quality of life and adaptation to new demands for quality took place. At the same time our analysis pays attention to what is called 'contested modernization.' The growth of well-being sometimes assumed unbridled forms, giving rise to entirely new problems of quality of life. Whereas in 1850 the field of food and health was dominated by malnutrition, the last decades have confronted us with obesitas as a new social problem.

2.4   Well-being 'Later': 1850 Versus 2010

Capitals or resources for 'later' are the theme of the second dashboard of the CBS monitor. From today's perspective, there was an enormous potential to develop the resources available in 1850. This is a simple conclusion in retrospect, but how was this viewed at the time?

The indicators of natural capital, as is to be expected, show a modest per capita consumption of energy and raw materials in 1850 (Table 2.2). At the time, energy was extracted from domestic turf and to a lesser extent from coal – which was for the most part imported. Raw materials also included mineral sub-soil assets (clay, sand and earth) and agricultural produce. The available natural capital could have been more intensively and extensively exploited – even on the basis of technologies then existing.

The sustainability monitor reveals a dilemma here. Biodiversity and air quality were in good shape, compared to today. From today's perspective, intensive exploitation of natural capital suggests a negative and undesirable development, while the solution for poverty as an important sustainability issue requires precisely that. Natural capital demands a fundamental analysis. To what extent did the exploitation of natural capital result in the depletion of raw materials, the pollution of the environment and the destruction of 'nature?' To what extent were prior generations aware of this and were nature and environment seen as relevant issues?

History shows that the exploitation of natural capital could occur in many different ways and that it did not always imply sustainability problems, for example in the case of a circular economy. History also shows that some problems

Table 2.2 Dashboard well-being 'later', 1850 versus 2010

Theme	Indicator	Unit	±1850	±2010	Evaluation corresponding CBS methods
Natural capital					
Energy	Energy consumption	TJ/capita	0.03	0.17	⬇
Non-fossil raw materials	Gross domestic consumption	ton/capita	2.1	9.8	⬇
Biodiversity	MSA	% original biodiversity	73	(63)	⬇
Air quality	SO_2 emissions	kg SO_2/capita	1.3	4	⬇
	Greenhouse gas emissions	ton CO_2/capita	1.2	10.6	⬇
Water quality	Public water supply	m^3/capita	0	120	⬆
Economic capital					
Physical capital	Economic capital stock/capita	index: 1850=100	100	1046	⬆
Financial capital	Gross national debt	% gdp	194	59	⬆
Knowledge	Stock knowledge capital	index: 2010=100	0	100	⬆
Human capital					
Health	Life expectancy	years	37	81	⬆
Labor	Unemployment	% workforce.	6.4	5.0	⬆
Level of education	Schooling	years	3	(11)	⬆
Social capital					
Trust	Generalised trust	% population with adequate trust	?	67	⬆
Political Institutions	Democracy	democracy-index: 0–100	0.31	(39)	⬆

Legend

	Positive development
	Negative development

Note: The numbers in brackets are from J.L. van Zanden et al. (ed.). *How was life? Global well-being since 1820* (OECD Publishing 2014) and relate to the year 2000. Numbers for these indicators – measured according to the same methodology – are not available for 2010
Source: See note 23 of this chapter

could be easily eliminated in consequence of the transition from the use of coal to oil to natural gas (like air pollution as a result of SO_2 emissions, while others were more intractable (like the ever-increasing manure surpluses in the dairy industry and the consequent eutrophication of the soil and sub-soil). This study singles out the negative effects that have followed on the exploitation of natural capital in the Netherlands. This makes it possible to acquire more insight into the roots of today's sustainability problems like climate change and the loss of biodiversity. These problems do not stem from the most recent period, but are the result of long-term manipulation by humans of the natural environment. It is only by adopting an intergenerational perspective that we can assess the extent to which societal activities have really increased our well-being.

With economic, human and social capital exhaustion can be prevented by reinvesting in the associated resources. The figures show that these forms of capital exhibit definite improvement between 1850 and 2010. It is also important to investigate the quality of the investments. For example, one of the three themes within economic capital in the Monitor is physical capital. This consisted for the most part of 'classic' investments in buildings (houses, workplaces, churches, government buildings), in infrastructure (roads, waterways and land reclamation) and in tools, machines and means of transport (sailing ships, windmills, etc.).[28] The question is how much was invested in a modern economy, e.g. in railways and steam engines, textile and bread factories.

From today's perspective human capital in 1850 was of a low quality, among other things because of widespread poverty, low life expectancy and a low level of schooling. Nowadays we see a connection between these factors and low levels of economic growth and well-being. To what extent did contemporaries want to invest in the improvement of public health and schooling in order to improve well-being in the future?

Social capital, a measure of social integration and political culture, is represented in the Monitor by two indicators: the democracy index and generalized trust. These indicators – as far as can be ascertained – paint a negative picture. What influence did the problematic nature of social capital have on the issue of poverty?

2.5   Well-being 'Elsewhere': 1850 Versus 2010

Nowadays, largely thanks to the Brundtland Report, much more attention is paid to how countries pursuing their own well-being lay a claim on resources elsewhere in the world. This issue will also occupy us in this volume. That said, the third dashboard of the monitor – 'elsewhere' – is based on only a very limited set of indicators. There is still much uncertainty – especially for historical research – about indicators that get to the core of the problematic transboundary dynamics of well-being.

Here we will hew to the international framework and work with two indicators: development aid and import from developing countries (Table 2.3). 'Development aid' and 'developing country' are anachronisms for the nineteenth century and have no meaning for an analysis of the situation in 1850. The notion of 'developing country' refers to the large present-day differences in well-being between, for example, the Netherlands and other countries. Such differences also existed in 1850. As noted above, the Netherlands and Great Britain were then the richest countries in the world. Most of the world was significantly poorer than the Netherlands. In comparison with the Netherlands, global poverty must have been of a different order altogether. The majority of the world population must have lived around or under the poverty line.

---

[28] J.L. van Zanden en A. van Riel, *Nederland 1780–1914. Staat, instituties en economische ontwikkeling* (n.p. 2000), 346–351.

Table 2.3 Dashboard well-being 'elsewhere,' 1850 versus 2010

Theme	Indicator	Unit	±1850	±2010	Evaluation corresponding CBS methods
Material Welfare					
Consumption, income	Development aid	% gdp	–	0.8	⬆
Natural capital					
Natural capital	Import of raw materials	ton/capita	0.41	12.9	⬇

Legend

	Positive development
	Negative development

Source: See note 23 of this chapter

The evaluation of the third dimension will take account of this context. We will take Dutch imports as a point of departure (Table 2.3). In 1850 these amounted to about 1300 kiloton, or 0.4 ton per Dutchman compared to about 214,000 kiloton or 12.9 tons per Dutchman in 2010. Imports have increased dramatically, both absolutely and relatively. The increase is indicative of the globalization of which the Netherlands is a part. Where do the imports come from? How did they influence – insofar as this can be determined – resources elsewhere? In present-day terms: what was the 'footprint' of the flow of foreign goods to the Netherlands?

In this connection the Dutch colonies demand special attention. Recent reports reveal that in comparison with other EU countries the Netherlands imposes a heavy burden on the natural resources of the poorest countries.[29] It has been suggested that this passing of the buck can be partially explained by the Netherlands' colonial past.[30] In this book we investigate what role foreign regions, and in particular the colonies, played in the growth of Dutch well-being.

2.6   Natural Capital: 1850 Versus 2010

The results of the Monitor for 1850 and 2010 reveal that over the last century and half the Netherlands has depleted some of its available natural capital. Hence this volume will examine the development of natural resources in greater detail by means of an analysis of raw materials and their derivative material flows.

---

[29] CBS, *Sustainablility monitor of the Netherlands 204. Indicator report* (Den Haag 2015), 34–37.

[30] H. Langenberg and J.P. Smits, 'Invoer van grondstoffen uit LDCs: geworteld in koloniale tijden?', *Internationaliseringsmonitor 2015 (4)*, 43–57. See: https://www.cbs.nl/nl-nl/publicatie/2015/44/internationaliseringsmonitor-2015-vierde-kwartaal

Table 2.4 shows how big the increase in raw materials was between 1850 and 2010. For bio-raw materials there was a 13-fold increase and for mineral and fossil sub-soil assets the factor was even higher (respectively 71 and 65). Population growth in this period goes some way toward providing an explanation for this explosive increase. But even the per capita growth of raw materials remains huge. For bio-raw materials this increased by a factor of 2.8 and for mineral and fossil sub-soil assets by, respectively, 13 and 12. These figures also show the extent to which claims on natural capital, measured as volume of raw materials, has become more international. For example imports of fossil raw materials rose between 1850 and 2010 from 18% to an impressive 69%, while exports rose from 1% to 47%. The other raw materials also reveal a growing integration with the global economy.

The increase in the total number of raw materials is not evenly distributed over the entire period (Graph 2.1). It fluctuates over time due to the two world wars and the economic depression in the thirties. However, a clear break in the trend can be seen around 1960. Between 1850 and 1960, annual growth averaged about 2.5%. In the period 1960–1975 it increased by 5.1% annually and then weakened to an average of 1.0% between 1975 and 2010 (see also Table 22.1).

Table 2.4 Raw materials in the Netherlands 1850 versus 2010

	1850	2010	Ratio 1850:2010
Bio-raw materials:			
Gross available	5260 kton	67,020 kton	1:13
Bio/capita	1.7 ton/capita	4.0 ton/capita	1:2.4
% import	11%	31%	1:2.8
% export	6%	23%	1:3.8
Mineral sub-soil assets:			
Gross available	1350 kton	95,570 kton	1:71
Mineral/capita	0.45 ton/capita	5.8 ton/capita	1:13
% import	11%	58%	1:5.3
% export	2%	17%	1:8.5
Fossil sub-soil assets:			
Gross available	3060 kton	199,630 kton	1:65
Fossil/capita	1.0 ton/capita	12.0 ton/capita	1:12
% import	18%	69%	1:3.8
% export	1%	47%	1:47
Total raw materials:			
Gross available	9670 kton	372,220 kton	1:37
Raw materials/capita	3.1 ton/capita	21.9 ton/capita	1:7
% import	13%	59%	1:4.5
% export	4%	35%	1:8.8

Remark: Gross available = domestic production + imports
Source: F. Lambert, *Massastromen in Nederland. In de jaren 1850, 1913, 1970, 2010* (researchrapport Technische Universiteit Eindhoven, oktober 2016)

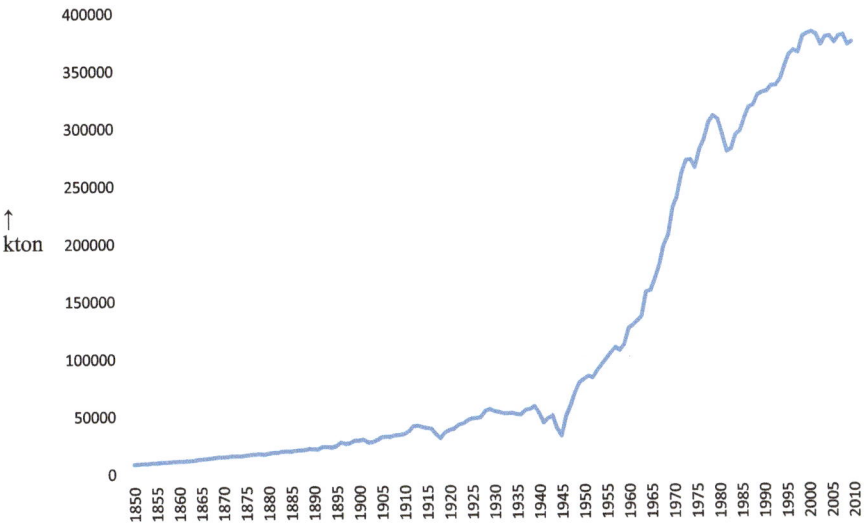

Graph 2.1 Total use of raw materials in the Netherlands, 1850–2010 (kton)
Note: Use = domestic production + import
Source: See note 10 of Chap. 24

2.7   The Great Transformation, the Tradeoff and the Fundamental Questions

The fifties and sixties of the twentieth century are also a special period for the development of economics, well-being and sustainability. Graph 2.2 shows the relationship between the growth of GDP per capita and the quality of life after 1850.[31] These figures show that economic growth and the development of well-being do not always go hand in hand. In the period from 1850 up to 1880 we see that well-being lags a little behind economic growth. After that time until the sixties of the twentieth century, however, the expansion of the quality of life is stronger than one might expect on the basis of economic growth. After that we see that economic growth has been accompanied by ever smaller increases in welfare. How can we explain these differences? Which factors determine the extent to which growth in GDP translates into increases in welfare or not? The relationship between growth in

---

[31] Technical explanation: Figure 2.1 compares the growth of GDP with an unweighted average of indicators for the quality of life 'here and now.' All the time series data are standardized, so that they all have the same bandwidth, preventing series that can increase without limit (like consumptive expenditure) from exerting more influence than series that by their nature move within a more limited bandwitdth (as for example average life-expectancy, being constrained by demographic limits). Due to this standardization technique, the graph exhibits deviations within a limited bandwidth.

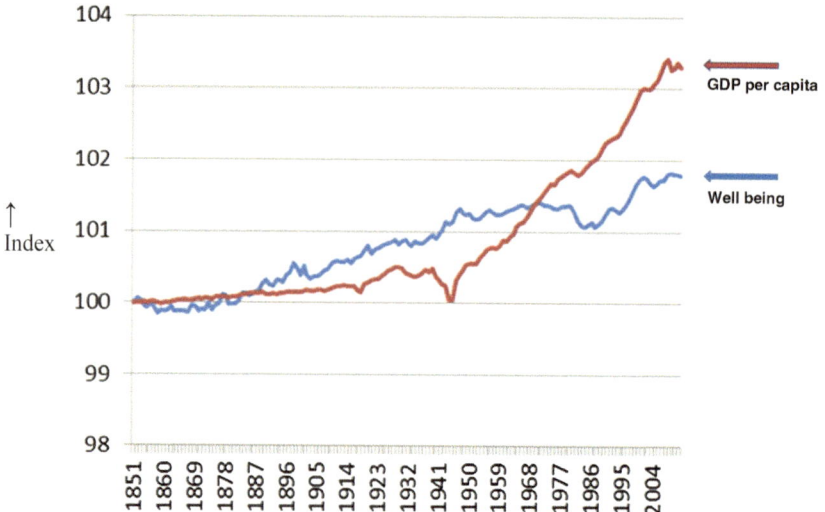

Graph 2.2 Well-being effects of economic growth in the Netherlands, 1850–2010 (compressed and composite index 1850 = 100)
Technical explanation: See note 5 of Chap. 22

GDP and increases in well-being appears to be very complex. In the rest of this volume the nature of this relationship will be investigated from period to period.

In addition to an examination of the development of well-being we are also interested in this study whether the increase in quality of life has been accompanied by overexploitation or even depletion of Dutch or foreign resources. From a present-day perspective the greenhouse gas problem is one of the greatest sustainability challenges.

In an economic system largely based on the exploitation of fossil fuels, every form of economic development is accompanied by CO_2 emissions. Still, important differences among the various periods can be ascertained. For example after 1850 economic growth is clearly associated with an increase in energy consumption (Graph 2.3). After the Second World War, however, energy consumption increases more than ever before. The same applies to greenhouse gas emissions, which means that the sustainability norm for CO_2 emissions around 1970 is exceeded (see also Graph 22.3).[32] How can this sudden turnabout be explained? Which economic activities were responsible?

Next to these general tradeoffs we also focus on more specific tradeoffs, for example that between biodiversity and efforts to achieve nutritional security or that between the quality of water, soil and air and the processes of upscaling and increas-

[32] J.P. Smits, 'De omvang en oorsprong van de milieuschade 1910–1995', in: R. van der Bie en P. Dehing, red., *Nationaal goed. Feiten en cijfers over onze samenleving (ca) 1800–1999* (Den Haag-Heerlen 1999), 235–254. Zie: https://www.cbs.nl/NR/rdonlyres/8EC09284-EF4A-4C0A-8FA3-CA95AB7ED4B1/0/nationaalgoed.pdf

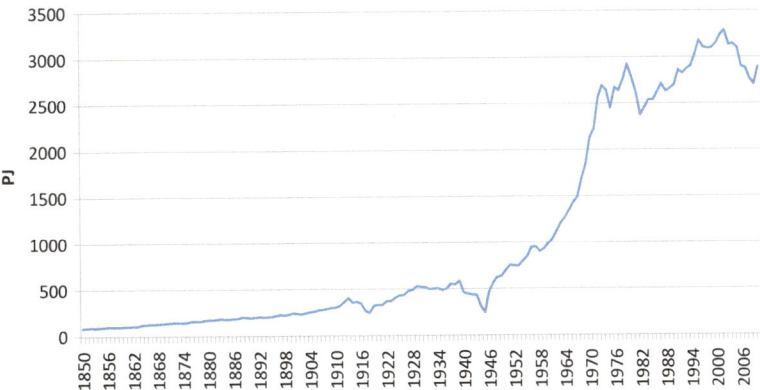

Graph 2.3 Total energy consumption in the Netherlands, 1850–2010 (PJ)
Source: H. Hölsgens, *Energy transitions in the Netherlands: Sustainability challenges in a historical and comparative perspective* (Groningen 2016), 11–12 and B. Gales and H. Hölsgens, 'Energy consumption in the Netherlands (1800–2012), in: H. Hölsgens, *Energy transitions in the Netherlands: Sustainability challenges in a historical and comparative perspective* (Groningen 2016), appendix I

ing consumption of energy. Issues like these are central to the upcoming chapters of this volume. We continue to focus on the choices historical actors made in solving their sustainability problems, on the technologies that they employed and above all on the institutional context that helped or perhaps hindered them in finding solutions.

2.8   The Structure of This Book

In this volume we distinguish three periods: 1850–1910, 1910–1970, and 1970–2010. Periodisation is of great importance because the relationship between economic growth and the development of well-being and sustainability is not linear. It always involves complex relationships that have to be studied in their specific institutional context.

This choice of periods has to do with the availability of data for the years 1850, 1910, 1970 and 2010. In addition, these years are seen as important watersheds in the social-economic literature.[33] An economic and technological perspective also justifies the choice for these periods.[34]

---

[33] See for example: A. Maddison, *Dynamic forces in capitalist development. A long-run comparative view* (Oxford 1991). Maddison uses 1850, 1913 en 1973 as turning points.

[34] J.P. Smits, H. de Jong en B. van ark, 'Three phases of Dutch economic growth and technological change, 1815–1997', *Groningen Growth and development Centre Research Memorandum GD 42* (Groningen 1999). See: http://www.ggdc.net/publications/memorandum/gd42.pdf

The period 1850–1910, certainly for the Netherlands, counts as that of the classical first industrial revolution with steam, textiles and the factory system as important features. The period 1910–1970 may at first sight appear to be anything but homogeneous. From a political perspective this is certainly the case; the period was characterized by a major economic depression in the interwar period and by two world wars. From a technological and economic perspective, however, it can be seen as a single period, shaped, among others, by a second industrial revolution rooted in electricity, chemistry and large-scale industrial enterprises.[35]

The first years of the 1970s can be seen as a turning point in many respects. Technologically speaking, the system of the second industrial revolution had by then all but exhausted its potential.[36] By the 1980s the contours of a new techno-economic regime could be discerned, a regime that emphasized information and communication technology, services and small and medium sized businesses.[37] Finally, at the outset of the 1970s it was becoming apparent that in an ecological sense we were reaching the 'limits to growth.' More and more attention was paid to the problem of the inordinate claims of economic growth on natural resources.[38]

For each period we review the results of the well-being monitor and present detailed data about natural capital. We consistently analyze developments in terms of the analytical quadrants of political and economic institutions, the societal 'midfield,' and the knowledge infrastructure.

To start off with, however, we shall examine in greater detail the situation around the middle of the nineteenth century, aiming to sketch an adequate frame of reference for subsequent periods. How did the Netherlands exploit its natural capital? How should we understand the quality of life around 1850? What were the most important issues in regard to well-being and sustainability from a present-day *and* a contemporary perspective?

We conclude the book with a summary of the historical development and a preview of 2050. Can we learn from history with a view to a sustainable future?

---

[35] R.J. Gordon, 'US economic growth since 1870: one big wave?', *American Economic Review* 89 (2) 1999, 123–128. Gordon argues that this 'one big wave' was impelled forward by, among other things, a second industrial revolution.

[36] J.P. Smits, 'Technology, productivity, and welfare', in: J. Schot, A. Rip and H. Lintsen, eds., *Technology and the making of the Netherlands. The age of contested modernization, 1890–1970* (Zutphen 2010), 454–455.

[37] Smits, Technology, productivity, and welfare, 455–456.

[38] The birth of this modern environmental consciousness is marked by among other things Rachel Carson's study of the DDT crisis and the way in which poisonous substances can penetrate deeply into ecosystems in a short span of time: Rachel Carson, *Silent Spring* (Harmondsworth 1962). Extremely influential in the early 1970s was the Report of the Club of Rome, 'limits to growth': Meadows et al., *The limits to growth: a global challenge* (n.p. 1972).

Literature

Bieleman, J. (1992). *Geschiedenis van de landbouw in Nederland 1500–1950*. Amsterdam: Boom.

Burger, A. (1996). 'Dutch patterns of development: economic growth and structural change in the Netherlands'. *Economic and social history in the Netherlands, 7*, 109–132.

Carson, R. (1962). *Silent spring*. New York: Mariner Books.

Centraal Bureau voor de Statistiek (2015a). *Sustainablility monitor of the Netherlands 2014:Indicator report*. Den Haag: Statistics Netherlands.

Centraal Bureau voor de Statistiek (2015b). *Internationaliseringsmonitor 2015: Vierde kwartaal*. Den Haag: Statistics Netherlands.

Gordon, R.J. (1999). 'US economic growth since 1870: one big wave?' *American Economic Review, 89*(2), 123–128.

Griffiths, R.T. (1980). *Achterlijk, achter of anders? Aspecten van de economische ontwikkeling van Nederland in de 19ᵉ eeuw*. Amsterdam: VU Uitgeverij.

Hayami, Y and Godo, Y. (2001). *Development economics: From the poverty to the wealth of nations*. Oxford: Oxford University Press.

Hoffmann, R.C. (2014). *An environmental history of medieval Europe*. Cambridge: Cambridge University Press.

Horlings, E. and Smits J.P. (1996). 'Private consumer expenditure in the Netherlands, 1800–1913'. *Economic and social history in the Netherlands, 7*, 15–40.

Human Development Report (2010). *The real wealth of nations: Pathways to human development*. United Nations Development Programme.

Knibbe, M.T. (2007). 'De hoofdelijke beschikbaarheid van voedsel en de levensstandaard in Nederland, 1807–1913'. *Tijdschrift voor Sociaal-Economische Geschiedenis, 4*, 71–107.

Kooij, van der T. P. (1931). *Hollands stapelmarkt en haar verval*. Zutphen: Paris.

Lambert, F. (2016). *Massastromen in Nederland. In de jaren 1850, 1913, 1970, 2010*. Eindhoven: Technische Universiteit Eindhoven, research report, open access.

Langenberg, H and Smits, J.P. (2015). 'Invoer van grondstoffen uit LDCs: geworteld in koloniale tijden?' In Centraal Bureau voor de Statistiek. *Internationaliseringsmonitor 2015: Vierde kwartaal*. Den Haag: Statistics Netherlands, 43–57. See: https://www.cbs.nl/nl-nl/publicatie/2015/44/internationaliseringsmonitor-2015-vierde-kwartaal.

Lintsen, H. (2005). *Made in Holland: Een techniekgeschiedenis van Nederland [1800–2000]*. Zutphen: Walburg Pers.

Maddison, A. (1991). *Dynamic forces in capitalist development. A long-run comparative view*. New York: Oxford University Press.

Meadows, D.H. et al. (1972). *The limits to growth: A report for the Club of Rome's project on the predicament of mankind*. New York: Universe Books.

Meere, J.M.M. de (1982). *Economische ontwikkeling en levensstandaard in Nederland gedurende de eerste helft van de negentiende eeuw*. Den Haag: Nijhoff.

North, D.C. (1990). *Institutions institutional change and economic performance*. Cambridge: Cambridge University Press.

Rooy R. de (2014). *Ons stipje op de waereldkaart: De politieke cultuur van modern Nederland*. Amsterdam: Wereldbibliotheek.

Smits, J.P., Jong, J. de, and Ark, B. van (1999). 'Three phases of Dutch economic growth and technological change, 1815–1997'. *Groningen Growth and development Centre Research Memorandum GD-42*.

Smits, J.P. (2010). 'Technology, productivity, and welfare'. In J.Schot, A. Rip and H. Lintsen (Ed.), *Technology and the making of the Netherlands: The age of contested modernization, 1890–1970* (pp. 454–455). Zutphen: Walburg Pers.

Wal, R. van der (2003). *Of geweld zal worden gebruikt: Militaire bijstand bij de handhaving en het herstel van de openbare orde 1840–1920*. Hilversum: Verloren.

Vries, J. de (1959). *De economische achteruitgang van de Republiek in de achttiende eeuw.* Leiden: Stenfert Kroese.

Vries, J. de (1974). *The rural Dutch economy in the golden age, 1500–1700.* New Haven: Yale University Press.

Vries, J. De (1978). *Barges and capitalism.* Wageningen: AAG Bijdragen.

Vries, J. de (1984). *European urbanization 1500–1800.* London: Harvard University Press.

Zanden, J.L. van (1991). *Arbeid tijdens het handelskapitalisme: Opkomst en neergang van de Hollandse economie 1350–1850.* Bergen: Octavo Uitgeverij.

Zanden, J.L. van (1993). 'The Dutch economy in the very long run. Growth in production, energy consumption and capital in Holland (1500–1805) and the Netherlands (1805–1910)'. In Szirmai, A., Ark, B. van and Pilat, D. (Eds.): *Explaining economic growth. Essays in honour of Angus Maddison* (pp. 267–283). Amsterdam: Elsevier.

Zanden, J.L. van and Riel, A. van (2000). *Nederland 1780–1914: Staat, instituties en economische ontwikkeling.* Amsterdam: Balans.

Zanden, J.L. van et al (2014). *How was life? Global well-being since 1820.* Paris: OECD Publishing.

Well-being and Sustainability Around 1850: A Search for a Frame of Reference

Chapter 3
Natural Capital, Material Flows, the Landscape and the Economy

Harry Lintsen

Contents

Abstract In this study, natural capital occupies a special place. After all, the way natural capital is exploited is crucial for issues of well-being and sustainability. This chapter analyses the exploitation of natural capital in the Netherlands around 1850. It deals with three categories of resources produced by natural capital.

First, organic resources (including grains, potatoes, cattle and milk) and the associated supply chains in agriculture and foods (farming systems, food production and food consumption).

Second, mineral subsoil resources (in particular sand, clay and gravel) and the associated supply chains in construction (housing construction, water management, and infrastructure).

This chapter is written by Harry Lintsen with contributions by Fred Lambert, Rick Hölsgens and Ben Gales.

© The Author(s) 2018
H. Lintsen et al., *Well-being, Sustainability and Social Development*,
https://doi.org/10.1007/978-3-319-76696-6_3

Finally, fossil subsoil resources (in particular turf and coal) and the associated energy supply chains (industry and households).

The exploitation of natural capital took place in a variegated landscape possessing great biodiversity and an agricultural economy that provided plenty of room for population growth.

Keywords Natural capital · Agriculture · Foods · Construction · Energy · Biodiversity · Economy

3.1   Recalcitrant Sand Grounds

It was always hard work on the Dutch sand grounds. 'The rooster is the alarm clock of the farming family', as an official report on farm life puts it:

> Work begins... in the early morning; crop farmers start in the summer at 4 and work until 8 o'clock in the evening, in the winter from 4 in the morning to 6 or 7 o'clock in the evening; grass or dairy farmers begin work an hour later in the winter and are done an hour earlier.[1]

Making manure was the core of farm work on the sand grounds.[2] This demanded heavy labour and was therefore men's work: cutting peat sods on the heath, loading the wagon, mixing the peat with manure from the stable, mucking out the stable, transporting manure to the fields and spreading it out over the fields. Adequate fertilizing of the soil was the key to survival on the sand grounds. Tilling the soil as well as harvesting was also men's work: ploughing, harrowing, delving, mowing, threshing the grain, etc. That was also the case for a variety of tasks like cutting wood, preparing flax and everything to do with the horse or the ox. Evenings were devoted to making brooms, repairing halters for the horses, knotting ropes for the cows, sharpening knives and so on.

Women's work consisted of taking care of the cow, the calf, the pig and the chickens. The farmer's wife was responsible for milking and processing the milk, in particular churning it into butter. She also helped with harvesting, maintained the vegetable patch and provided meals for her husband and the children. Making and repairing clothes was reserved for the evenings. In North-Brabant much of her time

---

[1] J. Zeehuisen, 'Statistieke bijdrage tot de kennis van den stoffelijken en zedelijken toestand van de landbouwende klasse in het kwartier Salland, provincie Overijssel', in: J.L. Van Zanden, *'Den zedelijke en materiële toestand der arbeidende bevolking ten platten lande'. Een reeks rapporten uit 1851* (Historia Agriculturae XXI, Groningen 1991). Original source: *Tijdschrift voor Staathuishoudkunde en Statistiek* 6(1851), 380.

[2] For a description of farming on the sand grounds see A.H. Crijns en F.W.J. Kriellaars, *Gemengd landbouwbedrijf op de zandgronden in Noord-Brabant 1800–1885,* (Tilburg 1987); J. Bieleman, *Boeren op het Drentse zand 1600–1910. Een nieuwe visie op de 'oude' landbouw* (Wageningen 1987); J. Bieleman, *Boeren in Nederland. Geschiedenis van de landbouw 1500–2000* (Amsterdam 2008) en G. van den Brink, *De grote overgang. Een lokaal onderzoek naar de modernisering van het bestaan. Woensel 1670–1920* (Nijmegen 1996).

was taken up with preparing slops for the cattle: three times a day in the winter and twice in the summer.[3] For the slops, the farmer's wife used cut tuber greens and tubers, corn spurrey (greens, hay, chaff and husks), potatoes, carrots, rape siliqua, flax husks, clover, cut hay, chopped straw, sometimes buttermilk and when the mixture was cooked also 'cake water' (made from rape cakes soaked in water), while oat or rye flour was often also added to the brew. 'Without fodder no good cattle, no good manure, no good harvest, no profit...' thus was its importance described.

> This important part of the enterprise rests largely on the woman, who from early morning to late evening toils for her cows and thus devotes little attention to her own home, because the time is needed in or for the stable: that is where she lives, there is her sole pride.[4]

Work was apportioned by sex but also by age. It was normal for children to be burdened with supporting tasks. From an early age they contributed by gathering wood, picking berries, digging up thistles and in some cases herding cattle in the fields. At the age of ten boys joined their fathers in helping to prepare the manure and work the land, while the girls helped the mother with her work.

On the bigger farms (on the sand grounds farms of more than 5 hectares) both the farmer and his wife had their hands full running their farm.[5] Usually they had the help of one or more domestic farmhands and maids. That depended on the size of the farm and the number of eligible children. But the sand grounds were dominated by smallholders with farms of less than 5 hectares. Many owned only a small plot of land of a hectare or less on which they cultivated potatoes and some vegetables and kept a cow, pig, goat or a few chickens. This was often inadequate to make a living. So next to running their own little enterprise they worked on other farms or derived income from other activities. Threads were spun to order, textiles woven, mats braided etc. The whole family took part. A reporter from the province of Drenthe described the smallholder as follows:

> He has very few needs and mostly a bit of land near his habitation and a Frisian sheep in the stable, sometimes a cow from Drenthe. He tills his soil in his free time, assisted by his wife and children; the sheep's wool serves for clothing; the *vest* jacket and the *five-shaft* coat are usually woven in the village itself; he is used to heavy labour...[6]

The sand grounds belong to the natural capital of the Netherlands. Natural capital occupies a special position in this study. It is viewed as the basis for quality of life (see Chap. 1). It is clear that natural capital does not bestow its favours without a struggle. It has enormous potential in the Netherlands, but its exploitation demands

---

[3] Bieleman, *Boeren in Nederland*, 266.

[4] Cited from: Van den Brink, *De grote overgang*, 125. Original source: W. van Iterson, *Schets van de landhuishoudkunde der Meijerij* (Groningen 1868), 85.

[5] Crijns en Kriellaars, *Gemengd landbouwbedrijf op de zandgronden in Noord-Brabant*, 78 en 85.

[6] P.W. Alstorphius Grevelink, 'Zedelijke en materiële toestand der arbeidende bevoking ten platten lande. Nopens Drenthe (1850)', in: J.L. Van Zanden, *'Den zedelijke en materiële toestand der arbeidende bevolking ten platten lande'. Een reeks rapporten uit 1851* (Historia Agriculturae XXI, Groningen 1991). Original source: *Verslag van het verhandelde op het vijfde Nederlandsche Landhuishoudkundig Congres, gehouden te Leiden van den 10den tot den 14den junij 1850* (Leiden 1850), Appendix no 13, 111.

tremendous efforts. This chapter develops the theme of natural capital. How can we characterize Dutch natural capital in the middle of the nineteenth century? What products did it help to produce and what services were provided? What were the consequences for nature? How did it contribute to economic growth? The next chapter examines the quality of life that was achieved in the Netherlands on the basis of that natural capital.

We begin with a characterization of natural capital and we do that in terms of colonised ecosystems. These are natural ecosystems that have been influenced by humans with the aim of producing products and services (see Chap. 1). We can characterize colonised ecosystems in the Netherlands in various ways, depending on the system boundaries we choose. Here we take water, air and soil as a point of departure. How was nature colonised around 1850? And to what degree were those systems sustainable?

3.2 Natural Capital: Water, Air and Soil

The Netherlands could (and can) be perceived as a large colonised water system, although with the reservation that in the nineteenth century the influence of the Dutch on that system was still limited. 110 billion cubic meters of fresh water annually flowed into the country, in part via the rivers and partly as precipitation.[7] Most of this (about 80%) left the country directly as it flowed into the sea. The rest evaporated (about 20%). Less than 1% was used in agriculture, by households and in industry. Even that miniscule percentage still represented a useful material flow of millions of tons of water. The Dutch population would have consumed something like 2.0 to 2.6 million tons of water annually (directly as well as indirectly as a constituent of food).[8]

Water played a key role in the Netherlands in a number of ways. As in other countries it was an elixir of life for the population in the form of drinking water. In the nineteenth century there was much ado about its quality. We will return to this point. Moreover it was an essential factor of production in agriculture (including fisheries) and industry. In addition water in the Netherlands uniquely fulfilled two other roles, namely as a medium for transportation and trade and for territorial defence.[9]

---

[7] *De Bosatlas van Nederland waterland* (Groningen 2010), 57. Figures refer to the present-day situation and will be of the same order of magnitude as in 1850. Only the percentage of water use in agriculture, households and industry has been adjusted from 4.5% at present to less than 1% at the earlier epoch. This seems realistic in view of the significantly smaller population and scope of economic activity.

[8] Adults need at least 2 to 2.5 liters of water per day (including water in foods); children older than one year 1.5 to 2 liters per day. We assumed an average of 1.8 to 2.3 liters. For a population of 3.1 million persons in the Netherlands around 1850 this amounts to 2.0 to 2.6 million tons of water (including water in foods).

[9] See: C. Disco and H.W. Lintsen, 'Het nijvere verbond', in: J. Schot, H.W.Lintsen, A. Rip and A.A. Albert de la Bruhèze (eds.), *Techniek in Nederland in de twintigste eeuw* (Zutphen 1998–

Since medieval times, inner waters, lakes and large natural waterways played an essential role in regional, interregional and even international trade. In the course of the seventeenth century a dense network of tow-canals came into existence on which tow-barges operating on accurate schedules maintained regular connections among important cities and smaller towns. In the first half of the nineteenth century a number of new canals like the Zuid-Willemsvaart and the North-Holland Canal (*Noord-Hollandsch Kanaal*) were added. From this perspective, water was the necessary condition for the riches gathered by, in particular, the province of Holland. Water was the most important means of transport for commodities. It was from water that the Netherlands, that is, the Republic of the Seven United Provinces, derived its power as a trading nation.

Water was also the ally that helped defend the commercial position and the accumulated riches against foreign aggression. Various defensive water-lines were constructed in order to defend the economic and political heart of the Netherlands – namely the province of Holland and in particular Amsterdam.[10] Water-lines operated on a simple principle: one opened the inundation sluices,[11] flooding the polders and creating a treacherous undeep water surface with numerous invisible ditches. The enemy could advance only along the more elevated dikes and roads, but this was blocked by a system of strategically positioned forts and sconces.

It is nonetheless important to realize that the colonisation of nature regularly misfired and that nature could easily become counterproductive: the sea threatened the land with storm surges, the rivers regularly flooded broad swaths of countryside as they overtopped their levies, lakes, their surfaces whipped by heavy winds, could also cause flooding. The low-lying portion of the Netherlands struggled with a surfeit of water while the higher portion ran the risk of water shortages. Water was not only an ally but a water-wolf that had to be tamed.

In the first half of the nineteenth century colonised ecosystems were Janus-faced. That certainly applied to air as a source of life for humans, plants and animals. As a factor of production it played a crucial role in industrial heating and chemical processes. It was used in households to heat and cook. If we focus on the atmospheric and meteorological aspects of air, then we soon encounter weather and the extensive use of wind as a source of energy. Wind supplied the energy for thousands of windmills and sailing ships.

But air also had a problematic side. Cold and poor air quality caused flu, tuberculosis and other epidemic diseases. On calm days the windmills languished, in stormy weather ships perished, during extreme weather entire harvests were lost and human existence was hard.

2003), part I, 55–63; G.P. van de Ven (ed.), *Leefbaar laagland. Geschiedenis van de waterbeheersing en landaanwinning in* Nederland (Utrecht 1993); A. Bosch and W. van der Ham, edited by H.W. Lintsen, *Twee Eeuwen Rijkswaterstaat 1798–1998* (Zaltbommel 1998).

[10] De oude Hollandse Waterlinie (1673), de Grebbelinie (1742), de nieuwe Hollandse Waterlinie (1815), de Stelling van Amsterdam (1880) en de IJssellinie (1951).

[11] Sluices by means of which low-lying ground could be inundated to form a defensive perimeter.

The soil and subsoil is the third element in our survey of colonised ecosystems. Underground we find resources like coal, oil, gas and salt, of which in the mid-nineteenth century only coal was known to exist. This was mined in modest amounts in South Limburg. The top layer of the Dutch delta exhibits a variegated structure. The usual dichotomy is between the high and the low Netherlands. The high Netherlands comprises the loess grounds in the south and the sand grounds in the south and east of the country. The low Netherlands derives its soil structure from complex interactions among a rising sea level, the course of the large rivers, the formation of thick packets of peat and of fossil beach ridges and dunes. The presence of sea clay, river clay, sand and peat makes for a rich variety in soil types. For any given location the type of soil, the situation and the elevation co-determine what kind of agricultural production is possible and what kinds of fuels and minerals can be extracted. This connects the useful material flows that characterize the Netherlands around 1850 with the structure of the ground. We now provide a brief sketch of the colonisation of the soil and subsoil.

3.3 Agriculture and Nutrition: Organic Raw Materials

3.3.1 Mixed Farming on the Sand Grounds

The exploitation of the Dutch soil led to a variety of modes of agricultural extraction.[12] This was due not only to the variety of soil types and their relative fertility. Tradition was also an element, especially styles of commercialization. This led to the dominance of mixed farming on the sand grounds, with a central role for the crop field.[13] Manure, cattle and crop-rotation were essential to maintain fertility.[14] Manure and cattle have already been mentioned (see the introduction to this chapter). A popular choice in crop rotation on the sand grounds was rye because it did well in poor soil. In Twente it was rotated with so-called catch crops like red clover, tubers and corn spurry (*spergula arvensis*) that were planted after the rye harvest and used

---

[12] See for agriculture in the Netherlands: J.L. van Zanden, *De economische ontwikkeling van de Nederlandse landbouw in de negentiende eeuw 1800–1914* (dissertation Wageningen 1985); Bieleman, *Boeren in Nederland*; Crijns en Kriellaars, *Gemengd landbouwbedrijf op de zandgronden in Noord-Brabant*; Bieleman, *Boeren op het Drentse zand*.

[13] It is important to note that mixed farms on the sand grounds did not comprise a single type of system, but that they differed across time and region in the Netherlands. Hence it was possible to invest more heavily in cattle husbandry and dairy farming. Bieleman, *Boeren in Nederland*, 260–267.

[14] Cattle could be pastured on stubble fields, grasslands and heather meadows. They could also be kept in stalls, like the deep-litter stalls of North Brabant and fed on hay, tubers, spurrey and other fodder. This practice became more common at the beginning of the nineteenth century and contributed to the increase of manure production and ultimately the size of the harvests. Van Zanden, *De economische ontwikkeling van de Nederlandse landbouw*, 179. Another factor in increasing the size of the harvests was improved tillage thanks to intensive weeding and the use of catch crops.

as cattle fodder and to combat weeds. As 'green manure' these crops also contributed to the restoration of soil fertility. In the province of Drenthe the rotation consisted of, for example, winter and summer rye, but also included potatoes, flax, oats and barley. In other regions, for example in North Brabant, buckwheat often replaced summer rye in the system of rotation, but potatoes and oats were also planted. A well-planned crop rotation scheme could significantly reduce the acreage that lay fallow. The way in which soil fertility was maintained differed through time and from place to place. The soil and agricultural expert, Winand Staring, distinguished six different systems on the sand grounds in 1869.

Moors, grasslands and forests were generally reckoned to the common grounds. They served as sources of peat and firewood, as pastures for sheep and so forth. The danger was overuse and the spread of sand drifts.[15] Adequate management was therefore of supreme importance. The common grounds were collectively owned or open to collective use. The owners were local farmers or, as in North Brabant and Limburg, the municipalities. Sometimes the national state was proprietor. Collective management was known by a variety of Dutch names such as *marken, maalschappen, holtelingen, gemeenten, boerschappen,* or *buurschappen.* They all enforced numerous proscriptions aimed at regulating overuse and ensuring survival of these vulnerable areas.[16] It was stipulated who could use the common grounds and in what way they were to be used. A board of governors decided on releasing grounds for cultivation.

Farming on the sand grounds, which consisted largely of smallholdings, was in many cases primarily oriented to satisfying the direct needs of the farmer and his family.[17] It was also essential for realizing a certain surplus in order to pay taxes. And in case there was anything left over it was possible to buy goods like salt, soap, vinegar and tobacco and to invest in land, cattle and agricultural tools. The surplus was realized by selling farm products or performing work for others. In this way the farm touched on the market economy, albeit in a very limited way. The first half of the nineteenth century nonetheless witnessed an increasing orientation to the national and even international market. The sand grounds in the eastern Netherlands pioneered this development.

---

[15] After digging out the peat sods a sandy subsoil remained that could regain fertility after a time if not used too intensively. These limits were violated if peat extraction became excessive. Another excess was the pasturing of flocks of sheep, a practice that created tracks of trampled and denuded paths through the heather.

[16] T. de Moor, *The dilemma of the commoners. Understanding the use of common-pool resources in long-term perspective* (Cambridge 2015). See also chapter 8 of our study.

[17] See: Van Zanden, *De economische ontwikkeling van de Nederlandse landbouw,* 145–199; Bieleman, *Boeren op het Drentse zand;* Van den Brink, *De grote overgang,* 184–192. H. Knippenberg and B. de Pater, *De eenwording van Nederland* (Nijmegen 1990), 98–102.

3.3.2   Grassland Farming in the Low-Lying Peat Marshes

Exploiting the Dutch soil demanded not only heavy labour, but also cautious interventions, much knowledge, careful procedures and sound management. That was not only the case for the sand grounds. In the low-lying peat marshes of the Western Netherlands, for example, grassland farming was developed.[18] This specialized mode of farming centred on the grass and the cow. Worldwide, these areas had the highest density of cows per hectare. The quantity of milk per cow was among the highest in the world. This was due among other things to astute cattle breeding, the quality of winter fodder and the hygienic preparation of the dairy products. The growth of grass was fundamental to this type of farming and was in large measure dependent on the water level.

Managing water levels was thus a major priority and this was accomplished within the institutional framework of the polders. Authority lay with the landowners. Management was in the hands of a polder or water board council composed of polder masters or dike reeves (*heemraden*). Relationships, interests and procedures were laid down in regulations. By the twelfth century, cooperation among polders had already led to the formation of regional water boards. These were responsible for the construction and maintenance of hydraulic works that were shared by all the polders, for example a common drainage basin (*boezem*) or a sea-dike. These were often called *hoogheemraadschappen*.[19]

3.3.3   Regional Variaty

In this way every region had its own specific struggle with the soil and its specific institutions. The clay grounds in the central river zone were tough and nearly impermeable to water. The clay embankments were well-suited to crop farming, the low-lying grounds (so-called *komgronden*) by contrast, were unprofitable. Weed pollution was one of the big problems. There was the recurrent danger of dike breaches. Seepage water plagued farming in the lower lying fields. Water boards lacked the means to control the big rivers.

In the northern and southwestern coastal zones extensive dike-building had deprived the sea of its casual access to the land and had marked the start of a continual struggle between the sea and the inhabitants under the leadership of the water boards. Behind the dikes, the banks of former tidal creeks, mud flats and pools had changed into fields and grassland. New land was added to the old. Both old and new land regularly had to be surrendered to the sea again. Crop growing with a mixed farm was the dominant type, but this included all kinds of variations depending on the quality of the soil and the market conditions.

[18] Bieleman, *Boeren in Nederland*, 210–234.

[19] Van de Ven, *Leefbaar Laagland*, 69–70.

3.3.4 *Market Orientation*

Market orientation also influenced how the soil was worked. Compared to farming in the eastern and southern provinces, farming in the coastal provinces (Zeeland, Holland, Friesland and Groningen) was strongly commercialized. Agriculture was therefore also more specialized. Grassland farming supplied Dutch cities and foreign countries with dairy products. North Holland was the prime cheese producing region; Friesland and South Holland were important centres of butter production.[20] Making butter and cheese was the work of the farmers' wives, their daughters and the maidservants on the farm. Merchants bought butter and cheese at the farms, traded large consignments at weekly markets in nearby cities where the farmers also sold their wares, and shipped large quantities abroad, in particular to England. At the outset of the nineteenth century a quarter of London's butter and cheese market was in Dutch hands.

The mixed farm in the coastal provinces produced wheat, oats and potatoes, often destined for the market and not for use on the farm itself. The farm also produced industrial crops like rape, flax and barley. Madder was a popular industrial crop in Zeeland and South Holland.[21] The roots provided a red dye used in dying and printing textiles. Extraction took place in small batches according to traditional practice in drying ovens called *meestoven*, usually the collective property of a number of farmers. The main elements of the process were the drying of the madder roots, their cleaning and finally pulverization. The powder was transported in large wooden barrels to the staple market in Rotterdam. Textile printers and dyers in England, the United States and Germany were important customers. Dutch madder was held in high regard and it was protected by a web of laws that, for example, specified the percentage of impurities that a certain quality of madder could contain.

Specialized, commercial agriculture also had consequences for the social structure of the countryside.[22] An important difference with the sand grounds was the size of farms. Large farms (more than 20 hectares) with their hired hands and maidservants dominated in the sea clay regions in the provinces of Zeeland, Friesland and Groningen. These were also dominant in the river clay regions but less so. Smallholders were all but non-existent. Instead, there was a large population of landless farm workers and day-labourers, that was totally dependent on paid work. They earned most in the summer, less in winter, and only on days that they worked. In part they were seasonal migrant workers that relieved the shortage of farm labour at harvest and mowing time. In the provinces of Drenthe, Overijssel and

---

[20] See for butter: M.S.C. Bakker, 'Boter', in: H.Lintsen et al. (ed.), *Geschiedenis van de techniek in Nederland. De wording van een moderne samenleving 1800–1890* (Zutphen 1993), part I, 103–133.

[21] J.W. Schot and E. Homburg, 'Meekrap en garacine', in: H.Lintsen et al. (ed.), *Geschiedenis van de techniek in Nederland. De wording van een moderne samenleving 1800–1890* (Zutphen 1993), part IV, 222–239.

[22] Bieleman, *Boeren op het Drentse zand*. Van Zanden, *De economische ontwikkeling van de Nederlandse landbouw,* 315–325

North Brabant they were in the minority. There the bulk of the rural population was composed of smallholders with one or two horses and farmworkers/smallholders without a horse but sometimes with a parcel of land and a few cows. Large agricultural enterprises with permanent farmhands also dominated in the peat marsh regions of the polders in North and South Holland with their specialized grassland farming. Inasmuch as this kind of farm was less subject to seasonal fluctuations in labour needs, these regions lacked a population of itinerant and occasional farmworkers.

Commercial farms also contracted more of their production out to specialized enterprises. Work formerly done by the farmers themselves, like spinning and weaving, were now undertaken by specialized businesses and persons. In consequence the commercial farming regions had a more differentiated occupational structure and a larger contingent of specialized craftsmen.

3.3.5 Food Supply

Dutch soil produced a cornucopia of agricultural products. These products were not subject to extensive processing.[23] Potatoes and vegetables were pretty much cooked as is. Wheat had to be ground and bread had to be baked. Milk was drunk straight from the cow, utilized in porridge, churned and turned into butter and cheese. Consumption of dairy products was, however, limited. There was little variety in foodstuffs, at most a few hundred different products. The Dutch above all ate what was locally produced. Culinary culture differed according to social class, region, season and between town and country.

Food processing was done on a small-scale, traditional basis, largely in the household. Sources of energy were classic and consisted of muscle power (of humans, horses and dogs), wind energy throughout the country and water power in some areas, and warmth provided by peat, wood or coal. Tools were made of wood and metal, and simple enough that the local carpenter or smith could make, repair or imitate them. Craft knowledge was transmitted from father to son, from mother to daughter, from master to apprentice and was further developed in practice. The distance between producers and consumers was small. The Dutchman bought directly from the farmer or the baker, at day or week markets or in small speciality shops. Food supply was above all bound to the rules of the local occupational communities and the municipal governments.

In addition to agricultural products, the soil and especially the subsoil of the Netherlands also produced fuel and minerals. We will now turn to these.

---

[23] H. Lintsen et al., *Made in Holland. Een techniekgeschiedenis van Nederland [1800–2000]* (Zutphen 2005), 38.

3.4   Building Materials and Construction: Mineral Resources

Clay, sand and gravel were the voluminous building materials for hydraulic engineering, road building and housing and building construction. But little could be done with only clay, sand and gravel. Many other building materials were necessary and these came in large part from abroad.[24] Wood was an important building material. Oak and pine were supplied from Germany via the Rhine. Scandinavia, the Baltic Sea harbours and Russia supplied softwoods like pitch pine and fir. Rocks were obtained from Germany (sandstone) and Belgium (basalt). Mortars were imported from the Belgian Meuse basin (chalk) and the German Eifel (trass cement). Fired shell chalk, another mortar, could be produced in the Netherlands itself by trawling for shells along the coast. With these and a few other materials like iron and asphalt, the Netherlands was able to carry on.

Many Dutchmen were in practice hydraulic engineers, because the local population, especially the farmers, were responsible for the maintenance of dike sections, digging drainage ditches, building simple sluices and so forth. This was inadequate for larger infrastructural works (canals, river works, coastal defences and such).[25] Here so-called polder workers commonly did the heavy labour of digging and transporting clay and sand. They came from the countryside and arrived *en masse* with their spades and wheelbarrows when a major work was set to start. Complex hydraulic projects involved all kinds of trades and technical specialists like carpenters, masons and engineers.[26]

Road building also depended on the local population, and in particular on farmers, workers and contractors. Local roads were simple and were made of clay, sand or peat. Sometimes they were surfaced with rubble or shells.[27] In addition there were hardened roads generally built under central government supervision. First the

---

[24] W. van Leeuwen, 'Woning- en utiliteitsbouw', in: H. Lintsen et al. (ed.), *Geschiedenis van de techniek in Nederland. De wording van een moderne samenleving 1800–1890* (Zutphen 1993), part III, 197–198.

[25] W. van Leeuwen, 'Waterbouw', in: H. Lintsen et al. (ed.), *Geschiedenis van de techniek in Nederland. De wording van een moderne samenleving 1800–1890* (Zutphen 1993), part III, 231–236.

[26] Building shipping locks and sluices was above all work for carpenters and masons. Wood was much used because the lock was primarily a wooden construction with wood for the lock gates, the sills, the foundations, 'eel planks' (contiguous wooden pilings placed at right angles to the lock axis to prevent seepage of water under the sills), etc. The construction site also sported other trades like pile drivers, smiths, laborers, contractors and sometimes even an engineer. Every type of hydraulic infrastructure had its own specialists. In dike construction, for example, willow workers made fascine mattresses that served to protect shorelines under water. A fascine mattress was a rectangular grid of bundles of brushwood filled in with reeds or willow twigs. On completion, the mattress was towed to its destination, ballasted with heavy stones and sunk to the bottom.

[27] A worn-in wagon track determined the profile of the road. The clay and peat roads were transformed into impassable mud-mires in the winter and after periods of heavy rain in other seasons. The sand roads were reasonably passable in winter but became loose sand in the summer.

existing ground was excavated and replaced by a layer of sand. On top of this a hardened road was built, usually composed of Dutch clinker bricks.[28]

Water management and infrastructure in the Netherlands were traditionally the responsibility of local interests (water boards and farmers) municipal and provincial governments and sometimes private investors. In the first half of the nineteenth century the national government, and in particular King William I, initiated the construction of comprehensive road and waterway networks and other infrastructural works. Provinces, municipal governments and water boards generally played a subsidiary role. The works contributed significantly to Dutch state formation.

In housing construction, new houses were built with a skeleton of brick load-bearing walls, wooden floors and beams, slate or baked roof tiles and forged iron wall anchors.[29] These techniques were applied to a wide variety of housing ranging from workers' homes, through middle-class homes, to mansions, urban villas and country estates. There was also a wide variety of non-residential buildings in the government sector (prisons, university buildings, covered markets, assembly halls), the private sector (among others warehouses, office buildings, workshops and hotels) and the semi-private sector (for example churches, theatres and exhibition halls). While these buildings were more or less complex – some might even be considered daring – they were for the most part traditional stone and wood structures. Other materials were limited to chalk, glass, paint, forged iron and stone.

Carpenters and masons were the most important craftsmen. They were part of a rich palette of craftsmen that could be encountered at a construction site, such as painters, stone masons, glaziers, decorators and slate roofers. There were also unschooled workers like porters and pile drivers. Part of the construction work force was employed by suppliers of materials, like sawmills and brick factories. The organization of a construction project was in the hands of a contractor who had learned by doing, generally as a carpenter or mason. Big projects might be managed by an architect or engineer.

Housing for workers comprised the majority of buildings. They hardly deserved the name. In the cities they took the form of slums built of stone and wood. On the edges of cities, in the countryside and in smaller towns one found numerous huts and underground burrows.[30] Walls were erected of lath, pine branches, and stones and filled in with clay, loam and horse manure. Roofs were covered with reeds. Burrows were dug in the ground. Huts were built of dried peat. The fireplace was a depression in the ground. Smoke escaped through a hole in the roof.

---

[28] Rainwater could easily drain through the layer of sand and the hard bricks did not sink into the subsurface after rainfall. The total width of the road was 8 to 10 meters, of which about half was paved.

[29] W. van Leeuwen, 'Een complexe sector', in: H. Lintsen et al. (ed.), *Geschiedenis van de techniek in Nederland. De wording van een moderne samenleving 1800–1890* (Zutphen 1993), part III, 192.

[30] See among others: A. van der Woud, *Koninkrijk vol sloppen. Achterbuurten en vuil in de negentiende eeuw* (Amsterdam 2010), 82–84.

3.5   Energy: Fossil Fuels

The extensive peat bogs that covered large parts of the Netherlands were a ready source of peat turf. Turf - a dried form of peat – represented a revolution in the domain of energy that had taken place as early as the late middle ages. From that time the country burned its own ground in order to produce, to cook and to warm itself.[31] The annual average temperature was 9 ° C and the winters were often bitterly cold. Turf was used in households for heating and cooking. Industry too was an important market. Beer breweries, chalk ovens, cotton printing mills, dye works, bleaching works and brick factories were the biggest users. Their fuel was turf and that would remain the case for centuries.

Dutch cities had a long history of dependency on turf.[32] Originally it was easily accessible in nearby surrounding peat bogs.[33] By the eighteenth century production had moved to the northeast.[34] The scale at which this took place required so much capital and was fraught with such risks that companies (*compagnieën*) were formed. These were consortiums, partnerships, in which anyone could invest and become stockholder, participant, partner or associate. Because it took a considerable time before these investments returned a significant profit, the companies were dominated by very wealthy investors from the province of Holland. Groningen was an exception. The city succeeded in retaining control over turf extraction in its region.

---

[31] See for this among others: T. Stol, 'Turfwinningslandschap', in: S. Barends, *Het Nederlandse Landschap. Een historisch-geografische* benadering (tenth revised edition, Utrecht 2010), 114–131.

[32] Economic prosperity in the Netherlands in the seventeenth century was in part due to the presence of this cheap source of energy in the immediate surroundings. Without turf the Golden Century would have been a more difficult business. Many activities were able to flourish thanks to this brown gold. Alternative fuels were available abroad, but at a higher price. Turf survived the Golden Century. Historians are of divided opinions on the meaning of turf in the Golden Century. According to J.W. de Zeeuw, turf was essential for economic success. See: J.W. de Zeeuw, 'Peat and the Dutch Golden Age. The historical meaning of energy attainability', *AAG Bijdragen* 21(1978), 3–32. J.L van Zanden takes a middle position. Other scenarios with less turf and more coal and firewood also led to a successful economic development. See: J.L. van Zanden, 'Werd de Gouden Eeuw uit turf geboren. Over het energieverbruik in de Republiek in de zeventiende en achttiende eeuw', *Tijdschrift voor Geschiedenis* 110 (1997), 484–499. Van Zanden writes not so much about a more difficult but rather about a different story for a Golden Century without turf.

[33] Originally peat extraction took place above the level of groundwater, so-called dry extraction. Once this supply was used up, wet extraction had to be employed. Working from a boat or a plank, the peat was dug out of the water with a dredging hook and dumped in a boat. The peat was subsequently unloaded on a laying field or dam, spread out, compacted, dried in the open air, pressed into forms, piled up, dried some more and finally sold as turf. Wet extraction was practiced until the seventeenth century above all in South Holland and the western part of Utrecht, in addition to the northern provinces.

[34] This displacement was part of the spread of the peat colonies. These began in Flanders and spread from there to northwest Brabant (thirteenth century), the Valley of Gelderland and Friesland (sixteenth century), Groningen (seventeenth century) and the Peel region (nineteenth century). Techniques and organization improved with each displacement.

Extraction was systematically executed with main canals, locks and bridges and additional infrastructure in the form of parcels, ditches and so forth. The turf was extracted mainly 'in the dry.' The top layer, the *bonkaarde* or *bolster*, had little value as fuel, but was later mixed with the infertile sand ground that was a common by-product of turf extraction. The *bonkaarde* provided an improved soil structure with better hydrological properties. It was in fact the point of departure for the system of mixed farming, with cattle providing the manure. Urban waste also served as fertilizer. The turf ships transported this commodity as return freight.

Turf production reached its zenith in the nineteenth century. Expansion of turf extraction in the north eastern provinces of Drenthe and Overijssel was made possible by the digging of a new canal, the *Dedemsvaart* (1809). A bit later the Peel region spanning the border between the southern provinces of North Brabant and Limburg was likewise made accessible for turf extraction by a new canal called the *Zuid-Willemsvaart* (1826). Official statistics indicate that in 1850 the Netherlands burned 2.5 million tons of turf.[35] This comes down to 750 kg of turf per capita. To this must be added the turf that residents of the countryside extracted for their own use from nearby peat bogs. Ultimately turf lost out as a fuel to ever-cheaper coal.

3.6   Natural Capital and Material Flows

Around 1850, the exploitation of the Dutch soil and subsoil produced an overseeable number of raw materials. Our point of departure is the three main categories of this study: bio-raw materials, subsoil fossil assets, and subsoil mineral assets. A reconstruction on the basis of official statistics reveals the following pattern (see Table 3.1).[36] The three categories together totalled an estimated 8390 kilotons in 1850 (domestic production). Bio-raw materials accounted for 4.670 kton (56%), subsoil fossil assets for 2520 kton (30%), and subsoil mineral assets for 1200 kton (14%).

Within these categories, six products were especially prominent: cereals, potatoes and dairy products (bio-raw materials), turf (subsoil fossil assets) and clay and sand (subsoil mineral assets). The three agricultural products totalled 3430 kton (amounting to 41% of total domestic production). Turf totalled 2500 kton and

---

[35] See Table 3.1. The caloric value of 2,5 million tons of turf is somewhere between 15,000 and 38,000 TJ. The caloric value of moist turf is 6 MJ/kg and of dry turf 15 MJ/kg. The caloric value (in other words the specific energy or energetic value) is a measure of the energy-content of a fuel. Hölsgens calculates an energy consumption of turf in 1850 of 22,000 TJ (five-year average). H. N.M. Hölsgens, *Energy transitions in the Netherlands. Sustainability challenges in a historica land comparative perspective* (Dissertation RU Groningen 2016), Appendix I, 213.

[36] The table provides only a rough indication. The figures provide an order of magnitude that is estimated and checked in different ways. It is probable that a more accurate calculation of the figures will not result in remarkable shifts in most of the (sub)categories. This is not the case for some categories like vegetables from vegetable gardens and brushwood, because these are not officially registered. In subsequent analyses we shall have to keep such unofficial material flows in mind.

Table 3.1 Raw materials in the Netherlands around 1850 in kilotons

		Domestic production	Import	Export	Available
Bio raw materials	Crops/market garden	2,040	290	200	2,130
	Potatoes	790	10	20	780
	Cereals	670	180	100	740
	Beans	80	<5	10	80
	Fodder beets etc.	60	<5	<5	60
	Oil seeds & oil	40	40	10	70
	Misc. Agr. Prod.	400	60	60	400
	Cattle farming	2,220	10	80	2,150
	Milk	1,970	<5	<5	1970
	Living animals	250	10	80	180
	Eggs	<5	<5	<5	<5
	Miscellaneous	<5	<5	<5	<5
	Fishing & hunting	210	<5	30	180
	Fish	130	<5	30	100
	Shells	80	<5	<5	80
	Forestry	200	290	10	480
	Timber	150	290	10	430
	Firewood	50	<5	<5	50
FOSSIL Sub-soil assets	Fossil sub-soil assets	2,520	540	30	3,040
	Turf	2,500	10	<5	2514
	Coal	20	530	30	518
Mineral Sub-soil assets	Mineral sub-soil assets	1,200	150	20	1,330
	Clay	650	20	10	660
	Sand	500	<5	<5	500
	Gravel	40	<5	<5	40
	Stone	<5	80	<5	80
	Salt	<5	40	<5	40
	Ore	10	<5	<5	10
	Total	**8,390**	**1,280**	**370**	**9,300**

Remark: Sand for heightening ground and dike clay are not included in the table
Source: F. Lambert, *Massastromen in Nederland. in de jaren 1850, 1913, 1970, 2010* (research rapport Technische Universiteit Eindhoven, oktober 2016)

accounted for 30% of the grand total. For clay and sand this amounted to 1150 kton and 14%. The remaining 15% consisted of a great number of other products: oil seeds, beans, wood, gravel and so on.

The Netherlands was for the most part self-sufficient. Of all the raw materials available to the Netherlands (domestic production plus import, totalling 9670 kton), only a small portion came from outside the country (1280 kton, 13%). In terms of weight this was above all coal (about 500 kton, or 6%), wood (about 300 kton, or 3%) and cereals (about 200 kton, or 2%). Nevertheless, the fact remains that the

Netherlands depended on other countries for its coal, while the foreign share of timber (for construction) and cereals (for food) was substantial.

This pool of raw materials gave rise to overseeable material flows.[37] The Dutch population directly consumed an important part of the agricultural produce and that virtually all the fossil raw materials (turf and coal) were directly burned (in households and industry). That took place without much preliminary processing. A smaller portion of the agricultural produce was processed in the food industry by grain millers, bakers, butchers and other artisans, while shell-chalk ovens transformed shells into quicklime and sawmills processed wood for use in constructions and buildings. Only a small portion was used directly, for example as firewood. Direct end-use also applied to some extent to mineral raw materials (for example the use of sand in building construction). On the other hand, a significant portion of the clay was used by the brick industry to form and bake bricks. As far as export goes, agricultural and food products dominated: wheat, cattle, sugar, butter and cheese.

In addition to these major flows there were a great number of smaller material flows – not visible in the table and the figure – for example imports of coffee, tea and cane sugar from the colonies (especially the Dutch Indies). When these lesser flows are taken into account, domestic agriculture suddenly appears to deliver a greater variety of products, like hops and madder. Then too, there were raw materials that were never registered despite their importance. Sand for raising the ground level and clay for dikes represented a considerable mass in certain periods but were not registered in the statistics.[38] These materials were usually obtained in the immediate vicinity and were, among other things, used to build railway embankments and

---

[37] See for the details of material flows in 1850 the open-access publication: F. Lambert, *Massastromen in Nederland. In de jaren 1850, 1913, 1970, 2010* (researchrapport Technische Universiteit Eindhoven, oktober 2016)., 11–17.

[38] The most important infrastructural works around 1850 were:

– Railways

1845 construction line Driebergen – Arnhem, Dutch Rhine Railway, 50 km
1847 construction line The Hague – Rotterdam, 25 km
1853 construction line Maastricht – Aachen, 40 km
1854 construction line Essen (B) – Zevenbergen/Etten Leur, 48 km
1855 construction line Utrecht – Rotterdam, 60 km
Railway construction: 23 km per year average
Embankments about 8 m wide – 1 m high: 184.000 m^3 (sand to make elevation) * 1600 kg/m^3 = ca. 300 kton.

– Reclamation Haarlem Lake (1840–1852):

Dikes Haarlem Lake – the encircling dikes (60 km) are completed in 1848 (7.5 km per year). 7.5 km per year 15 m wide, 2.50 m high = ± 280,000 m^3 (dike clay) * 1800 kg.m^3 = ± 500 kton.

– A conservative estimate for the period around 1850 comes out at about 800 kton of sand and dike clay.
– Not included are:

Excavations due to differences in elevation (railways).
Expansion of cities and towns.

Table 3.2 Contribution of economic clusters to GNP and employment rates in the Netherlands around 1850 (percentages)

	GNP (1850) (fl. 560.9 million)	Employment (1849) (n = 1,270,764)
Agriculture and foods	36	44
Construction and housing	8	7 (construction)
Textiles and clothing	7	10
Trade, distribution, transport	30	14
Other	19	35
Total	100%	100%

Source: J.P. Smits, E. Horlings and J.L. van Zanden, *Dutch GNP and its components, 1800–1913* (Groningen 2000), Table 4.5 and Table B.1

canal dikes. Growing vegetables in one's own vegetable patch and collecting brush wood for firewood were also not included in the statistics. Another such category was the domestic fabrication of clothing using wool or flax grown on the farm.

Two main flows dominated economic activity: first, agriculture-food-food consumption and, second, mineral raw materials-mineral processing-construction. The most important points of production were the farm, the brick factory and the construction site. These locations processed the bulk of raw materials. The farm produced above all potatoes, cereals, milk, butter and cheese. The brick factory delivered bricks. At construction sites building materials were transformed into houses, roads, dikes and other constructions. From this perspective, the Netherlands was above all an agricultural and building country.

Table 3.2 in part reflects this conclusion. The share of agriculture and foods in the GNP was 36% and in employment 44%. For construction and housing these percentages were respectively 8% and 7%. Trade, distribution and transport also constituted an important category with 30% of the GNP and 14% of the employment. Their economic contribution was however also realized to a large (but unknown) extent within the categories of agriculture, foods and construction. One would hardly expect textiles and clothing with their modest mass flow to be so prominent in the table with respectively 7% and 10%. Despite their relatively low weight, they scored high in the statistics in terms of money and expended labour.

In general the supply chains of raw material production, processing, use, export and transport were short. The statistics do not show the fact that the supply chains of bio-raw materials and mineral subsoil assets were predominantly closed cycles. At the end of the chain, agriculture and food produced animal and human faeces. These were collected, mixed with fireplace ashes and organic waste (derived from vegetables, fruit and vegetable patches), made into compost and sold to farmers as fertilizer. In this way urban manure contributed to the fertility of the soil. The collection of manure and faeces was unremarkable in a period in which many materials were re-used.[39] Many Dutch people, for example, bought their clothes

---

[39] Lintsen et al. (ed.), *Made in Holland* (Zutphen 2005) 56–58. See also: A. Oudejans, *Categorie één.. Dierlijke afvalverwerking door de eeuwen heen*, 27–32.

second-hand at the market. Once worn out, the clothing was used in the production of paper. Bones were collected to make glue. Building materials from demolished houses were used anew, while building rubble served as pavement material and to heighten building plots.

Matters were otherwise in the case of fossil subsoil assets.[40] Here we find a predominantly linear supply chain. Turf and coal were transformed into heat upon combustion.[41] Nowadays coal is reckoned to be a non-renewable resource. In this study we shall do the same for turf, even though the formation of peat as the basis for turf production is an immeasurably more rapid process than the formation of coal, namely several decades.[42] The total energy consumption of the Netherlands for 1850 can be estimated at 88,500 Tera joule (TJ), including not only turf and coal, but also muscle power, wood, wind and water as sources of energy. 36,200 TJ, or 41%, were supplied by turf and coal; the rest (52,300 TJ or 59%) from renewable sources. Discounting human muscle power, the share of renewable energy was 52%.[43]

The economy of the Netherlands in those days could justifiably be regarded as circular. To what extent was it also sustainable in terms of today's concepts? We answer this question in Chap. 6, 'Well-being and sustainability around 1850: the frame of reference.'

The natural capital of the Netherlands is described here in terms of the colonisation of water, air and ground (soil and subsoil). On the one hand the colonisation had consequences for the country's dead and living nature. This is the topic of the next section. On the other hand the colonised ecosystems produced raw materials and the material flows that derived from them. These were the foundations for economic development (see the concluding section) and the quality of life (see the next chapter).

---

[40] H. Hölsgens, *Energy transitions in the Netherlands. Sustainibility Challenges in a historical and comparative perspective* (dissertation Rijksuniversiteit Groningen 2016), 5–13 and appendix I: Energy consumption in the Netherlands (1800—2012).

[41] Ashes from burned turf were employed as an agricultural fertilizer. This can be seen as a form of recycling.

[42] Turf production was much smaller than the total energy consumed. Contemporaries hence regarded turf as a non-sustainable energy source (i.e. as non-renewable). See Hölsgens (2016) forthcoming in *Tijdschrift voor Economische Geschiedenis.*

[43] For 1850: total consumption (88,466 TJ), total human muscle power (12,661.9 TJ), total consumption excluding human muscle power (75,803.9 TJ). This new total includes 36,162 TJ/48% turf and coal and 39,642 TJ/52% renewable. Information R. Hölsgens (see previous note).

3.7 A Variegated Landscape

3.7.1 Biodiversity

In this study natural capital occupies a special position with respect to other forms of capital (economic, human, and social). It is the basis of production, consumption and well-being. Aside from this utilitarian function, natural capital also has an intrinsic value. This was referred to in the introduction as ecological well-being and is a measure of the quality of living and non-living nature. This intrinsic value does not depend on the contribution of natural capital to the well-being of humans. Nonetheless it is often associated with quality of life. Positive ecological well-being is considered good for fresh air, pure water, plant pollination, sustainable agricultural production, storage of CO_2, et cetera.

What was the impact of the colonisation of nature on ecological well-being in the Netherlands? In the sustainability monitor, ecological well-being is reduced to biodiversity, expressed by the indicator 'Mean Species Abundance' (MSA). What does this indicator tell us? MSA starts with a natural ecosystem that has not yet been disturbed by human activity, and takes the species diversity in that situation as its point of reference.[44] Intervention by humans causes a decline in biodiversity, that is, in the original variety of species. A region with an MSA of 100% means that biodiversity there is the same as the natural ecosystem. A region with an MSA of 0% is equivalent to a thoroughly colonised ecosystem, such that the original biodiversity has completely disappeared. The simplest method of calculation of an MSA is based on the land use of a given region. On this approach the transformation of a natural ecosystem into grassland reduces biodiversity by a factor of 0.3; into a field of crops by a factor of 0.7; and into an urban area by a factor of 0.95. More nuanced calculations are not possible for the nineteenth century due to a lack of data. It is however certainly remarkable that the heathlands, also known as the 'wastelands,' do not form a separate category in the method of calculation and are treated as a natural ecosystem.[45] This despite the fact that in the Netherlands an agricultural landscape was created on the basis of grazing the wastelands and the use of peat sods to improve their sandy soils. In 1850 about a third of the country consisted of heathlands. In the conclusion we will return to this shortcoming.

---

[44] J.L. Van Zanden et al., *How was life? Global well-being since 1820* (OECD n.pl. 2014), 182. This source is based on: R. Alkemade, M. van Oorschot, L. Miles, C. Nellemann, M. Bakkenes and B. ten Brink, 'GLOBIO3: a framework to investigate options for reducing global terrestrial biodiversity loss', *Ecosystems* 12(2009), 374–390 and K. Goldewijk, 'The HYDE 3.1 spatially explicit database of human-induced global land-use change over the past 12,000 years', *Global Ecology and Biogeography* 20(2011), nr. 1, 73–86. Voor de definitie van MSA: Alkemade et al., 'GLOBIO3', 375.

[45] The terms 'heather' or 'moorland' (*heide*) do not appear in the publications cited in the previous note. The publication by Alkemade et al. (2009 GLOBIO3, 379) attributes 'scrublands' (e.g. steppe, tundra or savannah) to the category of natural ecosystems. It may be the case that they also include heathlands or moorlands in this category. In some settings heathlands may be a natural ecosystem but that is not the case in the Netherlands.

We can estimate the MSA for the Netherlands around 1850 at 73%. What does this figure mean? It would indicate that the Netherlands possessed 73% of the biodiversity that existed before human influence in the various parts of the country became noticeable. We will return to this at the end of this section. First we want to give an impression of the big changes in the ecosystems of the sand grounds and the maritime zones. The Dutch forests, at least what was left of them, will be treated separately.

3.7.2   The Sand Grounds

Agriculture had thoroughly transformed the landscape of the sand grounds.[46] When human influence was still minimal – in the sixth and fifth millennium BC – the sand grounds were covered with extensive forests of, among other species, Scottish pine, birch, oak and alder.[47] The agricultural landscape of 1850 revealed little of this. What remained were isolated bits of forest, especially in the stream valleys and in remote spots, bits whose survival was constantly threatened. They were thoroughly integrated into the farming system. Cattle were tended there. Their wood was gathered and used for fuel (heating and cooking) or served as building material (furniture, clogs, tools, machines, load bearing constructions for farm buildings, etc.)

Crop fields, heath, grazing pastures and hayfields had taken the place of the forests. By today's standards the agricultural system had a rich ecology.[48] Heather and cross-leaved heath dominated the heath-fields. Birches and oaks were scattered here and there. Orchids, gentian and bog asphodel prospered, depending on local soil conditions. Bluegrass (*Cirsio dissecti-Molinietum*), an unusually biodiverse kind of poor meadow and hay land that existed in the infertile damp meadows in the stream valleys, sported plants that are now extremely rare: plants like the devil's bit, marsh violet, sundew and meadow thistle. The low-lying, saturated or wet stream dales were home especially to the forests, dominated by alder, willow and ash and in which one could find forest plants like the wood anemone, lily of the valley, and blueberry.

A characteristic element of the landscape was the wooded banks. They were built along paths, around fields and along pastures. As an enclosure the wooded bank was impermeable to cattle, as fencing it enabled the farmer to mark the extent of his

---

[46] The nature of the original Dutch landscape, that is the landscape before human influence made itself felt, is a matter of controversy. Was the Netherlands largely forested or was it an open landscape with large herds of grazing animals? In this section we assume a forested landscape. For the debate see: L. Kooijmans, 'Holland op z'n wildst? De Vera-hypothese getoetst aan de prehistorie', *De Levende Natuur* 113(2012), nr. 2, 62–66.

[47] For an exhaustive description of the landscape of the sand grounds see: J.A.J. Vloet, 'Zandlandschap', in: S. Barends et al., *Het Nederlandse Landschap. Een historisch-geografische benadering* (tenth revised edition, Utrecht 2010), 132–161; Thijs Caspers, *De hand van de mens in het landschap* (Haaren 1992); *Natuur in Noord-Brabant. Twee eeuwen plant en dier* (Haaren 1996).

[48] For this see: Caspers, *De hand van de mens*, 49–56; *Natuur in Noord-Brabant. Twee eeuwen plant en dier* (Haaren 1996), 28–71.

property and as means of production the bank supplied fruits, berries, firewood and wood for construction. A wooded bank consisted minimally of a deep ditch with a contiguous earthen bank on which grew trees, bushes, plants and herbs. Small mammals (mice, weasels and stoats) found forage there or used it as cover during their forays (the badger). Many birds found the wooded bank a congenial biotope because it provided cover, good nesting sites and a wealth of insects, fruits and berries. Wooded banks ran like green veins through the sand grounds and provided sharp contrasts in the landscape: the leafy environs of many villages and stream valleys over against the broad empty spaces of the heath meadows.

The heath meadow was a vulnerable component of the landscape.[49] It was constantly threatened by overexploitation. Heath meadows were once themselves the result of overexploitation. After forests had been felled and the land exhausted by farming, heather was one of the few species that could survive on the poor sand grounds that remained. The cutting of heather sods for the production of compost and manure exposed the bare sandy subsoil.[50] Where things got out of control and the sand got the upper hand, sandstorms threatened fields and villages and sand deserts like those on the Veluwe and in Noord Brabant (the Loon and the Drunen Dunes).[51] The question is to what extent the farming community lost its grip on the process. The danger lurked in growing population pressure and the consequent increase in cultivated acreage and more intensive use of the common lands, including the heath meadows. That was exactly the situation in the first half of the nineteenth century, with its rapid increase in rural population. At some places this did indeed cause problems. Nonetheless there are no indications that in general things took a dramatic turn for the worse.[52]

3.7.3   The Maritime Zones

Prior to the arrival of the first farmers in the fourth and third millennium BC, the northern and southwestern maritime zones bore a strong resemblance to the present day Wadden Sea and the parts of the Wadden islands now outside the dikes.[53] The

---

[49] The heath meadows belonged to the so-called communal lands or the 'commons'. This is the subject of an extensive literature, see: T. de Moor, *The Dilemma of the Commoners: Understanding the Use of Common Pool Resources in Long-Term Perspective* (Cambridge 2015).

[50] Another form of excessive exploitation was the pasturing of sheep, a practice that left traces of trampled and denuded paths across the heather.

[51] This was a troublesome phenomenon for the farmer. Crop fields and pastures were buried in sand drifts. Even buildings in the villages were threatened. To keep the advancing sands at bay, wood banks had to be raised, grey hair grass planted, ditches dug etc. Around 1850 the problem seemed to be reasonably under control. Caspers, *De hand van de mens*, 60–61.

[52] J. Renes, *Landschappen van Maas en Peel* (Maastricht 1999), 203–204. See also Chap. 8 of our study.

[53] For the following see: L. Hacquebord, 'Noordelijk zeekleilandschap', in: S. Barends, *Het Nederlandse Landschap. Een historisch-geografische* benadering (tenth revised edition, Utrecht

extensive tidal flats consisted of higher mud and sand ridges, tidal creeks, accreted silt from shore erosion, salt swamps and dunes. Species diversity was low because few plants could cope with the extreme conditions of the salty environment and the changing water levels. The tidal flats became the province of salt-tolerant plants common to brackish environments.[54] Seaweed, glasswort, and herbaceous seepweed grew on the mudbanks (*slikken*, i.e. low-lying sediment ridges that were regularly flooded), while for example the salt meadows (*schorren*, or high sediment ridges that only flooded at spring tides or during storm surges) were the domain of sea poa and sea lavender. Large numbers of migratory, overwintering or resident birds like brent geese, waders, avocets and spoonbills found nourishment, rest, and breeding grounds here. Throughout most of the Middle Ages these maritime zones retained this characteristic landscape.

In the first half of the nineteenth century the sea-clay landscape must have been considerably more diverse than the formerly un-diked land due to the variation between wet and dry, salt and fresh, clay and sand, dike and polder. Mud banks and salt meadows, geese and avocets were still part of this landscape, but now joined to various other landscape elements. In addition to the fields and the meadows, for example, there were now also the brackish polders with their pools and watering holes.[55] They were used as extensive grasslands and were home to salt-tolerant plants like wild celery and brackish dropwort as well as salt-intolerant plants like orchids and the greater yellow rattle.

The dikes also exhibited a variegated vegetation that was sometimes so profuse that they were called 'flower dikes.' This was due in part to grazing by itinerant flocks of sheep. Other important factors were the composition of the dike and the pitch of the slope that determined the position of the plants relative to the sun. Typical plants were, among others, wild marjoram, tuberous pea, wild onion and eryngo. Sometimes dikes were planted with willows and poplars.

Bramble hedges of hawthorn, blackthorn, dog rose, or blackberries were quite common. They served as fencing for fields and as a congenial biotope for insects, amphibians, birds and other animals. The hedges provided them with shelter, food and nesting sites. And then every region had its own characteristic elements like the numerous lakes in Friesland and the *inlagen* in Zeeland, small polders containing reed marshes and ponds that lay between the sea dike and a 'secondary dike' that had been created for the purpose of salt mining.

2010), 16–31 and A.P. de Klerk, 'Zuidwestelijk zeekleilandschap', in: S. Barends, *Het Nederlandse Landschap. Een historisch-geografische* benadering (tenth revised edition, Utrecht 2010), 32–45.

[54] Additional information on flora and fauna can be found on different websites, among others: www.natuurinformatie.nl; www.natuurkennis.nl en www.zwinstreek.eu

[55] Information on flora and fauna can be found on different websites, including: www. hetzeeuwselandschap.nl; www.deturfhoeke.nl

3.7.4 The Lost Forest

A resource that the inhabitants of the delta had long since exhausted was the forest. Around 1850 forest acreage had been reduced to a minimum of 5% of the Dutch surface area. There was a time when large parts of the Netherlands had been covered with forests of many sorts: alluvial forests, swamp forests, heron forests, dune forests etc. In the course of time they fell prey to the expansion of agriculture, turf extraction, wars and storms. The forests that survived were constantly threatened by overexploitation. Forests produced fruits, served as pasture for cattle and were royal repositories of fuel and construction material. Certain species of trees had survived agricultural expansion better than others. On the sand grounds, for example, the summer oak's thick bark had enabled it to survive the slash and burn tactics used to eliminate unwanted vegetation in preparation for making new farmland. The oak was also more tolerant for the decimation of the woods and the impoverishment of the soil. Moreover it was treated with circumspection because it produced strong timber as well as oak bark for tanning hides.[56]

Wood was a crucial and versatile material in pre-industrial times.[57] The demand for wood was so great that 'wood thievery and wood vandalism' was considered a problem. It was one of the only concerns that a commission was able to voice about public morals in the North Brabant countryside in 1851.[58] Poplar wood was used for clogs, oak for construction framing and furniture, pine for – among other things – shoring in mineshafts, wood from *grienden* (swampy fields planted with willows) for woven materials, etc. Further differentiations were made. Thin willow branches from the *grienden* were used as wicker to make baskets, chairs and fences; thicker branches were used to make fascine mattresses for hydraulic constructions. Thick branches were also used for broom handles and the new shoots for barrel hoops. Firewood, a generic term for waste wood, unusable wood and dead wood suitable for burning, was in a class by itself. For some types of wood the domestic supply was adequate to the national demand, for example poplar and willow. Not only forests, but also wood banks and rows of trees along roads, fields and properties supplied the necessary material. Firewood partly satisfied the need for fuel, especially in the countryside but less so in the cities. The Netherlands was itself

---

[56] The oak also supplied acorns, a favourite food of pigs, that played a role in all the agricultural systems.

[57] See for the following: J. Radkau und I. Schäfer, *Holz. Ein Naturstoff in der Technikgeschichte* (Reinbek bei Hamburg 1987); J. Buis, *Historia Forestis. Nederlandse bosgeschiedenis* (Dissertation, Wageningen 1985); H. van Zon, *Duurzame ontwikkeling in historisch perspectief. Enkele verkenningen* (Nijmegen 2002), 55–85.

[58] See: Caspers, *De hand van de mens…*, 58. For the original text: A. Martini van Geffen, 'Zedelijke en materiële toestand der arbeidende bevolking ten platten lande. I. Nopens Noord-Braband', *Verslag van het verhandelde op het vijfde Nederlandsche Landhuishoudkundig Congres, gehouden te Leiden van den 10den tot den 14den junij 1850* (Leiden 1850), Appendix no. 12, 104. The report is included in: J.L. Van Zanden, *'Den zedelijke en materiële toestand der arbeidende bevolking ten platten lande'. Een reeks rapporten uit 1851* (Historia Agriculturae XXI, Groningen 1991).

largely able to compensate the shortage of fuel by extracting turf. The remainder was accounted for by the import of coal.

Timber for construction was another matter. In some regions it proved possible to re-use old timber or fell local trees, in particular, oaks. For the rest there was a centuries-old dependency on import. As early as the fourteenth century large quantities of wood were supplied via the Rhine. This Rhine trade peaked in the middle of the eighteenth century after which it rapidly declined. In the first half of the nineteenth century wood was imported above all from Norway and the Baltic Sea (Sweden, Finland, Estonia and the hinterlands of St. Petersburg): chiefly coniferous wood (pine and fir) and oak. The imported wood was used for constructing housing, public buildings, windmills, locks and bridges. Great quantities of logs were needed for pilings for among other things quays and buildings in peat soils. Shipbuilding was also a bulk consumer.

With the massive import of timber the Netherlands effectively relocated to foreign countries the problems it had with its own forests. The collapse of the Rhine timber trade was partly a consequence of the overexploitation of the forests along the upstream Rhine. Uncontrolled logging had transformed large parts of the Black Forest and other forested areas into lunar landscapes.[59] The intensity of logging far outstripped the norm for natural recovery of the forest. Next to nothing was done about replanting. And it was not only the Netherlands that was consumed by wood hunger but also the German states. Great quantities of wood in the form of charcoal were burned there in blast furnaces in order to make iron.

3.7.5 Prosperous Farming Landscape

Since the onset of agriculture and the building of dikes every small corner of the Netherlands has repeatedly been torn down and built up again.[60] On the sand grounds and in the coastal zones colonisation commenced far before the Christian era. In the peat bogs of the low Netherlands human intervention was delayed until around 1000.[61] In that epoch the peat packets still towered several meters above sea level. Reclamation and turf extraction caused soil subsidence and changed the landscape into marshes, pools, lakes but above all polders and reclaimed lakes. Did this imply a reduction in biodiversity? According to the MSA indicator, 73% of the original

---

[59] The role of other factors like increased consumption in Prussia and rising prices is unclear. It is also uncertain how big a problem the deforestation was. Van Zon speaks of a catastrophic decimation of the German forests (Van Zon, *Duurzame ontwikkeling*, 85). Radkau and Schäfer are a bit more reticent: The deforestation was not so rigorous that recovery was impossible (Radkau and Schäfer, *Holz*, 139).

[60] See for a history of the Dutch landscapes among others: S. Barends, *Het Nederlandse Landschap. Een historisch-geografische* benadering (tenth revised edition, Utrecht 2010), 114–131.

[61] See for the history of the peat landscape among others: G.J. Borger, 'Agrarisch veenlandschap', in: S. Barends, *Het Nederlandse Landschap. Een historisch-geografische* benadering (tiende herziene druk, Utrecht 2010), 62–79 and websites like www.geologievannederland.nl

species diversity still existed in 1850. The question is whether this is an adequate picture of the actual biodiversity. Three objections may be raised.

First of all, the percentage seems a bit exaggerated for a country that had lost most of its original natural ecosystems. Here we are not counting the 'wastelands' (the heathlands) as original nature as per the MSA calculation. It may be the case that it was not so much the species diversity that declined but rather the population size of the individual species. Only a few species like the wolf, the brown bear, the elk and the beaver are known not to have survived the advent of humans with their penchant for hunting and transforming animal habitats. The high percentage then correctly expresses the modest decline in the original species diversity. Natural ecosystems were radically reduced in scope or nearly extinguished, but around 1850 small parts of the Netherlands still consisted of nearly authentic peat bogs, marsh forests and dunes. In some instances one could even speak of a large-scale natural landscape like the tidal flats of the Wadden Sea.

In the second place, the MSA provides not a single clue about the appearance of new species. The introduction and spread of agriculture must have facilitated the proliferation of entirely new species. Around 1850 the Netherlands harboured a large number of agricultural systems. The agricultural expert Staring, cited above, was able to identify fourteen of them in addition to special sectors like tobacco, flower bulbs and fruit orchards. There must have been a considerable variety of colonised ecosystems, each with their specific flora and fauna. The resultant biodiversity is difficult to reconstruct, but the variety of ecosystems each with their specific species diversity must have increased. In addition to the original ecosystems, new ones had emerged: the agrarian sand, clay, peat, dune and loess landscapes and their further differentiation into for example northern, southern and river clay landscapes. They contained biotopes that burgeoned with life, like the wood banks, the bluegrass lands, the brackish polders and the bramble hedges. In these cases human intervention did not eventuate in degradation of nature but rather in another 'nature' and presumably a more diverse one.

In the third place, the intensive exploitation of the available natural capital subjected the ecosystems and their constituent species to constant pressure.[62] Ecosystems disappeared, like the primeval forests; they were threatened, like the heath meadows; or subject to irreversible processes, like the transformation of open water into fens.[63] Overexploitation and exhaustion of the soil was an ever-present threat. 'Wild' animals were persistently pursued and killed for their fur, feathers or

---

[62] J.L. van Zanden en S.W. Verstegen, *Groene geschiedenis van Nederland* (Utrecht 1993), 17–31.

[63] In general the agrarian systems of the time facilitated stable ecosystems, an important precondition for continuity in species diversity. That also held for a large-scale practice like the extraction of clay with a production of an estimated 650 kton in 1850, about 8% of the weight of all raw materials extracted from Dutch soil. This had a modest effect on the natural ecosystem. Extraction and processing into roof tiles and bricks took place in virtually every region where clay could be found. There was a concentration of factories in the floodplains of the rivers. Locally, extraction did disrupt the order of things. After a clay pit had been abandoned it frequently turned into a pond. At some sites along the river, thanks to sedimentation, these clay pits could be worked again after 50 years.

Table 3.3 Total number of governmental bounties for killed 'harmfull animals' 1852–1857

Type of animal	Numbers
Weasels	88,449
Polecats	26,711
Sparrow Hawks	16,626
Foxes	5861
Kites	5017
Goshawks	2828
Falcons	2787
Stoats	2245
Buzzards	1474
Beech Martens	974
Martens	675
Eagles	219

Remark: Names used in the sources are retained here

Source: J.L van Zanden and S.W. Verstegen, *Groene Geschiedenis van Nederland* (*Green History of the Netherlands*) (Utrecht 1993), 27. Original source: J. Wttwaal, 'Over het verlenen van premiën voor het dooden van wezels en ander zoogenoemd schadelijk gedierte' (On the offering of bounties for the killing of weasels and other so-called harmful vermin') *De Volksvlijt* (1859), 317–345

meat but also because they threatened harvests and spoiled the hunt. Every province awarded bounties for the killing of polecats, sparrow hawks, foxes and other 'vermin.' (Table 3.3)

3.8   A growing Economy, a Growing Population[64]

The colonisation of nature had produced a rich variety of ecosystems. But did it also bring wealth to the Dutch? The colonised ecosystems were the original source of the three material flows identified in this book: biomass, fossil, and mineral. These provided the population with agricultural products, fossil resources (in particular turf) and building materials (among which sand and clay). These were processed, sold, consumed, used and exported. This enabled the import of yet other raw materials and products. What did this contribute to the development of welfare in terms of economic growth? We also look among the totality of material flows for

---

[64] An analysis of economic development in the first half of the nineteenth century can be found in J.L van Zanden and A. Van Riel, *Nederland 1780–1914. Staat, instituties en economische ontwikkeling* (Amsterdam 2000), 149–208. The following section is based on this book.

dominant industrial sectors. Which industrial sectors had the greatest impact on economic development? To find our way here we will first summarize economic developments prior to 1850.

By the outset of the nineteenth century the Dutch economy already passed through a long and difficult period of adjustment. The Dutch Golden Age with its glory days of economic prosperity between 1580 and 1670 was followed by a long period of decline of international competitiveness, loss of market share, and stagnation. This economic malaise persisted through the collapse of the Republic of the Seven United Provinces in 1795 and hit bottom during the French annexation (1810–1813). International trade ground to a halt thanks to Napoleon's economic blockade of England. In the province of Holland and the other maritime regions manufacturing all but collapsed. The national debt, already sizable by the end of the eighteenth century, had swollen out of all proportion due to the virtually permanent state of war during the Batavian and French period (1795–1813). The founding of the Kingdom of the Netherlands in 1813 was followed by a period of uneven economic development. The two decades after 1820 comprised a period of structural growth in the Dutch economy, after which a new period of relative stagnation set in (see Table 3.4).

Initially, economic development after 1820 was shaped by conditions in the agricultural sector. Agriculture had to cope with the aftershocks of the Napoleonic period: steeply rising prices for agricultural produce immediately after the war followed by a dramatic decline. The price of wheat in Groningen, for example, declined from fl. 12.97 per hectoliter in 1817 to fl. 3.59 in 1823. The price of rye also declined by more than 70%. Prices for grain remained low in subsequent years. Prices for potatoes, dairy products and meat also declined, but less steeply. Farmers responded in various ways. For the dairy farmers in the provinces of Holland, Utrecht and Friesland these were favourable years and presumably better than the years after 1820. On the sea- clay soils in the coastal provinces farmers in part abandoned the tillage of wheat for that of potatoes and commercial crops like madder and flax. Meadowland was 'torn' and transformed into crop fields. Formerly mixed farms thus gradually assumed the character of specialized crop farms, a process that had been going on in the province of Zeeland for a time and that now gained momentum in the province of Groningen as well. Farmhands were fired to save on labour costs. Presumably the farmer, his family and other residents of the farm had to work harder. Potatoes and commercial crops required more intensive tillage (weeding, delving, sowing in rows, et cetera). On the sand grounds, mixed farms began to concentrate on growing potatoes, husbanding cattle and producing butter. That was accompanied by intensification of crop farming and long work days.

The 1820s witnessed growth across the board in manufacturing and services. The recovery of the domestic market after the Napoleonic period stimulated a strong

Table 3.4 Growth in value added in the most important economic sectors, 1816–1850

	Growth 1816–1830	Growth 1830–1840	Growth 1840–1850
Agriculture	0.6	1.4	1.2
Industry, among which	2.7	2.4	0.5
Construction	1.4	0.6	−0.2
Clothing	3.2	−0.9	0.3
Textiles	5.0	6.2	−1.5
Foods	4.1	2.8	0.5
Services, among which	1.5	2.8	1.5
Internat. trade	0.1	5.9	3.1
Domestic trade	3.9	1.5	1.1
Maritime shipping	4.5	9.8	2.4
River shipping	1.0	5.8	0.7
Inland navigation	0.6	2.5	0.9
Gross Domestic Product GDP	1.6	2.3	1.1
Population	1.2	0.9	0.7
GDP per capita	0.4	1.4	0.8
Labour productivity in manufacturing[a]	0.2 (1807–1830)	0.7 (1830–1842)	−0.6 (1842–1860)

Remark: Value added and GDP in real prices
Source: J.L. van Zanden and A. van Riel, *Nederland 1780–1914. Staat, instituties en economische ontwikkeling* (*The Netherlands 1780–1914. State, institutions and economic development*) (n.p. 2000), Table 4.2, 153 and Table 6.3, 248. The figures for 1840–1850 differ between the tables. Here the data from Table 6.3 are used. The tables are based on: J.P. Smits, E. Horlings and J.L. van Zanden, *Dutch GNP and its components, 1800–1913* (Groningen 2000)
[a]Source: J.P. Smits, 'The determinants of productivity growth in Dutch manufacturing, 1800–1913', *European Review of Economic History*, 4 (2000), pp. 223–246, graph 1

expansion of the food sector, the construction industry, the clothing industry and turf extraction. In this decennium the food sector, comprising more than 60% of industrial production, exhibited the highest rates of growth for the first half of the nineteenth century. Growth in the service sector was above all attributable to international trade and international transport. This was due on the one hand to increasing trade with the Dutch East Indies and on the other hand to trade with the eastern neighbour, Prussia. Textiles formed the bulk of goods transported to Java, while sugar, coffee and other colonial products comprised the return trade. Rhine barges transported colonial wares to the Prussian hinterland and cereals and coals back to the Netherlands. The Dutch Trading Company (*Nederlandse Handelsmaatschappij, NHM*) coordinated these flows of commerce. The interdependence of shipping and the grain trade became evident in the years around 1820 when grain prices fell sharply and Rhine shipping declined by about 60% and shipping to European harbours by 40%.

Economic development in the 1830s was directly affected by the Belgian secession of 1830. Military expenditures demanded new and higher levels of taxation. In addition, there was a rise in prices of agricultural products and coal. This put an end to the rise of real income and the expansion of the domestic market. The building sector exhibited almost zero growth, while clothing production shrank and growth in the food sector was only modest. The sources of economic growth became narrower and were increasingly rooted in the so-called 'colonial complex.' The NHM reorganized the flows of commerce to and from the Dutch East Indies. They forcefully expanded the Dutch textile industry to compensate for the loss of Belgian production capacity. This required, among other things, mobilizing the countryside of the Twente region, where home weavers were put to work to produce huge quantities of textiles for Java. Maritime shipping also got a boost with the construction of great numbers of ships for the trade with Java. Sea trade and international commerce flourished.

All the agricultural sectors experienced a decisive upturn, particularly after 1835. The farmers on the sea and river clay, in the peat and on the sand all profited from higher prices for agricultural produce (including dairy produce). Common lands on the sand grounds were reclaimed and their productivity increased by working more intensively and improving the quality of the cattle and the manure. The viability of smallholdings in certain regions like Twente improved thanks to NHM policy. Home-weaving formed a welcome supplement to farming income and spread the risk of insolvency.

The 1840s were marked by stagnation on three fronts: manufacturing, services and agriculture. Industrial development stagnated due to a crisis in the 'colonial complex.' Trade with Java declined precipitously and dragged textiles and shipping in the Netherlands along with it. At the same time the government's gradual dismantling of protectionist measures confronted the NHM with increasing international competition. The 'colonial complex' settled into a long-term malaise. Agriculture, by contrast, initially performed well. The liberalisation of international trade at the outset of the 1840s provoked a further increase in the prices of agricultural produce. Agricultural productivity increased along with the prices. 1845 was the turning point for the agricultural sector. Potato blight caused several years of failed or strongly reduced potato harvests, while in 1846 the rye harvest failed as well. The domestic market stagnated. Stagnation in one sector was not compensated by improved performance in other sectors. Moreover labour productivity declined and real wages and purchasing power trailed behind.

Three complexes in large part shaped economic development in the first half of the nineteenth century: (1) agriculture and foods, (2) building materials and construction, (3) the colonial complex (including textiles and shipbuilding). The first two complexes were part of the three big material flows generated by the natural capital of the Netherlands. These interacted with the domestic market. Growing demand stimulated the supply of food and buildings. Increasing supply allowed for a greater demand. Increasing demand expressed itself in particular in a growing population. Natural capital (*and* the colonial complex) enabled a remarkable population growth in the first half of the nineteenth century.

Table 3.5 Growth of urban and rural population between 1795 and 1840

	Growth population 1795–1840 (%)	Growth rural population 1795–1840 (%)	Growth urban* population 1795–1840 (%)	Population 1840 (N)
Drenthe	83	81	86	72,484
Gelderland	55	46	82	345,762
Groningen	53	58	44	175,651
Utrecht	49	47	56	145,132
Overijssel	46	40	70	197,694
Noord-Brabant	45	44	50	378,437
Limburg	42	40	48	196,128
Friesland	41	43	35	227,859
Zuid-Holland	37	42	31	527,225
Zeeland	32	43	9	151,358
Noord-Holland	8	26	2	442,129
NETHERLANDS	38	44	28	2,859,859

*Urban refers to places with more than 2500 inhabitants in 1840

Economic development, albeit of a vacillating nature, was accompanied by a vigorous growth of the Dutch population in the first half of the nineteenth century. While in the entire eighteenth century the population grew by only 7%, in the 45 year period from 1795–1840 it increased by no less than 38% (see Table 3.5). Population in the countryside grew significantly faster than that in the towns, namely 44% compared to 28%. And the growth in the higher regions of the country was greater than in the low-lying portions. The agricultural system on the sand grounds appeared to have more than enough capacity to absorb population growth (in the province of Drenthe with a remarkable 81%). Peat marshes, heathlands, grasslands and forests continued to be reclaimed. Smallholders increased productivity by working the land more intensively or opted for more profitable alternatives like dairy produce. They also wove textiles at home for the *fabrikeur*, under the putting-out system.

In terms of the gross domestic product per capita, the Dutch economy grew throughout the entire period 1816–1850. The 1830s were the most advantageous (annual growth of ±1.4%) and the 1840s the least (annual growth of ±0.8%). That seems like a positive development. But what do these figures tell us about the quality of life of the population? This will be one of the topics of the next chapter.

Literature

Alkemade, R., M. van Oorschot, L. Miles, C. Nellemann, M. Bakkenes and B. ten Brink (2009). 'GLOBIO3: A framework to investigate options for reducing global terrestrial biodiversity loss'. *Ecosystems*, 12, 374–390.

Alstorphius Grevelink, P.W. (1991). 'Zedelijke en materiële toestand der arbeidende bevolking ten platten lande. Nopens Drenthe (1850)'. In J.L. van Zanden, J.L. van (Ed.), *'Den zedelijke en materiële toestand der arbeidende bevolking ten platten lande'. Een reeks rapporten uit 1851.* Groningen: Nederlands Agronomisch-Historisch Instituut. Original source: *Verslag van het verhandelde op het vijfde Nederlandsche Landhuishoudkundig Congres, gehouden te Leiden van den 10den tot den 14den junij 1850* (1850), appendix no. 13, 109–118.

Bakker, M.S.C. (1992). 'Boter'. In H.W. Lintsen et al. (Ed.), *Geschiedenis van de techniek in Nederland. De wording van een moderne samenleving 1800–1890* (pp. 103–133). Zutphen: Walburg Pers.

Barends, S. (2010). *Het Nederlandse landschap: Een historisch-geografische benadering.* Utrecht: Matrijs.

J.H. Beucker Andrea, J.H. (1991). 'Rapport ingediend voor het vijfde, Landhuishoudjundig Congres te Leyden, 11, 12, 13 junij 1850, betreffende een onderzoek naar den zedelijken en materiëlen toestand der arbeidende bevolking ten platten lande en van den middelen om dien zoveel mogelijk te verbeteren. [Friesland]'. In Zanden, J.L. van (red.), *'Den zedelijke en materiële toestand der arbeidende bevolking ten platten lande'. Een reeks rapporten uit 1851.* Groningen: Nederlands Agronomisch-Historisch Instituut. Original source: *Tijdschrift voor Staathuishoudkunde en Statistiek* (1851), 6, 156–200.

Bieleman, J. (1987). *Boeren op het Drentse zand 1600–1910: Een nieuwe visie op de 'oude' landbouw.* Het Goy: Hes en de Graaf Publishers.

Bieleman, J. (2008). *Boeren in Nederland: Geschiedenis van de landbouw 1500–2000.* Amsterdam: Uitgeverij Boom.

Bosch, A., Ham W. van der, and Lintsen, H.W. (1998). *Twee Eeuwen Rijkswaterstaat 1798–1998.* Zaltbommel: Europese Bibliotheek.

Borger, G.J. (2010). 'Agrarisch veenlandschap'. In S. Barends, *Het Nederlandse landschap: Een historisch-geografische benadering* (pp. 62–79). Utrecht: Matrijs.

Brink, G. van den (1996). *De grote overgang: Een lokaal onderzoek naar de modernisering van het bestaan Woensel 1670–1920.* Nijmegen: SUN.

Caspers, T. (1992). *De hand van de mens in het landschap.* Haaren: Stichting het Noordbrabants Landschap.

Caspers, T. (1996). *Natuur in Noord-Brabant: Twee eeuwen plant en dier.* Haaren: Stichting het Noordbrabants Landschap.

Crijns, A.H. and F.W.J. Kriellaars (1987). *Gemengd landbouwbedrijf op de zandgronden in Noord-Brabant 1800–1885.* Tilburg: Stichting Zuidelijk Historisch Contact.

Disco, C. and Lintsen, H.W. (2003). 'Het nijvere verbond'. In J. Schot, H.W. Lintsen, A. Rip and A.A. Albert de la Bruhèze (Ed.), *Techniek in Nederland in de twintigste eeuw* (Vol. I, pp. 55–63). Eindhoven: Stichting Historie der Techniek.

Goldewijk, K. (2011). 'The HYDE 3.1 spatially explicit database of human-induced global land-use change over the past 12,000 years'. *Global Ecology and Biogeography*, 20(1), 73–86.

Hacquebord, L (2010). 'Noorderlijk zeekleilandschap'. In S. Barends, *Het Nederlandse landschap: Een historisch-geografische benadering* (10th revised edition, pp. 16–31). Utrecht: Matrijs.

Hölsgens, H. N. M. (2016). *Energy Transitions in the Netherlands: Sustainability Challenges in a Historical and Comparative Perspective.* Groningen: University of Groningen, SOM research school.

Iterson, W. van (1868). *Schets van de landhuishoudkunde der Meijerij.* Groningen

Klerk, A.P. de (2010). 'Zuidwestelijk kleilandschap'. In S. Barends, *Het Nederlandse landschap: Een historisch-geografische benadering* (10th revised edition, pp. 32–45). Utrecht: Matrijs.

Knippenberg, H. and Pater, B. de (1988). *De eenwording van Nederland: Schaalvergroting en integratie sinds 1800.* Nijmegen: SUN.

Kooijmans, L. (2012). 'Holland op z'n wildst? De Vera-hypothese getoetst aan de prehistorie'. *De Levende* Natuur, 113(2), 62–66.

Lambert, F. (2016). *Massastromen in Nederland. In de jaren 1850, 1913, 1970, 2010.* Eindhoven: Eindhoven University of Technology (open-access research report)

Leenaers, H., Donkers, H. and Noordhoff Atlasproducties (2010). *De bosatlas van Nederland waterland.* Groningen: Noordhoff Uitgevers bv.

Leeuwen, W. van (1993). 'Een complexe sector'. In H.W. Lintsen et al. (Eds.), *Geschiedenis van de techniek in Nederland. De wording van een moderne samenleving 1800–1890* (Vol. III, pp. 191–195).Zutphen: Walburg Pers.

Leeuwen W. van (1993). 'Waterbouw'. In H.W. Lintsen et al. (Eds.), *Geschiedenis van de techniek in Nederland. De wording van een moderne samenleving 1800–1890* (Vol. III, pp. 233–249). Zutphen: Walburg Pers.

Leeuwen W. van (1993). 'Woning- en utiliteitsbouw'. In H.W. Lintsen et al. (Eds.), *Geschiedenis van de techniek in Nederland. De wording van een moderne samenleving 1800–1890* (Vol. III, pp. 197–231). Zutphen: Walburg Pers.

Lintsen, H.W. (2005). *Made in Holland: Een techniekgeschiedenis van Nederland [1800–2000].* Zutphen: Walburg Pers.

Moor, T. de (2015). *The dilemma of the commoners: Understanding the use of common-pool resources in long-term perspective.* Cambridge: Cambridge University Press.

Oudejans, A. (2012). *Categorie een: Dierlijk afvalverwerking door de eeuwen heen.* Mijnbestseller BV.

Renes, J. (1999). *Landschappen van Maas en Peel.* Maastricht: Eisma Edumedia.

Schot J.W. and Homburg E. (1993). 'Meekrap en garancine'. In Lintsen, H.W. et al. (Eds.), *Geschiedenis van de techniek in Nederland. De wording van een moderne samenleving 1800–1890* (Vol. IV, pp. 222–239). Zutphen: Walburg Pers.

Stol, T. (2010). 'Turfwinningslandschap'. In S. Barends, *Het Nederlandse landschap: Een historisch-geografische benadering.* Utrecht: Matrijs.

Smits, J.P., E. Horlings and J.L. van Zanden (2000), *Dutch GNP and its components, 1800–1913.* Groningen: University of Groningen

Woud, A. van der (2010). *Koninkrijk vol sloppen: Achterbuurten en vuil in de negentiende eeuw.* Amsterdam: Bert Bakker.

Ven, G.P. van de (2003). *Leefbaar laagland: Geschiedenis van de waterbeheersing en landaanwinning in Nederland.* Utrecht: Matrijs.

Vloet, J.A.J. (2010). 'Zandlandschap'. In S. Barends, *Het Nederlandse landschap: Een historisch-geografische benadering* (10th revised edition, pp. 132–161). Utrecht: Matrijs.

Zanden, J.L. van (1985). *De economische ontwikkeling van de Nederlandse landbouw in de negentiende eeuw, 1800–1914.* Leiden: Koninklijke Brill NV.

Zanden, J.L. van (1991). *'Den zedelijke en materiële toestand der arbeidende bevolking ten platten lande'. Een reeks rapporten uit 1851.* Groningen: Nederlands Agronomisch-Historisch Instituut.

Zanden, J.L. van and Verstegen, S.W. (1993). *Groene geschiedenis van Nederland.* Houten: Het Spectrum.

Zanden, J.L. van (1997). 'Werd de Gouden Eeuw uit turf geboren? Over het energieverbruik in de Republiek in de zeventiende en achttiende eeuw'. *Tijdschrift voor geschiedenis*, 110, 484–499.

Zanden, J.L., van en Riel, A. van (2000). *Nederland 1780–1914. Staat, instituties en economische ontwikkeling.* Amsterdam: Balans.

Zanden, J.L. van (2014). *How was life? Global well-being since 1820.* OECD.

Zeehuisen, J. (1991). 'Statistieke bijdrage tot de kennis van den stoffelijken en zedelijken toestand van de landbouwende klasse in het kwartier Salland, provincie Overijssel'. In Zanden, J.L. van (red.), *'Den zedelijke en materiële toestand der arbeidende bevolking ten platten lande'. Een reeks rapporten uit 1851*. Groningen: Nederlands Agronomisch-Historisch Instituut. Original source: *Tijdschrift voor Staatshoudkunde en Statistiek* (1851), 9, 375–408.

Zeeuw, J.W. de (1978). 'Peat and the Dutch Golden Age: The historical meaning of energy attainability'. *A.A.G. Bijdragen* 21: 3–31.

Zon, H. van (2002). *Geschiedenis en duurzame ontwikkeling: Duurzame ontwikkelingen in historisch perspectief.* Nijmegen: Netwerk Duurzaam Hoger Onderwijs.

Chapter 4
Quality of Life: A Poor and Vulnerable People

Harry Lintsen

Contents

Abstract Well-being is achieved by means of four resources (that is, the four capitals: natural, economic, human and social) of which natural capital is the basis. The previous chapter emphasised the natural capital of the Netherlands and the way this was exploited with the aid of the other capitals. This chapter asks what the outcome of this exploitation was in terms of well-being. What were the most important issues around well-being in the Netherlands at the middle of the nineteenth century? In terms of present-day norms for extreme poverty, around 21% of the population at that time lived in extreme poverty. From a present-day perspective, extreme poverty is among the most important issues in well-being around 1850.

A study of newspaper articles between 1830 and 1850 reveals that from a contemporary perspective too, extreme poverty was one of the most important societal issues of the time. The poor led not only a meagre, but also a vulnerable, existence. The latter also applied to a large part of the Dutch population. It had to cope with the elements in their extreme forms: heat waves, bitter cold, violent storms, heavy rains and hailstorms.

A component of well-being specific to the Netherlands as a country located in the delta of multiple rivers was the struggle against water. This was waged along three main fronts: the management of inner (fresh) water, the struggle against the sea, and the interminable fight with the rivers.

This chapter is written by Harry Lintsen with contributions by Lilianne Laan, Önder Nomaler, Martijn Anthonissen and Ben Gales.

Finally, by present-day lights, in the past all the cities in the Netherlands were filthy and polluted with organic waste, including human and animal faeces. This was in large part responsible for low life-expectancy and poor public health. From a present-day perspective this touched on an important aspect of well-being.

Keywords Poverty · Vulnerability · Water management · Hygiene

4.1 'How Can We Combat Pauperism?'

> Pauperism is the sickness that eats at the heart of all contemporary societies; it is the most fearsome epidemic that assaults the nations, and for which... appropriate means are lacking to heal the sickness once and for all...[1]

This was the opinion advanced in 1849 in a lead article by the editors of the *Arhemse Courant* as part of a short but heated debate on the causes of poverty. The provocation was an article in *Het Dagblad van 's-Gravenhage* titled 'How can pauperism be combated?'[2] Its author was Augustus Elink Sterk jr., son of a Lutheran minister, senior official in the Ministry of Finances and elected member of the Provincial Estates of South Holland. He was a well-known in Hague circles and published on a wide range of topics.

According to Elink Sterk, poverty was for '... nine tenths the victim's own fault... whether due to sluggishness or neglect, intemperance and dissolute behavior, rushing into unconsidered marriages without hope of means of support...' In short, because of 'chaotic living and lack of reflection.' The editors of the *Arnhemsche Courant* protested this portrayal.

> It is a grave untruth... One has to dig deeper if one wants to discover the source of the evil. Suppressing the means of exchange between nations and peoples, artificial industries kept alive only by main force and prohibitions, concentration of capitals in few hands, a production that consistently exceeds the needs of the users and in consequence too many hands and too little work – behold some of the causes ... of pauperism that has only become evident nowadays.[3]

Here in a nutshell are summarized the standpoints and arguments of contemporaries regarding the question of poverty.

Pauperism refers to quality of life, a concept that occupies a central position in this study with broad implications that relate, among other things, to health, domestic life, existential security and safety. Welfare is the core and has to do with the degree to which the population is able to meet its needs with scarce means, while well-being indicates how people perceive their quality of life. Well-being is realized with four resources (in other words the four capitals: natural, economic, human and social) of which natural capital is the foundation. The previous chapter dealt with natural capital in the Netherlands, its exploitation, the consequences for nature and

[1] Hoe is het pauperismus te stuiten', *Arnhemsche Courant* 28-08-1849.
[2] Hoe is het pauperismus te stuiten', *Dagblad van 's-Gravenhage* 13-08-1849.
[3] *Arnhemsche Courant* 28-08-1849.

the yields in terms of gross domestic income. The question for this chapter is: what is the well-being of inhabitants of the Netherlands in the mid-nineteenth century? Does pauperism belong to the 'most fearsome epidemic' of the Netherlands and with that to the most important problem of well-being and sustainability?

4.2   Poverty in the Netherlands

4.2.1   The Scope of Poverty[4]

From a present-day perspective, poverty ranks among the most important problems of the mid-nineteenth century. The historian Allen has developed a method to evaluate the living standard of the lowest classes in the past.[5] He starts with the disposable income that a person absolutely needs to survive. If he falls below this level then he will starve or die of cold. The cost of the minimum necessary consumption (C_{min}) is based on a 'basket,' containing the minimum quantities of food and fuel (for heating and illumination) and the minimum sums for clothing and housing. The magnitude of the entire sum depends on place and time. The cost of living is different for example in 1850 and in 1750 and for London and Amsterdam. We estimate C_{min} for the Netherlands in 1850 to amount to an average of 19 guilders per inhabitant per year (see Table 4.1).

In addition, Allen defines a basket that he somewhat euphemistically calls 'respectable.' The disposable income is a bit bigger and the individual can enjoy a menu with more variety (bread instead of porridge, better beer and a modest quantity of eggs, cheese and meat). He can also clothe himself better and enjoy more warmth and better housing. The 'respectable' basket serves to define the poverty line. Below this, people enter into the danger zone. They will often feel hungry, possibly have to dress in rags and in a moderately stringent winter will suffer bitter cold. The costs of living at the poverty line are C_{line} and these are estimated at 43 guilders per inhabitant per year (Table 4.1). That roughly corresponds to the line that the World Bank nowadays defines for extreme poverty.[6]

In 1850, with its four capitals, the Netherlands was able to generate a gross domestic income of 659 million guilders (Table 4.1). Of that amount, fl. 466 million was spent on private consumption. Had consumer expenditure in 1850 been equally distributed across all inhabitants, then all would have lived far above the poverty line: $C_{average}$, after all, amounted to 150 guilders/year. But because of social inequality the situation was entirely different. The gini – a measure of social inequality – in

---

[4] See for the calculations in this subsection: H. Lintsen, M. Anthonissen en B. Gales, *Berekening omvang extreme armoede in Nederland 1820–1913* (research report Eindhoven University of Technology, 2017).

[5] R.C. Allen, *Poverty lines in history, theory and current international practice* (Oxford 2013).

[6] The line of extreme poverty is set at $1.9 per day, i.e. $694 per person per year (constant prices 2016).

Table 4.1 Private consumer
expenditure and magnitude of
poverty 1850 (estimates)

	1850
Total private consumer expenditure	466 mln gld
Total population	3,098,000
$C_{average}$ (consumer expenditure per capita)	150 gld/year
C_{min} (costs consumption subsistence level)	19 gld/year
C_{line} (costs consumption poverty line)	43 gld/year
Gini in relation tot GDP	0.48
Gini in relation to consumer expenditure	0.41
Percentage of population under C_{line}	21%
Number Dutch under C_{line}	658,000

Remark: Consumer expenditure in constant 1913 prices. Al levels of consumption are determined for an 'average' inhabitant, taking account of the differences among men, women and children

Source: H. Lintsen, M. Anthonissen en B. Gales, *Berekening omvang extreme armoede in Nederland 1820–1913* (research report Eindhoven University of Techology, 2017)

relation to the consumer expenditure is estimated at 0,40 (1 is maximum inequality and 0 is maximum equality). This means that about 21% of the Dutch population lived under the poverty line (C_{line}), i.e. about 650,000 inhabitants of the Netherlands suffered extreme poverty.

On the basis of estimates for poor relief in 1850 it appears that 14% of the population was on the dole. This percentage does not, however, reveal how much was received *in natura* in the form of food and fuel. Moreover records of charitable organizations were sometimes missing. 1850 was not an especially problematic year. Quite the contrary. According to the papers, the number of the poor had decreased relative to previous years which were marked by the potato famine, high grain prices and cholera (Sect. 1.2). Our estimate of 21% poor therefore seems to accord with the magnitude of contemporary percentages (Table 4.2).

There are striking differences among the provinces. Drenthe, Overijssel and Gelderland scored low relative to the other provinces. Noord-Holland had the most poor in an absolute sense, more than 150,000 persons, while in a relative sense Friesland topped the list. There were also big differences between towns and the countryside: the number of the poor was higher in the city than in rural areas.[7]

---

[7] See for example the table in: J.L. Van Zanden, *Nederlandse landbouw in de negentiende eeuw 1800–1914* (dissertation Wageningen 1985), 218 and J.L. Van Zanden, *'Den zedelijke en materiële toestand der arbeidende bevolking ten platten lande'. Een reeks rapporten uit 1851* (Historia Agriculturae XXI, Groningen 1991), 36.

Table 4.2 Number of people on dole in the Netherlands in 1850

	Population	On dole	Percentage (%)
Drenthe	83,000	4896	5.9
Overijssel	216,000	16,206	7.5
Gelderland	365,000	34,844	9.5
Noord-Brabant	393,000	49,033	12.5
Limburg	204,000	32,589	16.0
Groningen	185,000	20,674	11.2
Friesland	248,000	47,982	19.4
Zeeland	160,000	22,330	13.9
Utrecht	153,000	25,263	16.5
Noord-Holland	472,000	90,740	19.2
NETHERLANDS (excl. Zuid-Holand)	2,513,000	344,557	13.7

Source: Nederlandsche Staatscourant 17-08-1851

Could poverty, as the most important factor in well-being, already have been solved at that time? Could, in other words, the level of extreme poverty in 1850 have been reduced to nearly nil? There were in principle three options. First, a further decrease in social inequality. This option was the bugaboo of the bourgeoisie: the possibility that the poor would rise up, that revolution would be propagated and the social order be attacked.[8] As noted above, extreme poverty would have disappeared in 1850 given an equal distribution of consumer expenditure (a gini of 1).

Table Numbers of those on poor relief and living at home[a] as percentage of the population in 1817, 1832 and 1850

Province	1817	1832	1850	1850 countryside
Drenthe	4.8	3.9	5.6	5
Overijssel	5.6	5.4	6.7	5
Gelderland	7.1	6.8	8.8	6
Noord-Brabant	10.2	6.5	11.9	10
Limburg	–	–	14.0	9
Groningen	5.2	5.9	9.0	7
Friesland	8.5	11.1	15.7	14
Zeeland	8.1	5.7	12.3	10
Utrecht	9.3	13.6	15.2	9
Zuid-Holland	11.3	10.6	15.7	10
Noord-Holland	21.9	15.3	23.0	–
NETHERLANDS	11.0 excl. Limburg	9.4 excl. Limburg	13.8	

[a]Poor relief distinguished between those on the dole *living at home, or the 'outside poor'* and the *'inside poor.'* The *inside poor* were those on poor relief that were permanently or temporarily housed in institutions: an orphanage, a hospital, an old age home, a widows' home and almhouse courts

[8]M.H.D. van Leeuwen, *Bijstand in Amsterdam ca. 1800–1850. Armenzorg als beheersings- en overlevingsstrategie* (Zwolle 1992), 123–125.

Table 4.3 Social inequality and economic growth if the poor in 1850 had been less than 5% of the population

	Situation 1850	In case of less social inequality	In case of greater economic growth 1820–1850
C_{line} (costs of consumption at poverty line)	43 gld/year	43 gld/year	43 gld/year
Portion of the population under C_{line}	21%	**<5%**	**<5%**
Gini in relation to GDP	0.48	**0.35**	0.48
Gini in relatie tot consumer expenditure	0.42	**0.30**	0.42
Growth of Gross Domestic Income 1820–1850	1.4%	1.4%	**3.3%**
Growth private consumer expenditure 1820–1850	1.2%	1.2%	**3.1%**

Source: H. Lintsen, M. Anthonissen en B. Gales, *Berekening omvang extreme armoede in Nederland 1820–1913* (research report Eindhoven University of Techology, 2017)

But even at today's degree of inequality (gini relative to consumer expenditure of 0.30) poverty at the time would in our estimation have been considerably less (Table 4.3).

The second option consisted of creating such a degree of economic growth that even with social inequality at the same level almost all the poor would be lifted above the poverty line. At the time this option was referred to as encouraging 'popular (or national) prosperity' (*volkswelvaart*).[9] In that case the total magnitude of consumer expenditure would have had to amount to 810 million guilders in 1850, or 261 guilders per inhabitant. This means that the growth of the Gross Domestic Product over the period 1820–1850 would have had to amount to 3.3%. This level of growth seems impossible with the then existing natural capital and the contemporary (agricultural) technology. Agriculture in the 1830s grew by only 1.4%. This in combination with the growth of industry and trade produced an increase of 2.3% in the GDP.

There was a third option, namely the exploitation of natural capital and populations elsewhere in the world. This was one of the aims of colonial policy. That policy was beholden to reap benefits for the development of 'national prosperity' in the Netherlands.[10] The construction of the 'colonial complex' certainly contributed to the decline in the level of poverty in the Netherlands, particularly in the 1830s, but not enough. This option would become a bone of contention after 1850.

---

[9] See the following Chaps. 5 and 6.

[10] J.L. Blussé, 'Koning Willem I en de schepping van de koloniale staat', in I. de Haan et al. (Eds.) *Een nieuwe staat: Het begin van het Koninkrijk der Nederlanden* (Amsterdam 2013), 169. Blussé notes that the originator of the Cultivation System, Johannes van den Bosch, was also concerned with improving life for the native population.

4.2.2 The Perception of the Poverty Question

In the mid-nineteenth century, many inhabitants of the Netherlands lived beneath the poverty line. In 1850 some 650,000 inhabitants of the Netherlands were extremely poor and lived from day to day. Their budgets approached the absolute minimum for survival. Poverty must have been visible everywhere in the Netherlands.

From today's perspective, the quality of life in those days – in terms of poverty – was one of the most important well-being and sustainability issues, if not the most important. The question is whether contemporaries also saw it this way. Was poverty an important issue for them? The answer is important if we are to avoid an anachronistic approach to well-being and sustainability. The elaborate system of poor relief suggests that poverty was an important theme, but was it also a problematic one? The system might have functioned so efficiently that contemporaries regarded the question as solved and felt themselves at liberty to turn their attention to other matters. What was the relationship of poverty to other societal issues?

One of the (few) ways to address this question is a systematic survey of newspapers, journals, pamphlets or ego documents. We limit our investigation here to all newspapers in the Netherlands from 1830 to 1850 and use as indicator those newspaper accounts in which 'misery' or a related concept appears. From a collection of more than 10,000 accounts we extracted an a-select sample (see Table 4.4).[11]

Poverty played a major role in accounts of 'misery' (29% of the articles). Much misery had to do with poverty and was often portrayed in heart-wrenching terms.

> …There I saw the Father or Mother struggling with death; there I saw the despairing Man or Woman, now gazing on the dying, now at the poor uncared-for brood. There I saw in another family again children surrounding the poor deserted Mother, weeping with her for the death of the loved Father, and the last bit of bread, charily divided…[12]

Along with 'Poverty,' 'Politics' pertains to the biggest category of newspaper accounts containing the term 'misery.' These accounts announced debates in parliament, political decisions, lawmaking, import duties and suchlike. These issues were nearly always discussed in combination with 'Poverty' or one of the other categories.

'Fate' is such a category (15% of the accounts). These articles recounted fires, floods, epidemic diseases, hurricanes, harsh winters and other natural violence and calamities. Some of the accounts made a connection to poverty. In these accounts misery was a question of fate and fate could lead to poverty.

'War' and 'insurrection' were other categories (respectively 13% and 11% of the accounts). Europe was restless in the 1830s and 40s. Belgium seceded from the Netherlands. Civil war threatened in Spain. In Portugal there was a struggle for power. Russia had occupied Poland. Turkey was at war with Egypt. France fought in Algeria. Great Britain conquered the Falkland Islands, etc. Several accounts made

---

[11] L. Laan, Ö. Nomaler en H. Lintsen, *Perceptie en kranten, 1830–1850* (research report Eindhoven University of Technology, 2017).

[12] 'St-Nicolaasfeest', *Rotterdamsche Courant* 09-12-1848.

Table 4.4 Theme in newspaper accounts between 1830 and 1850 in which 'misery' ('ellende') or a similar term appears (n = 346)

	Total	Domestic	Abroad
Poverty	101 (29%)	26	75
Politics	99 (29%)	31	68
Fate	52 (15%)	20	32
War	46 (13%)	19	27
Insurrection	38 (11%)	7	31
Morality	38 (11%)	20	18
Charity	36 (10%)	29	7
Miscellaneous	84 (24%)	35	49

Remark: The total exceeds the number of selected articles (n = 346) because a number of the articles was assigned to more than one category
Source: L. Laan, Ö. Nomaler en H. Lintsen, *Perceptie en kranten, 1830–1850* (research report Eindhoven University of Techology, 2017)

a direct connection with poverty. War and insurrections brought misery, chaos and poverty. Poverty could be the cause of wars and insurrections.

For the rest, the newspapers were a well-trodden podium for moralistic narratives; histories with a moral turn, a personal tragedy framed as a warning, an exciting serial with moralistic overtones or strong condemnation of behavior like opium trading or alcoholism (11% of the articles). These articles containing the term 'misery' fall into the category of 'morality.' These accounts also regularly made a connection to poverty. Poverty could be prevented by '…more frequent attendance at public religious services, quiet, stillness on Sunday evenings…'[13] An elevated moral sensibility inspired the charity that combatted poverty.

Finally, there was a great number of articles containing the term 'misery' that dealt with a rich variety of subjects (the category 'diverse' with 24% of the accounts). They dealt with, for example, misery due to religious tensions, misery in prisons, the misery of vulnerable groups in the overseas colonies and emigration in consequence of misery and need. Miscellaneous notices also fell into this category.

This analysis reveals poverty to have been the most important social question. Contemporaries saw it as a condition that brought much misery for individuals and social groups, both domestically and abroad. It had political aspects and was related to other social questions like morality and war. Poverty had to do with life's risks. Anyone could be a victim of fate. Chaos was ever-threatening and put existential security on the line. Contemporaries could have proclaimed poverty and vulnerability to be important, if not the most important, limits to well-being of their time.

To which contemporaries do we refer in this connection? Whose opinions were ventilated in the newspaper? Who was allowed to speak? Who read the paper? The newspaper was certainly not a 'podium for the people.' The newspaper was still for the most part the masculine domain of the upper middle class and the elite; the masses and the poor were not heard there. The authors were judges, writers and

---

[13] *De Noord-Brabander* 13-04-1844.

government officials. Newspaper readership was composed of the educated and the wealthy. These were the classes from which the officers of charitable organizations were recruited. They shaped the policies of and were responsible for poor relief. But to whom did they extend their helping hands? Who were the poor?

4.2.3 Poverty in the City and in the Countryside

In the city three categories of the poor could be distinguished.[14] First were the aged, the sick and the invalids. Around 1850 they comprised a significant portion of the urban population.[15] Depending on the institution, percentages ranged from 30% to 60%. Living with old age, illness and invalidity was in any case a heavy burden in those days. For those lacking financial resources it was punishing.

The employable poor formed another category. This could be divided into a group of workers holding often temporary jobs at the lowest wages and the group of minor craftsmen, small shopkeepers and the lowest-ranking civil servants.[16] The lowest-paid workers were referred to as day labourers, labourers or helpers or more specifically with, for example, porters, wagon loaders, barrow-men, and bag loaders. The lowest-paid female workers were called maids or more specifically washerwomen and cleaning ladies. Every city had a large class of unschooled and low-paid workers, who had a good chance of falling under the poverty line because of 'the ... difficulty, not to say the impossibility, for many heads of households to earn their keep by means of their labour...'[17] Having a trade, shop or job as civil servant was no guarantee of being able to avoid the trap of poverty. Tailors, seamstresses, pinmakers, spinners, vegetable sellers, teachers etc. often found it extremely difficult to keep their heads above water. Winter was an especially hard time. Business was slow, while the costs of heating and clothing increased.

---

[14] P. de Rooy, 'De armen hebt gij altijd met u: armenzorg en onderwijs', in: I. de Haan, P den Hoed and H. te Velde, *Een nieuwe staat. Het begin van het Koninkrijk der Nederlanden* (Amsterdam 2013), 221–222.

[15] See among others: Van Leeuwen, *Bijstand in Amsterdam ca. 1800–1850*; M. Prak, 'Overvloed of onbehagen. Armoede, armen en armenzorg in 's-Hertogenbosch, 1770–1850, in: J. van Oudheusden and G. Trienekens, *Een pront wijf, een mager paard en een zoon op het seminarie: aanzetten tot een integrale geschiedenis van oostelijk Noord-Brabant 1770–1914* ('s-Hertogenbosch 1993), B.P.A. Gales, L.H.M. Kreukels, J.J.G. Lujten and F.H.M. Roebroeks, *Het Burgerlijk Armbestuur. Twee eeuwen zorg voor armen, zieken en ouderen te Maastricht, 1796–1996* (Maastricht 1997), band 1 en 2. On poverty in earlier periods: H. Gras, *op de grens van het bestaan. Armen en armenzorg in Drenthe 1700–1800* (Zuidwolde 1989) and H.F.J.M. van den Eerenbeemt, *In het spanningsveld der armoede. Agressief pauperisme en reactie in Staats-Brabant* (Tilburg 1968). See also: C. Lis and H. Soly, *Armoede & kapitalisme in pre-industrieel Europa* (Amsterdam 1980) and A. de Swaan, *Zorg en de staat. Welzijn, onderwijs en gezondheidszorg in Europa en de Verenigde Staten in de nieuwe tijd* (Amsterdam 2004).

[16] Van Leeuwen, *Bijstand in Amsterdam ca. 1800–1850*, 30.

[17] Van Leeuwen, *Bijstand in Amsterdam ca. 1800–1850*, 189–190 and Prak, 'Overvloed of onbehagen', 26–34.

The third category of the poor were the social outcasts who were all but excluded from the labor process: beggars, drifters, paupers. They were most to be pitied. They were beyond the pale of poor-relief and were often chased away or locked op in workhouses.

For women, life was more difficult than for men and relatively speaking they turned more often to poor relief. They typically encountered difficulties when their husbands died or absconded and they were left to fend for themselves. Women earned less than men, even for similar work. Income from simple labour was far from adequate to maintain a family with children. In addition, women also typically had to run a household.

The category of the elderly and the sick and that of the beggars, drifters and paupers were also well-represented in the countryside. The category of the employable poor consisted of three groups: day-labourers, maids and hired hands, and small-holders. Day-labourers in 1850 were to be pitied.[18] In Zeeland there was a large population of such temporary labourers, an estimated 2/3 of the agrarian population. Their situation was lamentable: they barely had an income, were in poor health, and were poorly clothed and housed:

> When we see our eyes drawn to so many pale faces and emaciated bodies, and one sees the masses of weak and sickly beings... then assuredly we must admit that the material condition of the labouring class demands an investigation...[19]

Similar reports came from other provinces, describing day-labourers 'whose existence...is extremely uncertain, and most grievous, if illness prevents them from working.'[20] They often lived in shacks with a 'great number of children who, like little cave-dwellers, emerge half-naked from the ground.'[21]

---

[18]Van Zanden, *'Den zedelijke en materiële toestand der arbeidende bevolking ten platten lande'*. See for the analysis and discussion of sources: H. Lintsen, *De schraalheid van het bestaan, een reeks rapporten uit 1850* (research report Eindhoven University of Technology, 2017).

[19]'[Zeeland] Rapport naar aanleiding van een ingesteld onderzoek omtrent den zedelijken en materiëlen toestand der arbeidende en dienstbare bevolking ten platten lande', in van Zanden, *'Den zedelijke en materiële toestand der arbeidende bevolking ten platten lande'*. Originally published as a report to the annual meeting of the Zeeuwse Maatschappij van Landbouw (Zeeland Agricultural Association) at Tholen on June 7, 1849 (Middelburg 1849), 12.

[20]A. Martini van Geffen (1991), 'Zedelijke en materiële toestand der arbeidende bevolking ten platten lande. I. Nopens Noord-Brabant', in van Zanden, *'Den zedelijke en materiële toestand der arbeidende bevolking ten platten lande'*. Original source: *Verslag van het verhandelde op het vijfde Nederlandsche Landhuishoudkundig Congres, gehouden te Leiden van den 10den tot den 14den junij 1850* (Leiden 1850), Appendix no 12, 101.

[21]B.W.A.E. Sloet tot Oldhuis (1991), 'Statistieke beschouwing van den toestand der geringe plattelands bevolking op de Veluwe langs de Zuiderzee, (gemeenten Putten, Ermelo, Elspeet, Harderwijk, (Hierden), Doornspijk, Oldenbroek, Heerde en Epe)', in van Zanden, *'Den zedelijke en materiële toestand der arbeidende bevolking ten platten lande'*. Original source: *Tijdschrift voor Staathuishoudkunde en Statistiek* 9(1853), 291.

Hired hands and maidservants earned about the same as day-labourers. If that was all they could count on, they were condemned to an abject life.[22] But they enjoyed various advantages. They had more security because they were a permanent part of the farm. They were housed in the farm and could piggy-back on the farmer during prosperous times.[23]

Other inhabitants of the countryside that found it hard to make ends meet were the smallholders. Their small farms gave them a bit of security, but they had to exert themselves to the utmost to keep their heads above water. Discipline and frugality were the necessary ingredients; 'were this not the case, many farmers would soon sink into the estate of day-labourers or that of poverty.'[24]

Regional differences in rural poverty were large in 1850 and varied with the nature of the farm and the orientation to the market. Misery was no stranger, for example, to the clay-grounds in Zeeland. The mixed large farms there faced hard times. The wheat farmers suffered. The potato farmers were hard hit by the potato blight. This in the wake of a crisis in madder production due to strong competition from French madder. Farmers had to cut back on the high costs of wages and other expenditures like those for the blacksmith and the shopkeeper. Hence it was not only the farmers, their workers and the day-labourers who suffered from the agricultural malaise.

The livestock farms in Holland, by contrast, did rather well. There were fewer day-labourers. The farmers on the large farms generally employed permanent personnel for the regular work and had recourse to migrant labour during peak times in the summer.

On the sand grounds the smallholders were less dependent on the market, more self-reliant, less specialized and more flexible. If the market allowed, they could work their land more intensively or switch to other products. Moreover these small farmers were able to earn income from sources other than agriculture. That was enough to clothe themselves and to eke out an existence, albeit on an extremely sober footing.

The differences among the regions did not, however, hide the fact that hired hands, maidservants and smallholders lived around the poverty line and that day-

---

[22] M. van Geffen, 'Noord-Brabant' Bijlage no 12, 101–102. According to the Brabant report, there was an annual surplus of about five guilders with which to clothe a family of five children, meet miscellaneous household expenses, buy firewood, pay school fees etc.

[23] A report on Zeeland noted that maids and hired hands were well housed - on the farm or in the village – and that from the farmer 'they could obtain…at reasonable prices…grains, while most of them were granted a small plot of land for a modest rent on which to grow potatoes…' [Zeeland]…' 13.

[24] J. Zeehuisen (1991), 'Statistieke bijdrage tot de kennis van den stoffelijken en zedelijken toestand van de landbouwende klasse in het kwartier Salland, provincie Overijssel', in van Zanden, 'Den zedelijke en materiële toestand der arbeidende bevolking ten platten lande'. Original source: Tijdschrift voor Staathuishoudkunde en Statistiek 6(1851), 380.

labourers regularly fell below it. 'Over against the other members of society they are slaves…' as one investigator summarized the existence of agricultural workers (hired hands and maidservants, smallholders and day-labourers):

> '…slaves in body and spirit, labouring, if they have work, from early morning to the evening to earn a meagre bit of bread for themselves and theirs…having almost no hope of ever raising themselves to the status of independent citizens in society…'[25]

The poor had not only a meagre, but also a vulnerable existence. The smallest setback led to big problems and immersed them in a bitter struggle for survival. But not only the poor were vulnerable. Other large groups of the population were in the same boat. Prosperity, property, well-being and health could be threatened in numerous ways. Economic development was one of the factors that made life in the city and the countryside unpredictable and vulnerable. We discussed the fickle economic circumstances in Chap. 3. Here we will focus on factors associated with natural capital: life with natural forces, the struggle against the waters, and living in a delta. Natural capital was (in combination with human and economic capital) the most important source of subsistence. At the same time it was the source of an existential uncertainty that also played a part in shaping the quality of life.

4.3   Cold, Heat and Storm[26]

Well-being was heavily influenced by weather and the seasons. Every season brought its own problems. In the winter the theme was the 'cold.' Harsh winters were a common and regular occurrence in the Netherlands from the fifteenth up to and including the nineteenth century – a period also known as the little ice age.

---

[25] J.H. Beucker Andrea, 'Rapport ingediend voor het vijfde, Landhuishoudjundig Congres te Leyden, 11, 12, 13 junij 1850, betreffende een onderzoek naar den zedelijken en materiëlen toestand der arbeidende bevolking ten platten lande en van den middelen om dien zoveel mogelijk te verbeteren. [Friesland]', in van Zanden, *'Den zedelijke en materiële toestand der arbeidende bevolking ten platten lande'*. Original source: *Tijdschrift voor Staathuishoudkunde en Statistiek* 6(1851), 166.

[26] An exhaustive investigation into the history of the weather in the Netherlands has been undertaken by J. Buisman. This has resulted in a fascinating series with numerous data, observations, anecdotes, announcements etc. At the time of writing of the present volume, five volumes of Buisman's study have been published, covering the period from about 800 to 1750. J. Buisman, *Duizend jaar weer, wind en water in de lage landen* (Franeker 1995–2006), volumes 1 t/m 5. Mr. Buisman has generously provided us with the manuscripts for the years 1800–1850. Page references refer to the manuscript. Thanks to the chronological references, when the time comes citations and data will be easily traceable in the volumes to published.

In the first volume Buisman provides a description of the threats that natural forces presented to the survival of the Dutch in the middle ages, see: J. Buisman, *Duizend jaar weer, wind en water in de lage landen* (Franeker 1995) Vol. 1, to 1300, 97–100. We presume that up to 1850 not much had changed. Buisman's research does not in any case suggest that we should assume that the situation improved. After 1850 conditions would improve dramatically thanks to industrialization and modernization.

People were always wondering how harsh the winter would be and how long it would last. Did one have sufficient turf or wood to survive these months? Would food supplies be spoiled by the frost? The long and harsh winter of 1830 had already set in by the end of the previous November. On February 3 a farmer from Wirdum wrote that.

> ... the former bitter frost, ever-increasing up to now, is accompanied by a stiff east wind and clear skies. The frost has penetrated into the houses all the way to their hearths... This condition of frost and bitter cold will greatly deepen the misery of numerous people, in view of the general lack of turf and fuel, the need for food and shelter especially among the common folk. Potatoes that many still have in storage, and up to now have kept from the frost, will no longer be able to be saved from it...[27]

If during the winter months the supply of fuel was insufficient, a search for brushwood was undertaken or the few pieces of furniture were burned. There was no money to buy extra, expensive fuel. Commercial and industrial activity came to a standstill and unemployment rose precipitously. Appeals to poor relief were already frequent in normal winters, in harsh winters even more so: 'During harsh winters the worn out day-labourers, and those burdened by a large family and without work, are supported by the poor relief.'[28]

The coming of spring was joyously celebrated. At last there was light and mild weather. Nature came alive again and economic activities began to resume. What remained was the danger of frost. Now it was not the danger of freezing to death, but damage to crops for the farmer in the countryside and for the city dweller with his vegetable patch. Moreover if the frost held on too long, peat extraction was delayed which gave the turf too little time to dry, diminishing the quality.[29] The period of the so-called Ice Saints (saints with name days between 11 and 15 May) was taken to mark the definitive transition to summer; folk wisdom held that 'It can freeze until May, with the Ice-Saints it goes away.' But spring might hold another unpleasant surprise, namely long and heavy periods of rain. In such extremities, the windmills in the low Netherlands could no longer pump all the water away, especially when the rain was accompanied by a long windless period. Polders became swampy fields or were flooded. Farms were cut off from the world. Cows had to remain in the barn. Summer grain was lost. Sowing had to be postponed.[30]

---

[27] J. Buisman, *Duizend jaar weer, wind en water in de lage landen,* manuscript, 571 (to be published, presumably volume 7, 1825–1875.

[28] F.H.C. Drieling, 'Verslag over den toestand der arbeidende klasse, vooral ten plattelande, in de provincie Utrecht, en over de middelen tot verbetering van hunne zedelijke en stoffelijke welvaart', in van Zanden, *'Den zedelijke en materiële toestand der arbeidende bevolking ten platten lande'.* Original source: *Tijdschrift voor Staathuishoudkunde en Statistiek* 8(1853), 28.

[29] Hölsgens (2016), forthcoming.

[30] In the worst cases the dikes were breached. '... Much damage was done on May 14, 15 and 16 [of 1844] by the hard wind. In the new polder named Eijerland the summer barley and also a great quantity of oats have been beaten down by the hard wind and there are more places in Holland where much damage has been done. Due to the floodwaters on Kamper Island the houses were submerged to the attic and on Ameland hundreds of hay-wagons floated away...' Buisman, *Duizend jaar weer, wind en water...,* manuscript, 557, (to be published, presumably volume 7, 1825–1875).

The summer too had threats in store. Heavy thunderstorms or heat could destroy harvests. That led to food shortages in a matter of months. If these were not capably managed, famine ensued. Dry periods also affected cattle:

> Due to the extreme drought, moisture escapes not only from the earth, but also to such an extent from ditches and pools that they dry up, and if this continues, the animals can no longer be maintained on the land. Among farmers there is no greater burden than cattle wandering aimlessly over the fields.[31]

Shortages of water, grass and hay could lead to massive starvation of cattle.[32] Sometimes in dry summers farmers were confronted with plagues of mice that destroyed supplies of hay and grain. Friesland had such a summer in 1832. The plague was so bad that children on one farm caught 230 mice in a few days.[33]

The fall was among other things the period of the autumn-storms. These could wreak havoc in the countryside: crops that were lost; chimneys that came down; roof tiles that flew through the air; roofs that were lifted up; houses that collapsed; runaway windmills that caught fire from the heat of friction; polders that flooded (because now there was too much wind for the windmills). The first November storm of the nineteenth century set the tone. In Woensel (Noord Brabant) the church tower blew over and in Haren 43 of the 60 houses were damaged. Crop failure was a regular occurrence. In addition to bad weather, crop diseases were also to blame, such as the infamous potato blight of 1845.

4.4 The Vulnerable Dutch Delta[34]

The Dutch struggled heroically to survive in their delta. Nature's elements had to be resisted in their most extreme forms: heat waves, bitter cold, extreme storms, heavy and prolonged rainfall and hailstorms. The clothing and housing of large parts of the

---

[31] Buisman, *Duizend jaar weer, wind en water...*, manuscript, 557, (to be published, presumably volume 7, 1825–1875).

[32] An example, although from the eighteenth century: In the second half of July in 1750 a heat wave occurred that was '...so unbearably hot that many people and animals very quickly died from the heat. The animals in the fields suffered so much from this heat that throughout our entire fatherland they produced half as much milk as usual ... and the fish in the water died by the thousands of pounds...it seemed as though the water was everywhere strewn with dead fish.
Cited in: J. Buisman, *Duizend jaar weer, wind en water in de lage landen*, (Franeker 2006), Vol. 5, 1675–1750, 833.

[33] Buisman, *Duizend jaar weer, wind en water...*, manuscript, 557, (to be published, presumably volume 7, 1825–1875).

[34] This section is based on: G.P. van de Ven, *Leefbaar laagland. Geschiedenis van de waterbeheersing en landaanwinning in Nederland* (Utrecht 1993); A. Bosch and W. van der Ham, edited by H.W. Lintsen, *Twee Eeuwen Rijkswaterstaat 1798–1998* (Zaltbommel 1998); A. Bosch, *Om de macht over het water. De nationale waterstaatsdienst tussen staat en samenleving 1798–1849* (Zaltbommel 2000, dissertation); A. van Heezik, *Strijd om de rivieren. Tweehonderd jaar rivierenbeleid in Nederland of de opkomst en ondergang van het streven naar de normale rivier* (dissertation TUDelft 2007); H.W. Lintsen and N. Disco, 'Het nijvere verbond', in: H.W. Lintsen et al.,

population were ill-suited to the struggle. Bodies had a hard time of it. Moreover there was the perpetual war against the waters. This was fought on three fronts: management of 'inner water,' the fight against the sea and the incessant struggles with the rivers. The plodding toil in the Dutch mud was a war of attrition, one that repeatedly came to a premature and inglorious end.

The management of inner water entailed two kinds of risks, namely the risk of soil subsidence and that associated with lakes and ponds. Extraction of peat by drainage initiated an unintended process of soil subsidence that could not be stopped. Hydraulic engineering projects like polder drainage and reclamations accelerated the process. Increased habitation and continuing investments behind the dikes and in the polders exacerbated the consequences and risks of calamities. Lakes and ponds, of which there were hundreds at the beginning of the nineteenth century, were the second problem in managing inner water.[35] Especially problematic were large bodies of inner water, like the Haarlemmermeer, where the wind could blow unhindered for long distances. But for all that there were few fatalities and material losses were not dramatic.

The risks in these two cases did not in the first place concern a lack of safety. The big problem was an excess of water. This came home to farmers when business as usual (e.g. sowing in spring) became impossible. Sometimes, however, too much water became catastrophic. Extremely heavy rainfall, exceedingly high 'outer water' or long periods of no wind preventing the windmills from draining enough water could all, separately or in combination, cause damage to crops or cattle.[36]

Made in Holland. Een techniekgeschiedenis van Nederland [1800–2000] (Zutphen 2005), 75–93; H.W. Lintsen and A. van Heezik, 'In gevecht met de rivieren', in H.W. Lintsen et al., *Made in Holland. Een techniekgeschiedenis van Nederland [1800–2000]* (Zutphen 2005), 95–113.

[35] A brief and lucid description of the problems of polders and reclamations can be found in T. Stol. *Wassend water, dalend land. Geschiedenis van Nederland en het water* (Amsterdam 1993), particularly 21–58; 73–98.

[36] 'Danger' was not the motivation for the many reclamations. It was all about acquiring new land. The reclaimed land consisted for the most part of fertile clay. Investments could be attractive because the reclamations subsequently produced cereals, vegetables, meat and dairy products. Cereal and ground prices of course had to be favorable. Hence there was a clear relationship between the prices of cereal crops and the number of new reclamations. Initially private parties risked the investments. By the end of the eighteenth century governments were also taking initiatives.

King William I turned the reclamation of the Haarlemmermeer into a state project. The gigantic enterprise, he argued, provided work for thousands of families both in the short term (reclamation and the creation of an agricultural infrastructure) as well as in the long term (agriculture). It also produced an enormous amount of agricultural acreage to feed the kingdom and grow export products. The reclamation could also be financed from the sale of land. Public safety and loss of land due to erosion by waves were certainly important arguments for the reclamation. But they were not decisive in the deliberations. It was more the case that society no longer accepted the risks of flooding in the political and economic heart of the Netherlands and the destruction of agricultural products, capital goods and infrastructure. There was also opposition to the reclamation of the Haarlemmermeer. The lake was an important drainage catchment for the Rijnland Water Board, into which excess water from nearby polders could be pumped, whence it would gradually make its way to the sea. Where would that water be stored in the future? Similar problems with the

Danger from the sea was an especially impressive feature of Dutch history due to the storms accompanied by storm surges and claiming thousands of victims like the Saint Elisabeth Flood of 1421 and the All-Saints flood of 1570.[37] In 1825 there was another serious storm surge. Large parts of Groningen, Friesland and Overijssel were submerged. 380 people died.[38] The brackish water presumably caused additional indirect casualties. The following summer malaria erupted. In the Frisian countryside mortality exceeded normal levels by 4000 deaths.[39]

Storm surges were rare. Storms, on the other hand, occurred with great regularity. Damage was mostly local and the number of casualties small, but over the years losses could be significant and the number of deaths add up to hundreds. Acute loss of land did not occur on a large scale, but was a piecemeal affair. The old village of Egmond aan Zee, for example, gradually disappeared into the sea, church and all, between 1700 and 1850.

In general, invocations of the 'waterwolf' are assumed to refer to the struggle against the sea. But nature's elements could also wreak havoc in the region of the large rivers and cause heavy flooding.[40] These were above all a result of harsh winters, frozen rivers, sudden thaws and ice floes. Under these conditions, large masses of ice blocked the flow of water. In the eighteenth century river floods occurred with great frequency: 1726, 1740, 1751, 1757, 1781, 1784, 1799. In the first half of the

---

management of drainage catchments was expected for other reclamations. They required extra facilities. S.J. Fockema Andrea, 'Wat er aan de droogmaking van de Haarlemmermeer voorafging', in *Med. Der Kon. Ned. Akademie van Wetenschappen, afd. Letterkunde*, 18 (1955), 379–428; H.W. Lintsen, R.A. Lombaerts and R. Moerenhout, 'De droogmaking van het Haarlemmermeer: Wind of stoom', in: M.L. ten Horn-van Nispen, H.W. Lintsen and A. J. Veenendaal: *Nederlandse ingenieurs en hun kunstwerken. Tweehonderd jaar civiele techniek* (Zutphen 1994).

[37] Surprisingly, the last big storm surge before 1850 took place in 1717. The northern part of the country was hardest hit. In Groningen 2200 died and 37,000 head of cattle were lost. In the entire coastal area of the Netherlands, Germany and Scandinavia, 14,000 dead were mourned. Then for more than a century it remains 'quiet' on the coast. The storm surge of 1825 was regarded as a serious one. But precisely what *was* a period of 'quiet' with no storm surges? A heavy storm like that in January 1808 didn't make it into the history books. The epicenter was in Zeeland. The island of Walcheren suffered greatly. The Westkappele sea-dike was damaged. Near Domburg part of the dunes were blown away. Vlissingen and Middleburg were partly flooded. People died in Vlissingen, including a number of children. Household attributes, clothing and goods floated everywhere. Houses collapsed. Quays disintegrated. Walls were dislocated. Similar scenes were enacted in Zeeuws Flanders and other islands. More than 50 people died. Cattle were lost. And of course the storm tormented ships at sea in the coastal waters. Buisman, *Duizend jaar weer, wind en water...*, manuscript, 431 en 469, (to be published, presumably vol. 6, 1750–1825)

[38] Buisman, *Duizend jaar weer, wind en water...*, manuscript, 431 en 469, (to be published, presumably vol. 6, 1750–1825).

[39] Buisman, *Duizend jaar weer, wind en water...*, manuscript, 431 en 469, (to be published, presumably vol. 6, 1750–1825).

[40] See for the following: Lintsen and Van Heezik, 'In gevecht met de rivieren', 97–100; Van Heezik, *Strijd om de rivieren*, 44–52; Van de Ven (ed.), *Leefbaar laagland*, 227–232; Bosch and Van der Ham, *Twee Eeuwen Rijkswaterstaat*, 55–58; Bosch, *Om de macht over het water*, 177–184.

nineteenth century similar dramas occurred in 1805, 1809, 1820, 1827 and 1850. Before 1850 river floods were never absent from the collective memory. They were frequent occurrences.

That said, it is quite surprising how few deadly casualties there were. The river flood of 1809 was the most catastrophic ever and caused 275 deaths. The Dutch were in part able to defend themselves against these disasters. They generally inhabited the higher portions of the landscape. It was also the case that due to the many dikes the polders flooded one after the other in a gradual process. The inhabitants had time to get themselves and their cattle to safety. Moreover every farm had a rowboat that could be used to save people and cattle in case of a flooded polder.[41]

Inner water, the sea and the rivers all harbored specific risks for living in a delta. Water was also a problem in another respect. Life in the subsiding low portion of the Netherlands was increasingly confronted with saline water and the deterioration of water quality. Agriculture suffered. Drinking water supply became problematic. This makes water quality another important theme. The theme was broader than just increasing salinity. Water polluted by feces and organic waste was also part of the problem.

4.5 Organic Waste As the Biggest Environmental Problem[42]

As we now know, water is an important medium of contamination for infectious diseases. Bacteria and viruses flourish in water. Most of them are hardly a problem for humans. But among them are notorious perpetrators of disease. Some are responsible for innocent-sounding infections like diarrhea. But well into the twentieth century they were life-threatening for many inhabitants of the Netherlands, especially those with a poor condition. Other infectious diseases have never lost their fearsome auras, for example cholera and paratyphus. All these infectious diseases are transmitted in large part by drinking polluted water. But spoiled or contaminated food is also an important source of infection. The same goes for bathing or washing in polluted water. This knowledge was all but lacking in the first half of the nineteenth century. Only a small vanguard – a new generation of medical

---

[41] P. van Dam, *De amfibische cultuur: een visie op watersnoodrampen* (VU Amsterdam 2010, inaugural lecture).

[42] H. van Zon, 'Openbare hygiëne', in: H. Lintsen et al. (ed.), *Geschiedenis van de techniek in Nederland: De wording van een moderne samenleving 1800–1890* (Zutphen 1993), volume II, 49–55. Further: H. van Zon, *Een zeer onfrisse geschiedenis: Studies over niet-industriële vervuiling in Nederland, 1850–1920* (dissertation, Rijksuniversiteit Groningen 1986); A. van der Woud, *Koninkrijk vol sloppen: Achterbuurten en vuil in de negentiende eeuw* (Amsterdam 2010); H. Buiter and H. Lintsen, 'De stad, de stank en het water', in: H. Lintsen et al., *Made in Holland. Een techniekgeschiedenis van Nederland [1800–2000]* (Zutphen 2005), 55–73.

doctors, engineers and architects: the 'hygienists' as they would later be called – were aware of the problem.[43] Their activities had as yet little impact. The problems were most visible in the cities.

By today's standards all Dutch cities in the past were nothing short of filthy. Various studies have provided us with graphic images of the stench and the filth. Markets, squares and streets were strewn with vegetable waste, meat and cadavers and the feces of cows and pigs. Hide tanners were infamous for the harmfulness of their 'fumes and effluents.' Paints and other substances spread foul odors in the open air. In general, waste products of home-workers and workshops polluted many locations in the city.

That was also true of household garbage. Almost all households dumped their garbage in the street, on the city ramparts, or in the canals. Domestic sewers led directly to the street or the nearest water. Cesspools were often poorly maintained and rarely emptied. City dwellers dumped their feces in gutters and canals. The situation was most abysmal in alleyways and slums. Not only because of the massive 'impurity and filth' but also because of the dampness of the dwellings and the 'different households living in very close proximity.'[44]

Though all cities were dirty, there was still an important distinction. In many cities in the low-Netherlands the situation was worse than in the high-Netherlands. And that had everything to do with water management. Water was a defining aspect of the Dutch situation. By today's standards, the low-lying Netherlands of the nineteenth century was a gruesome world: soggy polders, sluggish flows of water, crowded slums, a glut of organic waste and heavily polluted surface and ground water.

In the 1850s the western part of the country (North and South Holland, Zeeland and Utrecht) had the highest mortality rates.[45] Whether we look at large or small cities, town or countryside, men, women or infants – in all cases death was a more immanent presence in this region. It is noteworthy that in the west there was hardly any difference between the city and the countryside. Apparently the problem of water quality – closely associated with increasing salinity – pervaded the entire region and influenced rates of mortality everywhere.

In the rest of the country there was a clear difference between town and country. Organic wastes and polluted water (in combination with food quality) contributed to higher mortality rates in the city due to diarrhea, cholera and other diseases of the digestive tract. Moreover, mortality due to tuberculosis, flu, measles, whooping

---

[43] E.S. Houwaart, 'Medische statistiek', in: H. Lintsen et al. (ed.), *Geschiedenis van de techniek in Nederland. De wording van een moderne samenleving 1800–1890* (Zutphen 1993), Vol. II, 19–45.

[44] Quotes from: *Verzameling van stukken, betrekkelijk de aanstelling eener Commissie van Geneeskundig toevoorzicht te Amsterdam* en daarin opgenomen een serie *Rapporten* (Amsterdam 1797). For a commentary on these reports see: van Zon, 'Openbare hygiëne', 49–55. Further: van Zon, *Een zeer onfrisse geschiedenis. Studies over niet-industriële vervuiling in Nederland, 1850–1920*, 29–32.

[45] See for the following, F.W.A. Poppel, *Stad en platteland in demografisch perspectief: de Nederlandse situatie in de periode 1850–1960* (Voorburg 1984) internal report nr.19, ch. 2, Sterfteverschillen tussen stad en platteland, 7–21.

cough etc. was significantly higher in the city. These respiratory diseases could spread more easily in the cities due to the high population density.

From today's perspective, the contamination of the environment with organic waste presents itself as the biggest environmental problem of society in the past. Whether contemporaries also saw things this way is another matter to which we shall return.

4.6   Poverty, Vulnerability and Sustainability

The quality of life in the Netherlands was *the* problem of well-being and sustainability in the mid-nineteenth century and that for two reasons. In the first place a considerable part of the population lived around and under the poverty line (an estimated 21%). In the second place existence was vulnerable due to the open economy, natural forces, the risks of the delta and epidemic diseases. All inhabitants of the Netherlands shared in this vulnerability but the rich were better able to protect themselves than the poor. They could attempt to escape from epidemics, flee the city and closet themselves in their country estates. But they could not escape altogether. That is evident from the differences in life-expectancy.

The average life-expectancy at birth in the period 1840–1851 was 37 years (men 36.1 year and women 38.5 years). A sample of the Dutch population in the period 1840–1859 distinguished among the life expectancies of different social classes. It is striking that the upper-bourgeoisie had the lowest life expectancy (about 30 years), while farmers possessed the highest (about 43), followed by the small-bourgeoisie (with about 39 years), skilled workers (with 38 years) and unskilled workers (36 years). The differences in life expectancy disappeared by the age of 15. Presumably the low figure for the upper classes had to do with breast feeding. Upper class babies received less breast feeding than other babies, were fed more often with porridge and milk, looked blooming and healthy, but had significantly less resistance to diseases.

Rich and poor did not differ so much in the struggle for existence, in which the poor would inevitably get the worst of it, but in the struggle for the basic necessities of life, which the rich could acquire more easily and more amply. In these respects they did not face big uncertainties, did not have to occupy themselves with scraping an existence together, were well-clothed and lived in relative comfort. Their quality of life was in those respects considerably higher than that of the poor.

Poverty and vulnerability had many faces in the first half of the nineteenth century. They differed per period, per region and per season. They had another face in the city than in the countryside. They were experienced differently by smallholders than by day-labourers and skilled workers. Much of the differences nevertheless disappeared in the catch-all category of 'pauperism.' That was how the elite and the middle classes referred to it. They recognized it as a problem and asked themselves how to deal with it.

Literature

Allen, R.C. (2013). 'Poverty lines in history: Theory and current international practice'. *Discussion paper series*. Department of Economics, University of Oxford.

Beucker Andrea, J.H. (1991). 'Rapport ingediend voor het vijfde, Landhuishoudjundig Congres te Leyden, 11, 12, 13 junij 1850, betreffende een onderzoek naar den zedelijken en materiëlen toestand der arbeidende bevolking ten platten lande en van den middelen om dien zoveel mogelijk te verbeteren. [Friesland]'. In J.L. van Zanden (Ed.), *'Den zedelijke en materiële toestand der arbeidende bevolking ten platten lande'. Een reeks rapporten uit 1851.* Groningen: Nederlands Agronomisch-Historisch Instituut. Original source: *Tijdschrift voor Staathuishoudkunde en Statistiek* (1851), 6, 156–200.

Blussé, J.L. (2013). 'Koning Willem I en de schepping van de koloniale staat'. In I. de Haan, P. den Hoed and H. te Velde (Eds.), *Een nieuwe staat: Het begin van het Koninkrijk der Nederlanden* (pp. 145–171). Amsterdam: Bert Bakker.

Bosch, A., Ham W. van der, and Lintsen, H.W. (1998), *Twee Eeuwen Rijkswaterstaat 1798–1998.* Zaltbommel: Europese Bibliotheek.

Bosch, A. (2000). *Om de macht over het water: De nationale waterstaatdienst tussen staat en samenleving, 1798–1849.* Zaltbommel: Technische Universiteit Eindhoven.

Buisman, J. (1995–2006), *Duizend jaar weer, wind en water in de lage landen.* Volumes 1 t/m 5. Franeker: Uitgeverij Van Wijnen.

Buisman, J., *Duizend jaar weer, wind en water in de lage landen.* Volume 7: 1825–1875. manuscript to be published. Franeker: Uitgeverij Van Wijnen.

Buiter, H. and H.W. Lintsen (2005). 'De stad, de stank en het water'. In H.W. Lintsen et al. (Eds.), *Made in Holland: Een techniekgeschiedenis van Nederland [1800–2000]* (pp. 55–74). Zutphen: Walburg Pers.

Dam, P. van (2010). *De amfibische cultuur: Een visie op watersnoodrampen.* Inaugural lecture, Vrije Universiteit Amsterdam.

Drieling, F.H.C. (1991). 'Verslag over den toestand der arbeidende klasse, vooral ten plattelande, in de provincie Utrecht, en over de middelen tot verbetering van hunne zedelijke en stoffelijke welvaart'. In J.L. van Zanden (Ed.), *'Den zedelijke en materiële toestand der arbeidende bevolking ten platten lande'. Een reeks rapporten uit 1851.* Groningen: Nederlands Agronomisch-Historisch Instituut. Original source: *Tijdschrift voor Staathuishoudkunde en Statistiek* (1853), 8, 405–452.

Eerenbeemt, H.F.J.M. van den (1968). *In het spanningsveld der armoede: Agressief pauperisme en reactie in Staats-Brabant.* Tilburg: Stichting Brabants Historisch Contact.

Fockema Andrea, S.J. (1955). 'Wat er aan de droogmaking van de Haarlemmermeer voorafging'. *Med. Der Kon. Ned. Akademie van Wetenschappen, afd. Letterkunde*, 18, 379–428.

Gales, B.P.A., et al. (1997). *Het burgerlijk armbestuur: Twee eeuwen zorg voor armen, zieken en ouderen te Maastricht, 1796–1996.* Maastricht: Stichting Historische Reeks Maastricht.

Gras, H. (1989). *Op de grens van het bestaan: Armen en armenzorg in Drenthe, 1700–1800.* Zuidwolde: Stichting het Drentse Boek.

Heezik, A. van (2007). *Strijd om de rivieren: Tweehonderd jaar rivierenbeleid in Nederland of de opkomst en ondergang van het streven naar de normale rivier.* Delft: Technische Universiteit Delft.

Houwaart, E.S. (1993). 'Medische statistiek'. In H.W. Lintsen et al. (Eds.), *Geschiedenis van de techniek in Nederland. De wording van een moderne samenleving 1800–1890* (Vol. II, pp. 19–45). Zutphen: Walburg Pers.

Laan, L., Ö. Nomaler en H. Lintsen (2017), *Perceptie en kranten, 1830–1850.* Eindhoven: Eindhoven University of Techology, open-access research report.

Leeuwen, M.H.D. van (1992). *Bijstand in Amsterdam ca. 1800–1850: Armenzorg als beheersings- en overlevingsstrategie.* Zwolle: Waanders'.

Lintsen, H.W., R.A. Lombaerts and R. Moerenhout (1994). 'De droogmaking van het Haarlemmermeer: Wind of stoom'. In M.L. ten Horn-van Nispen, H.W. Lintsen and A. J. Veenendaal (Eds.), *Nederlandse ingenieurs en hun kunstwerken: Tweehonderd jaar civiele techniek* (pp. 31–40). Zutphen: Walburg Pers.

Lintsen, H.W. and N. Disco (2005). 'Het nijvere verbond'. In H.W. Lintsen et al. (Eds.), *Made in Holland: Een techniekgeschiedenis van Nederland [1800–2000]* (pp..75–93). Zutphen: Walburg Pers.

Lintsen, H.W. and A. van Heezik (2005). 'In gevecht met de rivieren'. In H.W. Lintsen et al. (Eds.), *Made in Holland: Een techniekgeschiedenis van Nederland [1800–2000]* (pp. 95–113). Zutphen: Walburg Pers.

Lintsen, H., M. Anthonissen en B. Gales (2017), *Berekening omvang extreme armoede in Nederland 1820–1913*. Eindhoven: Eindhoven University of Techology, open-access research report.

Lintsen H. (2017), *De schraalheid van het bestaan, een reeks rapporten uit 1850*. Eindhoven: Eindhoven University of Techology, open-access research report.

Lis, C. and H. Soly (1980). *Armoede en kapitalisme in pre-industrieel Europa*. Antwerpen: Standaard Wetenschappelijke Uitgeverij.

Martini van Geffen, A. (1991). 'Zedelijke en materiële toestand der arbeidende bevolking ten platten lande. I. Nopens Noord-Brabant'. In J.L. van Zanden (Ed.), *'Den zedelijke en materiële toestand der arbeidende bevolking ten platten lande'. Een reeks rapporten uit 1851* (Appendix no 12, 101). Groningen: Nederlands Agronomisch-Historisch Instituut. Original source: *Verslag van het verhandelde op het vijfde Nederlandsche Landhuishoudkundig Congres, gehouden te Leiden van den 10den tot den 14den junij 1850*, appendix no. 12, 100–108.

Poppel, F.W.A. (1984). *Stad en platteland in demografisch perspectief: De Nederlandse situatie in de periode 1850–1960* (pp. 7–21). Voorburg: Nederlands Interdisciplinair Demografisch Instituut.

Prak, M. (1993). 'Overvloed of onbehagen: Armoede, armen en armenzorg in 's-Hertogenbosch, 1770–1850. In J. van Oudheusden and G. Trienekens (Eds.). *Een pront wijf, een mager paard en een zoon op het seminarie: Aanzetten tot een integrale geschiedenis van oostelijk Noord-Brabant 1770–1914* (pp. 7–44). Den Bosch: Stichting Brabants Regionale Geschiedbeoefening.

Rooy, P. de (2013). 'De armen hebt gij altijd met u: Armenzorg en onderwijs'. In I. de Haan, P. den Hoed and H. te Velde (Eds.), *Een nieuwe staat: Het begin van het Koninkrijk der Nederlanden* (pp. 221–222). Amsterdam: Bert Bakker.

Sloet tot Oldhuis, B.W.A.E. (1991). 'Statistieke beschouwing van den toestand der geringe plattelands bevolking op de Veluwe langs de Zuiderzee, (gemeenten Putten, Ermelo, Elspeet, Harderwijk, (Hierden), Doornspijk, Oldenbroek, Heerde en Epe)'. In J.L. van Zanden (Ed.), *'Den zedelijke en materiële toestand der arbeidende bevolking ten platten lande'. Een reeks rapporten uit 1851*. Groningen: Nederlands Agronomisch-Historisch Instituut. Original source: *Tijdschrift voor Staathuishoudkunde en Statistiek* 9(1853), 291, 283–311.

Stol, T. (1993). *Wassend water, dalend land: Geschiedenis van Nederland en het water*. Amsterdam: Kosmos.

Swaan, A. de (2004). *Zorg en de staat: Welzijn, onderwijs en gezondheidszorg in Europa en de Verenigde Staten in de nieuwe tijd*. Amsterdam: Bert Bakker.

Ven, G.P. van de (2003). *Leefbaar laagland: Geschiedenis van de waterbeheersing en landaanwinning in Nederland*. Utrecht: Matrijs.

Woud, A. van der (2010). *Koninkrijk vol sloppen: Achterbuurten en vuil in de negentiende eeuw*. Amsterdam: Bert Bakker.

Zanden, J.L. van (1985). *Nederlandse landbouw in de negentiende eeuw, 1800–1914*. Wageningen: University of Wageningen.

Zanden, J.L. van (1991). *'Den zedelijke en materiële toestand der arbeidende bevolking ten platten lande'. Een reeks rapporten uit 1851*. Groningen: Nederlands Agronomisch-Historisch Instituut.

Zeehuisen, J. (1991). 'Statistieke bijdrage tot de kennis van den stoffelijken en zedelijken toestand van de landbouwende klasse in het kwartier Salland, provincie Overijssel'. In J.L. van Zanden (Ed.), *'Den zedelijke en materiële toestand der arbeidende bevolking ten platten lande'. Een reeks rapporten uit 1851.* Groningen: Nederlands Agronomisch-Historisch Instituut. Original source: *Tijdschrift voor Staathuishoudkunde en Statistiek* (1851), 9, 375–408.

Zon, H. van (1986). *Een zeer onfrisse geschiedenis: Studies over niet-industriële vervuiling in Nederland, 1850–1920.* Groningen: Rijksuniversiteit Groningen.

Zon, H. van (1993). 'Openbare hygiëne'. In H.W. Lintsen et al. (Eds.), *Geschiedenis van de techniek in Nederland. De wording van een moderne samenleving 1800–1890* (Vol. II, pp. 49–55). Zutphen: Walburg Pers.

Chapter 5
Stagnation and Dynamism in Three Supply Chains: Agriculture and Foods, Building Materials and Construction, Energy

Harry Lintsen

Contents

Abstract At the time, extreme poverty could be fought by, among other things, economic growth. That demanded another approach to the exploitation of natural capital and accordingly to innovation in the three main supply chains.

In the agriculture and foods supply chain (one of the three main chains in this study, based on organic raw materials) experimentation with new techniques did take place (among other things the use of guano as artificial fertilizer) but this did not lead to practical innovations.

In the supply chain of building materials and construction (the second main chain based on mineral subsoil resources) the construction of a national road system amounted to an important innovation in road infrastructure. Hardly any innovations were undertaken in the fields of water management and housing construction. It is remarkable that little was done about the social problem of organic wastes, including human and animal faeces. Public hygiene was not one of the most important societal issues of the time.

In the energy supply chain (the third main supply chain based on fossil subsoil resources) innovations were equally lacking, in particular applications of steam power. Up to 1850 the Netherlands did not industrialise on the basis of steam and coal.

Keywords Innovation · Agriculture · Foods · Construction · Water management · Infrastructure · Hygiene · Energy · Industrialisation · Steam technology

© The Author(s) 2018
H. Lintsen et al., *Well-being, Sustainability and Social Development*,
https://doi.org/10.1007/978-3-319-76696-6_5

5.1   The Conspiracy[1]

On January 31 1827 Lodewijk Cantillon submitted a request to King William I for permission to build a steam-powered gristmill in Amsterdam, the first in the Northern Netherlands. Cantillon was a grain merchant from Hasselt, which town at the time belonged to the Southern Netherlands. The procedure required consultation with the surrounding neighbors. These were dead set against the project because of the danger of 'explosion of the boiler, the danger of fire and other unpredictable disasters...' They also feared that the value of their real estate would plummet due to the 'thick clouds of coal smoke, the bad smell, the noise of the machine, even the multiplication of rats and other vermin attracted by the grain...'[2]

The wind-powered grist-millers also objected. They feared 'the total ruin' of Amsterdam's wind-powered milling industry, that counted 34 windmills and fed 170 households. The steam gristmill would have 'unhappy consequences' for the millers with their 'so dearly bought buildings,' but also for the government because the windmill industry paid significant taxes and these could then no longer be levied.[3]

The city of Amsterdam also opposed the plan. It observed that the 'location where the construction is requested is the most beautiful and prestigious part of that quarter of the city.'[4] Steam engines did not belong in the girdle of canals but in lesser neighbourhoods. The city also supported the grievance of the millers. Amsterdam did not need competition for the existing milling industry. Let Cantillon apply his energy and inventiveness in a better way, was the city's final word.

But King William, proponent of a liberal economy that he was, granted Cantillon's request, albeit after Cantillon had found a new location and submitted a new request. To attract customers, Cantillon tried to cut his milling rates to an absolute minimum. The going rate in Amsterdam was about 39 eurocent for milling one hectoliter of wheat and almost 38 eurocent for an hectoliter of rye. Cantillon asked about 33 eurocents and 23 eurocents, respectively. The windmillers retaliated at once. They began to undercut Cantillon's prices. The bakers were loyal to the windmillers and refused to let Cantillon grind their grain. Cantillon tried to turn the tide and to come

---

[1] See for the Cantillon episode: H. Lintsen et al., *Made in Holland: Een techniekgeschiedenis van Nederland [1800–2000]* (Zutphen 2005), 33–36; H. Lintsen and M. Bakker, 'Meel', in: H. Lintsen et al. (Eds.), *'Geschiedenis van de techniek in Nederland. De wording van een moderne samenleving 1800–1890* (Zutphen 1993), Part 1, 70–101, in particular 81–84; J.L van Zanden and A. Van Riel, *Nederland 1780–1914: Staat, instituties en economische ontwikkeling* (Amsterdam 2000), 178–185.

[2] *Gemeentearchief Amsterdam, Archief Secretarie-Afdeling Financiën*, 1827, nr. 725, Brief van 17 eigenaren aan de burgemeester van de stad Amsterdam (*Letter of 17 proprietors to the mayor of Amsterdam*), 13 March 1827.

[3] *Gemeentearchief Amsterdam, Archief Secretarie-Afdeling Financiën*, 1827, nr. 1119, Brief van zeven molenaars aan de Koning (*Letter of seven millers to the King*).

[4] *Gemeentearchief Amsterdam, Archief Secretarie-Afdeling Financiën*, 1827, nr. 927, brief van de commissaris over de publieke werken aan de wethouder van financiën, 28 maart 1827 (*Letter of the commissioner of public works to the alderman of finances, 28 March, 1827*).

to terms with the windmillers. But to no avail, they flatly refused. After this, the steam gristmill drops out of sight. Cantillon had failed.

In retrospect Cantillon's failure was a misfortune for the people of Amsterdam. His innovation would have lowered prices for flour and bread. That would have stimulated the consumption of bread as a staple food and improved the physical condition of the poor. About 30 years later the reformist Amsterdam physician Samuel Sarphati did succeed in decreasing the price of bread by 30%. To do this he had to found the first (steam-powered) bread and flour factory (see Chap. 8). This had far-reaching consequences for the quality of life of the poor or, in terms of this book, for well-being in the Netherlands. Why did Cantillon fail?

With his innovative application of steam power Cantillon was in a good position to compete with the windmillers. He could grind grain considerably cheaper, but only if he succeeded in working at full capacity. If he fell much below this, his fixed costs began to prevail and to raise the unit milling costs. The Amsterdam windmillers knew exactly how to exploit this Achilles heel of large-scale production. Since time immemorial the millers had formed a mighty cartel, first in the form of a guild and after the abolition of guilds in 1798 in the form of a cooperative association. This cartel disposed over large sums of money that it employed to wage a price-war with Cantillon (and later with other steam-powered gristmills). The windmillers (in coalition with the bakers) were able to hold out long enough to run the steam gristmills out of business.

Cantillon's initiative signified a disturbance of the existing agricultural and food supply chain. That supply chain fulfilled a crucial function in the development of popular welfare for it shaped the production and consumption of primary foodstuffs, in particular porridge, bread and potatoes. In that supply chain the exploitation of the soil (natural capital), the processing of agricultural produce and the consumption of foodstuffs all had specific characteristics. The windmillers, for example – at least that is what Cantillon's case suggests – were dominant actors in the grain, flour and bread chain, who influenced the price and therefore the distribution of food. In so doing they also influenced the quality of life. The analysis of these supply chains is the business of this chapter. The focus is on poverty and vulnerability as the most important issues in well-being and sustainability around 1850.

Supply chains are a sequence of successive activities at specific locations and within certain organizations. They refer to the institutional framework of natural capital and their derivative material flows. We have divided the material flows into three main categories:

1. Biomass (agriculture and foods). The broad variety of supply chains and material flows in this category pertained mostly to foods (potatoes, grain, sugar beets, oil seeds, beans etc.) and additionally to industrial products (like wool, madder, leather and starch). We will elaborate on the food chain.
2. Mineral assets (building materials and construction). The supply chain for building materials was rooted in the exploitation of mineral raw materials like clay, sand, gravel and ores of iron, tin and zinc. Processing of mineral raw materials took place in the earthenware industry, in mining, in the metalworking industry

and in construction. These products were used primarily in construction and for machines. We focus here on construction.

3. Fossil sub-soil assets (energy). The supply chains up to 1850 were limited to those for turf and coal. Here we concentrate on the coal supply chain.

We take these three categories as a point of departure for a brief analysis of stagnation and dynamism in the supply chains before 1850. How could changes in the supply chains of production and consumption have supported sustainable development in the Netherlands. Which interventions in natural capital and the derivative material flows could have improved the quality of life of especially the poor? To what extent did innovations by contemporaries lead to transitions in the three supply chains?

5.2 Agriculture and Foods: More of the Same

In the first half of the nineteenth century, well-being was primarily a matter of quality of life and poverty. Food was a crucial element in the problematic situation, affecting especially the poor. Would they have access to enough food of sufficient quality throughout the year and in subsequent years? The poor were exceedingly vulnerable and faced a daily struggle to scrape a meagre meal together. A graph of food consumption confirms this image.

The average number of calories available to the Dutch in the period 1815–1850 was about 2400 kcal per capita per day (see Graph 5.1). We can add a margin of about 10% for the calories contained in beverages, in particular beer that was consumed on a regular basis. Taken together, this exceeds the norm of 2000–2100 kcal/day, that is universally taken to be the absolute minimum necessary to stay alive.[5] The graph shows that the poor had a hard time of it. They ate no (or nearly no) meat. They fed themselves chiefly with potatoes, grain and dairy products. In the period under consideration, these foodstuffs delivered on average 2200 kcal per day. But there were large fluctuations in the first half of the nineteenth century. There were years with less than 2000 kcal/day. What the graph also hides is the unequal distribution of food. This makes the condition of the poor in certain periods even more problematic.

The first years of the century were a time of misery. This had everything to do with the political situation.[6] The Napoleonic wars and the continental system wreaked havoc with food production and the food trade. In the 1820s and 1830s there was on average sufficient food available for the Dutch population. Nonetheless,

---

[5] R.C. Allen, *Poverty lines in history, theory, and current international practice* (Oxford 2013), discussion paper series ISSN 1471-0498, 3.

[6] For this see: M.T. Knibbe, 'De hoofdelijke beschikbaarheid van voedsel en de levensstandaard in Nederland, 1807–1913', *Tijdschrift voor Sociale en Economische Geschiedenis* 4(2007), nr. 4, 71–107.

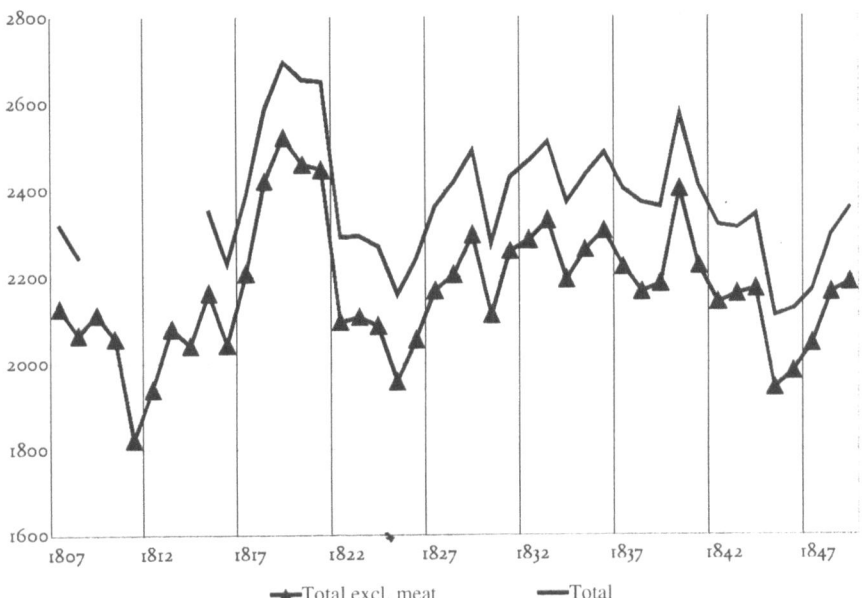

Graph 5.1 Daily amount of available kilocalories per capita including and excluding meat, 1807–1850
Source: Knibbe 2007

within this period there was at least one notoriously bad year, in which masses of the poor went hungry. 1845, 1846 and 1847 were again dramatic. Agricultural production declined in these years due to failed potato and rye harvests. Trade and industry stagnated. Under these circumstances the Dutch population was vulnerable. Food shortages ensued. The poor went hungry. For the first time that century the death rate outstripped the birth rate. The population declined.

But it was not only wars and crop failures that depressed food consumption. Between 1800 and 1850 the population increased by roughly 50% (from 2,120,000 to 3,012,000 persons). The Netherlands faced the challenge of feeding all these mouths. International trade also affected the food situation. Increasing foreign demand, especially from industrialising England, did not go unnoticed. Domestic food patterns changed to suit. The Dutch enjoyed fewer dairy goods because butter and cheese were increasingly exported.

But at the same time, trade also provided a certain degree of food security. The Netherlands was self-sufficient only with respect to potatoes. Grain, particularly wheat and rye, had always been imported. When the domestic harvests of these bread grains failed, imports increased.[7]

Nutrition was an important factor in the physical and mental condition of the Dutch population. While exact criteria for quantity and quality are hard to provide, data on length and weight in the nineteenth century suggest that by today's stan-

---

[7] Knibbe, 'De hoofdelijke beschikbaarheid...', 87.

Table 5.1 Average annual
growth of agricultural
production and productivity,
1810–1850

Agricultural production	0.6
Labour productivity	−0.2
Production per hectare	0.4
Total population	0.8

Source: J.L. van Zanden, 'Mest
en ploeg', in: H. Lintsen et al.
(ed.), 'Geschiedenis van de techniek in Nederland. De wording
va een moderne samenleving
1800–1890 (Zutphen 1992), Part
1, Table 2.2, 67

dards large parts of the population were insufficiently and poorly fed.[8] And not only
the poor, but also workers, carpenters, petty tradesmen, hired hands, in short 'the
people,' the 'masses,' 'the working population – in other words, the 'third estate.'

The food supply could have been improved by certain structural changes in the
food supply chain. Changes in the food supply chain were in any case endemic.
Markets, scarcity and price influenced the choices of farmers, entrepreneurs, tradesmen and consumers. The purchasing power of the population determined the composition of the daily menu and in this way influenced the supply chain. By means of
taxation and legislation the government influenced the production, sale and consumption of food. Many changes had a temporary character. We focus here on the
structural changes in the food supply chain and take as our point of departure innovations in food production and consumption. To what extent can we speak of innovations in the agricultural sector, food processing industry and food consumption?

The Netherlands had a difficult time feeding its own population with the natural
capital at hand. Agricultural productivity did not keep pace with population growth
as is clear from Table 5.1. Even in years of high grain prices, such as 1817, grain
production increased inadequately. In other words, there was little elasticity in the
grain supply. Agricultural production responded only marginally to changes in
demand and the development of agricultural prices.

Why did crop farming respond so feebly to the increasing demand?[9] An important factor was the surface area of the available agricultural land. This increased
inadequately. It is true that the Netherlands possessed extensive heath moors – the
'wastelands' – but these already played an important role in the agricultural system,
namely the production of fertilizer and this was a crucial bottleneck. Reclamation
and transformation into crop fields and grassland was possible. But then this bottleneck first had to be solved. That proved impossible before 1850. The consequence

---

[8] See for the relationship between nutrition and length/weight: J.J.A. de Beer, Voeding, gezondheid
en arbeid in Nederland tijdens de negentiende eeuw. Een bijdrage tot de antropometrische
geschiedschrijving (Utrecht 2001).

[9] See for the following: J.L. van Zanden, 'Mest en ploeg', in: H. Lintsen et al. (ed.), 'Geschiedenis
van de techniek in Nederland. De wording van een moderne samenleving 1800–1890 (Zutphen
1993), Part 1, 52–61.

was that agricultural production could only be augmented by increasing the productivity per hectare (farming lands plus 'wastelands'). That in fact took place and suggests the development of new farming techniques. To trace this we look at innovations in crop farming.

Inspired by reform movements in England and France in the second half of the eighteenth century, the Netherlands had launched an offensive to renew its agriculture just after the turn of the century. The first 'minister of agriculture' (1805) and the founding of provincial Agricultural Commissions were intended to promote 'scientific' agriculture. These initiatives had little effect. A second offensive in the 1830s and 40s was more successful. It included the founding of a number of provincial agricultural societies, the first Dutch Agricultural Congress (1846), experimentation with new farming methods and the publication of periodicals like *Vriend van de Landbouw* (*Friend of Agriculture*) including descriptions of experiments.

A characteristic of the movement was the dominance of local elites of lawyers, politicians, large landowners, and farming gentry. Smallholders did not participate in the agricultural societies. In 1850 the members of all the agricultural societies together totalled some 10,000, while the number of dirt-farmers and smallholders was many times greater. The movement, in pursuit of a 'scientific' agriculture, searched among other things for alternatives to the traditional method of fertilizing the soil: cattle manure supplemented by peat sods from the moors, urban waste and industrial waste products, including earth-foam (aphrite) from the genever distilleries, bone-meal, and fish waste).[10] In the 1840s a new fertilizer became available: guano. This was dried sea-bird manure found in meters-thick layers off the coast of Peru. Two years after successful experiments in England a number of farmers and large landowners in the Netherlands followed suit. The results were published in agricultural reports and in a journal like the *Tijdschrift ter bevordering van nijverheid* (*Journal for the promotion of industry*).

Not long after, practical experiments with guano became more widespread. The advantage of the new fertilizer was the concentration of active ingredients. Relatively small amounts (compared to classic fertilizers) of guano were already effective. That also meant lower transport costs and less work to fertilize the land. The problem was that results were mixed and that the debate did not resolve the question of the advantages of guano. Fundamental insight into the workings of fertilizer was lacking. Justus von Liebig had only just published his *Die organische Chemie in ihrer Anwendung auf Agricultur und Physiologie* (*Organic Chemistry in its Application to Agriculture and Physiology*) (1840). His theories still wanted diffusion and widespread acceptance. Moreover, developments in the price and quality of guano left much to be desired. Prices climbed while quality declined due to

---

[10] 'Scientific' agriculture for example also embraced the idea that farming had to be industrialised. This included among other things a shift toward mechanisation, steam technology and large-scale enterprises. This is the context in which J.P. Amersfoordt, lawyer and gentleman farmer at the Badhoeve in the Haarlemmermeer Polder, experimented with a steam plow in the 1860s. It became crystal clear that the plow, mounted on a steam locomobile managed by three men and three boys, was not an option for a farmer with only a few hectares of land.

import of inferior guano from other regions and to malversations. Large landowners working on an experimental basis and gentry-farmers could deal with these disadvantages, but for the small farmers the new alternative was hardly appealing. They had few reserves and were not in a position to take risks. The diffusion of guano remained limited, even after 1850.

Moreover, in the first half of the nineteenth century, the smallholders on the sand grounds had been able to increase the yield per hectare. Among other things, they succeeded in fighting weeds more effectively and in improving the manure production of their cows. Their reform strategy consisted of incremental innovation. Though this strategy demanded more labour, that was the one thing in plentiful supply. A growing population and increased employment went hand in hand.

In addition to innovation in agriculture, innovation in the food-processing industry could also have improved popular nutrition. Various initiatives had been taken in other countries.[11] Steam power had penetrated into various production processes as a substitute for classical sources of energy (wind, water, and horse). It was applied in grain milling, bread baking and beer brewing. Moreover, different industries exhibited all kinds of branch-specific innovations.

In the grain milling industry, for example, in addition to implementing new milling methods, entrepreneurs also devised new installations for purifying the grain and for cooling the flour as well as transport mechanisms for moving the grain and intermediate products from one stage to another. Equipped with these innovations a grain mill assumed the character of a flour factory. A number of these factories already functioned in France, England, the United States and Austria; some in combination with a mechanized bread bakery so that one could speak of a bread and flour factory.[12]

There were certainly efforts to innovate, as the example of Cantillon and the steam grain mill shows, but the majority failed. Innovations in the food processing industry were limited to a small number of branches among which sugar refining. In the 1830s this branch of industry was still in thrall to small-scale traditional manufacture; by the 1860s large-scale industrial production had become the norm.[13] Frontrunners invested in new purification methods, steam-heating and vacuum kettles. Steam engines were used to power the pumps. The big investments in expensive machines and building returned a profit only if huge amounts of raw sugar could be processed. The *Nederlandse Handel-Maatschappij* (Dutch Trading Company) was responsible for the supply from the Dutch East Indies and also extended large credits to entrepreneurs.

---

[11] See for an extensive treatment of innovations in the food processing sector: 'Landbouw en voeding', H. Lintsen et al. (ed.), '*Geschiedenis van de techniek in Nederland. De wording van een moderne samenleving 1800–1890* (Zutphen 1992), Part 1, 37–277 in particular the chapters 'Voeding in Nederland' (by M. Bakker, 38–51), 'Meel' (by H. Lintsen and M. Bakker, 70–101), 'Boter' (by M. Bakker, 102–133), 'Bier' (by H. Schippers, 170–213), 'Suiker' (by M. Bakker, 214–251) and 'Techniek en voeding in verandering' (by M. Bakker, 252–277).

[12] Small-scale craft enterprises survived this development by shifting to the fabrication of specialty products like candy.

[13] Bakker, 'Suiker', 214–251.

Despite the limited number of innovations in the food processing industry there must have been some productivity increases. The growth of the added value in the food processing industry was about 2.7% (1816–1850), while the work force in this sector grew by 1.9% (1807–1849)[14] (Table 3.4). But an important part of the growth derived from simply adding more grain-millers, bakers, butchers etc. In this way the food processing industry easily succeeded in keeping up with population growth.

Were there changes in patterns of consumption? Food comprised the biggest portion of the budget of an average Dutch household, 55–60%.[15] This had remained constant for decades. The most important shift in consumption was a negative one. Butter and cheese appeared less and less frequently – as we noted above – on the Dutch 'menu.' These were shipped in huge quantities to foreign consumers. The food package therefore seemed to be getting more meagre. It is difficult to get this clearly in focus. It is true that potatoes and grain products (in particular bread) were the staples of the Dutch diet, but the latter varied throughout the country and depended strongly on the region and residence in the city or the countryside. Many people, moreover, had a vegetable patch in which they grew all kinds of vegetables. The quality of the diet will have been far from adequate, but it was not an issue at the time. The debate concerned the quantity. The social issue was not good food, but enough food for the population.

Structural changes were limited to a very few links in the food supply chain. There was no question of any kind of structural improvement in the food situation. In fact, the situation around 1850 looked quite gloomy – no novelty in the history of the Netherlands. The daily fare of the poor part of the population was meagre, of

---

[14] **Table** Growth of value added and of the workforce in agriculture and the food processing industry in the first half of the nineteenth century.

	Growth 1816–30	Growth 1830–40	Growth 1840–50	Growth first half of the nineteenth century (estimate)
Agriculture value added	0.60	1.43	1.20	0.7 (1816–1850)
Agriculture workforce				0.7 (1807–1849)
Food processing value added	4.18	2.75	0.52	2.7 (1816–1850)
Food processing workforce				1.9 (1807–1849)
Population	1.16	0.93	0.71	1.0 (1816–1850)
				0.8 (1807–1849)
GDP per capita	0.42	1.41	0.81	0.8 (1816–1850)
GDP	1.58	2.34	1.52	1.8 (1816–1850)

Source: van Zanden and van Riel, *Nederland 1780–1914,* table 4.2, 153

[15] E. Horlings and J-P. Smits, 'Private consumer expenditure in the Nethetlands, 1800–1913, *Economic and Social History in the Netherlands* 7(1995), Graph 3, 24.

poor quality and from time to time insufficient. Initiatives to improve this sustainability issue met with little success.

5.3   Building Materials and Construction: King and State

The supply chain of building materials was also directly related to the issue of poverty. The chain also touched on other themes of well-being and sustainability. The building materials supply chain can be divided into three categories: housing construction, hydraulic engineering and the construction of infrastructural works. Housing construction (including commercial and public buildings) concerns a broad range of buildings, including public housing for the poor. How problematic was the housing situation of the poor? Were there initiatives to improve their lot? And housing construction raises the issue of hygiene. Poor public health was closely associated – as we saw – with poor hygiene in houses and cities. What happened in this domain around 1850? Hydraulic engineering was concerned with the vulnerable delta, the struggle against water. What policies were developed in order to deal with this important sustainability issue? The construction of infrastructural works (roads, waterways and railroads) has a direct relationship with the sustainability issue. These infrastructures were important for trade and the economy. Did they lay a foundation for prosperity robust enough to solve the issues of poverty and vulnerability? We arrive at the following overview.

5.3.1   Public Housing and Public Health

About the housing of the poor and the hygienic measures we can be brief. By today's standards housing and hygiene were very problematic. This conviction was not shared by the contemporaries of 1850, with the exception of the hygienists, a small intellectual elite of engineers, architects and physicians.[16] The hygienists developed a judicious program to improve housing. But in the years around 1850 this came to nothing.

Typical for the time is the fate of an 1854 report by the Royal Institute of Engineers, *Report to the King, on the requirements and design of workers' hous-*

---

[16] For the vanguard role of hygienists, engineers and architects, see: H. van Zon, 'Openbare hygiëne', in: H. Lintsen et al. (ed.), *Geschiedenis van de techniek in Nederland. De wording van een moderne samenleving 1800–1890* (Zutphen 1993), Part II, 57; E.S. Houwaart, 'Medische statistiek', in: H. Lintsen et al. (ed.), *Geschiedenis van de techniek in Nederland. De wording van een moderne samenleving 1800–1890* (Zutphen 1993), Part II, 27–28. For the perception of contemporaries see the results of research in the internet newspaper archive Delpher in H. Lintsen and T. Lintsen, *Opvattingen over hygiëne 1830–1850* (research report Eindhoven University of Technology, 2017).

ing.[17] It was written by a commission consisting of three architects, a physician and an engineer of the national Public Works Agency (*Rijkswaterstaat*). It is not clear what prompted the report, but possibly the members of the commission had been inspired by the British Society for Improving the Condition of the Labouring Classes, founded in 1844, and by the London World Fair of 1851, at which Prince Albert's Model-homes were on exhibit.[18]

Be that as it may, the commission's work was the first effort to arrive at an assessment of workers' housing, or more broadly, the 'less fortuned estate', in short, the poor. Though it was based on a sample of neighbourhoods in six cities (Amsterdam, Rotterdam, The Hague, Utrecht, Arnhem and Delft) the conclusions, in view of what subsequent investigations revealed, will not have been much different for other cities. The picture they painted was, in a word, shocking.

The poor lived packed into small, primitive quarters with little light and fresh air and with holes in roofs and floors. The rooms were inevitably damp, drafty, filthy and full of smoke. They were sometimes half buried in the ground and were accessible only from narrow alleyways, into which sunlight barely penetrated, where mounds of garbage lay strewn about and the faeces of the residents remained for days on end. The poor lived like animals:

> Alas! The burrows of the people – and no other name suits the homes of many in the less fortuned estate – are not seldom inferior to the places reserved for the housing of animals: the primary requirements for life and health are absent; everything seems arranged to maintain the moral life led in those burrows at the animal level, and hence those burrows, as inexhaustible sources of depravity, inhibit all mental and above all moral development, and stand in the way of the progress of an important class of society...[19]

The poor city-dweller was a caveman. Other sources reveal that the condition of the rural poor was not much different, except that his home was not built of bricks and wood, but of clay, peat, or loam with a roof of reeds or branches. It is estimated that at least 130,000 houses in the Netherlands (of a total of about 620,000) answered to this description in whole or in part.[20]

---

[17] 'Verslag aan den Koning, over de vereischten en inrigting van arbeiderswoningen, door eene commissie uit het Koninklijk Instituut van Ingenieurs', *Het Tijdschrift van het Koninklijk Instituut van Ingenieurs*, 1854–1855, 50–75, with 15 illustrations.

[18] A. van der Woud, *Koninkrijk vol sloppen: Achterbuurten en vuil in de negentiende eeuw* (Amsterdam 2010); H. Buiter and H.W. Lintsen, 'De stad, de stank en het water', in: H.W. Lintsen (Ed.), *Made in Holland: Een techniekgeschiedenis van Nederland [1800–2000]* (Zutphen 2005), 90.

[19] 'Verslag aan den Koning, over de vereischten en inrigting van arbeiderswoningen', 4.

[20] Hole, shack, hovel and slum are qualitative terms. Around 1850 the government did little or nothing in the way of investigating housing quality. Such investigations are of a later date. Dwellings that did not meet specific criteria could be endowed with such labels. Criteria shifted in the course of time. Having a privy in or near the dwelling, being connected to a public water supply, or having heating could be criteria, just like light, air and space as the traditional indicators of quality. These norms could be set ambitiously or be light as a feather. In the latter case hovels were dwellings that should actually already have been demolished; in short, dwellings that later appeared on the list of uninhabitable buildings and where the uninhabitability was no longer at issue. It was also possible to opt for a stringent norm which left the better part of the available housing beyond the pale. It goes without saying that the figures pertinent to each approach would differ wildly.

It is notable that the commission for the first time introduced norms to which workers' housing was to conform: sufficient space, sufficient light and fresh air with potable water and sanitary facilities. This was translated into a series of designs. It is typical of the period that nothing was done with them. King William II was not the kind of person to be concerned about these kinds of issues. The problem was moreover not the business of the central state, but of the municipalities. What is even more telling is that the bourgeoisie was anything but concerned about housing the poor. The same held for poor relief. A room and roof above one's head were considered enough.

The investigation into workers' housing can also be seen as a new approach to the problem of hygiene. This too was of little interest to the bourgeoisie.[21] It viewed hygiene from the classic vantage point of cleanliness: a clean home, a spotless kitchen, a scrubbed sidewalk, clean clothes, clear water. Hygiene also included a 'calm cheerful disposition ...' and 'a firm trust in God's Providence.' Much attention was paid to pure air and smelly fumes because according to the miasma theory infectious diseases were communicated chiefly through the air. Rotting substances, spoiled food and stagnant water had to be avoided. Homes needed to be ventilated, canals flushed and garbage cleaned up. Hygiene in public spaces was subject to stringent regulations but these were quite ineffective. Local governments lacked the wherewithal to monitor their observance. The bourgeoisie had an inadequate sense of public hygiene as a serious social problem and treated the issue '... in general in an inattentive, imprecise, careless, yes, reckless fashion...' as an inventor of ino-

---

We derive an estimate of the number of hovels around 1850 from the following calculation:

- In 1856–1857, 362,159 dwellings and 12,160 warehouses were taxed. 257,706 dwellings were not taxed. Total: 632,025 dwellings and warehouses. The tax officials counted 619,865 dwellings. Of these, 42% was not taxed and 58% was taxed.
- Using other sources, Duyndam calculated: 680,000 dwellings in the Netherlands. This is a difference of about 60,000 dwellings, or ±9%.
- Points of departure in this study:

Number of dwellings: 619,865
Number of inhabitants in the Netherlands 3,115,421
Average number of persons per dwelling: 5
Non-taxed dwellings are in large part qualitatively inferior dwellings.

- Non-taxed means a rent-value below fl. 18/year for 257,706 dwellings, or 42%

Around 1850 about 7% is paid for rent.
Maximum annual income for residents of non-taxed dwellings: fl. 257 per dwelling

- Maximum annual income per capita in non-taxed housing fl. 51
- 42% of the inhabitants has on average an annual income less than fl. 51.
- 21% of the inhabitants of the Netherlands lives below the poverty line of fl. 43/year consumptive expenditures. This amounts to 658,000 inhabitants in 131,600 dwellings (on average 5 persons per dwelling). We assume that these are certainly qualitatively poor dwellings.
- Conclusion: at least 130,000 and at most 260,000 dwellings are of poor quality.

[21] See H. Lintsen and T. Lintsen, *Opvattingen over hygiëne 1830–1850* (research report Eindhoven University of Technology, 2017).

dores (non-smelling privies) and *urinoirs* complained.[22] To be sure there were complaints about smells, but more in terms of nuisance than as a threat to health. Something similar applied to drinking water from wells, canals and rivers. It was considered to be of poor quality and the population therefore drank plenty of beer. The complaints were mostly about the smell and the taste. The citizens were little concerned about public health. Cholera was one of the few diseases that frightened governments and citizens and got them thinking about radical changes. The disease broke out in 1848, but after a few years the unrest had passed.

The new approach advocated by the hygienists put a high value on scientific research.[23] They were the first to gather medical statistics. They stressed the salience of drinking water and fecal wastes. Proposed solutions were piped water and sanitary facilities in the home and public hygienic facilities on the street. Government should assume responsibility for hygiene and provide the necessary means and personnel. The commission appointed by the Royal Institute of Engineers was an exponent of a new generation of the bourgeoisie that would begin to exert influence only after the mid-nineteenth century.

5.3.2 Water Management

Hydraulic engineering faced an entirely different challenge. It had to respond to the risks associated with living in a delta. Storms and river floods regularly threatened chattels and goods of the Dutch. New impulses for dealing with this sustainability problem emerged in the wake of new relationships in the domain of water management.

For centuries water management had been the business of water boards, local government and provinces. During the Batavian and French period (1795–1813) the contours of a unified state were forged and the central state acquired supervisory authority over the entire domain of water management as well as direct responsibility for a number of sea and river works of national importance.[24] This introduced a new actor into the sphere of water management. Between 1813 and 1840 this coincided with the person of King William I. The *Rijkswaterstaat* was his executive agency. Major projects required enormous amounts of tax money, persuasive power and credibility. These the King could deliver. The *Rijkswaterstaat* was able to mobilise and organise knowledge, large quantities of materials and masses of labourers.

---

[22] De Noord-Brabanter: staat- en letterkundig dagblad 07-07-1949.

[23] E.S. Houwaart, 'Professionalisering en staatsvorming', in: H.W. Lintsen et al. (Eds.), *Geschiedenis van de techniek in Nederland: De wording van een moderne samenleving 1800–1890* (Zutphen 1993), Part II, 82–85.

[24] H.W. Lintsen, *Ingenieurs in Nederland in de negentiende eeuw: Een streven naar erkenning en macht* (proefschrift Technische Universiteit Eindhoven 1980), 45–64; A. Bosch and W. van der Ham, onder redactie van H.W. Lintsen, *Twee eeuwen Rijkswaterstaat 1798–1998* (Zaltbommel 1998, eerste druk), 33–41.

Perennial floods, as we have seen, were among the biggest problems of the period. In fact these had been one of the main reasons for founding the *Rijkswaterstaat* in 1798. But however urgent the problem might have been, solutions were not forthcoming. The main impediment was a shortage of river mouths. Close to the sea the river system formed a kind of bottleneck that encouraged enormous obstructions when the rivers were high and clogged with ice floes. But the construction of new river mouths was a gigantic enterprise which even the King and the *Rijkswaterstaat* refused to risk undertaking.[25]

Another impediment was the nature of the river. The Dutch river was a so-called 'green river,' a river with islands, sandbanks, brushwood, reeds, swamp forests, and other vegetation along the river banks. It did not possess a single channel, but a braided system of channels. The narrow channels frustrated a rapid discharge at high water and encouraged the formation of ice blockages and ice dams when ice floes clogged the rivers. Possible solutions were the subject of chronic debates.[26]

Some hydraulic engineers looked to lateral diversions using overbank floodways. These consisted of a number of dike sections that had been lowered to allow the river at high stages to spill into a floodway. This created artificial floods and directed the excess water to another river or to a downstream section of the same river. In other words, the system of channels was temporarily expanded.

Other engineers regarded the preservation and elaboration of the 'classical green river' as a thorn in their sides. They viewed the lateral diversions as a 'soothing plaster on a filthy stinking wound.' According to them the problem was only being displaced, while a fundamental approach was needed, namely 'normalization' of the rivers. A 'normalized' river no longer flowed through a system of channels, but through one continuous channel maintained at fixed dimensions and in itself able to discharge the necessary water and ice into the sea.

The two governmental commissions appointed by William I could not arrive at a consensus. Added to this was the fact that the King did not prioritize the rivers and preferred to spend money on canals and other plans. After the war with Belgium between 1830 and 1838 funds for hydraulic projects were in any case lacking.

A big project backed by the King that did succeed was the reclamation of the Haarlemmermeer. Public safety, the continual loss of land and the destruction of goods were important arguments for the reclamation. Nonetheless these were not of overriding importance for the definitive decision to eliminate the threat constituted by the lake. More to the point was the fact that the reclamation enacted the politics of welfare-promotion pursued by William I. The gigantic enterprise provided work for thousands of families. It also delivered an enormous expanse of new farming land with which to feed the realm and grow export products. In 1837 the King appointed a governmental commission to deliberate on the reclamation. In 1839 a

---

[25] H.W. Lintsen and A. van Heezik, 'In gevecht met de rivieren', in: H.W. Lintsen et al., *Made in Holland: Een techniekgeschiedenis van Nederland [1800–2000]* (Zutphen 2005), 98.

[26] Zie voor het navolgende: A. van Heezik, *Strijd om de rivieren: Tweehonderd jaar rivierenbeleid in Nederland of de opkomst en ondergang van het streven naar de normale rivier* (dissertation TUDelft 2007).

law was adopted mandating a reclamation by the state. Thousands of workers constructed the encircling dike and the accompanying drainage canal – also intended as a waterway for inland navigation. Three enormous steam engines (the most powerful in the world at that time) – the Cruquius, the Lijnden and the Leeghwater – pumped the lake dry. The job was completed by 1853.

5.3.3 Infrastructure[27]

King William's big ambition was to improve the welfare of his people. The theme recurred in all his King's speeches. The state represented the public good and was beholden to pursue the greatest happiness for the greatest number. The impoverished Netherlands had to be revived with diligence and industry and with useful knowledge and projects. The development of infrastructure was a crucial feature of his politics. This was essential for prosperity, but also for unity. Prosperity demanded unification of the country, politically, economically and culturally. Decent infrastructure was one of the preconditions. The King went to work with a will.

It is true that thanks to its extensive network of waterways the Netherlands already had one of the best transport infrastructures in the world, but that was true only for the western part of the country. The eastern Netherlands were poorly integrated. Moreover, other countries like England and Belgium were busy improving their infrastructures with roads and canals. And then there were also the railways as revolutionary means of transport. The harbours of Amsterdam and Rotterdam were threatened with the loss of their unique position due among other things to competition from Antwerp and Hamburg.

During the reign of William I, for the first time, a coherent network of roads with international connections came into being. Initially the national state had instigated the project. Provinces, municipalities and private capital followed with refinements to the network, that ultimately connected all regions of the country.[28]

William I also invested in the construction of canals. This did not eventuate in a coherent network. Every canal had its own purpose. The North-Holland Canal and the Voornse Canal, for example, were dug to improve access to the maritime harbours of, respectively, Amsterdam and Rotterdam. The Dedemsvaart and the Willemsvaart were constructed to facilitate the transport of turf. The Zuid-Willemsvaart served to incorporate the eastern part of North Brabant and a part of Limburg into the rest of the waterways system. Canal construction was above all an extension of William I's economic politics with its high priority for international trade followed by the promotion of agriculture, industry and jobs.[29]

---

[27] Zie voor deze paragraaf: G. Mom and R. Filarski, *Van transport naar mobiliteit: De transport-revolutie [1800–1900]* (Zutphen 2008).

[28] William I's policy extended initiatives already started in the French and Batavian period (1795–1813). Mom and Filarski, *Van transport naar mobiliteit*, 108.

[29] Mom and Filarski, *Van transport naar mobiliteit* 112–115, 123.

Only two railways were built under William I. One between Amsterdam and Rotterdam and the so-called Rhine Railway. The first was intended as an attractive alternative for the existing goods and passenger transport between the two harbours. The second, intended to connect Amsterdam with the German hinterland, was the anxious answer to the railway line between Antwerp and Cologne.

Many projects encountered resistance and engendered heated debates. The construction of a road, canal or railway disrupted the status quo. The choice of a route threatened local interests, strained property rights and disturbed long-standing practices. The route could also be contested because it literally bypassed municipalities and deprived them of the benefits of the new infrastructure. Some actors had problems with new technologies. Many of the above issues came to the fore with the construction of the railway between Amsterdam and Rotterdam. Opponents considered the railway completely superfluous. Existing infrastructure was more than adequate. They could not imagine that the population had any need for this modern mode of transport. The railways would barely show a profit. Quite the contrary, the railways would seriously harm local interests.

> Stable owners, barge haulers, shipping agencies, certifiers, delivery services, ships' carpenters, shipbuilders would naturally suffer a large and important loss. And then we do not even speak of stage-coach entrepreneurs and everything associated therewith.[30]

The choice of a route had to take the interests of national defense into account; owners of land needed for the right of way bargained with the railway company for extra stations, etc. It was only royal will-power that endowed the Netherlands with two railways before the middle of the century.[31]

Did all these infrastructural projects augment national welfare, as the King so ardently desired? The question is difficult to answer. Some doubts can, however, be raised.[32] Economically speaking, the maritime canals were a failure. Canals like the North Holland Canal, the Marken Canal and the Zederikkanaal were barely used and provided far too little revenue to justify the costs of construction. On the other hand, turf canals like the Dedemsvaart and regional access canals like the Zuid-Willemsvaart did have the desired social and economic effects. The railways also exhibited a similar dualism. The railway between Amsterdam and Rotterdam was a success from the very start, while the Rhine Railway became a drama and financially speaking a bottomless pit.[33]

The impact of new roads leads to some interesting speculations. The main roads in the region bounded by Amsterdam, Rotterdam and Utrecht were sufficiently profitable from a socio-economic point of view. But what about the situation in the peripheries? It is noteworthy that road transport in the (relatively sparsely settled) provinces of Gelderland, Overijssel and Drenthe increased significantly.[34] The same

---

[30] Quoted in: Mom and Filarski, *Van transport naar mobiliteit* 129.

[31] Mom and Filarski, *Van transport naar mobiliteit,* 149.

[32] Mom and Filarski, *Van transport naar mobiliteit,* 381–383.

[33] Mom and Filarski, *Van transport naar mobiliteit* 361–378 in particular 373–378.

[34] J.L. van Zanden, *De economische ontwikkeling van de Nederlandse landbouw in de negentiende eeuw, 1800–1914* (dissertation Universiteit Wageningen 1985), 148–149.

held for North Brabant and Limburg. In Drenthe, for example, the number of farm-wagons and carts increased from about 2880 in 1818 to 9400 in 1854. Transport costs declined by 50% to possibly 75%. Small farmers were able to earn extra money in the transport business. Markets were more accessible. Agricultural products found markets over larger areas. The cottage textile industry flourished. Earlier we concluded that after 1820 poverty in the Netherlands declined precipitously and proposed as cause de-urbanization and the rise of the country-provinces, in particular on the sand grounds. The prerequisite would then be the improved accessibility of those regions thanks to the construction of a rural road system.

If this analysis holds water, then the development of infrastructure did stimulate national welfare, but in a different way than imagined by William I. It was not the powerful cities of Amsterdam and Rotterdam that contributed to national welfare but the countryside and the small cities on the sand grounds.[35] Economic growth was rooted not in industrialization and large-scale production on the English model, but in a commercializing agriculture and cottage industry in the former periphery. That said, it remains the case that William I's efforts did lay the groundwork for a transport revolution in the following period, a revolution that also affected national welfare.

Around 1850 housing construction, water management and infrastructure were the three relevant themes of the building materials supply chain in relation to sustainability. It was only in the domain of infrastructure that dramatic changes took place. These had an unexpected effect on the question of poverty.

To conclude, we examine the third supply chain, that of the fossil raw materials. What changes took place here?

---

[35] Using a gravity model, J. de Vries analysed the historical structure of the Dutch 'Randstad' (urban agglomeration in the western Netherlands). He applied this to the passenger traffic serviced by the Hollandsche Yzeren Spoorweg Maatschappij (Holland Iron Railway Company) in 1857–1861 and contrasted this pattern with that of the tow-barges of the seventeenth and eighteenth centuries. The balanced urban system of the Golden Century - in reality a collection of highly autonomous cities and towns - had disappeared. Amsterdam was originally dominant but stagnated as far as passenger traffic was concerned. That made room for other cities. Passenger traffic on the railroad showed that Rotterdam and subsequently The Hague were secondary centers and that they moreover bound relatively many smaller cities to themselves. In 1855 Amsterdam had only one satellite city, Haarlem, while Rotterdam had four. This development had roots reaching back to the period before 1800. The period 1700–1750 was that of the (Holland) urban system in decline. The period 1750–1800 that of the dual city system, the rise of rivals to Amsterdam. J. de Vries, 'Barges and capitalism: Passenger transportation in the Dutch economy, 1632–1839', in: *A.A.G. Bijdragen* 21 (1978), 252–351.

5.4   Energy: Coal and Steam[36]

The promotion of national well-being could be pursued along different paths. There were three options, as we showed in the previous chapter: diminution of social inequality, economic growth with the aid of domestic natural capital and economic growth with the aid of natural capital elsewhere, in particular in the Dutch East Indies. William I's infrastructural policy, in particular on the point of roads, enabled a combination of the three options. The sand grounds and the high moors – previously relegated to the periphery – could now develop and be integrated into the national economy and the colonial complex. The rapidly growing population in those regions experienced less social inequality than the urban population.

Economic growth with the aid of natural capital could take place in the first half of the nineteenth century in yet another way, namely by radical changes in the supply chain of fossil raw materials. The output of an economy is in its most elementary form reducible to a function of capital and labour.[37] Both capital and labour require inputs of energy to realise an output. For labour (human and animal force) the source of energy is food. In an agricultural economy the production of food is limited by the availability of fertile ground and the transformation of sunlight into agricultural produce. For capital, sources of energy are the sun, the wind, water and fossil and nuclear raw materials. These can, in principle, break the bonds of an economy based on labour and land. Wind, water and turf already do this in the pre-industrial phase. For centuries, the Netherlands had already been exploiting wind power and turf on a large scale – to great economic effect. The breakthrough of the nineteenth century is the use of the fossil fuel coal in combination with the steam engine.[38]

Prominent contemporaries in the first half of the nineteenth century were already convinced of the crucial role of coal and steam. Anthony Hendrik van der Boon Mesch, professor of chemistry and later president of the University of Leiden, put it like this in 1843:

> The Government and the Nation, we see both...exerting their powers, to the end of recovering as much as possible of what has been lost and to expansion into a field hitherto unexplored... The introduction of steam engines...breathes new life and importance into many manufactures. Much of what was considered to be possible only in England, is now happening here.[39]

---

[36] See for this section: H. Hölsgens, *Energy transitions in the Netherlands: Sustainability challenges in a historical and comparative perspective* (dissertation Rijksuniversiteit Groningen 2016), chapter 1, 1–18 and chapter 2, 19–60; H.W. Lintsen, 'Een land met stoom', in: H. Lintsen et al. (ed.), *Geschiedenis van de techniek in Nederland. De wording van een moderne samenleving 1800–1890* (Zutphen 1993), Part VI, 191–209

[37] Hölsgens, *Energy transitions in the Netherlands*, 4–6.

[38] On the limitations of land and labour in relation to economic growth: E.A. Wrigley, *Energy and the English Industrial Revolution* (Cambridge 2010); Sieferle writes about the 'subterranean forest' when writing about fossil sources of energy: R. Sieferle, *The subterranean forest: Energy systems and the Industrial Revolution* (Cambridge 2001).

[39] Quoted in: Hölsgens, *Energy transitions in the Netherlands*, 9.

Table 5.2 Energy use in the Netherlands per source of energy in 1850

	TJ	Percentage
Human and animal	33,500	38
Turf	21,900	25
Coal	14,300	16
Wood	9800	11
Wind and water	9000	10
Total	88,500	100

Source: B. Gales and H. Hölsgens, 'Energy consumption in the Netherlands (1800–2012)', in: H. Hölsgens, *Energy transitions in the Netherlands. Sustainability challenges in a historical and comparative perspective* (dissertation Rijksuniversiteit Groningen 2016), Table A.I.1, 211

Table 5.3 Number of prime movers in industry around 1850 by type of prime mover

	Number	Percentage
Windmills	3050	53
Horse treadmills	1930	34
Watermills	470	8
Steam Engines	290	5
Total	5470	100

Source: H. Lintsen, 'Een land met stoom', in: H. Lintsen et al. (ed.), *Geschiedenis van de techniek in Nederland. De wording van een moderne samenleving 1800–1890* (Zutphen 1995), vol. VI, Table 7.1, 192

In the Netherlands, wind and turf had long-ago occasioned a breakthrough. The economic success of the Golden Century was in part due to the application of the Dutch windmill and the use of cheap turf. The future lay, according to Van der Boon Mesch, in imitation of England, in the steam engine and coal. In his time William I was certainly not alone in his efforts to modernise industry.

The total consumption of energy in the Netherlands in 1850 can be estimated at 88,500 Tera joule (TJ) (Table 5.2). Humans and animals provided the most energy, 33,500 TJ (or 38%), followed by turf with 21,900 (25%). Coal took third place with 14,300 TJ (16%) even before wood as a source of energy (9800 TJ or 11%) and wind and water (9000 or 10%). But these statistics give a wrong impression of the role of coal. In that year coal was used not so much for steam production to power steam engines, but as a source of heat in households and production. This type of use did not occupy centre stage in the process of industrialization.

The number of steam engines in Dutch industry was quite modest in 1850, 290 of a total of 5740 prime movers (5%). The rest consisted of windmills (53%) horse treadmills (34%) and a small percentage of water mills (8%) (Table 5.3). Why did

Table 5.4 Costs of depreciation, interest, maintenance, labour and fuel of different sources of energy in industry, 1825–1850

Place/region	Overijssel	Overijssel	Amsterdam	Amsterdam	Amsterdam
Technology	Human power	Water power	Wind power	Steam (max. 20 hp)	Horse
Period	1825–1850	1825–1850	1825–1850	1843	1825–1850
Depreciation, interest and maintenance		0.54	0.51	0.68	0.32
Wages	5.00	0.02	0.39	0.19	0.34
Fuel		0.33		0.79	2.00
Total	5.00	0.91	0.90	1.66	2.75

Source: H. Lintsen, 'Een land met stoom', in: H. Lintsen et al. (ed.), *Geschiedenis van de techniek in Nederland. De wording van een moderne samenleving 1800–1890* (Zutphen 1995), vol. VI, Table 7.1, 198

not the supply chain coal mining-coal burning-steam production-steam power develop? What are the reasons for the limited use of steam technology?

Economics provides part of the answer.[40] For this we compare the steam engine to classical prime movers and regard them as completely interchangeable in the process of production (Table 5.4). In the first half of the nineteenth century the windmill and watermill were the cheapest, after that the steam engine (assuming an engine of about 12 hp) and finally the horse treadmill and the human body. Compared to the windmill and watermill the steam engine was an expensive source of power. Operating costs were high, among other things because of the cost of coal.[41] A horse suffered from even higher operating costs especially because of the cost of food. But humans were dearest. Considering human food as fuel, and the fact that acquiring this fuel consumed a large part of the worker's wages, he easily became the most expensive prime mover. Nonetheless, humans were hard to replace. Human workers were also clever in utilizing tools and simple machines and could be employed at many places in the process of production. The horse's position was considerably weaker. He would be the first to be ejected from the process of production and managed to retain a position only in transport. It should also be noted that the costs of running steam engines depended strongly on the region. Due to the lack of decent waterways, coal was considerably more expensive in Twente than in Amsterdam. In Twente cottage industries long competed successfully with the textile factory.

The entrepreneur who introduced steam technology prior to 1850 was almost always 'doomed' to higher production.[42] The steam engine demanded the presence or the creation of a sufficiently large market. In most cases this requirement was not met. That brings us to different kinds of reasons for the limited application of steam

---

[40] Lintsen, 'Een land met stoom', 195–198.

[41] Taxes were levied on coal until 1863. That made coal more expensive than turf. It was true that the entrepreneur could be refunded but that required a long march through the tax bureaucracy. Even without taxes the fuel costs of steam engines were still considerable.

[42] Lintsen, 'Een land met stoom', 199–202.

power in industry. In the countryside, poor transport infrastructure long hindered the expansion of local markets into regional or national markets. Urban markets were dominated by informal cartels of old crafts. The legal system was oriented to small-scale production and hindered large-scale production processes.

The economic structure is another explanatory factor. The Netherlands lacked economically exploitable raw materials like coal and iron ore. Significant mining was lacking, as opposed to England and Belgium. This branch of industry was a crucial domain of application for steam technology. Demand for steam engines to pump water, ventilate mine shafts and transport mine workers and ore stimulated the iron industry and machine engineering.

Mining, the iron industry, machine engineering and the textile industry (and in its wake the sulphur industry) were the dynamic sectors in the industrialization of England and Belgium. These had an enormous impact on the other industrial sectors. The Netherlands lacked mining of any importance. The textile industry was based on cottage production. The metalworking sector was small in scope. Machine engineering consisted of a small nucleus of machine builders. Industrialisation had not started in 1850 and thus neither was the promotion of national welfare on the basis of industrialisation.

Literature

Allen, R.C. (2013). 'Poverty lines in history: Theory and current international practice'. *Discussion paper series.* Department of Economics, University of Oxford.

Bakker, M. (1992). 'Boter'. In H.W. Lintsen et al. (Eds.), *Geschiedenis van de techniek in Nederland: De wording van een moderne samenleving 1800–1890* (Vol. I, pp. 103–133). Zutphen: Walburg.

Bakker, M. (1992). 'Suiker'. In H.W. Lintsen et al. (Eds.), *Geschiedenis van de techniek in Nederland. De wording van een moderne samenleving 1800–1890* (Vol. I, pp. 215–251). Zutphen: Walburg.

Bakker, M. (1992). 'Techniek en voeding in verandering'. In H.W. Lintsen et al. (Eds.), *Geschiedenis van de techniek in Nederland: De wording van een moderne samenleving 1800–1890* (Vol. I, pp. 253–277). Zutphen: Walburg.

Bakker, M. (1992). 'Voeding in Nederland'. In H.W. Lintsen et al. (Eds.), *Geschiedenis van de techniek in Nederland: De wording van een moderne samenleving 1800–1890* (Vol. I, pp. 39–51). Zutphen: Walburg.

Beer, J.J.A. de (2001). *Voeding, gezondheid en arbeid in Nederland tijdens de negentiende eeuw. Een bijdrage tot de antropometrische geschiedschrijving.* Utrecht: PhD-study University Utrecht.

Bosch, A., Ham W. van der, and Lintsen, H.W. (1998). *Twee Eeuwen Rijkswaterstaat 1798–1998.* Zaltbommel: Europese Bibliotheek.

Buiter, H. and H.W. Lintsen (2005). 'De stad, de stank en het water'. In H.W. Lintsen (Ed.), *Made in Holland: Een techniekgeschiedenis van Nederland [1800–2000].* Zutphen: Walburg Pers.

'Verslag aan den Koning, over de vereischten en inrigting van arbeiderswoningen, door eene commissie uit het Koninklijk Instituut van Ingenieurs', *Het Tijdschrift van het Koninklijk Instituut van Ingenieurs*, 1854–1855, 50–75.

Gales, B. and H. Hölsgens (2016). 'Energy consumption in the Netherlands (1800–2012)'. In Hölsgens, H. (Ed.), *Energy transitions in the Netherlands: Sustainability challenges in a historical and comparative perspective (dissertation)*. Groningen: University of Groningen, SOM research school.

Heezik, A. van (2007). *Strijd om de rivieren: Tweehonderd jaar rivierenbeleid in Nederland of de opkomst en ondergang van het streven naar de normale rivier (dissertation)*. Delft: Technische Universiteit Delft.

Hölsgens, H. (2016). *Energy transitions in the Netherlands: Sustainability challenges in a historical and comparative perspective (dissertation)*. Groningen: University of Groningen, SOM research school.

Horlings, E. and J.P. Smits (1995). 'Private consumer expenditure in the Netherlands, 1800–1913'. *Economic and Social History in the Netherlands*, 7, 15–40.

Houwaart, E.S. (1993). 'Medische statistiek'. In H.W. Lintsen et al. (Eds.), *Geschiedenis van de techniek in Nederland: De wording van een moderne samenleving 1800–1890*. Zutphen: Walburg.

Houwaart, E.S. (1993). 'Professionalisering en staatsvorming'. In H.W. Lintsen et al. (Eds.), *Geschiedenis van de techniek in Nederland: De wording van een moderne samenleving 1800–1890* (Vol. II, pp. 81–92). Zutphen: Walburg.

Knibbe, M.T. (2007). 'De hoofdelijke beschikbaarheid van voedsel en de levensstandaard in Nederland, 1807–1913'. *Tijdschrift voor Sociaal-Economische Geschiedenis*, 4, 71–107.

Lintsen, H.W. (1980). *Ingenieurs in Nederland in de negentiende eeuw: Een streven naar erkenning en macht (dissertation)*. Eindhoven: Technische Universiteit Eindhoven.

Lintsen, H.W. (1995). 'Een land met stoom'. In H.W. Lintsen et al. (Eds.), *Geschiedenis van de techniek in Nederland: De wording van een moderne samenleving 1800–1890* (Vol. VI, pp. 191–210). Zutphen: Walburg.

Lintsen, H.W. and M. Bakker (1992). 'Meel'. In H.W. Lintsen et al. (Eds.), *Geschiedenis van de techniek in Nederland: De wording van een moderne samenleving 1800–1890* (Vol. I, pp. 71–101). Zutphen: Walburg.

Lintsen, H. (2005). *Made in Holland: Een techniekgeschiedenis van Nederland [1800–2000]*. Zutphen: Walburg Pers.

Lintsen, H.W. and Heezik, A. van (2005). 'In gevecht met de rivieren'. In H.W. Lintsen (Ed.), *Made in Holland: Een techniekgeschiedenis van Nederland [1800–2000]* (pp. 95–113). Zutphen: Walburg Pers.

Lintsen, H. and T. Lintsen (2017), *Opvattingen over hygiëne 1830–1850* Eindhoven: Eindhoven University of Technology, open-access research report

Mom, G. and R. Filarski (2008). *Van transport naar mobiliteit: De mobiliteitsexplosie (1895–2005)*. Zutphen: Walburg Pers.

Schippers, H. (1992). 'Bier'. In H.W. Lintsen et al. (Eds.), *Geschiedenis van de techniek in Nederland: De wording van een moderne samenleving 1800–1890* (Vol I, pp. 171–213). Zutphen: Walburg.

Sieferle, R. (2001). *The subterranean forest: Energy systems and the Industrial Revolution*. Cambridge: The White Horse Press.

Vries, J. de (1978). 'Barges and capitalism: Passenger transportation in the Dutch economy, 1632–1839', in: *A.A.G. Bijdragen* 21, 252–351.

Woud, A. van der (2010). *Koninkrijk vol sloppen: Achterbuurten en vuil in de negentiende eeuw*. Amsterdam: Bert Bakker.

Wrigley, E.A. (2010). *Energy and the English Industrial Revolution*. Cambridge: Cambridge University Press.

Zanden, J.L. van (1985). *De economische ontwikkeling van de Nederlandse landbouw in de negentiende eeuw, 1800–1914 (dissertation)*. Wageningen: Universiteit Wageningen.

Zanden, J.L. van (1992). 'Mest en ploeg'. In H.W. Lintsen et al. (Eds.), *Geschiedenis van de techniek in Nederland: De wording van een moderne samenleving 1800–1890* (Vol. I, pp. 53–69). Zutphen: Walburg.

Zanden, J.L. van en Riel, A. van (2000). *Nederland 1780–1914. Staat, instituties en economische ontwikkeling*. Amsterdam: Balans.

Zon, H. van (1993). 'Openbare hygiëne'. In H.W. Lintsen et al. (Eds.), *Geschiedenis van de techniek in Nederland: De wording van een moderne samenleving 1800–1890* (Vol. II, pp. 47–79). Zutphen: Walburg.

Chapter 6
Well-being and Sustainability Around 1850: The Frame of Reference

Harry Lintsen

Contents

Abstract The previous chapter reported a dearth of innovation regarding the exploitation of natural capital around 1850. This chapter deals with the dynamics of the institutional quadrants at the time (see Chap. 1). In a number of respects the 1840s marked the start of a new phase. 'Civil society' awoke, mainly thanks to the contribution of younger generations of Netherlanders. Due to the abdication of King William I, the political institutions required a makeover. Economic institutions were under a great deal of pressure due to the emerging liberal climate and the liberalisation of the world market. In the domain of technology, new institutions blossomed with the emergence of civil and mechanical engineers and other professional groups. These developments had not yet led to fundamental social change. The Netherlands remained a mercantile capitalist, colonial and agricultural nation.

This is the context in which the well-being monitor for 1850 must be placed. This monitor is the 'benchmark' for this study, the standard against which the monitors for 1910, 1970 and 2010 are evaluated.

On the basis of the monitor and from both a contemporary and a present-day perspective, three important sustainability problems can be discerned: material welfare (poverty), the institutional environment (political instability) and social capital (little trust in political institutions). In addition, from a present-day perspective a series of issues is problematic: poor public health, nutrition and lower-class housing

© The Author(s) 2018
H. Lintsen et al., *Well-being, Sustainability and Social Development*,
https://doi.org/10.1007/978-3-319-76696-6_6

(the personal characteristics), insufficient innovations (economic capital), a lack of qualified labour (human capital) and the immoral "culture system" in the Dutch East Indies (trans border effects). The state of water management also gave cause for concern. Both from contemporary and present-day perspectives the Dutch delta was vulnerable.

Keywords Civil society · Politics · State · Economy · Technology · Engineers · Monitor

6.1 The Society of Benevolence[1]

Johannes van den Bosch returned from the Dutch East-Indies in 1812 a wealthy man. He had advanced rapidly in his military career and had supervised a plantation. Back in the Netherlands after the fall of Napoleon he immediately applied for positions in the army and government. King William I promoted him after several years to the rank of major-general. Despite this he had very different, almost grandiose, ambitions. He was going to abolish poverty in the Netherlands. The 'French period' had impoverished the Netherlands. Trade had ground to a halt. To make matters worse, harvests in 1816 and 1817 had failed. Poverty was endemic.

Van den Bosch proposed a unique plan: a national organization for the poor. The urban poor would be put to work reclaiming the 'wastelands' in Drenthe and Overijssel; they would become farmers, produce food, become self-reliant and make a useful contribution to society. The idea of an organization for the poor was not new. It was often seen as an ideal and cheap solution for pauperism. The pauper would no longer be a burden on society, but would instead produce value that could be used to finance the organization. However, this solution generally proved illusory. Subsidies were almost always necessary.[2]

Van den Bosch would have been aware of the doubtful economic viability of organizations for the poor. But he was convinced of his success, even if he did need money to buy land and dwellings for the 'colonists.' To this end he founded the Society for Benevolence in 1818. The bourgeoisie was enthusiastic. Thousands con-

---

[1] See for this section: S. Jansen, *Het pauperparadijs: Een familiegeschiedenis* (Amsterdam 2008), 39–46, 52–56. P. de Rooy, *Ons stipje op de waereldkaart: De politieke cultuur van Nederland in de negentiende en twintigste eeuw* (Amsterdam 2014), 63–69.
[2] A. de Swaan, *Zorg en de staat* (Amsterdam 1989), 55–60.

tributed. Local committees popped up everywhere. The King became patron and his son, Frederik, chairman of the board.

Initially things looked promising. The first colony, not coincidentally called Frederiksoord, was ready by 1818 to receive the first families. This was celebrated in verse:Brothers! Glad and awake

Singing to the field
Where labour awaits us.
In former times when in hovels,
We hid from the light...
Now it's different.[3]

Van den Bosch optimistically announced that 6 weeks had been time enough to 'raise (the families) from their downcast state' and decided to create another two colonies in addition to Frederiksoord, to be known as Willemsoord and Wilhelminaoord. These were reserved for the 'decent' poor. In addition he built two penal colonies, Ommerschans and Veenhuizen. Here abandoned children found a home, but also beggars, vagabonds and drifters – in short all the troublesome poor that were difficult to handle.[4]

Reality proved more complex. The land was poor and produced less than had been expected. The cities had a hard time convincing their poor to move to distant Drenthe. Police were called in to supervise forced emigration. The cities refused to surrender the occupants of their children's homes. The colonists – amounting to some 11,000 in the early 1840s – chafed at the discipline. That was especially true of the 'unregenerate' paupers. But even the 'decent' poor could not get used to the rules made by the Society for Benevolence for living in a colonist's dwelling and to the rhythm of labour on the land. Financially speaking, the initiative could only be kept afloat by artifice and improvisation. In 1859 the state assumed responsibility for the colonies.

Van den Bosch and his Society for Benevolence belonged to the domain of 'civil society.' Contemporaries were preoccupied with poverty and more generally the national welfare in many different ways. In the previous chapter we analysed the initiatives from the perspective of the three material flows: biomass (agriculture and foods), mineral substances (building materials and construction), fossil substances (energy). In this final chapter we focus on the institutional quadrants. In addition to the domain of 'civil society' these include the state, economy, and technology (see Chap. 1). This perspective brings cross-sections into relief. It enables us to evaluate the institutions relevant to all three material flows one by one. The evaluation of the institutions also provides insight into societal dynamics in the mid-nineteenth century. In addition, the institutional environment is an important theme in the sustainability monitor.

---

[3] See for this and the following quote: Jansen, *Het pauperparadijs*, 46.

[4] In 1822 Van den Bosch also founded the Society of Benevolence for the Southern Netherlands and under this flag founded the free colony of Wortel and the unfree colony of Merksplas.

After that we return to the CBS sustainability monitor. By now we have collected sufficient information in this part of the book to be able to provide a summary of well-being and sustainability in the Netherlands around the middle of the nineteenth century.

6.2 Deficient Dynamism: Citizens, Government, Entrepreneurs and Researchers

6.2.1 'Civil Society'

'Civil society' was intensely involved with the issue of poverty and it had done this for centuries in the form of poor-relief. In contrast to Van den Bosch's centrally organized project, poor-relief was traditionally decentralized and segmented. Estimates are that some five thousand local organizations implemented relief, mostly on a religious basis.[5] Local governments exercised some surveillance over the churchly organizations or sometimes possessed their own public agencies. The national government was minimally involved. The upshot was an enormous variety in poor relief. Usually the organizations supplied sums of money and goods like bread, clothing, turf and other necessities. Some also provided medical aid, education for the children, housing for the elderly and so on. The substance of poor relief differed from administration to administration.

Taking care of the poor was one of the churchly duties. It provided moral peace of mind or was exercised in the full conviction of brotherly love. It was also a bourgeois strategy of control. Trade, industry and agriculture had need of an army of reserve labour. The size of the labour market fluctuated in the course of a year and across the years. A generous supply of cheap labour was a precondition for the Dutch mercantile capitalist agricultural economy. Poor relief was the means to this end. It also provided social stability. By providing minimal support in normal times and extra support in difficult years it prevented riots and revolt among the poor.

For the poor, the dole was part of their survival strategy. They were unable to make ends meet on the basis of labour alone. The winters were the worst. Old age and the child-rearing years (up to age seven) were difficult periods in the life cycle. Disaster struck with sickness and economic recessions. Charity never covered the costs of living. There was always need of supplementary sources of income from labour, loans, help from neighbours etc. Begging and vagrancy could also provide incomes but those sources were risky, including the risk of expulsion from poor relief. These activities were socially unacceptable, were a nuisance and were judged as criminal acts. Theft was of course completely beyond the pale.

[5] P. de Rooy, 'De armen hebt gij altijd met u: Armenzorg en onderwijs', in: I. de Haan, P. den Hoed and H. te Velde, *Een nieuwe staat: Het begin van het Koninkrijk der Nederlanden* (Amsterdam 2013), 222.

A big problem of poor relief was the levying of costs. The poor residing in a given municipality were the least problematic. They were supported by local organizations. Struggles about levying costs arose around the poor originating from other municipalities. Poor relief organizations and municipalities persistently tried to foist these costs off on other municipalities. Towns and cities wanted to rid themselves of the wandering poor as quickly as possible. Some municipalities, for example, appointed waggoners who wasted little time in transporting vagrants and vagabonds beyond the municipal limits. A solution to these problems could have been found in a national approach, but the proprietary and churchly organizations had no faith in this option.

The institution of poor relief was stable. The organizations were conservative. Once in a while an effort was made to change things, as in the case of Van den Bosch. 'Civil society' was in any case characterized by a conservative bent. This extended to other societal issues like public health, education and industrialisation. After the turbulent periods of the Enlightenment and the French Revolution the impetus to social change weakened and the intensity of public debate declined.[6] The dominant role of the state in the person of William I did not help matters. Citizens only became active in questions that threatened their own interests, as we saw with the first railways and the first steam grain-mills.[7] Maintaining the status-quo was often the goal. Virtue aimed at economic recovery and a penchant for change within the existing frameworks were the prime characteristics of 'civil society.' Jan Rudolf Thorbecke, constitutional reformer and the first prime-minister after the 'bourgeois revolution' of 1848, characterized the period as follows:

> In the swings and numerous changes that had afflicted our State since 1795, political zeal and political conviction were weakened, if not extinguished, even among the flower of the nation... Not participation but abstinence seemed to be the patriotic duty.[8]

It took until the 1840s before a new generation of Netherlanders were able to liberate themselves from the scourge of political indifference and to propose initiatives that would reverberate in politics, policy and economy. Provincial agricultural committees initiated the first national Congress for Rural Economy (*Landhuishoudkunde Congres*) in 1846. A new occupational group, engineers, founded the Royal Institute of Engineers in 1847. In 1849, local medical associations founded the Dutch Society for the Promotion of Medicine.

---

[6] R. Aerts, 'Het ingetogen vaderland: huiselijkheid, maatschappelijke orde en publieke ruimte', in: I. de Haan, P. den Hoed and H. te Velde, *Een nieuwe staat: Het begin van het Koninkrijk der Nederlanden* (Amsterdam 2013).

[7] 'Civil Society' had a weekly opportunity to present its interests directly to the king. Every Wednesday there was an open invitation to all citizens to have an audience with the king. Many took advantage of this standing invitation. J. Koch, *Koning Willem I, 1772–1843* (Amsterdam 2013), 294.

[8] De Rooy, *Ons stipje op de waereldkaart*, 47.

6.2.2   King and State

The nation belonged to the King, that is how William I regarded political relations.[9] The political elite and the bourgeoisie thought no differently. Popular sovereignty had failed miserably during the French Revolution. The King had been responsible for restoring order and stability. His authoritarian style of governance was firmly rooted in the Constitution of 1815 and was long tolerated.

William I was authoritarian, but also enlightened. He regarded himself as the creator of the modern Netherlands and worked on this ideal with irrepressible energy. He was to become known as the 'merchant king' because of his meritorious service to trade or the 'canal king' because of the construction of numerous canals with a total length of 500 km. He might with equal justification have been called the 'road king' because during his reign a coherent network of roads was created; or the 'industrialist king' because he envisioned the creation of a modern industry based on the incorporation of modern technology. Must we see his reign as an authoritarian intermezzo between the Republic and the constitutional monarchy? From the perspective of modernisation his reign can be viewed as one big experiment.[10]

That experiment entailed promotion of the national welfare in addition to the development of the economy and the earning of money as the most important goals.[11] These goals were synonymous with combatting poverty and justified infrastructural projects and industrial subsidies. They were also part of the construction of the 'colonial complex.'[12] The Netherlands Trading Society (*Nederlandsche Handel-Maatschappij, NHM*) founded in 1824 at the King's behest, played a crucial role. The Society was established to organize trade with the Dutch East-Indies. It would buy and trade preferably Dutch products that would be transported in preferably Dutch ships that had been built if at all possible in the Netherlands. The Cultivation System was also part of the complex. Van den Bosch, he of the poor-colonies in Drenthe, introduced in the Dutch East Indies that with which he had started in Drenthe: forced agricultural labour. In this case it concerned the Javanese farmer. The NHM organised the flow of goods that this system generated.

The aim of the 'colonial complex' was to generate income for the Dutch state and for the stockholders of the NHM. Among the latter's largest stockholders were William I and a small coterie of Amsterdam investors, along with many smaller ones. For William I, the NHM was also a vehicle for stimulating various industries, creating jobs and fighting poverty. The cotton industry in Twente, for example, flourished thanks to the NHM. The Society founded weaving schools, acquired the most modern looms and put hundreds of smallholders, home weavers and their children to work. This policy, in combination with tariffs and the rural road network, was able

---

[9] Koch, *Koning Willem I*, 575.

[10] Koch, *Koning Willem I*, 203. See above all: J.L van Zanden and A. van Riel, *Nederland 1780–1914: Staat, instituties en economische ontwikkeling* (Amsterdam 2000), chapter 3, 109–148.

[11] Koch, *Koning Willem I*, 572.

[12] Koch, *Koning Willem I*, 43–48, 142–148.

to resist competition from British mechanized cotton production. Even after the collapse of the East Indies economy in the 1840s, the Twente textile industry managed to survive and in a later phase make the transition to factory production.

In the Dutch East-Indies the situation was different. The colony was seen by William I and the Dutch state as a source of profit. The 'colonial complex' produced immense incomes. Millions of Javanese farmers laboured to reduce the number of the poor in the Netherlands and augment the assets of the rich.[13] Profits from the Indies were also used to finance part of the enormous national debt and to pay for much of the war with secessionist Belgium in 1830.

The reign of William I must have contributed substantially to public well-being in the Netherlands, directly due to the policy of the NHM and indirectly with the development of infrastructure that enabled agriculture in formerly peripheral regions to flourish. As far as we can tell, his policies helped to realize a sharp decline in the percentage of poor in the first half of the nineteenth century to about 21% in 1850. That was achieved at the cost of the Javanese farmers who worked under degrading and inhumane conditions comparable to slavery.[14]

Public well-being was not the only sustainability issue on which William I exerted a big influence. Another was the problem of water management. The struggle against water could in his opinion be effectively conducted only by a strong state supported by a professional corps of hydraulic engineers. State formation, in other words, was the prerequisite for building a safe delta. But around 1850, with the exception of the reclamation of the Haarlemmermeer, the King, the state, and the corps of engineers could not yet claim any great successes. The large rivers and the sea remained persistent threats to safety.

William I's reign ended dramatically in 1840 with government finances in total disarray. About 60% of the state's income went to pay instalments and interest on the national debt.[15] In the following decades he was much maligned and his remarkable experiment ignored. The abdication of William I marked the start of a turbulent political period. The government and parliament initiated a program of reforms culminating in 1848 in a peaceful revolution led by Thorbecke and marked by the introduction of new constitution. The failure of the King's rule was due among other things to old economic institutions. These now demand our attention.

---

[13] De Rooy writes about ten million Javanese who augment the welfare of two and a half million Netherlanders. He is then speaking of the beginning of the cultivation system. De Rooy, *Ons stipje op de waereldkaart*, 67. Termorshuizen speaks of an average of 60–75% of all Javanese small farmers in the period 1836–1860. T. Termorshuizen, 'Indentured labour in the Dutch colonial empire, 1800–1940', in: G. Oostindie (ed.), *Dutch colonialism, migration and cultural heritage* (Leiden 2008), 266.

[14] E. Hondius, 'Het slavernijverleden achter de Hollandse horizon', in: I. de Haan, P. den Hoed and H. te Velde, *Een nieuwe staat: Het begin van het Koninkrijk der Nederlanden* (Amsterdam 2013), 186.

[15] Van Zanden and van Riel, *Nederland 1780–1914*, 219.

6.2.3   Economy and Trade[16]

William I's experiment was first of all inspired by the unification of the Southern and Northern Netherlands in 1815. This would be an ideal synthesis between a South that was developing industrially and a North oriented to commerce. According to the King, in this configuration the Netherlands was now exceedingly well-placed to become a modern industrial-capitalist nation. However it quickly became clear that this was an uneasy combination. There were, for example, differences of opinion about the introduction of a uniform tax system. The South wanted import and export rights in order to protect its domestic industry and as little taxation as possible on basic necessities in order to stimulate the domestic market. The preferences of the North, desiring as few obstacles to trade as possible, were diametrically opposed. These kinds of contradictions ultimately led to the secession of the Southern Netherlands and the founding of the Kingdom of Belgium in 1830.

Other efforts to modernize the economy ran into various kinds of opposition. It proved impossible to create a large domestic market. The Dutch market was strongly fragmented, among other things due to the autonomy of the cities, local excise taxes and the inter-provincial and local tolls. Local markets were moreover dominated by guild-like organizations in craft manufactures, retail trade, transport and fisheries. Groups of entrepreneurs and craftsmen chronically resisted changes that threatened their rights or competitive position. In this way the historical legacy remained firmly anchored in the institutions.

The historical legacy also caused trouble in other quarters. Amsterdam's commercial elite kept on claiming its former hegemonic position in international trade, despite the fact that the glory days of Amsterdam's staple market had long since passed. In the wake of the Belgian secession the king was *nolens volens* forced to concentrate his policy on the Northern commercial elite, its international commerce in colonial products and associated industries like sugar refining, shipbuilding and textiles (largely based on cottage industry and not on the factory system). This was not an economy that could generate strong impulses in the direction of a modern industrialisation process. Another spanner in the works was the role of agriculture. Agriculture was traditionally a strong sector and had comparative advantages relative to other countries. This sector absorbed entrepreneurial initiatives and investments, while industrialisation demanded efforts to build up key sectors like machinery building.

William I had to tack among all kinds of interest groups: Southern industrialists, Northern commercial elites, organized trades, groups of entrepreneurs, provinces, municipalities etc. It often proved impossible for him to maintain his progressive momentum. He was also not always consistent in his politics. The destruction of old and rigid structures in different economic sectors therefore proved extremely difficult. Even after William I abdicated the throne in 1840, it would be another two

---

[16] See for this subsection: Van Zanden and Van Riel, *Nederland 1780–1914*, 115–121; 203–208.

decades before industrialisation in the Netherlands took off. The institutions in the economy proved to be extremely resilient.

6.2.4   Technology and Science[17]

The traditional craft mode of production dominated within the institutional framework of industry in the mid-nineteenth century. Production was modest in scale and was carried on in a workshop, shed or dwelling. Manual labour with hand tools dominated this mode of production. To the extent machines were used, these were invariably simple in nature, traditionally constructed of wood and wrought iron, driven by the workman himself, a horse, a windmill or a watermill. Fabrication consisted of unique products or small series. The craftsman or craftswoman was the central figure.

Small scales and traditional craft production also held for agriculture, trade and services. Knowledge was of a traditional craft nature. It was largely based on experience that was accumulated and transmitted in practice. For some trades, particularly in construction and metalworking, there was supplementary schooling at drafting schools and drawing academies.

Besides the traditional craft knowledge infrastructure there was a modest infrastructure of learned societies. The most important of these dated from the eighteenth century, like the Holland Society for Sciences (*Hollandsche Maatschappij der Wetenschappen*) of 1752, the Batavian Fellowship for Experimental Philosophy (*Bataafsche Genootschap der Proefondervindelijke Wijsbegeerte*) of 1769, and the Society for the Promotion of Industry (*Maatschappij ter Bevordering van Nijverheid*) of 1797. They were a meeting place for merchants, physicians, army officers, entrepreneurs, scientists, hydraulic engineers and other groups within the bourgeoisie. The societies occupied themselves with scientific and technological topics, but regularly also with societal issues. They held competitions for treatises on poverty, national welfare, water management and industrialization.

The universities were independent of the societies. They were above all educational institutions for lawyers, physicians and ministers.[18] The universities were

---

[17] See for this sub-section: H.W. Lintsen with contributions by M. Davids, 'Een revolutie in kennis', in: H.W. Lintsen et al., *Made in Holland: Een techniekgeschiedenis van Nederland [1800–2000]* (Zutphen 2005), 293–314 and G. Verbong, 'Techniek, beroep en praktijk', in: H.W. Lintsen et al. (ed.), *Geschiedenis van de techniek in Nederland: De wording van een moderne samenleving 1800–1890* (Zutphen 1993), part V.

[18] A. Maas, 'Civil Scientists: Dutch Scientists between 1750 and 1875'. In: *History of Science* XIVIII (2010), pp. 75–103. The Organic Law (*Organiek Besluit*) of 1815 apportioned mathematics and the natural sciences to separate faculties. Research became a task of the universities, next to teaching. Initially it was not an obligation for professors, although it was highly encouraged. In the course of the century it would become a more prominent part of the academic job description and an important element in teaching. Scientists were less often seen as self-made researchers and more often as academically educated persons.

socially oriented. Science had to produce knowledge useful for technology, industry, seafaring and so forth. While this did not have to be immediately applicable knowledge, it certainly had to be knowledge possessing a clear link to socially useful matters. Most professors at the beginning of the nineteenth century could be regarded as socially engaged scientists. They placed great value on social service and engaged in all kinds of social activities. For example, they gave lectures to non-academic publics like industrialists, farmers, pharmacists and teachers.[19]

In the first half of the nineteenth century new types of knowledge structures were developed in two domains. The first was the state domain of military defence and water management. To educate army officers and hydraulic engineers, the state founded military schools around 1800, that eventually crystallized in the Royal Military Academy at Breda in 1829. The program of studies for engineers for the national Department of Waterways and Public Works (*Rijkswaterstaat*) was transferred to the Royal Academy for Engineers in Delft. The professionalisation of the domain was further consolidated by the founding of the Royal Institute of Engineers in 1847.

In the second place a new kind of knowledge infrastructure developed in the industrial domain of machinery construction. The central figure here was the mechanical engineer. He was the quintessentially modern bearer of knowledge in the age of steam and iron. His profession was nonetheless rooted in craft practices. By dint of years of practice he acquired skills in milling, drilling, planing, riveting and other mechanical operations. But due to the technological dynamism of his domain, his practical preparation ultimately proved inadequate By means of courses, literature, lectures and study trips abroad he managed to acquire a basis with which to appropriate international developments.

A gradual transformation of the technological knowledge domain took place in the shadow of this traditional craft knowledge infrastructure.[20] Knowledge became less subjective and personal. Craft knowledge became more objectified and was codified in books, journals and study materials. All kinds of practical knowledge was inventoried, investigated, described and internationally exchanged. Authors made efforts to explain technical phenomena and laws in technical processes. But while these changes had only a limited impact on contemporary issues in agriculture, industry and other technical domains, they would eventually provide new impulses for industrialisation, public well-being and other social issues.

In many respects the 1840s marked the start of a new era. 'Civil society' woke up, partly thanks to the contributions of younger generations of Netherlanders. William I's departure required renewal of existing political institutions. Venerable

---

[19] B. Theunissen, *'Nut en nog eens nut': Wetenschapsbeelden van Nederlandse natuuronderzoekers, 1800–1900* (Hilversum 2000), 190–193.

[20] This is Joel Mokyr's term. See his book: *The gifts of Athena: Historical origins of the knowledge economy* (Princeton 2002).

economic institutions were challenged by the new liberal climate and the ongoing liberalisation of the global market. In the domain of technology new institutions flowered with the emergence of engineers, especially mechanical engineers, and other professional groups. These developments did not yet lead to fundamental social change. The Netherlands remained a mercantile capitalist, colonial and agricultural nation.

At the same time, the 1840s brought a series of shocking calamities: failed harvests, epidemics and worst of all cholera. Poverty increased and showed its ugliest face. Disease was omnipresent and as in the case of cholera exhibited awful images of rapid decay. The popular mood was restless and at times downright grim. This is the context in which the sustainability monitor for 1850 must be placed.

6.3 The Monitor of 1850

The middle of the nineteenth century is the starting point for our investigation into well-being and sustainable development. To chart this dynamic relationship we make use of a measuring system developed at Statistics Netherlands, the sustainability monitor. As explained in Chap. 1, this monitor has three dashboards: the well-being of a people in the 'here and now,' the resources preserved for future generations ('later') and the transboundary effects of domestic activities on other peoples ('elsewhere'). Twenty-four indicators serve to track trends in the three dimensions. Every indicator is associated with an important theme and is assigned a value for each period. An historical 'benchmark' is an important feature of this approach. This is the frame of reference against which further developments can be set out. The 'benchmark' in this study will pertain to the period around 1850. Above, we argued that in order to interpret the results of such a measurement a context is necessary. The preceding chapters have provided this context. The 'benchmark' can now be briefly summarized (Table 6.1).

6.3.1 Well-being 'Here and Now'

A proper interpretation requires distinguishing between a present-day perspective and that of contemporaries regarding the themes of well-being and sustainability. Without this distinction we would be hard pressed to acquire a proper understanding of the developments. As far as the dimension of well-being 'here and now' is concerned the conclusion in regard to the situation around 1850 is clear from a present-day perspective: this was problematic in many respects (Table 6.1). The indicators reveal a situation in which a significant part of the population (21%) lived below the

Table 6.1 Monitor well-being of 1850 from a contemporary and a present-day perspective

Dashboard well-being 'here and now'					
Theme	Indicator	Unit	1850	Perspective 1850	Present day perspective
Population	Number inhabitants	million	3.1		
Material welfare and well-being					
Consumption, income	Consumptive expenditures per capita, constant prices	Index (1850=100)	100	−	−
	Income inequality, general	Gini coefficient 0–1	0.48	+	−
	Gender income inequality	% difference hourly wage M/F	?	+	−
Subjective well-being	Satisfaction with life	Score 0–10	?	?	?
Personal characteristics					
Health	Life expectancy	year	37	+	−
Nutrition	Height (military conscripts)	cm	165	−	−
Housing	Housing quality	% slums	30 á 50	+	−
	Public water supply	m³/capita	0	+	−
Physical Safety	Victims of murder	number per 100.000 inhabitants.	0.8	+	+
Labour	Unemployment	% workforce.	6.4	−	−
Education	Level of education	years	3	O	−
Free time	Free time	hours per week.	?	+	−
Natural environment					
Biodiversity	MSA	% original biodiversity	73	+	+
Air quality	SO_2	kg SO_2/capita	1.3	+	+
	Greenhouse gas emissions	ton CO_2/capita	1.2	+	+
Water quality	Public water supply	m³/capita	0	+	−
Institutional environment					
Trust	Generalised trust	% population with adequate trust	?	−	−
Political Institutions	Democracy	Democracy-index 0–100	0.3	−	−

(continued)

Table 6.1 (continued)

Dashboard well-being 'later'					
Theme	Indicator	Unit	1850	Perspective 1850	Present day perspective
Natural Capital					
Energy	Energy consumption	TJ /capita	0.03	+	+
Non-fossil fuels	Gross domestic consumption	ton/capita	2.1	+	+
Biodiversity	MSA	% original biodiversity	73	+	+
Air quality	SO_2 emissions	kg SO_2/capita	1.3	+	+
	Greenhouse gas emissions	ton CO_2/capita	1.2	+	+
Water	Public water supply	m³/capita	0	+	−
Economic Capital:					
Physical capital	Economic capital stock/capita	index (1850=100)	100	O	−
Financial capital	Gross national debt	% gdp	194	−	−
Knowledge	Stock knowledge capital	Index (2010=100)	−	+	−
Human Capital:					
Health	Life expectancy	years	37	+	−
Labour	Unemployment	% workforce	6.4	−	−
Educational level	Schooling	years	3	O	−
Social Capital:					
Trust	Generalised trust	% population with adequate trust	?	−	−
Political institutions	Democracy	democracy index 0–100	0.31	−	−

Dashboard well-being 'elsewhere'					
Theme	Indicator	Unit	1850	Perspective 1850	Present day perspective
Welfare					
Consumption, income	Development aid	% GDP	−	+	−
Natural capital					
Natural capital	Import of raw materials	ton/capita	0.4	+	?

Legend

+	Not problematic or not problematized
−	Generally acknowledged as problematic
O	Under discussion: different opinions about the scale and nature of the problems
?	Unknown

Note: The signs − and O in the column of 1850 are the then important themes. The column of the contemporary perspective indicates with the sign − which current themes would now be regarded as problematic. For the justification of the evaluation in the table, see the main text

poverty line. The poor found it immensely difficult to provide themselves with the basic necessities of life. They were poorly housed or forced into vagrancy. They were ill-clothed and could barely protect themselves against rain and dampness. Their food was meagre and lacked variation. They were able to consume just sufficient calories to maintain their physical bodies and to supply a bit of labour. The poor and their children were susceptible to all kinds of diseases due to the poor hygienic circumstances in which they lived. They were extremely vulnerable. They frequently suffered hunger and cold. Many could not survive without aid. Their life expectancy was low. These general characteristics hide the fact that there were large differences in poverty between town and country, between the low and high Netherlands and among different regions. Such differences have been described in this section of the book.

Contemporaries saw poverty as one of the most important issues, but they set their norms for poverty considerably lower than we would nowadays. They made few demands in regard to the quality of housing for the poor, the quality of their food or their hygienic circumstances. These themes were barely problematized by contemporaries and certainly not in terms that we would recognize. This is hardly surprising considering the available knowledge, the changed context and the changed attitudes with respect to housing, nutrition, hygiene and health. For contemporaries, the struggle against poverty aimed at survival for the poor and not at improving their quality of life. It was sufficient to provide for a minimum of basic necessities. The shifting norms in regard to poverty after 1850 appear to have had a great impact on the sustainable development of the Netherlands.

Another aspect of quality of life is the natural environment of the Dutch (and not just of the poor). From a present-day perspective, one aspect of the situation around 1850 was extremely problematic: the pollution of the human environment with organic waste, especially organic pollution of surface water and the lack of good drinking water. This extensive environmental (or hygienic) problem was understood only by a small and little-influential vanguard of professionals (especially physicians and engineers: the hygienists) as a social problem that required political action.

From a present-day perspective, the situation in other respects (biodiversity, greenhouse gas emissions, air quality) was at worst only mildly problematic. For example the CO_2 level at that time was far below the norm that we now try to achieve (Fig. 22.3). It will come as no surprise that at the time such issues were neither problematized nor on the political agenda.

An aspect that was problematic both from a present-day and contemporary perspective was the institutional context. It is true that this aspect has not been elaborated on in the preceding text and was only mentioned in relation to poverty, nonetheless the thesis seems defensible. After the departure of William I, who ruled as an enlightened despot, the constitutional order had to be established anew. That process engendered much disquiet and was still not brought to closure around 1850. Also under William I there were few social movements in 'civil society' that engaged with important social issues. An active and powerful 'societal midfield' able to provide a political counterweight to the king on important issues like housing, nutrition and health, was lacking.

An important theme for which indicators are lacking in the sustainability monitor is hydraulic safety. The Dutch delta was perennially threatened by the 'water wolf.' The problematic rivers were high on the agenda. From a present-day perspective that was certainly justified. The rivers of those days were not designed for a safe discharge of floodwaters and large quantities of ice. Life in the region of the big rivers was full of risks.

6.3.2 Well-being 'Later'

The resources for future generations – the dashboard for 'later' – is split into natural, economic, human and social capital. To what extent does the present-day perspective on these resources differ from that of the mid-nineteenth century?

In our eyes, natural capital around 1850 – in terms of depletion and environment – is a nuanced story. Various indicators like greenhouse gas emissions and energy consumption are (far) under the present-day norm. Moreover, 50% of the energy was provided by a variety of renewable resources like wind, water, muscle power and wood. The degree of biodiversity is hard to ascertain but as far as we can tell, it must have been high. These were in any case concerns that did not arise in those days.

At the same time, it is true that turf and to a lesser extent coal, were used in great quantities. Both belong to the class of finite and exhaustible resources. Nowadays this would give us pause. There were some in those days who were likewise concerned. They were worried about the rate at which turf was extracted and warned of the 'ultimate disappearance of our fens.' Some toyed with schemes for growing timber to head off a possible future fuel shortage. Others were less worried and pointed to the immense reserves of coal, especially in England.[21]

Natural capital was vulnerable, among other things for flooding, sand drifts and exhaustion of the soil. Contemporaries acknowledged this vulnerability. Another big problem was the widespread water and air pollution due to organic wastes. Contemporaries barely or only incidentally saw this as a problem. The bourgeoisie often complained of garbage and stench, but framed this above all in terms of a nuisance. Only a small and as yet uninfluential group of hygienists framed the issue as a social problem that demanded political action.

For contemporaries, economic capital that provided the wherewithal to exploit natural capital was an extremely important theme, though any kind of consensus about the nature of the issue was lacking. From a present-day perspective it is clear that modernisation of the economy and transport was a bitter necessity for popular well-being and the fight against poverty. In other countries major transformations were underway. The future of the country was at stake. But differences of opinion and contradictory interests dominated the debate. Contemporaries debated on the improvement of the rivers, the construction of railways, the introduction of the

---

[21] Zie Hölsgens (2016), forthcoming in *Tijdschrift voor Sociale en Economische Geschiedenis*.

steam engine and the use of new kinds of fertilizer. Initiatives to improve industry and infrastructure regularly met with opposition. In addition a solution had to be found for the enormous national debt. Well-being for future generations was on the public and political agenda, but there were serious differences of opinion on the way this might be achieved.

Human capital was not a big issue in those days. The condition of the paupers and poor workers was worrisome, but the concerns did not go beyond maintaining the poor as available labour power. Upgrading human capital by investing in housing, hygiene, better nutrition and health care was – barring scattered initiatives – not an option often considered. Upgrading schooling was in discussion. With the wisdom of hindsight such investments should have had a high priority. Moreover it would have been wise to invest in new forms of knowledge like mechanical engineering and the professionalization of existing domains of knowledgeable expertise like hydraulic engineering.

To conclude with social capital. This was problematic both from a contemporary mid-nineteenth century perspective as well as from a present-day perspective. Trust in political institutions had to be regained and 'civil society' to be empowered. The sustainability monitor neglects one theme that is an important component of social capital: namely social inequality. We have shown that social inequality has a major impact on popular welfare and the prevalence of poverty. Around 1850 contemporaries did next to nothing about inequality in terms of consumptive expenditures. From a present-day perspective it was imperative to renew social relationships in addition to relationships concerned with economic and human capital.

6.3.3   Well-being 'Elsewhere'

This important dimension – the transboundary effect – is underrepresented both in the sustainability monitor and in historical research. The monitor for 1850 provides just one indicator: the import of goods from abroad. Subsequent investigation should have systematically inventoried the effects on popular welfare in the affected countries. Alas, we shall have to content ourselves with an impression.

Imports consisted chiefly of four types of goods: grain from the Baltic, wood from the Baltic and the Rhinelands, coal from England and colonial goods from the Dutch East Indies. It is not known what effects the grain trade with the Netherlands had on the exporting countries. It is possible that for countries like Finland and Estonia the trade may in some periods have contributed to serious shortages and famine. This has not yet been looked into. It is known that the wood trade with the Netherlands contributed to the decline of tree populations in the Black Forest. The situation in other countries is unknown. It is unclear whether the 'scientific forestry' that was then emerging had achieved new equilibria in the ecosystems. Of the English coal mines it is known that by present-day standards they were notoriously unsafe and that every year they claimed hundreds of victims. The cultivation system in the Dutch East Indies was unacceptable by today's standards because of forced labour, the impositions on the population and the monopoly of the Netherlands Trading Company (NHM). But from today's perspective this can also be relativised. Pressure on foreign natural capital in terms of quantities of imported raw materials by weight was in fact quite modest by today's standards. Moreover the cultivation system initially bestowed various benefits on Java like a money economy, a kind of property register and a certain increase in welfare.[22]

But from the perspective of those days all these concerns were not at issue. Foreign trade was not problematised in the Netherlands in terms of popular welfare elsewhere or excessive demands made on foreign natural capital.

The evaluation of transboundary effects should also be approached from another angle. As we have seen, exports abroad had negative effects on the situation in the Netherlands. Exports of butter and cheese were an example. They were responsible for an impoverishment of the Dutch diet, particularly among the poor.

To summarize, on the basis of the monitor we can distinguish three important sustainability issues from both contemporary and present-day perspectives: material welfare (poverty), the institutional environment (political instability) and social capital (marginal trust in political institutions). There is, however, the caveat that the two temporal perspectives exhibit big differences in how the issues are interpreted. In addition, from a present-day perspective the situation in 1850 is problematic in yet other ways: the personal characteristics (among other things the poor health, diets and housing of a significant part of the population), economic capital (a dearth of innovations), human capital (a lack of high-quality labour power) and the transboundary effects (among others the unethical cultivation system in the colonies). The monitor ignores an important sustainability problem, namely the water management situation. Both from a contemporary and present-day perspective the Dutch delta was vulnerable.

What shifts does the sustainability monitor show for the second half of the nineteenth century?

Literature

Aerts, R. (2013). 'Het ingetogen vaderland: Huiselijkheid, maatschappelijke orde en publieke ruimte'. In I. de Haan, P. den Hoed en H. te Velde (Eds.), *Een nieuwe staat: Het begin van het Koninkrijk der Nederlanden* (pp. 251–273). Amsterdam: Bert Bakker.

Blussé, J.L. (1984) 'Labour takes root: Mobilization and immobilization of Javanese rural society under the cultivation system'. *Itinerario*, 8(1), 77–117.

Blussé, J.L. (2013). 'Koning Willem I en de schepping van de koloniale staat'. In I. de Haan, P. den Hoed and H. te Velde (Eds.), *Een nieuwe staat: Het begin van het Koninkrijk der Nederlanden* (pp. 145–171). Amsterdam: Bert Bakker.

---

[22] L Blussé, 'Koning Willem I en de schepping van de koloniale staat', in: I. de Haan, P den Hoed and H. te Velde, *Een nieuwe staat: Het begin van het Koninkrijk der Nederlanden* (Amsterdam 2013), 167–170; L. Blussé, 'Labour takes root: Mobilization and immobilization of Javanese rural society under the cultivation system', *Itinerario* 8–1(1984), 77–117.

Hondius, E. (2013). 'Het slavernijverleden achter de Hollandse horizon'. In I. de Haan, P. den Hoed and H. te Velde (Eds.), *Een nieuwe staat: Het begin van het Koninkrijk der Nederlanden* (pp. 183–187). Amsterdam: Bert Bakker.

Jansen, S. (2008). *Het pauperparadijs: Een familiegeschiedenis*. Amsterdam: Balans.

Koch, J. (2013). *Koning Willem I, 1772–1843*. Amsterdam: Bert Bakker.

Lintsen, H.W. (2005). Een revolutie in kennis'. In H.W. Lintsen et al. (Ed.), *Made in Holland: Een techniekgeschiedenis van Nederland [1800–2000]* (pp. 293–314). Zutphen: Walburg Pers.

Maas, A. (2010). 'Civil Scientists: Dutch Scientists between 1750 and 1875'. *History of Science XIVII*, 48(1), 75–103.

Mokyr, J. (2000). *The gifts of Athena: Historical origins of the knowledge economy*. Princeton: Princeton University Press.

Oostindie, G. (Ed.) (2008). *Dutch colonialism, migration and cultural heritage*. Leiden: KITLV.

Rooy, P. de (2013). 'De armen hebt gij altijd met u: Armenzorg en onderwijs'. In I. de Haan, P. den Hoed and H. te Velde (Eds.), *Een nieuwe staat: Het begin van het Koninkrijk der Nederlanden* (pp. 221–222). Amsterdam: Bert Bakker.

Rooy, P. de (2014). *Ons stipje op de waereldkaart: De politieke cultuur van modern Nederland*. Amsterdam: Wereldbibliotheek.

Swaan, A. de (2004). *Zorg en de staat: Welzijn, onderwijs en gezondheidszorg in Europa en de Verenigde Staten in de nieuwe tijd* (6th edition). Amsterdam: Bert Bakker.

Theunissen, B. (2000). *'Nut en nog eens nut': Wetenschapsbeelden van Nederlandse natuuronderzoekers, 1800–1900*. Hilversum: Verloren B.V.

Verbong, G. (1993). 'Techniek, beroep en praktijk'. In H.W. Lintsen et al. (Eds.), *Geschiedenis van de techniek in Nederland. De wording van een moderne samenleving 1800–1890* (Vol. V, pp. 11–19). Zutphen: Walburg.

Zanden, J.L. van and A. van Riel (2000). *Nederland 1780–1914: Staat, instituties en economische ontwikkeling*. Amsterdam: Balans.

Part I: The Great Transformation 1850–1910

Chapter 7
The Point of Departure Around 1850:
The Turn of the Tide

Harry Lintsen

Contents

Abstract Fundamental changes in politics, economy, technology and 'civil society' took place in the Netherlands in the second half of the nineteenth century. The eminent politician Thorbecke guided the nation through the constitutional reforms around 1850. A new constitution put an end to the power of the king and shifted political power to parliament. The Dutch economy modernised thanks to the liberalisation of trade, an entrepreneurial spirit and other new economic conditions. It was moreover embedded in a new culture that regarded technological innovations almost by definition as social progress. A dynamic 'civil society' was populated by emergent professionals, including engineers and hygienists.

In this chapter, with the help of the well-being monitor, we explore the changes in quality of life 'here and now,' 'later,' and 'elsewhere' for the period 1850–1910. Extreme poverty began to decline significantly, while the burden on natural capital and the natural environment increased. The question is whether this increased burden was problematic and whether it was in fact problematized. We subsequently focus on natural capital, that in this study is seen as the basis of well-being. We make an inventory of some of the important shifts in the production of raw materials and the derivative material flows in the period between 1850 and 1910. This overview isolates the main themes that are worked out in greater detail in the following chapters.

This chapter is written by Harry Lintsen with contributions by Fred Lambert, Jan-Pieter Smits and Martijn Anthonissen.

H. Lintsen et al., *Well-being, Sustainability and Social Development*,
https://doi.org/10.1007/978-3-319-76696-6_7

Keywords Constitution · Liberalisation · Hygienists · Engineers · Monitor · Well-being · Natural capital

7.1   Interpreter of Freedom

'While the notables regale themselves with partridges, snipes, hares, ducks etc. etc. that are all non-taxed…', according to the newspaper *Interpreter of Freedom* in 1840, '…the working man, as a consumer of meat and bacon, has to pay heavy taxes…bread is also heavily taxed due to the law on milling and this, as is the plan, will also be increased by ten percent…'.[1]

The chief editor of the paper was Eillert Meeter of Groningen.[2] As son of a barber he belonged to the petit bourgeoisie. He had attempted to make a career in the army. His failure in that endeavour he ascribed to his lowly origins. 'The hierarchical chain of society, which has to be climbed by everyone in normal times, is a hindrance to me.' His radical pronouncements, in which he took the side of day-labourers and workers, regularly got him into trouble. In 1841 the High Court condemned him to 4 years in prison and he fled to Paris. This meant the end of the *Interpreter of Freedom*.

Articles from the inflammatory paper were regularly reported in the national press. These were restless years in the Netherlands. William I had abdicated the throne and had left for Berlin to be married to his Catholic sweetheart – to the accompaniment of a torrent of scorn from the protestant part of the populace.[3] William II, his son and successor, was beside himself because his silver wedding anniversary was disrupted. The Amsterdam elite was also furious about the financial chaos that William I had left behind. The people were angry and accused the former king of enriching himself and leading a dissolute life with his lady friend. The common man threatened to become the victim of tax increases on food and other daily products that were already heavily taxed and that kept the cost of living high.

The state was all but bankrupt. Successive budget cutbacks by governments had brought little relief. Finally in 1843 the state finances were brought to heel thanks to interventions by the Minister of Finance, Floris van Hall. As a member of Amsterdam's elite he was able to pressure that city's financial and banking establishment into tendering a 'voluntary' loan under favourable conditions. This significantly reduced the interest payments on the national debt.

---

[1] Tolk der Vrijheid', *Arnhemsche Courant*, 28-11-840.

[2] M.J.F. Robijns, 'Meter, Eillert', *Biografisch Woordenboek van het Socialisme en de Arbeidersbeweging in Nederland* 3(1988), 132–135. Latest edition: 26-08-2002.
    http://hdl.handle.net/10622/40FE3915-C651-4BEC-95EF-9E8C802406D4 (externe link)

[3] J. Koch, *Koning Willem I, 1772–1843* (Amsterdam 2013), 548.

Nonetheless, this did not suffice to quell societal unrest. Quite the contrary, dissatisfaction increased throughout the 1840s with the potato famine and rising food prices. Political unrest in foreign countries also played into the hand of the political opposition. The time was ripe for radical reforms. The constitutional order would be revised, the tax system revamped and economic life would be regulated. With this, issues of well-being acquired an entirely different context.

7.2   The Reforms[4]

7.2.1   Thorbecke

In 1848 King William II bowed to popular pressure and appointed a commission to propose a new constitution. The commission commenced its labours on March 17th, under the chairmanship of the lawyer Thorbecke. Ten days later a proposal was ready. Five months after that, parliament saw fit to ratify the new constitution. This provided for direct elections, ministerial responsibility and more parliamentary powers. It was in any case enough to quell disquiet in the country. The political elite, that up to then had turned a pragmatic blind eye to the radical opposition, now abandoned the populist leaders and suppressed an uprising in Amsterdam.[5]

Thorbecke would also preside over the first cabinet to be formed under the new constitution. As Minister of Internal Affairs he successfully shaped a new political landscape. The Provincial Law of 1850 put an end to the ancient estate-based representation in the Provincial Estates. The Law on Municipalities of 1851 put an end to the autonomy of cities and towns that up to then had enjoyed numerous rights, among them the right to veto vital infrastructural projects and thereby strongly influence decision-making by the provinces and the central state. In the newly established framework, the state delegated societal tasks to the provinces and cities. These acquired incomes via a system of earmarked transfers from the public funds. Municipal taxes were formally abolished in 1865. Up to then they had functioned to protect the local markets. Their abolition meant the end of impediments to free market exchange among the cities and between the cities and the countryside. This measure also reduced the cost of living in the cities.

A series of national taxes were also eliminated, including excise taxes on the slaughtering of pigs and sheep (1852), excise taxes on milling (1855) and excise

---

[4] This section is based on: J.L. van Zanden and A. van Riel, *Nederland 1780–1914: Staat, instituties en economische ontwikkeling* (Amsterdam 2000), 'Chapter 5. Het liberale offensief, 1840–1870', 209–235.

[5] P. de Rooy, *Ons stipje op de waereldkaart: De politieke cultuur van Nederland in de negentiende en twintigste eeuw* (Amsterdam 2014), 86–89.

taxes on turf and coal (1865). These had been a burden on the day-to-day existence of particularly the poor and the workers. They also inhibited the modernisation of production. As we shall see, the tax system included numerous regulations to prevent fraud. These measures were the source of considerable bureaucratic red tape. They were aimed in part at small businesses and small-scale production and hindered the coming of the factory and mass-production. The tax reforms put an end to this situation.

These years also witnessed a radical liberalisation of international trade. Under British leadership, a long period of international protectionism was ushered out the back door. The elimination and simplification of various import duties and laws in Great Britain led among other things to the liberalisation of shipping and the import of agricultural produce. Germany eased restrictions on the import of cattle and other products. France abolished measures that strongly protected its industry. The Netherlands participated in the international movement, lowered its import duties to an absolute minimum and put an end to transit and shipping duties at almost all the Rhine and IJssel tolls. The complete freedom of Rhine navigation – blocked for decades by the Dutch state, much to the exasperation of Germany – had finally been achieved.

In the 1840s and 1850s government finances had been put in order. National bankruptcy had just been avoided. The interest burden, especially the interest on the national debt, declined as a percentage of state expenditures from more than 60% in 1840 to less than 40% in 1860. This provided an opportunity for the state to lower taxes, assume new duties and initiate new projects. Starting in the 1850s, river floods were tackled with great zeal. Budgets for education increased significantly after the 1860s. Transport infrastructure was stimulated by the Railway Law (1860). A legal arrangement in 1860 determined that the state would construct the New Waterway (to the harbour of Rotterdam) and private initiative the North Sea Canal (to the harbour of Amsterdam).

The stabilisation of the state finances was possible thanks to the colonial profits from the Cultivation System and the activities of the Dutch Trading Company (NHM). At the outset of the 1840s the so-called credit balance (*batig slot*) of the colonial Cultivation System that accrued to the Dutch state had decreased to a record low. This was followed by a period of recovery. During the 1840s, 39% of the state's tax income came from the colonial credit balance. During the 1850s that rose to 53%.[6] The Dutch East Indies, and in particular Java, paid for a large part of the Dutch state budget. The Dutch railway network, for example, was financed in large part by colonial profits.[7] This had considerable adverse effects. The creation of more profits led to increased pressure on the Javanese population and the neglect of rice cultiva-

---

[6] Van Zanden and van Riel, *Nederland 1780–1914*, tabel 5.1, 223.

[7] G. Mom and R. Filarski, *Van transport naar mobiliteit: De transportrevolutie [1800–1900]* (Zutphen 2008), 216.

tion. In the early 1840s it had already contributed to the outbreak of a famine. The failure of the rice crop in 1849 and 1850 again caused serious food shortages on Java.

7.2.2   Looking for a New Political Culture

The political revolution of the 1840s released a lot of social energy. According to the prestigious literary journal *De Gids*, Dutch society had now 'broken for good with that epoch of feebleness and slackness, of lack of ambition…[of a] Netherlands sunken into a deathlike slumber…'[8] *De Gids* was not referring to William I, who with his inexhaustible energy had initiated countless projects, but to the effete political culture that prevailed during his reign. His autocratic rule had done anything but encourage public debate on social issues. Though both public and politicians were now mobilized, it was still unclear how the new energy should be channelled.

For example, while the constitution guaranteed freedom of religion, the Pope's official declaration of the Netherlands as a church province initiated a shadowy political process.[9] In 1853, under Thorbecke's leadership, the government took a principled stance and allowed the declaration. King William III, who had no constitutional say in this question, nonetheless used the occasion to support the opposition. 200,000 Netherlanders (out of a total of 3 million!) had been prepared to sign a petition against the 'conspiracy of Rome' and had found in the King a willing ear. In response, the government asked, and was granted, permission to step down, but unrest remained. Due to the acerbic debates and political turbulence in this period, popular trust in political institutions was at a low ebb.

This in itself is hardly surprising, because with the new constitution a number of fundamental questions had to be answered. It was true that parliament now occupied a key position in the new democracy and that the parliamentarian was the chosen representative of a district. But who or what did parliament represent? It was in any case elected by only a small part of the population (initially by some 80,000 male citizens); but even then the question remained whether the parliamentarian had a mandate from his district or whether he was a delegate who represented the interests of the district. Did all the nuances of opinion have to be present in parliament? Or were parliamentarians the advocates of interests? Did something like the 'common interest' exist? A derivative question was who represented the poor and how the interests of the workers, of public health and of public housing were to be represented in parliament.

---

[8] 'Staatkundige beschouwingen', *de Gids* 13(1849) II, 487. Cited in: A. van der Woud, *Een nieuwe wereld: Het ontstaan van het moderne Nederland* (Amsterdam 2006), 15.

[9] De Rooy, *Ons stipje op de waereldkaart*, 75–77.

7.2.3   The Promise of Technology

The new political culture provided new chances for young professionals, including engineers, physicians and agronomists. An ambitious generation of engineers, for example, strained at the leash in their impatience to tackle great works in the interest of the Netherlands and popular well-being. 'A newly born nation,' as they described the situation around 1850.[10] The plans for the North Sea Canal, the railway network and the reclamation of the Zuiderzee were all ready and would put the Netherlands on the map again. Plans for river improvements could be carried out forthwith and would guarantee a safe delta. For years the engineers had been forced to wait patiently. The sorry state of the public finances prohibited the execution of large projects. One of the few large projects carried out in this period, the reclamation of the Haarlemmermeer, had not brought civil engineers the recognition they had hoped for.[11] In view of the constitutional reforms, the engineers hoped to acquire a more central social position.

The time also seemed ripe for a greater role for the professional. Technology and science held great promise: '...*everything* has become possible for our descendants...', wrote the Haarlem instrument maker and aficionado of technological innovation, W.M. Logeman in 1854.[12] His enthusiasm was rooted in a process that had the entire western world in its grip.

The first World Exposition held in London in 1851 embodied the belief in progress and exhibited its promise to a public of millions that descended on London from far and near.[13] The main exhibition hall was in itself already an imposing example of technological prowess. It had been built in less than a year from iron and glass and was so transparent that it was justly named the 'Crystal Palace.' It was filled with impressive machines and installations for flour mills, sugar refineries, the machinery industry, breweries and many other branches of industry. The powerful ambiance of mechanical engineering was combined with the more refined radiance of the industrial arts. These exalted crafts exhibited an unbelievable collection of ten thousand luxurious items including cabinets inlaid with mother of pearl, exquisite fabrics, artistic rugs and decorated pianos. Many countries were represented, but it

---

[10] E. Berkers, *Technocraten en bureaucraten: Ontwikkeling van organisatie en personeel van de Rijkswaterstaat, 1848–1930* (Zaltbommel 2002), 37–38.

[11] H.W. Lintsen, R.A. Lombaerts and R. Moerenhout, 'De droogmaking van het Haarlemmermeer: Wind of stoom', in: M.L. ten Horn-van Nispen, H.W. Lintsen and A. J. Veenendaal, *Nederlandse ingenieurs en hun kunstwerken: Tweehonderd jaar civiele techniek* (Zutphen 1994). The government civil engineers employed by the Rijkswaterstaat had played only a marginal role in the decision-making on the reclamation of the Haarlem Lake, were passed over when it came to leading and executing the project and made no contribution to the British design of the steam engines and pumps for the reclamation (at the time the largest steam engines in the world).

[12] Ln [W.M. Logeman], 'Eeuwigdurende beweging', *Praktisch Volks-Almanak* 1(1854), 164–167. Cited in: A. van der Woud, *Een nieuwe wereld*, 15.

[13] M. Bakker, 'De geest van Crystal Palace', in: H.W. Lintsen et al. (Eds.), *Geschiedenis van de techniek in Nederland: De wording van een moderne samenleving 1800–1890* (Zutphen 1993), deel VI, 13–16.

was above all England that could now demonstrate to the world that it was the planet's leading nation.

The spirit of the Crystal Palace also haunted the Netherlands. The country adopted the culture that accompanied the belief in progress. Two rather modest expositions in 1847 and 1849 were followed by a whole series of national and regional expositions. In part they exhibited a broad selection of technological and industrial novelties. Sometimes they were also thematically oriented to, for example, agricultural machines, gas engines, or colonial products. The Palace for Popular Diligence [*Paleis voor Volksvlijt*] an imposing exposition hall built in 1864 in Amsterdam, became the Dutch ode to progress. The expositions also stimulated a culture of congresses. All kinds of organisations seized the opportunity to meet foreign colleagues and launch new initiatives. A Dutch journalist wrote in response to the founding of the International Workers Association during the World Exposition of 1862: 'We not only expect that its civilised culture will penetrate into all corners of the world, but also that this will give rise to useful cosmopolitan institutions, that must improve the material and moral condition of the European countries...'[14]

The new culture also included a variety of periodicals that kept professionals and the public at large informed about developments in the Netherlands and especially in foreign countries.[15] The general public was served by magazines like the *Practical Peoples' Almanac* (1854), *Treasure of Health* (1858), *Contemporary Questions* (1875) and *Nature* (1881). A professional readership had its own periodicals. *Transactions of the Royal Society of Engineers* (1848) were aimed at engineers and other technical experts. *The Economist* (1851) aimed to reach all those interested in domestic and foreign economics. *The Observer* (1866) was directed to 'architects, engineers, manufacturers, contractors and foremen.' *Gas* (1881) was the periodical of the directors of gasworks. The trade journals were the expression of the emergence of a professional mid-field that had begun to organize itself around academic curricula and professional associations.

The modernisation of the Dutch economy that took place in the second half of the nineteenth century, was not only the result of the liberalisation of trade, free enterprise and other economic variables, but most certainly also of a culture that viewed technical innovations almost by definition as social progress.[16] But technology and science promised more than merely modernisation and economic development.[17] In the course of centuries they had professed their ability to abolish poverty, end hunger and bring welfare for the masses.

---

[14] J.W. del Campo a.k.a. Camp, *Verslag der Wereldtentoonstelling te Londen in 1862* ('s Gravenhage 1864), 436. Cited in: Bakker, 'De geest van Crystal Palace', 19.

[15] Van der Woud, *Een nieuwe wereld*, 16.

[16] See also: D. van Lente, *Techniek en ideologie: Opvattingen over de maatschappelijke betekenis van technische vernieuwingen in Nederland, 1850–1920* (Groningen 1988). The minimal resistance to technology in Dutch politics and society is striking.

[17] See, among others: H.W. Lintsen, *Made in Holland: Een techniekgeschiedenis van Nederland [1800–2000]* (Zutphen 2005), 15–17.

Was this a reasonable claim? Using the monitor for well-being we first explore the changes in quality of life for the period 1850–1910. We then change our focus to the natural capital that is viewed as the basis for well-being in this study. We make an inventory of several important shifts in the production of raw materials for the period 1850–1910 and of the derivative material flows. The overview provides the main themes that will be worked out in the rest of this book section.

7.3 Well-being 'Here and Now': Less Extreme Poverty, 1850 Versus 1910

The Dutch population nearly doubled between 1850 and 1910, increasing from 3.1 million to 5,9 million inhabitants. It may be considered a remarkable achievement that despite this growth well-being also increased (Table 7.1). In 1910, on average, every Netherlander had twice as much to spend as in 1850. Nevertheless it was not the case that extreme poverty – at the time the most important issue in well-being – had disappeared. It is estimated that about 6% of the Netherlanders still lived in bitter poverty. As a percentage this represented a significant decline relative to 1850 (21%). In absolute terms it still included about 350,000 fellow countrymen. Extreme poverty must have remained quite visible on the streets.

The decline of extreme poverty did not – as far as we can tell – occur gradually. Graph 7.1 uses estimates to chart the trend between 1850 and 1913. In the mid-nineteenth century poverty had increased again after a period of decline, in consequence of among other things the crisis of the colonial complex, the potato crisis, failed harvests and foreign wars (Chap. 3). But after these crises poverty continued to increase. The tide turned only in the 1860s after which extreme poverty continued to decline until 1890. The initial increase after the crises and the subsequent decline demand an explanation. As we noted earlier, two factors played an important role in extreme poverty: economic growth and income inequality. Economic growth creates an increase in consumptive expenditures and less inequality a more equitable distribution of the latter among the population. Both factors will demand our attention in the following chapters.

On the basis of various indicators we can conclude that with increasing economic welfare the situation of the Dutch in the areas of health, nutrition, hygiene and education also improved. In one respect the personal situation had worsened, namely housing. At the outset of the twentieth century, many Netherlanders – estimates suggest some 60% – lived under poor conditions and occupied dwellings that by today's standards were too small. After 1850, public housing would become a prominent item in the debate on well-being.

From today's perspective the indicators for 1850 exhibit a negative trend relative to those for 1910 in regard to the natural environment and natural capital: a lower value for MSA (the indicator of biodiversity), higher values for SO_2 and greenhouse gas emissions (the indicators for air quality) and a higher consumption of energy

Table 7.1 Dashboard well-being 'here and now', 1850 versus 1910

Theme	Indicator	Unit	±1850	±1910	Corresponding CBS methods
Population	Size	million inhabitants	3.1	5.9	
Material welfare and well-being					
Consumption, income	Consumptive expenditures per capita/constant prices	index (1850=100)	100	200	⇧
	General income inequality	Gini coefficient 0–1	0.48	0.47	⇔
	Gendered income inequality	% difference in hourly wages M/W	?	?	?
Subjective well-being	Satisfaction with life	score 0–10	?	?	?
Personal characteristics					
Health	Life expectancy	years	37	55	⇧
Nutrition	Height	cm	165	173	⇧
Housing	Housing quality	% slums	30 á 50	60	⬇
	Public water supply	m³/capita	0	19	⇧
Physical safety	Murder victims	number per 100.000 inhabitants	0.8	0.4	⇧
Labour	Unemployment	% workforce	6.4	2.0	⇧
Education	Educational level	years	3	5.8	⇧
Free time	Free time	hours / week	?	?	
Natural environment					
Biodiversity	MSA	% original biodiversity	73	54	⬇
Air quality	SO_2	kg SO_2/ capita	1.3	4.6	⬇
	Greenhouse gas emissions	ton CO_2/capita	1.2	3.8	⬇
Water quality	Public water supply	m³/capita	0	19	⇧
Institutional context					
Trust	Generalised trust	% population with adequate trust	?	?	?
Political institutions	Democracy	democracy-index 0–100	0.3	9.5	⇧

Legend

⇧	Positive development
⬇	Negative development
⇔	Not positive/not negative
?	Unknown or irrelevant

Source: See note 23 of Chap. 2

Graph 7.1 Percentage of
the population living under
the line of extreme poverty,
1850–1913
Source: Appendix 4.1

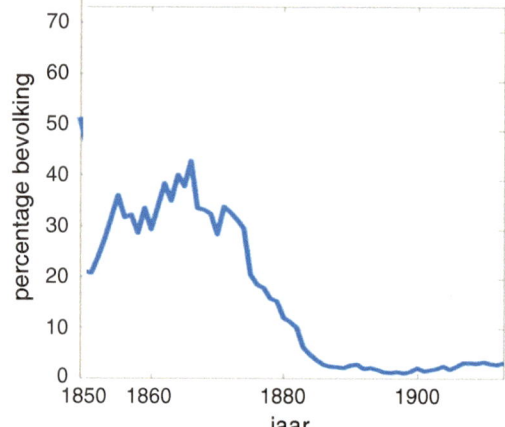

and raw materials (the indicators for the exploitation of natural capital) (Tables 7.1 and 7.2). The monitor here shows that trade-offs are inherent to changes in well-being. The increase in well-being here was based on more forceful exploitation of natural capital. In 1913 the Netherlands had access to almost four times more raw materials than in 1850, partly from domestic and partly from foreign sources. This enabled the Dutch to feed themselves better, to produce more, to clothe themselves better, to increase trade, achieve better water management etc. This was accompanied by a decrease in biodiversity and in air quality. The question is to what extent this is problematic. How must the 'gains' and 'losses' be weighed according to present-day norms? But the contemporary perspective is also of importance: what did contemporaries think of the trade-offs that took place? These questions guide the research informing this book section.

7.4   Well-being 'Later': An Economy Under Steam, 1850 Versus 1910

The quality of life for future generations depends on the resources left them by preceding generations. Resources are included in the monitor under four types of capital: natural, economic, human and social capitals. Natural capital has a special position. Its exploitation with the aid of the other three capitals is the basis for a given quality of life. Natural capital has already been briefly discussed. The other three capitals – as the monitor shows – are in much better shape in 1910 than they were in 1850.

The stock of economic capital per Netherlander was twice as high in 1910 as it was in 1850, the national debt had shrunk by more than half and measures had been taken to get a modern knowledge infrastructure off the ground. The monitor uses investments in research and development (R&D) as an indicator for the supply of

Table 7.2 Dashboard well-being 'later', 1850 versus 1910

Theme	Indicator	Unit	±1850	±1910	Corresponding CBS methods
Natural Capital					
Energy	Energy consumption	TJ/capita	0.03	0.05	⬇
Non-fossil resources	Gross domestic consumption	ton/capita	2.1	3.8	⬇
Biodiversity	MSA	% original biodiversity	73	54	⬇
Air quality	SO_2 emissions	kg SO_2/capita	1.2	3.8	⬇
	Greenhouse gas emissions	ton CO_2/capita	0	19	⬇
Water quality	Public water supply	m³/capita	1.3	4.6	⬆
Economic capital:					
Physical capital	Economic capital stock/capita	index (1850=100)	100	141	⬆
Financial capital	Gross national debt	% gdp	194	71	⬆
Knowledge	Stock knowledge capital	index (2010=100)	–	<0.5	⬆
Human capital:					
Health	Life expectancy	years	37	55	⬆
Labour	Unemployment	% workforce	6.4	2.0	⬆
Level of education	Schooling	years	3	5.8	⬆
Social Capital:					
Trust	Generalised trust	% population with adequate trust	?	?	⬆
Political institutions	Democracy	democracy index 0–100	0.3	9.5	⬆

Legend

⬆	Positive development
⬇	Negative development
⬌	Not positive/not negative
?	Unknown or irrelevant

Source: See note 23 of Chap. 2

modern knowledge capital. Its extent is unknown. However, in 1850 there were hardly any private or semi-governmental laboratories; around 1910 at least 40 could be counted.[18] According to the monitor, the volume of human capital had also increased in 1910 relative to 1850.

The condition of the Dutch population had improved, educational levels had increased and there was a surfeit of work. In terms of political participation, social capital was more robust in 1910 than in 1850. Trust in political institutions will have to be estimated qualitatively. In our estimation this was not particularly high in

---

[18] J. Hutter, 'Nederlandse laboratoria 1860–1940, een kwantitatief overzicht', *Tijdschrift voor de Geschiedenis der Geneeskunde, Natuurwetenschappen, Wiskunde en Techniek* 9(1986), nr.54, 150–174, graph 1.

1850, but around the turn of the century there was also much political turbulence, as we shall see.

The question here is to what extent investments in the three capitals by successive generations between 1850 and 1910 led to an improvement in the quality of life of each generation itself and to a better point of departure for the next generation. Around 1850 the Netherlands still had, in various respects, a classical economy dominated by agriculture, craft production and by trade in colonial and domestic products. A central question of this section is: How did the modernisation of the economy and the indices of well-being develop in successive decades?

7.5   Well-being 'Elsewhere': Colonial Profit, 1850 Versus 1910

What claims did the Netherlands stake to natural capital elsewhere in the world? To answer this question, the monitor provides only one usable indicator for this period, namely imports of raw materials. This indicator, however, reveals much. In 1913 the Netherlands imported 8 á 9 times more raw materials per capita than in 1850. Of all the raw materials (in kilotons) that were at the disposal of the Netherlands at the beginning of the twentieth century, about half was imported! In 1850 that was only 13%.[19] The Netherlands had become considerably more dependent on foreign raw materials, especially stone (100%, primarily limestone to make cement), coal (88%), gravel 67%) and grain (68%) (Table 7.5). The same dependency held for a series of industrial products like metal products (86%), artificial fertiliser (81%), cokes (64%) and wooden planks (61%). Dependence made the Netherlands vulnerable. Did that worry contemporaries? That was one side of the coin. The other side was the influence of these imports on well-being elsewhere.

In this connection the relationship to the colonies once again demands our attention. Above we noted the dependence of the state finances around 1850 on the colonial credit balance. We also pointed out the effects on the lives of the Javanese population. In the same period the Cultivation System came under increasingly heavy fire. In 1860 the writer Multatuli (pseudonym for Eduard Douwes Dekker) would publish the *Max Havelaar*, an indictment of the Cultivation System, forced labour and the corrupt government in the Dutch East Indies. To what extent did this kind of opposition lead to reforms? (Table 7.3)

---

[19] The total weight of available raw materials in 1850 (bio, mineral and fossil) was 9700 kton, of which 1300 kton was imported. For 1913 these figures were respectively 40.000 kton and 20.400 kton (see appendix 2.1).

Table 7.3 Dashboard well-being 'elsewhere' 1850 versus 1910

Theme	Indicator	Unit	±1850	±1910	Corresponding CBS methods
Material Welfare					
Consumption, income	Development aid	% gdp	–	–	?
Natural capital					
Natural capital	Import of raw materials	ton/capita	0.4	3.4	⬇

Legend

⬆	Positive development
⬇	Negative development
⬌	Not positive/not negative
?	Unknown or irrelevant

Source: See note 23 of Chap. 2

7.6   Natural Capital and Material Flows, 1850 Versus 1910

Natural capital is the basis of the quality of life, that is the presupposition of this study. Three varieties of natural capital can be distinguished: bio-raw materials, mineral subsoil resources and fossil subsoil resources. These are the origin of three material flows, that can be roughly denoted as agriculture and foods, construction and building materials and energy. Along these lines, the link with the most important problems of quality of life in the second half of the nineteenth century can be directly laid. The material flow 'agriculture and foods' was decisive for the food situation of the Dutch population: famine, malnutrition, food quality, food distribution and food security. The material flow 'construction and building materials' was decisive for public health (the construction of public hygienic facilities), public housing (the construction of dwellings for the poor and workers) and the maintenance of a safe delta (the struggle against the sea, inner water and the rivers). The material flow 'energy' was decisive for providing heat to the households of the poor and workers in preparing meals and fighting the cold. In addition, the same time the three material flows were part of an economic system that generated surplus value and economic growth that enabled the achievement of a higher level of well-being, in particular for the poor and the workers. At the same time natural capital is part of a process of trade-offs. From a present-day perspective, more intensive exploitation means increasing depletion of natural resources.

What changes did natural capital and the material flows in the period 1850–1910 undergo? We can identify four. First of all, we noted above that at the outset of the twentieth century the Netherlands commanded significantly more raw materials than in 1850, about four times as much. This came down to twice as much per capita

Table 7.4 Raw materials in the Netherlands, 1850 versus 1913 in kilotons

	1850	1913	Ratio 1850:1913
Bio raw materials:			
Gross available	5260 kton	14,740 kton	1:2.8
Bio/capita	1.7 ton/capita	2.4 ton/capita	1:1.4
% import	11%	22%	1:2.0
% export	6%	14%	1:2.3
Mineral subsoil resources:			
Gross available	1350 kton	8040 kton	1:6.0
Mineral/capita	0.45 ton/capita	1.3 ton/capita	1:2.9
% import	11%	33%	1:3.0
% export	2%	8%	1:4.0
Fossil subsoil resources:			
Gross available	3060 kton	17,430 kton	1:5.7
Fossil/capita	1.0 ton/capita	2.8 ton/capita	1:2.8
% import	18%	81%	1:4.5
% export	1%	32%	1:32
Total raw materials:			
Gross available	9670 kton	40,210 kton	1:4.2
Raw materials/capita	3.1 ton/capita	6.5 ton/capita	1:2.1
% import	13%	50%	1:3.8
% export	4%	21%	1:5.3

Remark: Gross available = domestic production + imports
Source: F. Lambert, *Massastromen in Nederland. In de jaren 1850, 1913, 1970, 2010* (researchrapport Technische Universiteit Eindhoven, oktober 2016).

(Table 7.4). Mineral and fossil subsoil resources in particular were much more intensively exploited, their use increasing by a factor of five or six. In the second place, the Netherlands were ever more tightly integrated into an international economy. Imports of raw materials increased by nearly 400% and exports by a factor of five. Imports of processed goods also increased substantially (Tables 7.4 and 7.7). In the third place, industrial processing of raw materials more and more became the norm: In 1850 26% of the bio-raw materials were industrially processed as opposed to 51% in 1913; for mineral raw materials the figures were respectively 54% against 60%. Finally, in terms of volume, entirely different raw materials and products dominated the Dutch economy in 1913 in comparison with 1850 (Tables 7.5 and 7.8).

This last shift illustrates the modernisation of the economy that took place in this period. Coal replaced turf as the most important fossil subsoil resource. Metal products and machinery had joined the top-ten list of manufactured goods. Both developments thus represented the mechanisation of production. Gravel and stone (marl) had conquered a place among the top ten raw materials and concrete products filled a slot in the top ten processed goods. This reflects the rise of concrete technology

Table 7.5 Ten most prominent raw materials (in kton) and the percentage imported, 1850 versus 1913

		1850	1913	1850	1913
		Raw material (kton)	Raw material (kton)	Import (%)	Import (%)
Bio raw materials	Total, of which	5,260	14,740	11	22
	Milk	1,970	3,210	0	0
	Grain	840	2,830	21	68
	Potatoes	800	2,650	2	2
	Lumber	440		66	
	Living cattle	260		3	
	Fish	130		2	
	Sugar beets		1,710		3
Mineral subsoil resources	Total, of which	1,350	8,040	11	33
	Clay	670	3,450	2	2
	Sand	500	1,750	0	9
	Gravel		1,200		67
	Stone		1,060		100
Fossil subsoil resources	Total, of which	3,060	17,430	18	80
	Turf	2,510	1,600	1	6
	Coal	550	15,610	97	88
	Total	**9,670**	**40,210**	**15**	**50**
		8,670	35,090	12	45

Source: F. Lambert, *Massastromen in Nederland. In de jaren 1850, 1913, 1970, 2010* (researchrapport Technische Universiteit Eindhoven, oktober 2016)

and of concrete as new building material. As far as agriculture and foods are concerned, various raw materials (like cattle and fish) and processed products (like beverages and meat) disappeared from the top ten. They were replaced only by sugar beets. This is illustrative of the slow but certain decline of the relative importance of agriculture and foods in the Dutch economy. To be sure, this sector also modernised, as illustrated by the arrival of artificial fertiliser and sugar beets (in support of a new industry, the beetroot sugar industry) in the top ten.

In 1850 the Netherlands faced a major challenge. As the monitor 1850–1910 shows, much remained to be done in the area of well-being. Extreme poverty was widespread and had become the most important issue of well-being in all of Dutch history. The satisfaction of primary needs exhibited serious shortcomings. At the same time the population continued to grow and to dwell in an unsafe delta.

The Netherlands sought a solution for these issues in a new mode of exploitation of natural capital, 'here' in the Netherlands and 'elsewhere.' Changes in material flows point to this. The question is to what extent this brought new problems to the fore. The monitor suggests these might be found in the domain of the natural environment and natural capital. The question in this context is then: in what way are the dynamics of the material flows related to changes in well-being and sustainability. In this analysis we take bio-materials (agriculture and foods), mineral materials (construction and building materials) and fossil materials (energy) as our point of departure (Tables 7.6, 7.7 and 7.8).

Table 7.6 Percentage of raw materials and subsoil resources industrially processed, 1850 versus 1913

	1850	1913
	Industrial processing of raw materials and subsoil resources	Industrial processing of raw materials and subsoil resources
Bio-raw materials	20%	51%
Mineral subsoil resources	54%	60%
Fossil subsoil resources	0%	7%

Remark 1: 'Industry' excludes extraction of mineral and fossil subsoil resources
Remark 2: Mineral resources such as clay and sand directly used in building activities (in for example dikes) are not treated as processed materials
Source: F. Lambert, *Massastromen in Nederland. In de jaren 1850, 1913, 1970, 2010* (researchrapport Technische Universiteit Eindhoven, oktober 2016)

Table 7.7 Import and export of processed products in kton and percentages, 1850 versus 1913

	1850	1913
Total import	1480 kton	26,000 kton
Import processed products	190 kton	6060 kton
Import processed products (% of total)	13%	23%
Total export	570 kton	11,420 kton
Export processed products	170 kton	3110 kton
Export processed products (% of total)	30%	27%

Source: F. Lambert, *Massastromen in Nederland. In de jaren 1850, 1913, 1970, 2010* (researchrapport Technische Universiteit Eindhoven, oktober 2016)

Table 7.8   Ten most prominent processed products (in kilotons) and the percentages imported and exported, 1850 versus 1913

		1850	1913	1850	1913	1850	1913
		Processed products (kton)	Processed products (kton)	Import (%)	Import (%)	Export (%)	Export (%)
Food processing industry	Bread	320	860	0	0	0	0
	Flour	220	750	0	30	0	10
	Beverages	170		5		7	
	Meat	110		2		4	
	Grain chaff	100		0		0	
	Sugar	80		100		75	
	Fodder grain		1260		8		13
Lumber industry	Wood processing	210	1450	6	61	1	0
	Wood waste	150		0		0	
Mineral processing industry	Coarse ceramics (e.g. bricks)	670	3100	2	10	5	0
	Chalk	70		5		4	
	Concrete products		580		0		0
Chemical industry	Cokes		1330		64		50
	Artificial fertiliser		880		81		30
Metals and machine industry	Metal products		950		86		4
	Machine building		610		20		6

Source: F. Lambert, *Massastromen in Nederland. In de jaren 1850, 1913, 1970, 2010* (researchrapport Technische Universiteit Eindhoven, oktober 2016)

Literature

Bakker, M. (1995). 'De geest van Crystal Palace'. In H.W. Lintsen et al. (Eds.), *Geschiedenis van de techniek in Nederland: De wording van een moderne samenleving 1800–1890* (Vol. VI, pp. 13–26). Zutphen: Walburg.

Berkers, E. (2002). *Technocraten en bureaucraten: Ontwikkeling van organisatie en personeel van de Rijkswaterstaat, 1848–1930*. Zaltbommel: Europese Bibliotheek.

Hutter, J. (1986). 'Nederlandse laboratoria, 1860–1940: Een kwantitatief overzicht'. *Tijdschrift voor de Geschiedenis der Geneeskunde, Natuurwetenschappen, Wiskunde en Techniek*, 9(54), 150–174.

Koch, J. (2013). *Koning Willem I, 1772–1843*. Amsterdam: Bert Bakker.

Lente, D. van (1998). *Techniek en ideologie: Opvattingen over de maatschappelijke betekenis van technische vernieuwingen in Nederland, 1850–1920*. Groningen: Wolters.

Lintsen, H.W., R.A. Lombaerts and R. Moerenhout (1994). 'De droogmaking van het Haarlemmermeer: Wind of stoom'. In M.L. ten Horn-van Nispen, H.W. Lintsen and

A.J. Veenendaal (Eds.), *Nederlandse ingenieurs en hun kunstwerken: Tweehonderd jaar civiele techniek* (pp. 31–40). Zutphen: Walburg Pers.

Lintsen, H.W. (2005). *Made in Holland: Een techniekgeschiedenis van Nederland [1800–2000]*. Zutphen: Walburg Pers.

Mom, G. and R. Filarski (2008). *Van transport naar mobiliteit: De mobiliteitsexplosie (1895–2005)*. Zutphen: Walburg Pers.

Robijns, M.J.F. (1988). 'Meter, Eillert'. *Biografisch Woordenboek van het Socialisme en de Arbeidersbeweging in Nederland*, 3, 132–135.

Rooy, P. de (2014). *Ons stipje op de waereldkaart: De politieke cultuur van modern Nederland*. Amsterdam: Wereldbibliotheek.

Woud, A. van der (2006). *Een nieuwe wereld: Het ontstaan van het moderne Nederland*. Amsterdam: Bert Bakker.

Zanden, J.L. van and A. van Riel (2000). *Nederland 1780–1914: Staat, instituties en economische ontwikkeling*. Amsterdam: Balans.

Chapter 8
Agriculture and Nutrition: The Food Revolution

Harry Lintsen

Contents

Abstract The chapter analyses the fundamental changes in the agriculture and foods supply chain between 1850 and 1910 and investigates the consequences for the food supply, in particular for the poor.

Initially, agriculture profited from the liberalisation of international trade. The mixed crop tillage farms in the region of the large rivers and on the sand grounds commercialised and specialised themselves. After 1880, cheap, especially American, grain imports cast Dutch agriculture into a profound crisis. In part because of this crisis a number of innovations were introduced, like the use of artificial fertiliser and the founding of agricultural cooperatives. In addition, common lands were to disappear and large tracts of heathland were to be reclaimed.

The 1860s proved a turning point for the food processing industry. The revival of the domestic market in these years was a key factor. Also, a number of sectors oriented to foreign markets like the potato starch and the sugar beet industry flourished. The steam engine gained ground at the cost of horse-mills and windmills. Moreover new sectors like the margarine and the dairy-processing industry were established.

The modernisation of agriculture and the food processing sector had contributed to the improvement of the food situation. That also resulted from changes in the tax structure, whereby taxes on food were lowered and from increased welfare.

© The Author(s) 2018
H. Lintsen et al., *Well-being, Sustainability and Social Development*,
https://doi.org/10.1007/978-3-319-76696-6_8

Quantitatively there was sufficient food at the beginning of the twentieth century, also for the poor. Potatoes and grain were still the main menu of the majority of the populace. The problem now shifted to food quality.

Keywords Innovation · Modernisation · Agriculture · Artificial fertiliser · Cooperatives · Commons · Agricultural crissis · Foods

8.1   The First Flour and Bread Factory[1]

In 1851 the Amsterdammer Samuel Sarphati left for a study trip to Brussels, Paris and London.[2] As physician to the poor he had seen much poverty and misery at first hand and with this trip he wanted among other things to acquaint himself with the magnitude of poverty in those cities, their public hygiene, education, public health care and their struggles against poverty. In this endeavour he was supported by Thorbecke. Sarphati was an admirer of Thorbecke. Thorbecke, for his part, had great sympathy for the ideas and forcefulness of this kindred spirit. In London, Sarphati visited the first World Exposition and was deeply impressed by all those robust machines and technical innovations from foreign countries. He was particularly interested in the set of machines for milling grain and baking bread: steam engines, grain purifiers, bucket conveyors, air-cooled millstones, kneading machines, baking ovens etc. Upon his return he joined forces with the Delft professor S.A. Bleekrode and the agronomist W.C.H. Staring to found the 'Association for Public Diligence' in order to stimulate national industry and improve popular well-being.

During one of the meetings, Sarphati proposed to set up what would be the first flour and bread factory in the Netherlands. The goal was to bring good and cheap bread to market and hence to improve the nutrition of the poor and the workers. Amsterdam's philanthropists and industrialists found common ground in a new joint stock company. In 1857 the factory with steampower began production. By means of this and other projects, Sarphati mounted the social ladder from his origins in an obscure Portuguese-Jewish family to a place among the social elite of the capi-

---

[1] For this see: H.W. Lintsen et al., *Made in Holland: Een techniekgeschiedenis van Nederland [1800–2000]* (Zutphen 2005), 33–35. The case is based on: H.W. Lintsen and M.S.C. Bakker, 'Meel', in: H.W. Lintsen et al. (eds.), *Geschiedenis van de techniek in Nederland: De wording van een moderne samenleving 1800–1890* (Zutphen 1992), part I, 71–101; J.L van Zanden and A. van Riel, *Nederland 1780–1914: Staat, instituties en economische ontwikkeling* (Amsterdam 2000), 178–185.

[2] L. Hagoort, *Samuel Sarphati: Van Portugese armenarts tot Amsterdamse ondernemer* (Amsterdam 2013), 219–225, 231–232.

tal. Amsterdam still honours Sarphati with a Sarphati Street, a Sarphati Square and a Sarphati Park, the latter containing an imposing bust of the man himself.

Amsterdam's example was quickly followed. Eight years later, in 1865, there were already ten mechanized flour mills and eleven flour and bread factories in the Netherlands. This sector was clearly booming business. Whence this sudden upsurge? Was Sarphati the first Dutchman to make acquaintance with the industrial fabrication of flour and bread? That would seem a serious underestimation of the entrepreneurs of the time. Many travelled around the world and had seen similar factories in France, England, America and various other countries. Grain merchants, above all, had been particularly interested but had made no use of their knowledge. The situation was even more curious in view of the fact that the flour and bread factory, as an innovation, was by then about a century old. The origins of the idea lay in Paris in 1760. How was it possible that Dutch industrialists only picked up on the idea after such a long period?

The opening of the first Dutch flour and bread factory had everything to do with a change in the tax laws: in 1855 the Law on Milling was rescinded. History shows that technical development and legislation are closely related. This is certainly a case in point. The Law on Milling regulated taxation of flour. This was an ancient tax, originally in the form of local ordinances. After the founding of the kingdom of the Netherlands in 1813 these were succeeded by national legislation. Every kilo of rye or wheat flour was taxed at the grain mill and the government saw to it that nobody milled without permission.

Of course millers did everything in their power to avoid paying these taxes. The government was therefore compelled to take strict precautions and to maintain a small army of tax officials. Millers were required to submit a detailed floorplan of their mills, sheds, storage facilities and dwellings. In the course of a possible search these would aid the tax official in locating secret storage spaces and possible smuggling routes for grain on which no taxes had been paid. These tax laws not only affected the peace of mind of the miller, but had consequences for the entire conduct of the business.

Large-scale production was impossible, because the law demanded that every sack of grain in every mill had to be traceable. Grain was in fact milled sack by sack. Flour could not be purified, not re-milled for more delicate kinds of bread, not traded by the miller and so on. In short, the stringent milling law was adapted to traditional small-scale production, frustrating for the miller, but equally so for anyone who wanted to start a flour and bread factory. This was openly acknowledged even by the government and parliament.

The law drew criticism for another reason as well: it taxed a staple element of the popular diet. But the tax income was hardly trivial for a national government struggling with big debts and budget deficits. Only after the government succeeded in getting its financial house in order, did it become possible to rescind the law, a decision taken in 1855.

The elimination of the Milling Law in any case affords an explanation for the sudden rise of industrial-scale flour and baking companies: Sarphati had seen and seized the opportunities offered by a liberalising market and had been followed by others in short order. What remains to be explained is the spectacular growth of this sector in the subsequent decade. We will come back to this below.

The flour and bread factory illustrates the transformation of the Netherlands in the 1840s and 50s. The innovation was the initiative of an emergent midfield of active and socially engaged citizens who had great faith in technology. It was also made possible by a government that was re-inventing itself and that wanted to create conditions for the modernisation of economy and technology.

The new societal context would influence the entire supply chain of agriculture and foods. This supply chain belonged to the core of the Dutch economy. Agriculture had just experienced two decades of reasonable prosperity, but saw itself confronted around 1850 with the potato disease and failed harvests. The food processing industry was to a great extent locally organised and based on a craft system of production. Consumption consisted chiefly of potatoes and cereals and was, for a large part of the population, just adequate to remain alive.

Sarphati's innovation was one of the many innovations that transformed the food supply chain between 1850 and 1910. These would fundamentally alter the food situation. What did these changes mean for the nutrition of the population, in particular for the poor and the workers? The example of Sarphati suggests that this would have improved. Is that so? We shall deal separately with the various links in the supply chain (agriculture, the food-processing industry and food consumption).

8.2   The Modernisation of Agriculture

8.2.1   Prosperity, Crisis and Innovations[3]

Between 1840 and 1870 agriculture profited abundantly from the liberalisation of international trade. Dutch agriculture had clear comparative advantages over that of other countries. Grassland farming had long enjoyed a high level of productivity. Nowhere was milk production per cow as high as in the provinces of Holland and Friesland. This manifested itself in a flourishing export trade especially to industrialising England with its almost inexhaustible demand for 'luxury' foods like butter, cheese and meat. In addition, Dutch agriculture produced several other successful export products like the dyestuff madder.

Trade liberalisation caused the crop farming sector in the area of the big rivers and on the sand grounds to change and become commercialized. The price of grain remained under that of dairy products. It became more attractive to increase the amount of livestock and to transform crop fields into pastures. Up to then, cattle

---

[3] See for this subsection: Van Zanden and van Riel, *Nederland 1780–1914*, 248–256; J.L. van Zanden, *De economische ontwikkeling van de Nederlandse landbouw in de negentiende eeuw 1800–1914* (dissertation Wageningen 1985), 246–252; J.L. van Zanden, 'Mest en ploeg', in: H. Lintsen et al. (eds.),*Geschiedenis van de techniek in Nederland. De wording van een moderne samenleving 1800–1890* (Zutphen 1992), part I, 53–69.

husbandry had been part of a circular production process, subservient to the production of manure and grain. Now, crop farming increasingly came to serve cattle husbandry and the production of butter and livestock. Cattle husbandry hence increasingly became part of a linear chain of production and consumption. Cattle fodder was imported in large quantities and imports of grain partly fed to livestock, while the more expensive dairy products were exported. Initially the manure was destined to improve soil fertility, but in the twentieth century the increase in livestock in combination with the introduction of artificial fertiliser would lead to the conundrum of manure surpluses.

Agriculture's international success also had consequences for the domestic market. Agricultural products became more expensive and the domestic demand stagnated. The cost of living in the cities increased. The introduction of flour and bread factories in the 1860s could not reverse this trend. Moreover, not everyone in the countryside profited equally from the agricultural prosperity. Employment increased thanks to the intensification of agricultural production, but it consisted above all of an increase in the number of hired hands, maids, day-labourers and the increasing recruitment of the families of smallholders.[4] These workers were barely able to pick the fruits of the increased prosperity and lived around or under the poverty line. The large farmers profited first and foremost. Income inequality in the countryside increased. It was one of the causes of increasing poverty in the Netherlands in the 1850s and 1860s.

The large farmers invested the accumulated wealth partly in innovations. On the clay-grounds the Arend-plough was introduced, a new kind of plough that reduced the necessary traction and thus economised on horses. In a number of polders steam-powered pumping replaced wind-powered pumping, a change that improved the manageability of groundwater levels. Other mechanical innovations, originating chiefly in Great Britain and the United States, included sowing and mowing machines, hay bailers and mechanical churns. They spread slowly, but did contribute to the increase of production per labourer and per hectare.

Around 1880 the agricultural prosperity came to an abrupt end only to terminate in a crisis that would last about two decades. It started with crop farming. Cheap cereals flooded the Netherlands and Europe. The United States had opened up vast tracts of agricultural land in its interior on which it cultivated wheat and corn. The costs of production were low because the land was in fact being over-exploited. Railways and steamships assured competitive transport. The railways connected the agricultural regions with the coast and steamships transported the grain to Europe. Similar developments took place in Russia and Argentina.

Dairy farming also ran into trouble. Dutch butter increasingly faced international competition from Denmark, France and Germany as did cheese from American suppliers. Moreover, butter began to be replaced by a cheaper substitute, margarine. Thanks to improvements in refrigeration technologies, trade in meat was able to replace trade in living cattle. The Netherlands played only a modest role in this international market.

---

[4] Van Zanden, *De economische ontwikkeling van de Nederlandse landbouw*, 331–332. Van Zanden refers to the increasing proletarianisation of the countryside and the increasing economic distance between farmer and labourer.

The crisis also had its positive sides. Foodstuffs became cheaper. Domestic demand increased. The cost of living declined. These developments contributed to the decline of poverty in the Netherlands. That said, this is not a sufficient explanation, because the decline had already set in at the end of the 1860s and not in the 1880s.

Another effect of the crisis was to accelerate the modernisation of agriculture. We will illustrate three aspects of this modernisation: the use of artificial fertiliser because it played a crucial role in this period – but also in the twentieth century – in the development of agriculture; the enclosure of the 'commons' because as a centuries-old corporative institution it was of essential importance for the management of the communal grounds and therefore of a large part of the Dutch landscape; and the emergence of the cooperative movement, because this was essential for the well-being of the small farmer and the smallholder.

8.2.2   Artificial Fertiliser and the End of the Closed Chain

At the beginning of the nineteenth century, foreign chemists and agronomists had discovered that nitrogen (N), potassium (K) and phosphor (P) were essential elements for plant growth. In 1840, Justus von Liebig, the 'father' of the artificial fertiliser industry, had summarized this knowledge, together with the results of his own investigations, in his book *Agricultur-Chemie*.[5] His insights rapidly diffused in agricultural circles throughout Europe and stimulated dedicated efforts to develop fertilisers that would enable plants to absorb these elements with ease.

Up to then, Dutch farmers used manure from stables, if possible, augmented with urban waste, urban night soil (human and animal faeces) and waste materials like slaughterhouse waste, rotten fish and ground bones. In the 1840s this was supplemented by a lively trade in Guano, dried manure from sea-birds that was especially rich in nitrogen.[6] After that things moved quickly in regard to the application of new fertilisers. There were various options.[7] 'Super Phosphate' contained phosphates and nitrogen compounds and was based on treating bone with sulphuric acid. It was imported chiefly from Belgium, England and Germany. Chili Saltpetre was a South American raw material composed among other things of the nitrogen compound sodium nitrate. Thomas slag powder, imported from Germany and Belgium, was a phosphate-rich artificial fertiliser made from finely ground blast furnace slag derived from iron ore and raw iron with a high phosphate content. Potassium Salt was mined in Germany. The fertiliser ammonium nitrate was a component of ammonium sulphate, derived from the wastewater of municipal gasworks.

For a long time, the Netherlands was highly dependent on foreign suppliers for its artificial fertiliser. The domestic fertiliser industry consisted in part of suppliers who mixed fertilisers produced elsewhere. It also included factories that produced

[5] E. Homburg, *Groeien door kunstmest. DSM Agro 1929–2004* (Hilversum 2004), 25–27.

[6] Van Zanden, *De economische ontwikkeling van de Nederlandse landbouw*, 252–254.

[7] Homburg, *Groeien door kunstmest*, 25–44, passim.

fertilisers using foreign raw materials, for example the super phosphate factories. Finally there were also small enterprises that fabricated fertilisers from domestic raw materials. Gasworks, for example, processed ammonia water into ammonium sulphate. Beet sugar factories used their earth foam and potassium to produce potassium salts. At the outset of the twentieth century, there were six super phosphate factories in the Netherlands that were able to produce at large scales against internationally competitive prices. In addition, there were another ten to twenty small miscellaneous fertiliser factories.[8]

Initially, the use of artificial fertiliser spread slowly in the Netherlands. After 1880 things accelerated. Around 1900 lead users applied between 100 and 170 kg per ha, but the Dutch average was much lower. The Belgians were the most energetic fertilisers, followed by Germany and the Netherlands. In the 1920s the Dutch would seize the lead and become by far the biggest users of artificial fertilisers (see Table 13.2).

At the outset, four factors played a role in the adoption of artificial fertilisers: the price of the fertilisers, the farmers' ability to bear the costs, the knowledge he had at his disposal and the type of agricultural zone.[9] Artificial fertilisers were expensive and it was long uncertain what the profit was. The pioneers were above all the large wealthy farmers able to bear greater risks and embedded in networks that gave them ready access to information. Some of them even performed elaborate experiments on test fields. The lead users also tended to be located in the fen-colonies in Groningen and Drenthe. These regions had a tradition of employing alternative fertilisers, particularly urban wastes. Regions like eastern Zeeuws Vlaanderen and the island of Goeree-Overflakkee were big users of urban waste. Farmers there cultivated sugar beets, a crop that required intensive fertilisation.

The sand grounds would eventually follow, even though there, because of the nature and the intensive cultivation of the soil, the issue of soil fertility was extremely urgent. In these regions, the declining cost of artificial fertilisers was an important incentive to adoption. The founding of cooperative purchasing associations was a contributing factor. Smallholders and dirt farmers first waited for the market to recover so that they could free up the necessary funds. They were further supported by so-called 'walking teachers' that educated them about the use of fertilisers. These teachers were part of a knowledge infrastructure that the government had been putting together since the onset of the agricultural crisis. Agricultural education was modernised. Every province was provided with teachers of agronomy that organised winter courses and provided farmers with information. Moreover, the government set up three additional experimental stations, in addition to the existing one at Wageningen. Among other things these institutes monitored the quality of fertilisers and experimented with them.

Artificial fertilisers did more than just increase the yield per hectare. Their use also meant the end of a production cycle, that is to say, a largely closed cycle of production and consumption. Agricultural products were formerly produced with

---

[8] Homburg, *Groeien door kunstmest*, 35.
[9] Van Zanden, *De economische ontwikkeling van de Nederlandse landbouw*, 255–262.

organic fertilisers derived from animals and humans, who after consumption contributed their organic wastes, or faeces, to agricultural production. Sunlight was the most important source of exogenous energy feeding the cycle. The adoption of artificial fertiliser stimulated the development of a linear chain, in which mineral and fossil raw materials – generally imported from abroad – were the beginning and in which the organic metabolic products produced at the end were only organic waste.

8.2.3   The Landscape and the End of the 'Common' Lands[10]

Another theme that was discussed throughout much of the nineteenth century in agricultural circles was the management of the 'commons' or the communal lands. At the outset of the nineteenth century, associations of owners managed at least a third of the national territory. They were known variously as commons, *marken, holtelingen* and *buurschappen*. They consisted mostly of heaths and moors, but also of lakes, swamps, meadows and woods. Most communal lands were on the sand grounds, but some also in the regions of river-clay and the high moors.

There were differences of opinion about the role of the common lands. Was the land more optimally exploited under collective or individual property rights? Proponents pointed to the functions fulfilled by the common lands. They served as a source of turf and firewood and to graze sheep, cows and pigs. Sometimes parcels of common land were cultivated. The common lands also provided heath sods for the production of manure compost. Agriculture could not exist without the common lands. Smallholders and dirt farmers with a small plot of land were extremely dependent on the commons.

Those pursuing the modernisation of agriculture spoke of a regressive remnant of the past that was maintained by 'short-sightedness, prejudice, blind egoism and the most egregious ignorance...'[11] According to them this was a far from optimal use of the lands:

'...Everywhere you can ... see the skinny cows of Pharao walking about, separated from the fat ones by a single drainage ditch: the first as representatives of the common pasture, the other as those of particular ownership.'[12]

---

[10] See for this subsection: T. de Moor, *The dilemma of the commoners: Understanding the use of common-pool resources in long-term perspective* (Cambridge 2015); A. van der Woud, *Het lege land: De ruimtelijke orde van Nederland 1798–1848* (Amsterdam 1998), 205–208, 213–237; van Zanden, *De economische ontwikkeling van de Nederlandse landbouw,* 152–165; Van Zanden and van Riel, *Nederland 1780–1914*, 158–162.

[11] Cited in van der Woud, *Het lege land*, 207. Original source: 'De landbouw op de Nederlandsche Zandgronden', *Mededeelingen en Handelingen van de Geldersche Maatschappij van Landbouw* III(1848), 10. Van der Woud endorses this standpoint. He speaks of a 'remarkable relic from the early Middle Ages' (206).

[12] Cited in van der Woud, *Het lege land*, 206. Original source: 'De landbouw op de Nederlandsche Zandgronden', *Mededeelingen en Handelingen van de Geldersche Maatschappij van Landbouw* III(1848), 4–5.

The opponents spoke of the 'wastelands,' lands that were barely profitable or had degenerated into bad lands due to poor management. That system had to be demolished and the *marken* organisations to be dismantled. The common lands had to be sold, so that farmers could get to work efficiently reclaiming them to the benefit of society.

In the twentieth century this theme would reappear on the political agenda in the guise of the 'tragedy of the commons.' The common lands were a metaphor for the problem of over-exploitation. Under a regime of common ownership farmers would above all pursue their own interests. Every farmer would cut as much heath sod as possible for his own fertiliser production or allow as many cows as possible to graze for his own milk production. If all farmers pursued this course, than over-cutting and over-grazing would be the consequence and the utility of the land would accordingly be seriously compromised. Similar phenomena occur at present with the over-fishing of the oceans and the pollution of the atmosphere by exhaust gases.

A recent study shows that the this metaphor is based on incorrect presuppositions. The collective management of the common lands was rather meticulous and provided a certain flexibility, for example if population pressure increased in the region, harvests failed and market demand fluctuated.[13] Three principles underpinned the management: utility, fairness and appropriateness. Utility referred to the degree to which the use of the common lands was sufficient for the owners. Fairness was related to the degree to which the owners participated in the economic use and the management of the lands. The third principle, appropriateness, referred to an ecological optimum: the ratio between the yields and the future exploitation of the lands. The application of the three principles led in many cases to long periods of sustainable management.[14]

Despite this the common lands would disappear in the nineteenth century. A regulation dating from 1810 that envisioned the dissolution of the mark associations, had little effect. An edict by William I in 1837 that attempted to breathe new life into this regulation was effective especially in those marks where large landowners and large farmers were in charge. Where small-scale farmers had more influence, the dismantling occurred later.[15] In the end, a law passed in 1886 would seal the fate of the common lands. In the background other factors played a role in arriving at this decision. Reclamation of the wastelands was the immediate goal, but that demanded the solution of the fertiliser problem, the elaboration of infrastructures, the improvement of water management and adequate legal registration.[16] By the end of the century private parties could enroll the services of the Nederlandse Heidemaatschappij (a professional reclamation firm) to advise them while in 1899 the government set up *Staatsbosbeheer* (State Forest Management) to reclaim state-owned lands.

---

[13] De Moor, *The dilemma of the commoners*, 110–120.

[14] De Moor, *The dilemma of the commoners*, 143–148.

[15] Van Zanden, *De economische ontwikkeling van de Nederlandse landbouw,* 162–163; Van Zanden en Van Riel, *Nederland 1780–1914*, 161.

[16] Van der Woud, *Het lege land*, 229.

Table 8.1 Land-use in the Netherlands, 1833–1913 (×1000 ha)

	1833	1913	Increase/decrease
Farmland	1895	2185	+ 290
'Wastelands'	907	515	− 392
Forest	169	258	+ 89
Roads and Railways	11	53	+ 42
Built up	25	48	+ 23
Total	**3007**	**3059**	**+ 52**

Remark: Due to (minimal) differences in the definition of various kinds of land-use the figures are not completely comparable with each other

Source: J.L. van Zanden and S.W. Verstegen, *Groene geschiedenis van Nederland* (Utrecht 1993), 65, table 4.1

Between 1833 and 1913 a bit less than 400,000 ha of 'wastelands,' some 12% of the Dutch territory, disappeared. That provided (including the reclamation of the Haarlemmermeer and other reclamations) 290,000 ha of extra farmland and 89,000 ha of extra forest. Planting forests was a long-term investment. The forest could be felled after 20 or 30 years and would leave an improved humus-rich soil behind. In many parts of the Netherlands the landscape acquired a rather different aspect. That caused little consternation at the time. Not so in the twentieth century when decisions on the remaining 'wastelands,' still some 515,000 ha, had to be taken (Table 8.1).

8.2.4 Small Farms and the Cooperative Movement

For a while it seemed as if the smallholder and dirt-farmer would become victims of the disappearance of the common lands. But in the end the opposite appeared to be the case. At the outset of the twentieth century the small farm and in particular the family farm was the dominant form of agricultural entrepreneurship. These small farms were moreover apparently in most cases capable of providing a reasonable existence – above the poverty line – for the family.[17] Oddly enough, the elimination of the mark organisation contributed to this. Smallholders and dirt-farmers were namely recompensed for the loss of their use of the common lands with small parcels of 'wasteland' or had become able to lease extra land. However that was still inadequate to build up a decent existence, in particular on the sand grounds in the east and south of the Netherlands,. We focus here on those regions.

An important cause of the relative prosperity of the smallholders was the commercialisation and specialisation of their farms. Thanks to improvements in infra-

---

[17] Exact figures are hard to come by. Van Zanden and others speak of de-proletarianisation of the countryside and an improvement in the situation of the smallholder and dirt-farmer. For the rural village Woensel (near Eindhoven) van den Brink was able to establish that the farmers (and a small upper stratum) were able to avoid the dynamics of structural poverty in this municipality in the period 1850–1920. G. van den Brink, *De grote overgang: Een lokaal onderzoek naar de modernisering van het bestaan. Woensel 1670–1920* (Amsterdam 1996), 110.

structure, farming on the sand grounds had become completely integrated into national and international trade. The emphasis was no longer on self-sufficiency.[18] Flax, hemp, barley and other products that formerly supplied the wants of the farm itself were no longer cultivated; flax and wool were no longer spun and woven for direct use on the farm. Activities that produced less income, for example timber production, were terminated. Emphasis was placed on cattle husbandry. Pastures were extended and improved. Parcels that supplied cattle fodder required less tillage because of cheap foreign fodder and the use of artificial fertiliser. Butter production was delegated to the dairy factory. Work focused on taking care of the milk cows, the pigs and the chickens. The most important raw materials were bought on the market, which is also where almost the entire produce was destined.[19]

An extremely important factor was also the rise of the cooperative movement at the end of the nineteenth century.[20] This institutional innovation commenced with the collective purchase of cattle fodder, artificial fertiliser, seeds and other raw materials. That not only meant considerable savings for smallholders, but also more control over quality. The communal use of machinery also provided economies of scale. Purchasing a steam-powered threshing machine was hardly attractive for a smallholder, but became profitable as a collective possession. The cooperatives expanded their scope of operations to also include the processing and sale of products. Cooperative auctions came into existence to sell and establish prices for products and transport cooperatives were set up to get the produce to export markets. Farmers established cooperative sugar factories to process sugar beets and cooperative dairies to process milk. In this way farmers attempted to avoid obstacles in the markets: the dealer as part of a cartel that demanded excessively high prices for raw materials, the monopolistic factory owner who underpaid for agricultural produce, the local store owner who exploited the farmer with compulsory purchases, etc.

Cooperatives developed later on the sand grounds than in the rest of the Netherlands. After 1890, however, in those regions too their popularity soared. The founding of the farmers' loan banks can be seen as the keystone of the movement. Access to credit was a major obstacle for farmers on the sand grounds, an impediment that hindered the modernisation of their enterprises. The cooperative banks accumulated the savings of the rural population in order to satisfy demands for credit. They turned out to be able to extend credit on favourable conditions, while keeping banking costs low, limiting the risks and offering a reasonable interest on savings.[21] The driving force behind the founding of the banks was the Catholic elite.

---

[18] Van Zanden, *De economische ontwikkeling van de Nederlandse landbouw,* 283–284.

[19] Small enterprises also survived because real wages in Netherlands rose and labour became an ever more scarce commodity. Large farmers encountered difficulties because the increasing wages of hired hands and maids threatened to make their enterprises insolvable. Small enterprises, in contrast, had the possibility of doing without hired labour. Van Zanden, *De economische ontwikkeling van de Nederlandse landbouw,* 335.

[20] See for the cooperative movement in agriculture: Van Zanden, *De economische ontwikkeling van de Nederlandse landbouw,* 273–281; Van Zanden and van Riel, *Nederland 1780–1914,* 365–376.

[21] However, H. Denweth, O. Gelderblom en J. Jonker doubt whether the mode of financing contributed much to the development of the small enterprises (including the smallholder), see: J. Jonker,

They were concerned not only with economic motives, '…to fight usury, support the farmer in his need…,' but also with Christian values, '… to promote thrift, brotherly love, industriousness and temperance…'[22]

At the beginning of the twentieth century Dutch agriculture was characterized by dense networks.[23] More than 50% of the farmers owning more than 1 ha of land was associated with a cooperative. In 1884 the provincial agricultural societies, long the association of choice for large landowners, large farmers and rural elites, had founded the Netherlands Agricultural Committee as a national umbrella organisation. The founding of the Netherlands Farmers' Union in 1896 led to Catholic farmers' unions in every province and to a rapid increase in the level of agricultural organisation. By 1913 the Netherlands Agricultural Committee, that by then had subsumed the Netherlands Farmers' Union, counted 130,000 members, some 65% of all farmers.

At the same time the government was also beginning to intervene in the sector. Up to the agricultural crisis it had remained passive. The 1886 report of the State Commission regarding the condition of agriculture changed all that. The government began investments in the agricultural sector with the creation of a knowledge infrastructure. This would mark the beginning of the well-known OVO-triptych in the twentieth century, the infrastructure for Education, Extension and Research for the agricultural sector.

Farmers' organisations became interwoven with the governmental apparatus, which laid the basis for a powerful agricultural lobby in the twentieth century.

The modernisation of agriculture was by no means a catastrophe for the smallholder and dirt farmer. He certainly had to work hard and employ his entire family to keep the farm going, but at the outset of the twentieth century the modernised sector was able to support the farming family at a level above the poverty line. Favourable economic circumstances also lent a helping hand. The agricultural crisis, meanwhile, had been bested.

What effects can we attribute to the modernisation of the food processing industry, the second link in the food supply chain?

8.3   The Modernisation of the Food Processing Industry

While agriculture flourished in the 1850s, the food processing industry fell on hard times. The cost of living had risen due to the liberalisation of international trade, which caused stagnation of demand in domestic markets.[24] Netherlanders ate less

---

'Welbegrepen Eigenbelang. Ontstaan en Werkwijze van Boerenleenbanken in Noord-Brabant, 1900–1920', *Jaarboek voor de Geschiedenis van Bedrijf en Techniek 5(1988), 188–206*; H. Deneweth, O. Gelderblom, J. Jonker, 'Micro-finance and the Decline of Poverty: Evidence from the Nineteenth-Century Netherlands', *Journal of Economic Development* 39(2014), no 1, 79–110. O. Gelderblom, *Waar hebben we de financiële sector eigenlijk voor nodig?* (Inaugural Lecture Universiteit Utrecht, 2015), TPEdigitaal 9 (2015), no 1, 45–46.

[22] Cited in: Van Zanden and van Riel, *Nederland 1780–1914*, 375.

[23] Van Zanden and van Riel, *Nederland 1780–1914*, 369–371.

[24] Van Zanden and van Riel, *Nederland 1780–1914*, 282.

Table 8.2 Number of power sources in the food processing industry, 1850–1890

	1850	1860	1880	1890
Steam engines	100	270	850	1310
Windmills	2150	2390	2280	1210
Horse mills	1640	1450	780	470
Water mills	290	290	90	60
Total	**4180**	**4400**	**4000**	**3050**

Remark: Figures rounded off to tens
Source: H. Lintsen, 'Een land met stoom', in: H. Lintsen et al. (eds.), *Geschiedenis van de techniek in Nederland. De wording van een moderne samenleving* (Zutphen 1995), part VI, table 7.9, 269–279

meat, resulting in less work for the slaughterhouses. More rye bread was consumed instead of the more expensive wheat bread, a shift that had consequences for bakers, millers and bread and flour factories. Beer brewers had to cope with declining beer consumption. The foods sector was also not very innovative. We can point to the small number of steam engines (270) over against the many windmills (2390) and horse treadmills (1450) in 1860 (Table 8.2). In this respect, the founding of the first bread and flour factory by Sarphati in this period was a ray of light. That was also the case for the first beetroot sugar factory in 1858, the start of an entirely new branch of industry.

The 1860s were a watershed for the food processing industry. The recovery of the domestic market was an important factor. In addition, a number of sectors oriented to the foreign market, like the potato starch and beetroot sugar industries, began to flourish. After this, the modernisation of the food processing sector proceeded apace. The steam engine gained terrain, first at the cost of the horse treadmills and from the 1880s on at the cost of the windmills. In addition, new branches of industry developed, like the margarine and the dairy industries.

An important characteristic of the modernisation of the food processing industry was the application of steam technology and associated modern apparatus like kneading machines, milk centrifuges and refrigerators. In some cases legal strictures had impeded the introduction of new production technologies, as we saw in the case of Sarphati's bread and flour factory. But other factors also played a role. The operating costs of the new process technologies were lower than those of the classic ones. Sarphati, for example, was able to sell his wheat bread in 1865 at an impressive 30% lower price than the bakers.

It did, however, mean that the entrepreneur was forced to realize a larger output. The new mode of production demanded the existence or creation of a sufficiently large market. We can again take Sarphati's factory as an illustration. The Amsterdam market was in principle sufficiently large, but it was dominated by the cartel of windmill operators and bakers that had successfully opposed earlier novelties. For centuries these two groups had been able to ply their trades under a benevolent providence, originally as part of the guild system, later in the form of cartels in protected urban markets. But the old corporatist institute lost the competitive struggle with the steam engine and disappeared from the Amsterdam scene.

The need for a large market applied not only to an industrialist who invested in large steam engines, but more particularly also to the entrepreneur with a small steam engine. This is why small-scale steam-powered enterprises were all but lacking in the countryside until well into the nineteenth century. Here windmills continued to produce for the local market. Their scale of production suited these circumstances. Due to limited transportation infrastructure, there was little threat of competition from outside. With the construction of a tramway network from the 1880s on, the countryside became accessible and the many remaining windmills quickly disappeared.

An entrepreneur could also create a larger market by creating a new product with the new process technology or by transforming a luxury commodity into a mass commodity. Again, Sarphati's factory can serve as an example. The demand for his white bread increased explosively. That was not only because of the low price. Wheat and white bread counted as a luxury, a sign of prosperity. Up to then it had graced the tables of the poor only on high holidays. On weekdays coarse breads and porridges were the common fare. But now finer bread came within the reach of the masses and they embraced the opportunity. White bread became a 'hype' in Amsterdam in the 1860s.[25]

At the end of the nineteenth century the biggest steam boilers could be found in the sugar refineries, the beet sugar factories, the bread and flour factories and the potato flour factories (Table 8.3)[26] Smaller steam installations were found in the beer, margarine and dairy industries. Steam technology was not reserved to new branches of industry or to classical industries producing at a larger scale. The classical small enterprises modernised as well. Many grain mills, hulling mills, roasting houses and distilleries invested in steam apparatus, but of course of limited size. The consequence was that at the end of the nineteenth century far and away the majority of steam engines were to be found in the food processing industry, but with a modest average horsepower (10.6 hp) that lay below the national average of 13 hp (Chap. 10, Table 10.6).

The introduction of steam provides a first impression of modernisation, but innovations in the food processing industry reached further than the steam engine and the machines connected to it.[27] New processes, for example the pasteurisation of milk to increase its shelf life, and new methods, for example the thorough and hygienic

---

[25] In other places too, but not everywhere in the Netherlands. In some regions like the provinces of Drenthe and Overijssel, white bread counted as something for 'fine folk' or for the ill and popular opinion favoured firm dark rye bread.

[26] M.S.C. Bakker, 'Voeding', in: H.W. Lintsen et al. (eds.), *Geschiedenis van de techniek in Nederland: De wording van een moderne samenleving 1800–1890* (Zutphen 1992), volume I, 48–51. See for innovations in the production of flour, butter, margarine, beer and sugar, various chapters in: H.W. Lintsen et al. (eds.), *Geschiedenis van de techniek in Nederland: De wording van een moderne samenleving 1800–1890* (Zutphen 1992), volume I.

[27] M.S.C. Bakker, 'Techniek en voeding in verandering', in: H.W. Lintsen et al. (eds.), *Geschiedenis van de techniek in Nederland: De wording van een moderne samenleving 1800–1890* (Zutphen 1992), volume I, 253–264.

Table 8.3 Most important users of steam in the food processing industry, 1890

	Average size of the steam boilers, rounded off to 100 (m²)	Number of enterprises	Total size of the steam boilers, rounded off to 100 (m²)
Flour factories, bread factories, flour and bread factories	±590	±100	59,000
Beet sugar factories, sugar refineries	±522	±36	18,800
Potato flour factories	100	28	2800
Margarine factories	±52	±50	2600
Beer breweries	±42	±105	4400
Butter factories, cheese factories, dairies	±28	±50	1400
Grain mills	<16	>200	3200
Roasting houses, distilleries	±14	±150	2100
Hulling mills	±8	±150	1200

Source: table based on table 1.4 in: M.S.C. Bakker, 'Voeding in Nederland', in: H. Lintsen et al. (eds.), *Geschiedenis van de techniek in Nederland. De wording van een moderne samenleving* (Zutphen 1992), part I, 49

cleansing of apparatus, were introduced. New instruments, among others the introduction of the thermometer, improved the reliability of production processes.

But the question is whether the nutrition of the lowest classes of the population improved as a result of the emergence of new process industries, large-scale modes of production and countless innovations.

8.4 Food Quantity and Food Quality

It is estimated that around 1900 Netherlanders consumed about one and a half times as much potato and grain as in the middle of the nineteenth century.[28] The modernisation of agriculture certainly was a contributing factor, particularly as regards the cultivation of potatoes. The Netherlands was self-sufficient in this staple and its

---

[28] The inventory of material flows in our investigation (see appendix 2.1) shows that the domestic supply of potatoes in 1913 was 1.4 times that in 1850, and of grain 1.7 times. Bakker, 'Voeding,' Table 1.1, shows that for the period 1897–1901, 1.8 times as many potatoes were consumed as in the period 1852–1856 (including cattle fodder), 3.0 times as much wheat and 1.1 times as much rye (including cattle fodder). M.T. Knibbe paints a somewhat different picture in his 'De hoofdelijke beschikbaarheid van voedsel en de levensstandaard in Nederland, 1807–1913,' *Tijdschrift voor Sociale en Economische Geschiedenis*, 4 (2007), nr. 4, graph 3. There, the *availability of calories* from the consumption of potatoes grosso modo remains the same between 1850 and 1890 and but that from bread cereals (wheat and rye) doubles.

yield per hectare had been greatly increased.[29] It is notable that the contribution of the modernisation of the food processing industry to the improvement of popular nutrition remained limited to the bread and flour factories.[30] In a number of cities these contributed to the decline in bread prices, though lower grain prices in the United States were probably the most important factor. Potatoes and cereals remained the staple foods of the largest part of the population. Sugar was a luxury; butter remained expensive and was first and foremost exported to England. Even margarine, introduced as a cheap substitute for butter, was mainly destined for export. Sugar, butter, margarine and meat did appear more often on the tables of the 'lesser classes', but they remained limited to a few grams per day.[31]

The modernisation of the agricultural and foods sector contributed in a largely indirect fashion to the improvement of the 'popular diet.' It improved the international competitive position of the Dutch economy. At the beginning of the twentieth century, the production and export of foodstuffs were still the most important sectors of the Dutch economy. The food processing industry had the biggest share in the added value of industry.[32] In this way, agriculture and foods made an important contribution to the growth of the gross domestic product and of the consumptive expenditures per Netherlander. Among other things, the decrease in food prices meant that the relative share of foods in household budgets slowly but surely declined. It decreased from about 60% around 1850 to 50% around 1910.[33]

After 1860 and up to the First World War, famines no longer occurred. Hunger stalked the poor from time to time, but ever less frequently and at increasingly smaller scales. Social movements were still active in supplying food to the poor. Amsterdam, for example, opened a People's and Children's Kitchen in 1887, where families could enjoy nutritious noonday meals for the price of 7 cents. The improved food situation was the most important cause of improvements in public health. Average height increased from 165 cm in 1850 to 173 cm in 1910 and the life expectancy of the Dutch from 37 to 55 years. That was an impressive accomplishment.

However the improvement in the food situation was above all a quantitative accomplishment. Food quality was another matter. Around 1900, politicians, physi-

---

[29] Van Zanden, *De economische ontwikkeling van de Nederlandse landbouw,* table 5.7. The yield per hectare in the period 1913–1918 is approximately two times higher than in the period 1854–1862.

[30] Bakker, 'Voeding in Nederland', 50.

[31] Bakker speaks of a few grams per day. Estimates on the basis of our research into material flows (appendix 2.1) gives the following result: sugar consumption per Netherlander per day 16 grams in 1850 and 25 grams in 1913 (by comparison: 47 grams in 2010), butter and cheese 31 grams in 1850 and 30 grams in 1913 (by comparison: 47 grams in 2010), margarine 0 grams in 1850 and 18 grams in 1913 (by comparison: 28 grams in 2010), meat 93 grams in 1850 and 113 grams in 1913 (by comparison: 221 grams in 2010).

[32] Van Zanden and van Riel, *Nederland 1780–1914,* table 8.9. See also tables 6.7 and 8.1.

[33] E. Horlings and J.P. Smits, 'Private consumer expenditure in the Netherlands', *Economic and social history in the Netherlands* 7 (1996), 15–40, graph 3. A.H. van Otterloo refers to a decline of 70% in 1850 to 50% in 1890. A.H. van Otterloo, 'Voeding'. in: J. Schot, H.W. Lintsen, A. Rip and A.A. de la Bruhèze (eds.), *Techniek in Nederland in de Twintigste Eeuw* III (Zutphen 2000), 241.

cians, nutritional experts and citizens engaged in heated debates on this issue.[34] The diets of the poor and workers was too one-sided, exhibited shortages of essential foodstuffs and contained too few calories. Knowledge about healthy nutrition, spoilage, proteins, fats, and carbohydrates was not widely disseminated. Exhibitions about nutrition were organised. Cooking teachers gave demonstrations for bourgeois ladies, servants, female workers and their daughters. These sorts of initiatives would result after the First World War in the adoption of nutrition as part of home economics curricula, food research at universities and research institutes, food counselling by expert bureaus and in 1953 the publication of the standardised food protocol, 'the disk of five.'

A separate question was that of the spoilage and adulteration of food. The sale of for example spoiled meat was not unusual, nor was illicit fiddling with food. Milk could be diluted with dirty water, flour mixed with alum, bread baked with poisonous copper sulphate, and candies doped with red lead to give them a cheerful red colour. This kind of fiddling was perennial, but it increased with the changes in the food supply chains. The scale of food production increased, it was distributed across numerous links and it became more anonymous. Producers attempted to generate trust in their products by introducing brand names that were supposed to represent quality. City governments intervened with inspectors and the establishment of food inspectorates, the first in Rotterdam in 1893. In 1919 the national government ratified the Law on Commodities.[35]

At the start of the twentieth century one important aspect of the issue of 'public nutrition' had been solved. Quantitatively there was enough food, for the poor as well. The problem had now shifted to food quality. Nonetheless, the First World War would confront the Netherlands with another problem, namely that of food security. Was the food supply sufficiently robust to withstand serious shocks such as a war?

Literature

Bakker, M. (1992). 'Techniek en voeding in verandering'. In H.W. Lintsen et al. (Eds.), *Geschiedenis van de techniek in Nederland: De wording van een moderne samenleving 1800–1890* (Vol. I, pp. 253–277). Zutphen: Walburg.

Bakker, M. (1992). 'Voeding in Nederland'. In H.W. Lintsen et al. (Eds.), *Geschiedenis van de techniek in Nederland: De wording van een moderne samenleving 1800–1890* (Vol. I, pp. 39–51). Zutphen: Walburg.

Brink, G. van den (1996). *De grote overgang: Een lokaal onderzoek naar de modernisering van het bestaan Woensel 1670–1920.* Nijmegen: SUN.

Deneweth, H., O. Gelderblom and J. Jonker (2014). 'Micro-finance and the decline of poverty: Evidence from the nineteenth-century Netherlands'. *Journal of Economic Development, 39*(1), 79–110.

---

[34] For this see: van Otterloo, 'Voeding', 240–242; Bakker, 'Techniek en voeding in verandering', 269–274.

[35] H.W. Lintsen (eds.), *Tachtig jaar TNO 1932–2012*, 108.

Gelderblom, O. (2015). 'Waar hebben we de financiële sector eigenlijk voor nodig?' (Inaugural lecture University of Utrecht). *TPEdigitaal*, 9(1), 37–49.

Hagoort, L. (2013). *Samuel Sarhpati: Van Portugese armenarts tot Amsterdamse ondernemer.* Amsterdam: Bas Lubberhuizen.

Homburg, E. (2004). *Groeien door kunstmest: DSM agro, 1929–2004.* Hilversum: Uitgeverij Verloren.

Horlings, E. and J.P. Smits (1995). 'Private consumer expenditure in the Netherlands'. *Economic and Social History in the Netherlands*, 7, 15–40.

Jonker, J. (1988). 'Welbegrepen eigenbelang: Ontstaan en werkwijze van boerenleenbanken in Noord-Brabant, 1900–1920'. *Jaarboek voor de Geschiedenis van Bedrijf en Techniek*, 5, 188–207.

Knibbe, M.T. (2007). 'De hoofdelijke beschikbaarheid van voedsel en de levensstandaard in Nederland, 1807–1913'. *Tijdschrift voor Sociaal-Economische Geschiedenis*, 4, 71–107.

Lintsen, H.W. et al. (1992–1995). *Geschiedenis van de techniek in Nederland: De wording van een moderne samenleving 1800–1890* (Vols. I-VI). Zutphen: Walburg.

Lintsen, H.W. and M. Bakker (1992). 'Meel'. In H.W. Lintsen et al. (Eds.), *Geschiedenis van de techniek in Nederland: De wording van een moderne samenleving 1800–1890* (Vol. I, pp. 71–101). Zutphen: Walburg.

Lintsen, H.W. (2005). *Made in Holland: Een techniekgeschiedenis van Nederland [1800–2000].* Zutphen: Walburg Pers.

Lintsen, H.W. (Eds.) (2013). *Tachtig jaar TNO.* Delft: TNO.

Moor, T. de (2015). *The dilemma of the commoners: Understanding the use of common-pool resources in long-term perspective.* Cambridge: Cambridge University Press.

Otterloo, A.H. (2000). 'Voeding'. In J.W. Schot, H.W. Lintsen, A. Rip and A.A. Albert de la Bruhèze (Eds.), *Techniek in Nederland in de twintigste eeuw, deel 3: Landbouw, voeding* (pp. 235–374). Zutphen: Walburg Pers.

Woud, A. van der (1998). *Het lege land: De ruimtelijke orde van Nederland 1798–1848.* Amsterdam: Contact.

Zanden, J.L. van (1985). *De economische ontwikkeling van de Nederlandse landbouw in de negentiende eeuw, 1800–1914 (dissertation).* Wageningen: Universiteit Wageningen.

Zanden, J.L. (1992). 'Mest en ploeg'. In H.W. Lintsen et al. (Eds.), *Geschiedenis van de techniek in Nederland: De wording van een moderne samenleving 1800–1890* (Vol. I, pp. 53–69). Zutphen: Walburg.

Zanden, J.L. van and Riel, A. van (2000). *Nederland 1780–1914: Staat, instituties en economische ontwikkeling.* Amsterdam: Balans.

Chapter 9
Building Materials and Construction: The Four Building Challenges

Harry Lintsen

Contents

Abstract The chapter analyses the radical changes in the supply chain of building materials and construction between 1850 and 1910. It investigates the consequences for the four building challenges in this period: public hygiene, public housing, the struggle against water and the development of road infrastructure.

The hygienists succeeded in getting the issue of urban pollution due to faeces and other organic waste on the societal agenda. Despite this, the effect of the movement on the health of the poor and workers in this period was still minimal. Realising the hygienist program demanded firm municipal policy and large investments. Moreover there was exuberant debate on the nature of the measures to be taken.

Public housing too became a significant political issue thanks to the efforts of the progressive bourgeoisie and socially conscious entrepreneurs. The most important result was the passing of the Housing Law of 1901, that provided the future framework for the condemnation, dispossession and improvement of dwellings.

In the field of water management, the government public works agency (Rijkswaterstaat), intensified its struggle against regularly recurring river floods. An impressive program of river normalisation was carried out under its leadership.

Successive governments would devote much attention to infrastructure with the aim of improving welfare, reinforcing the position of the maritime harbours and improving the accessibility of remote regions. The construction of a network of canals and railways guaranteed the spatial integration of the Netherlands and had far-reaching consequences for the modernisation of the economy.

Keywords Construction · Hygienists · Public hygiene · Public housing · Housing Law · River improvements · Infrastructure · Railways

© The Author(s) 2018
H. Lintsen et al., *Well-being, Sustainability and Social Development*,
https://doi.org/10.1007/978-3-319-76696-6_9

9.1   The Filthy Hole[1]

In 1864 Jacob van Niftrik, raised in the Ooipolder near Nijmegen, moved to Amsterdam. He went to work there as an engineer in the municipal Public Works Department. The first thing that struck him was how filthy the capital city was. The streets were muddy and the gutters clogged. Garbage accumulated around waste bins. The canals were repositories of waste. During walks taken to acquaint himself with the city he noticed 'in many neighbourhoods how pestilential the emanations from that garbage were.' It smelled in alleyways and side-streets, the canals reeked abominably and those approaching the many cellar dwellings and houses on inner courtyards were greeted by sour-smelling rotten air. In the course of an inspection, the effusion from 'The Hole' between the *Nieuwe Zijds Voorburgwal* and the *Kalverstraat* literally robbed him of consciousness. The neophyte municipal engineer was overwhelmed by the rotten air that drifted out of the dwellings and had to be taken to the broad canal behind the houses to be revived by his colleagues.

Amsterdam was no exception. The other Dutch cities also smelled and were filthy. But there was a difference between the cities in the low Netherlands, where due to the nature of the water the situation was worse, and the cities in the high Netherlands. (see Chap. 4). The issue of the filthy city has two interesting aspects. First, the conditions giving rise to the complaints were as old as the hills. Cities in the past had always been dirty, at least by present-day standards. Contemporaries once in a while complained and when even by contemporary standards things got out of hand, measures were sometime taken. But why did the number of complaints increase in the second half of the nineteenth century? In the second place it is striking that the cities were barely capable of improving the sanitary situation. Despite a permanent stream of alarming publications and chronic debate, for a long time after 1850 little seemed to improve. Was the issue so complicated? Where was the opposition?

From a contemporary perspective public health long remained an unimportant aspect of quality of life, though it certainly became so in the second half of the nineteenth century. It thus became incorporated in the great building challenges in the context of well-being. In our broad definition of well-being, health and human capital are important aspects of the quality of life and sustainable development. A hygienic city is a contributing factor to both. This is connected to another important building challenge, namely public housing. Public health demands sanitary facilities in dwellings. At the time, the dwellings of the poor and workers were not amenable to such facilities. A large part of the population lived in slums, sheds and hovels, that not only failed to satisfy the new hygienic norms, but also violated any number of other new emerging norms.

---

[1] The introduction is based on: H.W. Lintsen, *Made in Holland: Een techniekgeschiedenis van Nederland [1800–2000]* (Zutphen 2005), 55–56; H. Buiter, *Riool, rails en asfalt: 80 jaar straatrumoer in vier Nederlandse steden,* (dissertation Technical University Eindhoven 2005), 153–155; I. Jager, *Hoofdstad in gebreke: Manoeuvreren met publieke werken in Amsterdam, 1851–1901* (Rotterdam 2002).

In earlier chapters we noted two other large building challenges from the perspective of well-being: the struggle against water and the improvement of infrastructure. The Dutch delta was a vulnerable region. The land was threatened from all sides by water: the sea, enclosed waters and the rivers. In the struggle against water the rivers appeared to be the most immanent threat at the time. River floods had afflicted successive generations prior to 1850. In the 1850s and subsequent decades the issue would be joined and dealt with. Whence this sudden vigour and ambition and with what consequences? Improving infrastructure had already been a government priority under King William I. And under liberal domination the government continued to concern itself intensively with infrastructure. Proper infrastructure was seen to be essential to transportation, trade and national welfare. To what did this basic orientation lead?

In this chapter the four building challenges– in relation to the issue of poverty – provide a framework for further analysis.

9.2 Working on a Hygienic City[2]

In the first half of the nineteenth century, various reports had put the issues of public hygiene and health on the agenda, but only in an incidental way. It was the hygienists who after 1850 were the first to bring this issue to the attention of the public and who regularly succeeded in getting the problem onto the political agendas of municipalities and the national state. They were young physicians generally employed at a medical service for the poor, run by a municipality or one of the many churches.[3] They earned little and worked hard. Their work was barely acknowledged. In any case, the status of physicians was not in general very high. Cholera epidemics only made things worse. In the face of this disease, the profession appeared powerless. It is hardly surprising that the hygienists were in search of a new orientation for the profession of physician.[4]

---

[2] This section is based on: E. Houwaart, 'Medische statistiek', in: H.W. Lintsen, et al. (eds.), *Geschiedenis van de techniek in de negentiende eeuw* (Zutphen 1993), volume 3, 19–45; H. van Zon, 'Openbare hygiëne', in: H.W. Lintsen, et al. (eds.), *Geschiedenis van de techniek in de negentiende eeuw* (Zutphen 1993), volume 3, 47–79; E. Houwaart, 'Professionalisering en staatsvorming', in: H.W. Lintsen, et al. (eds.), *Geschiedenis van de techniek in de negentiende eeuw* (Zutphen 1993), volume 3, 81–92; H. Buiter, *Riool, rails en asfalt: 80 jaar straatrumoer in vier Nederlandse steden* (dissertation Technische Universiteit Eindhoven, 2005); E. Houwaart, *De hygiënisten: Artsen, staat en volksgezondheid in Nederland, 1840–1890* (dissertation Groningen 1991); H.W. Lintsen, *Wat is techniek? Een geschiedenis van menselijke secreten en discrete technieken* (Inaugural lecture TUE, 1992); H. van Zon, *Een zeer onfrisse geschiedenis: Studies over niet-industriële vervuiling in Nederland, 1850–1920* (dissertation Groningen 1986).

[3] Houwaart, 'Medische statistiek', 27–30.

[4] Physicians (including the hygienists) wanted not only to define a new practice for the profession of medicine but also to improve its position. With the aid of the Netherlands Society for the Promotion of Medicine (1849), for example, they succeeded in reducing the number of physicians per 100,000 inhabitants from 79 in 1850 to 41 in 1890 and thereby improving their financial position.

They derived inspiration from abroad. Progressive physicians in France, England and Germany saw themselves as the 'advocates of the poor.' It was their task to accurately chart and analyse abuses. They felt compelled to search for correlations among the number of ill, the spread of diseases, anatomical pathologies and the chemical composition of water, soil and air. Public health had to be expressed in figures describing age, sex, birth, death, illness, height, weight, nutrition etc. A scientific, statistical approach would be the basis of a new science, namely the science of public health, and of a new perspective on health, namely the maintenance of public health. Substandard hygienic circumstances were the evil genius of poor public health. The solution for hygienic problems lay in sewer systems, pure drinking water and other new technologies. In addition, it was important that the government, in particular the municipalities, acknowledged public hygiene as a governmental task, for which means and personnel had to be made available.

The constitutional reforms of 1848 gave the hygienists the chance to manifest themselves.[5] Physicians acquired voting rights. They became involved in elections for parliament, the provincial estates and the municipal councils. Some of them entered local and national politics. The hygienists also took steps to set up health boards that began to concern themselves with municipal governance. At the same time the hygienic program of this emancipated group of physicians began to enjoy increasing support. The movement became broader and began to include citizens, politicians, entrepreneurs and engineers. During and after the 1870s the issue became part of a more encompassing problem, the so-called 'social question,' that also included poverty, public housing and the question of labour.

That said, the effect of the hygienic movement on the health of the poor and the workers remained negligible in the nineteenth century.[6] The hygienic program required bold municipal policy and big investments. For this reason support for these policies among the bourgeoisie, politicians and administrators long remained far from sufficient. Municipal finances were inadequate. Consciousness of a public health problem and public health policy was limited. Two additional factors also played a role.[7]

First, for a long time there was no consensus among politicians and administrators about the exact causes of poor public health and of epidemic diseases, for example the cholera that caused so much commotion in 1848, 1853, 1859 and 1866.[8] Were these contagious diseases, that is to say, diseases spread by contacts from person to person? This perspective led to quarantine measures, the isolation of the ill and the disinfecting of dwellings. Were epidemics associated with so-called miasmas? This referred to particles in the air that made people sick. The *mal'aria*, literally bad air, that prevailed in the low Netherlands, was attributed to this cause.

---

[5] Houwaart, 'Professionalisering en staatsvorming', 84–85 and Houwaart, 'Medische statistiek', 28.

[6] Van Zon, 'Openbare hygiëne', 47–49.

[7] Lintsen, *Made in Holland*, 58–64.

[8] Houwaart, 'Medische statistiek', 25–26.

This perspective emphasized the necessity of fresh air, sunlight and much space. Others, again, pointed to causes like spoiled food, polluted soil or a profligate lifestyle. Drinking water too counted as a potential source of infection. Research in London during the cholera epidemic of 1853 had pointed to infected drinking water as the most important vehicle for the spread of the disease. Investigations like these influenced the debate in the Netherlands. The hygienists pointed to practices with faeces and polluted water as important (but not sole) causes of poor public health. Confusion reigned.

A second factor was that there was no consensus – in particular within the broad hygienist movement – about the solution for the problem. The hygienists were united in emphasising the importance of a supply of reliable and pure drinking water. They supported plans for drilling new and modern wells that became the dominant practice in the 1860s and they advocated the construction of piped water systems whose numbers increased rapidly from the 1880s on.[9] They also concurred in supporting plans to set up sanitation departments, to flush urban canals, fill in ditches and lay underground sewers. But they were quite at odds about the proper approach to the disposal of faeces.

There were three options: the barrel system, the Liernur system and the flushing system with a water closet.[10] The barrel system, introduced for example in Delft in 1871, consisted of the placing of toilet barrels in dwellings. A (municipal or private) enterprise regularly collected the barrels and processed the faeces into manure to sell to farmers. The Liernur system – invented by a Dutch officer of the army engineer corps, C. Liernur, was based on the transport of faeces from dwellings to large underground tanks through vacuum pipes. Manure was collected from the tanks and transported to farmers and garden farmers by small boats. The vacuum was produced with steam engines. Leiden (1871), Dordrecht (1873) and Amsterdam (1879) performed large-scale experiments with this system. The flushing system with water closets consisted of a system of sewers that was coupled to the public drinking water supply. For a long time the hygienic movement was unable to settle on the one best solution among these alternatives.

---

[9] Modern wells were drilled with a new method developed by the American Norton during the American Civil War. He drilled for potable water with an extensible telescopic tube fitted with a steel point with openings. This technology made it possible to drill to great depth at low costs. In the Hague, the city government together with the *Association for the Improvement of the Health Situation in The Hague*, a local association of hygienists, used this machine to install more than 25 public pumps that delivered water that was much more trustworthy than that in other wells. Lintsen, *Made in Holland*, 63.

[10] Van Zon, 'Openbare hygiëne', 62–77.

9.3   Striving for Public Housing[11]

As we saw above, in 1854 a commission of the Royal Institute of Engineers was among the first to address the issue of the housing of the poor and the workers. After that the Institute preserved a long silence on this point despite the miserable situation in the slums that the commission had reported. The Institute would develop into a conservative, liberal fortress that barely addressed the social question and that oriented itself above all to technology and the applied sciences. Despite this, the engineering world would be confronted with public housing by another route, namely via the municipal Public Works Departments, that employed engineers, architects and other technical professionals. Niftrik was one of them and also one of the engineers that developed ambitious plans. But one thing they did not do was to improve public housing.

Big cities, namely Amsterdam, Rotterdam, The Hague and Utrecht faced the challenge of accommodating their rapidly growing populations (Graph 9.1). Public works had a crucial role to play. They produced plans for urban development. Municipalities were in principle empowered to steer this development. Legislation was certainly inadequate, but they could get a lot done on the basis of building codes, land-use ordinances, the granting of concessions and the Public Nuisance Act.[12] Despite this, little was accomplished on this front. The question was whether

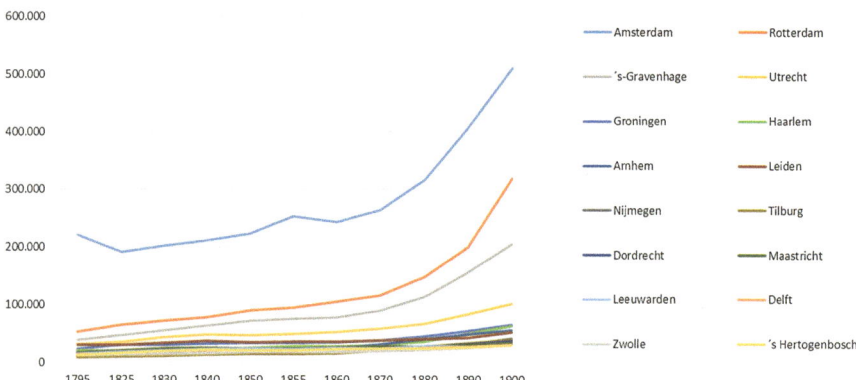

Graph 9.1 Growth of the big cities in the Netherlands, 1795–1900
Remark: Cities with a population greater than 30,000 in the year 1900 count as 'big' cities
Sources: *Statistisch Jaarboek*, various years. F.W. van Voorden. *Schakels in stedebouw; een model voor analyse van de ontwikkeling van de ruimtelijke kwaliteiten van de 19de-eeuwse stadsuitbreidingen op grond van een onderzoek in Gelderse steden* (Zutphen 1983)

---

[11] This section is based on: L. de Klerk, *De modernisering van de stad: De opkomst van de planmatige ontwikkeling in Nederland 1850–1914* (Rotterdam 2008); A. van der Woud, *Koninkrijk vol sloppen. Achterbuurten en vuil in de negentiende eeuw* (Amsterdam 2010); Buiter, *Riool, rails en asphalt.*

[12] De Klerk, *De modernisering van de stad*, 168.

Table 9.1 Municipal expenditures in guilders per capita of population, 1862–1907 (three year running average)

	1862	1877	1892	1907	1907 Cities	1907 Countryside
General Administration	1.50	1.23	1.44	1.97	2.72	1.44
Public Safety		1.07	1.77	1.77	3.27	0.72
Public Works	1.95	3.44	3.17	8.70	17.17	2.69
Education	1.00	2.38	3.78	5.14	7.71	2.69
Poor relief	0.98	0.97	1.22	1.69	2.74	0.94
Interest and amortisation	0.83	2.33	3.39	4.06	8.16	1.16
Investments		0.65	0.28	1.68	3.29	0.54
Other	0.75	0.21	0.48	0.66	1.22	0.26
Unknown	0.65					
TOTAL	7.57	12.27	15.43	25.64	46.27	10.81

Source: H. Knippenberg and B. de Pater, *De eenwording van* Nederland (Nijmegen 1988), 162 and 164

urban development should be determined by the market or by government.[13] The former approach was dominant until the end of the nineteenth century. And municipal finances also played a role. Municipalities had lost important sources of revenue due to the abolition of local taxes. At the same time they acquired new responsibilities, among other things in the area of education. Expenditures for Public Works that increased strongly after 1870, were continuously under pressure (Table 9.1).

The ambitious plans produced by departments of Public Works supplied much fuel for debates in municipal councils and among the broad public. They were regularly voted down or radically altered. The consequence was a fragmented policy.[14] The projects that survived were related to infrastructure (adaptation of water management, building new roads, connecting up to the railway network, building harbours) and public hygiene (laying down sewers, flushing the canals, filling in drainage ditches).[15] Building houses was considered to be a task for the market. An answer to the question of public housing thus had to be provided by private initiative.

One of the responses of private initiative was the so-called 'revolution' building, made possible by a new 'revolutionary' method of market oriented financing.[16] In this approach a land-developer bought up land in a particular area of a city and divided it up into parcels. He then sold the parcels to housing contractors, who took out a loan at a mortgage bank with as security the ground and the building plans. The bank provided the loan in installments after each phase of the building process (delivery of the land, driving of the first pile, etc.). That meant fast-paced construction due to the cost of interest and the fact that various parties needed to be paid – which also meant that the building materials became a kind of afterthought. The

[13] De Klerk, *De modernisering van de stad*, 269.

[14] De Klerk speaks of chaotic planning. De Klerk, *De modernisering van de stad*, 194.

[15] Buiter, *Riool, rails en asfalt*, 43–60; 106–119; 145–167.

[16] De Klerk, *De modernisering van de stad*, 238–239.

approach provided a record number of dwellings in a brief period of time for cities that were bursting at the seams, like Amsterdam (the Pijp and the Kinker-neighbourhoods), The Hague (the Schilderswijk) and Rotterdam (the Oude Westen). But the soundness of the buildings often left much to be desired. Building materials and construction were sometimes of inferior quality even to the extent that buildings sometimes collapsed. Even though the quality might not differ very much in comparison with the existing housing of the poor and workers, the dwellings drew heavy fire from hygienists and health commissions. The houses were too close together, lacked an adequate drainage system for faeces and often consisted of only one room, in which the occupants also cooked and slept.

That things could be otherwise was demonstrated by several socially engaged entrepreneurs. Gerard Adriaan Heineken, for example, had a series of dwellings built for his workers in the Pijp that had separate kitchens and small gardens. They were designed by the architect Gosschalk. Both men were members of 'Citizen's Duty' (*Burgerplicht*), a progressive-liberal political party in Amsterdam. Similar entrepreneurial initiatives were seen in other parts of the country as well. A famous example was the Agneten Park in Delft, built in the 1880s at the behest of Jacob Cornelis van Marken, director of the Dutch Yeast and Methylated Spirits Factory (*Nederlandsche Gist- en Spiritusfabriek*).[17] A number of his workers and higher personnel lived there. The firm also boasted a number of other social projects like a system of profit-sharing, an association building and an employee's council for consultation. In this way the entrepreneurs sought to encourage loyalty among their employees, to exhibit their involvement in the social question, and to project an image of themselves as progressive, modern businessmen.[18]

An entirely different kind of initiative was undertaken by socially responsible members of the upper classes. An example was the Jordaan Building Company Inc., set up in 1896 by social-liberals like the architect Jan van der Pek and the writer Helena Mercier. The company was an experiment intended to investigate whether the construction of affordable workers' housing on the basis of hygienic norms was possible. The experiment concerned one-room dwellings with sufficient daylight, a separate kitchen and a toilet. A housing inspector, inevitably female, collected the rent and checked to see if the occupants obeyed the rent contract, didn't hang the washing up in the room and kept the house clean. In practice it turned out that the homes were too expensive for many of the working-class families.

---

[17] E. Nijhof and A. van den Berg, *Het menselijk kapitaal: Sociaal ondernemersbeleid in Nederland* (Amsterdam 2012), 121.

[18] This section is based on: A. Bosch and G.P. van de Ven, 'Rivierverbetering', in: H.W. Lintsen et al. (eds.), *Geschiedenis van de techniek in Nederland: De wording van een moderne samenleving* (Zutphen 1993), volume II, 95–102; A. van Heezik, *Strijd om de rivieren: Tweehonderd jaar rivierenbeleid in Nederland* (dissertation Technische Universiteit Delft 2005); G.P. van de Ven (ed.), *Leefbaar laagland: Geschiedenis van de waterbeheersing en landaanwinning in Nederland* (Utrecht 2003); A. Bosch and W. van der Ham, with H.W. Lintsen, *Twee eeuwen Rijkswaterstaat, 1798–1998* (Zaltbommel 1998); A. Bosch, *Om de macht over het water: De nationale waterstaatsdienst tussen staat en samenleving, 1798–1849* (Zaltbommel 2000) and Lintsen, *Made in Holland*, 95–101.

The experiment in the Jordaan deeply influenced the debate on public housing in the 1890s. Socially responsible housing construction appeared to be possible only when supported by subsidies and legislation. The debate acquired momentum after a report on this issue by the Society for the Common Good (*Maatschappij tot Nut van 't Algemeen*), written by the social-liberal and lawyer Jan Kruseman, one of the founders of the Jordaan Building Company Inc. Inspired by this report, the progressive-liberal cabinet Pierson (1897–1901) submitted a proposal for a Habitation Act that was ratified by parliament in 1901. The act provided a framework for condemning dwellings and for dispossessing and improving them. It defined the role of municipalities in public housing and established the basic rules for providing subsidies for socially responsible housing construction. The issue of public housing had acquired entirely new dynamics.

9.4 The Improvement of the Rivers[19]

In the twentieth century, public hygiene and public housing would become one of the administrative tasks of municipalities. Municipal governments had already had some experience in the area of public hygiene. This had yet to happen in the case of public housing. The crucial importance of a tradition of policy-making and implementation of governmental tasks can be appreciated in the case of another building challenge for well-being in the second half of the nineteenth century: the problem of the rivers.

The Dutch lived in a vulnerable delta. They experienced that particularly in the polders, where flooding and high water were regular occurrences, in the coastal zones that were from time to time harassed by storm surges and in the regions of the big rivers where successive generations experienced serious flooding. The struggle against water in the polders and against storm surges was above all a task for the water boards and especially the large regionally consolidated water boards (*hoogheemraadschappen*). They were able to manage high water thanks to the introduction of steam-powered pumping stations that enabled them to exercise much better control of water levels in the polders. They combatted storm surges by building dikes, a strategy that in the nineteenth century was codified into a well-organized body of knowledge and resulted in a complete system of defensive works.[20] The rivers were the responsibility of the water boards and the national government. The scale of the problem was such that it had even been fundamental for the process of state formation and the founding of the *Rijkswaterstaat*, the national hydraulic and public works agency, in 1798.

---

[19] This section is based on: Bosch and van de Ven, 'Rivierverbetering', 95–102; van Heezik, *Strijd om de rivieren;* van de Ven, *Leefbaar laagland;* Bosch and van der Ham, with Lintsen (ed.), *Twee eeuwen Rijkswaterstaat 1798–1998*; Bosch, *Om de macht over het water* and Lintsen, *Made in Holland*, 95–101.

[20] Van de Ven (ed.), *Leefbaar Laagland*, 210.

The national government thus had a long tradition in caring for the rivers. This governmental task was uncontroversial. However, there was a problem: prior to 1850 the government and the Rijkswaterstaat had been singularly unsuccessful in combatting river floods. A certain indecisiveness caused in part by a controversy about how to come to grips with the problem lay at the bottom of this malaise. This came to an end in the 1850s.

In 1850 the leadership of the Rijkswaterstaat published a short and lucid report proposing a strategy for river improvement.[21] They rejected the concept of flood-ways, which could draw off an excess of river water along temporary alternative channels, and argued for the regulation and normalisation of the rivers, that is, the creation of a river with a bedding as straight as possible, of normalised dimensions and with a sizable opening to the sea. The report was a synthesis of ideas and foreign experiences accumulated over the preceding period. It came at a point in time that, thanks to the victory of the liberals in 1848, a new wind was blowing through the political landscape. And after the national treasury had also found its feet again, the government and parliament provided large sums of money. This made it possible to work on river improvements continually over a long period of time. For years there had been great pressure from an industrialising Germany that wanted better access to the sea. In Germany itself, massive efforts had already significantly improved the navigability of rivers. The Germans now demanded action from the Dutch.

A gigantic program was launched, one in which the Netherlands would invest hundreds of millions of guilders of tax revenues and to which successive generations of engineers, foremen, contractors, polder workers and other labourers would contribute. Rijkswaterstaat was in charge and occupied itself chiefly with the river beds. The water boards continued to concern themselves with the dikes, as they had done for centuries. In the end the entire river region would be transformed. Shores were rebuilt and consolidated with groynes and longitudinal embankments, meanders straightened and the river beddings widened. Secondary channels were eliminated by joining islands and sandbanks to the shores. New river mouths were created. Dikes were heightened, their width increased and their slopes rendered less steep, the outside faces reinforced with basalt and the floodplains cleared.

At the end of the this period hundreds of groynes and lateral dams had been constructed along the Waal, Nether-Rhine and Lek. At a number of places, meanders had been eliminated, the one at Wijk bij Duurstede taking 6 years. The New Merwede, a new Rhine river opening to the Hollands Diep had been in the works for 35 years and still the river was of inadequate depth. In 1896, after 33 years, Rotterdam had a new opening to the sea of the required dimensions, the New Waterway. The Meuse and the Waal (main branch of the Rhine in the Netherlands) had been separated and the Meuse provided with its own mouth, the Bergsche Maas (1883–1904). Extensive dredging was carried out, in particular after 1875. All this was possible only for a state that had the necessary organisational and financial prowess. Modern technology also played a crucial, if not decisive, role. Dredging machines, sand suction dredgers and bucket-excavators, all powered by steam, had made the large-scale works possible.

---

[21] Bosch and van de Ven, 'Rivierverbetering', 122–123.

Originally, safety had been the motivation for the river works. The rivers had first and foremost to be able to handle large volumes of water and ice. Subsequently attention shifted to navigability: achieving and maintaining a certain depth in the rivers and estuaries. In time it became normal usage to speak of the rivers as 'shipping channels.' Hence, the river works contributed in two ways to the positive development of well-being. The Netherlands would be considerably less vulnerable to ice jams and high water, while the rivers, as natural capital, functioned more effectively in transport and the economy.

9.5   The Infrastructural Revolution[22]

The development of infrastructure was also an age-old governmental responsibility. King William I had invested heavily in this domain. This continued under the new constitutional regime. Successive cabinets would devote considerable attention to infrastructure. And just as had been the case during William I's reign, the most important arguments for state intervention were the advancement of national welfare, the consolidation of the position of the maritime harbours and the improvement of access to cities and the countryside. But there was an important difference. Decision-making on the large projects took considerably more time after 1850 than in the previous period (Table 9.2).

In regard to infrastructural projects, the government and politicians were as usual approached by countless lobbyists: farmers' organisations, chambers of commerce, water boards, cities, private individuals and a diversity of interest groups. But King William I usually came to quick decisions, while governments and parliaments within the democratic order were busy evaluating interests, achieving compromises and coming to decisions. In this period some of the plans never made it across the finish line, or if they did, only after lengthy debates. For example, in 1849 an engineer of the Rijkswaterstaat published a two-volume work in which he proposed to drain and reclaim the Zuiderzee, the Frisian Wadden Sea and the Lauwerszee. He regarded the 'present point in time' eminently suited to realise his plans because of the 'new political life' and 'the simplified state administration.'[23] This idea would be followed by numerous plans for the total or partial reclamation of the Zuiderzee. Governments and parliament regularly returned to this question. Still, it would take until 1918 – after serious flooding in North Holland in 1916 – before a law mandating the closure and partial reclamation was adopted.

---

[22] This section is based on: G. Mom and R. Filarski, *Van transport naar mobiliteit: De transportrevolutie [1800–1900]* (Zutphen 2008); A.J. Veenendaal, 'Spoorwegen', in: H.W. Lintsen et al. (eds.), *Geschiedenis van de techniek in Nederland: De wording van een moderne samenleving* (Zutphen 1993), volume II, 129–163 and Lintsen, *Made in Holland*, 193–212.

[23] B.P.G. van Diggelen, *De Zuiderzee, de Friesche Wadden en de Lauwerszee, hare bedijking en droogmaking* (Zwolle 1849).

Table 9.2 Lead times (excluding execution) of large projects, 1813–1918 (in years)

Project	Law	Lead time
Period 1813–1815		
Great North Holland Canal	1819	7
Keulse Waterway	1821	4
Zuid-Willems Waterway	1822	7
Canal Brugge-Oostende	1822	5
Zederik canal	1824	6
Canal Gent-Terneuzen	1825	8
Reclamation Haarlemmermeer	1837	2
Period 1850–1918		
North Sea Canal	1863	14
New Waterway	1863	10
Merwede Canal	1881	14
Relocation Meuse mouth	1883	23
Closure Zuiderzee/Polders	1918	50

Remark: Lead time is the time from the moment that the responsible administration adds the project to its agenda to the start of the actual execution. The year indicates the point in time that the project is mandated by law. In that same year (or soon thereafter) the actual work is generally started

Source: H.W. Lintsen and M.L. ten Horn-van Nispen, 'Grote infrastructuurprojecten als belangenstrijd: niets nieuws onder de zon?', in: *Grote infrastructuurprojecten: inzichten en aandachtspunten, achtergrondstudies ten behoeve van de Tijdelijke Commissie Infrastructuurprojecten* (Den Haag 2004), 36–45

The national government was the decisive factor in the development of a national infrastructure after 1850.[24] After 1860 it took the construction of a national railway network to hand, after private initiative had failed to do so. In 1863 it committed itself to major works to improve maritime access to Rotterdam and Amsterdam: the execution by the state of the New Waterway and its subsidising of the North Sea Canal. In addition a number of other important canals were dug. From 1894 on, the national government also developed a subsidy system for a nationwide network of local narrow-gauge steam tramways.

An important factor was also that the state could count on popular support.[25] To be sure, there were heated discussions about the role of the state, the contribution of private enterprise, the design of the networks, the losers in policy making and so forth; nonetheless users, local administrators, chambers of commerce, the press and public opinion all in principle favoured a good national infrastructure. Transport companies eagerly used the new roads, canals, railways and tramways. Travellers were enthusiastic and chose the means of transport that promised the shortest travel

---

[24] Mom and Filarski, *Van transport naar mobiliteit*, 403–404.

[25] Mom and Filarski, *Van transport naar mobiliteit*, 404–405.

Table 9.3 Passenger transport. Development of speed and transport costs, 1850–1900

	Speed (km/hour)		Transport costs (cents/km)	
Means of transport	1850	1890–1900	1850	1890–1900
Tow barge	7	–	2.5	
Stage coach/omnibus on paved road	10–12	?	7.2–8.6	3.6–7.5
Steamship	10–15	10–15	2.3–3.0	1.0–1.5
Steam Railway	30–35	45–55	2.4–2.6	1.7–1.9
Steam Tram		12–14		1.75–3.0

Source: R. Filarski and G. Mom, *Van transport naar mobiliteit. De transportrevolutie [1800–1900]* (Zutphen 2008), 402

Table 9.4 Freight transport: Development of speed and transport costs, 1850–1900

		Transport costs (cent/tonxkm)	
Means of transport	Type of transport	1850	1890–1900
Maritime shipping		0.4–1.0	0.1–0.4
Inland shipping	Domestic shipping	1–4	0.4–1.0
	Rhine upstream	4.6	0.6–1.3
	Rhine downstream	2.0	0.3–0.5
	Small rivers in eastern Netherlands	9–14	
Horse and wagon	Unpaved road	28–35	
	Paved road	12–19	
Steam railway		6.3	1.8–1.9
Tram			ca.5

Source: R. Filarski and G. Mom, *Van transport naar mobiliteit. De transportrevolutie [1800–1900]* (Zutphen 2008), 403

time. Passenger transport had increased in speed and decreased in cost (Table 9.3). The costs of freight transport had also declined (Table 9.4).

9.6 The Balance

From a present-day perspective on well-being, the Netherlands faced four building challenges: public hygiene, public housing, water management and transportation infrastructure. At the time, the latter two were also acknowledged to be important building challenges and had been the responsibility of the state since the founding of the Kingdom (in 1815). The two former challenges were manoeuvred onto the political agenda by the 'societal midfield' (in particular the hygienists) and increasingly came to be seen as construction agendas for municipalities. Results were mixed.

Public Works departments of municipalities, particularly in the low Netherlands, regularly initiated construction of hygienic facilities (sewers, flushing of canals) and entrepreneurs were active in public housing ('revolution' building). But municipalities were incapable of formulating coherent policy in these areas. After the Law on Municipalities of 1851, they had to reinvent themselves before they could assume tasks in the domain of public hygiene, urban development and public housing and before they could provide the necessary means. Municipal governments were goaded into action by public opinion, social movements and a rapidly growing urban population. The financial situation proved an impediment. Many municipalities attempted to surmount their financial limitations by establishing a municipal gasworks (instead of granting concessions to a private firm) in order to generate extra income. A crucial factor was also that the national government created a legal framework for public health and public housing in order to legitimate, record and support municipal action.

The national government had a better track record with its building challenges than the municipalities. Its struggle against water had made the region of the large rivers much safer. Flooding as a result of ice jams ravaged the river region one more time in the nineteenth century, but that was in 1861when the normalisation project had only just been initiated. In the twentieth century, high river stages due to excessive rainfall and melting snow would once again overtax the rivers. For the first time in 1926 with extensive flooding as a consequence and – after a long period of relative calm – in 1993 when large parts of Limburg were flooded and in 1995 with a mass-evacuation of population from the river region as levees threatened to break. The national state also realised major works in the area of transport infrastructure and under its leadership coherent networks were created. This had two kinds of effects in relationship to well-being.

In the nineteenth century the Netherlands long remained 'an archipelago of regions and societies.' Until the 1860s large parts of the Netherlands were poorly accessible, barely opened up and inhabited by isolated and autonomous social communities. To be sure, the country was formally a unified state with a national legal system, but by no means a country characterised by a certain degree of economic, mental and cultural unification.[26] Every region had its own dialect, costume, products, dishes and local time.

In the first place, the transportation infrastructures enabled spatial integration and were an important prerequisite for the state formation of the Netherlands. Travel times, for example, decreased dramatically between 1850 and 1920. With the exception of Zeeuws Vlaanderen, the entire country could be reached from Utrecht within 5 h (Fig. 9.1). People and goods could circulate faster, ideas spread more quickly and social networks extend over greater distances. This was of course also abetted by the rise of newspapers, the telegraph and the telephone. New ideas about poverty, health, housing and labour penetrated more quickly into the furthest corners of the Netherlands. The activities of political parties, unions, health commissions acquired a broader reach. It was easier to mobilise social and occupational groups. Health became public health and housing became public housing. Both became a problem for the nation and hence also of the state.

---

[26] H. Knippenberg and B. de Pater, *De eenwording van Nederland* (Nijmegen 1990).

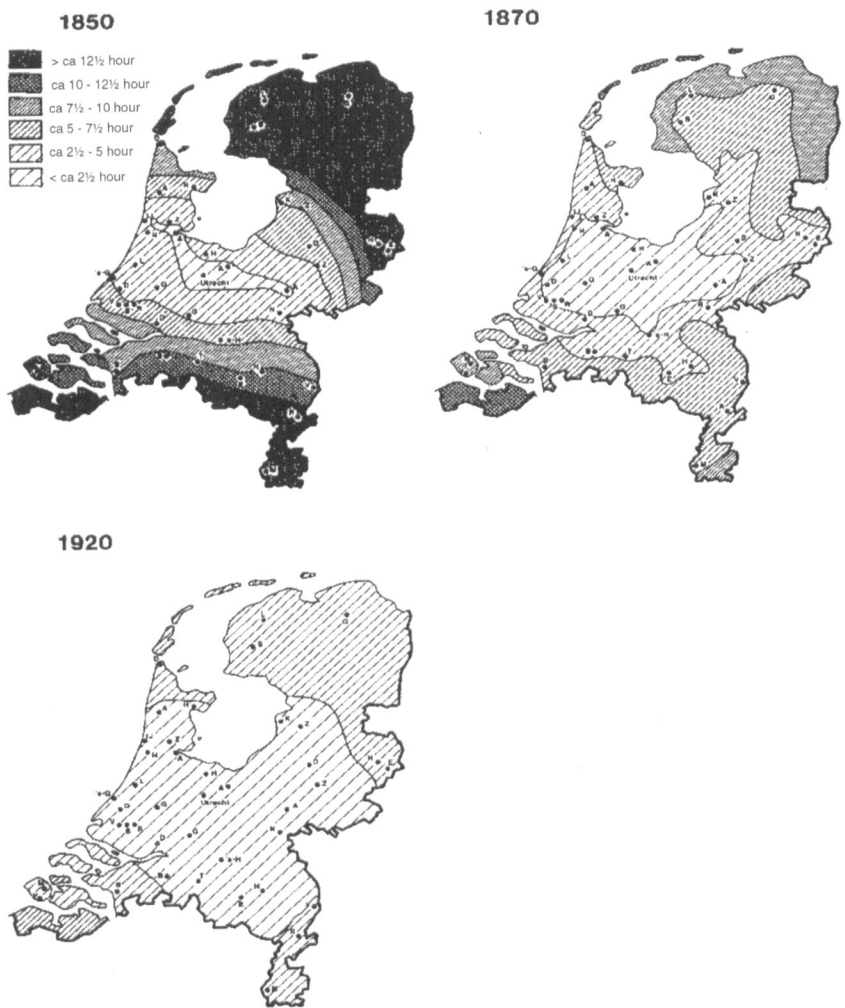

Fig. 9.1 Area in the Netherlands in 1850, 1870 and 1920 accessible from Utrecht within a given time

Source: A.J. Thurkow, J.D.H. Harten, H. Knippenberg et al., 'Bewoningsgeschiedenis', *Atlas van Nederland* (Den Haag 1984), part 2. See also: H. Knippenberg and B. de Pater, *De eenwording van Nederland* (Nijmegen 1990), 57

In the second place, spatial integration had far-reaching consequences for the modernisation of the economy. The modern economy was heavily dependent on scale, mass and speed. A key technology like steam could only develop on the basis of larger-scale production than was customary with existing technologies; it demanded a faster supply of raw materials and required larger markets. The opening up of the countryside, providing access to urban markets and making connections with foreign regions were important factors in this connection. Roads, canals, rivers, railways and local light rail made them possible.

Literature

Bosch, A. and G.P. van de Ven (1993). 'Rivierverbetering'. In H.W. Lintsen et al. (Eds.), *Geschiedenis van de techniek in Nederland: De wording van een moderne samenleving 1800–1890*. Zutphen: Walburg.

Bosch, A. (2000). *Om de macht over het water: De nationale waterstaatsdienst tussen staat en samenleving, 1798–1849*. Zaltbommel: Europese Bibliotheek.

Buiter, H. (2005). *Riool, rails en asfalt: 80 jaar straatrumoer in vier Nederlandse steden* (dissertation). Eindhoven: Technical University Eindhoven.

Bosch, A., W. van der Ham and H.W. Lintsen (1998). *Twee Eeuwen Rijkswaterstaat, 1798–1998*. Zaltbommel: Europese Bibliotheek.

Diggelen, B.P.G. van (1849). *De Zuiderzee, de Friesche Wadden en de Lauwerszee, hare bedijking en droogmaking*. Zwolle: Tjeenk Willink.

Heezik, A. van (2007). *Strijd om de rivieren: Tweehonderd jaar rivierenbeleid in Nederland of de opkomst en ondergang van het streven naar de normale rivier* (dissertation). Delft: Technical University Delft.

Houwaart, E. (1991). *De hygiënisten: Artsen, staat en volksgezondheid in Nederland, 1840–1890* (dissertation). Groningen: Groningen University.

Houwaart, E. (1993). 'Medische statistiek'. In H.W. Lintsen et al. (Eds.), *Geschiedenis van de techniek in Nederland: De wording van een moderne samenleving 1800–1890*. Zutphen: Walburg.

Houwaart, E. (1993). 'Professionalisering en staatsvorming'. In H.W. Lintsen et al. (Eds.), *Geschiedenis van de techniek in Nederland: De wording van een moderne samenleving 1800–1890*. Zutphen: Walburg.

Jager, I.W.M. (2003). *Hoofdstad in gebreke: Manoeuvreren met publieke werken in Amsterdam, 1851–1901*. Rotterdam: 010.

Klerk, L. de (2008). *De modernisering van de stad 1850–1914: De opkomst van planmatige stadsontwikkeling in Nederland*. Rotterdam: NAI Uitgevers.

Knippenberg, H. and B. de Pater (1988). *De eenwording van Nederland: Schaalvergroting en integratie sinds 1800*. Nijmegen: Sun.

Lintsen, H.W. (1992). *Wat is techniek? Een geschiedenis van menselijke secreten en discrete technieken* (Inaugural lecture). Technical University Eindhoven.

Lintsen, H.W. (2005). *Made in Holland: Een techniekgeschiedenis van Nederland [1800–2000]*. Zutphen: Walburg Pers.

Mom, G. and R. Filarski (2008). *Van transport naar mobiliteit: De mobiliteitsexplosie (1895–2005)*. Zutphen: Walburg Pers.

Nijhof, E. and A. van den Berg (2012). *Het menselijk kapitaal: Sociaal ondernemersbeleid in Nederland*. Amsterdam: Boom.

Veenendaal, A.J. (1993). 'Spoorwegen'. In H.W. Lintsen et al. (Eds.), *Geschiedenis van de techniek in Nederland: De wording van een moderne samenleving 1800–1890*. Zutphen: Walburg.

Ven, G.P. van de (2003). *Leefbaar laagland: Geschiedenis van de waterbeheersing en landaanwinning in Nederland*. Utrecht: Matrijs.

Woud, A. van der (2010). *Koninkrijk vol sloppen: Achterbuurten en vuil in de negentiende eeuw.* Amsterdam: Bert Bakker.

Zon, H. van (1986). *Een zeer onfrisse geschiedenis. Studies over niet-industriële vervuiling in Nederland, 1850–1920* (dissertation). Groningen University.

Zon, H. van (1993). 'Openbare hygiëne'. In H.W. Lintsen et al. (Eds.), *Geschiedenis van de techniek in Nederland: De wording van een moderne samenleving 1800–1890.* Zutphen: Walburg.

Chapter 10
Energy: A Revolution with Steam

Harry Lintsen

Contents

Abstract The chapter analyses the radical changes in the supply chain of fossil fuels between 1850 and 1910. It does this in relationship to the development of industry and investigates the consequences for the factory system, the environment and class relations.

The modernisation of the economy was an important precondition for the increase in welfare and the decline of extreme poverty. It involved agriculture and nutrition, infrastructure and transportation, industry and steam technology. Agriculture, foods, infrastructure and transportation were discussed in previous chapters. Industry and steam technology are the focus of this chapter.

The transition to coal and steam was accomplished in this period. It was typical of the Netherlands that the development of steam technology did not in the first place lead to the establishment of factories, but to the modernisation of workshops and small firms. Craft production remained dominant. To be sure, labour relations became more business-like due to liberalisation and sharp competition, especially in the cities, but it never came to social disruption or class conflict.

The emergence of steam technology raised questions about the safety of the vicinity and the nuisance inflicted on residents. A Steam Law and a Steam Inspectorate were relied on to minimise the risks of possible explosions. The nuisance law was intended to prevent irritating smoke, smell and vibrations as far as possible. However complaints about nuisance were almost never sufficient cause to refuse a permit. In most cases the interests of industry prevailed.

Harry Lintsen with contributions from Rick Hölsgens and Ben Gales.

© The Author(s) 2018
H. Lintsen et al., *Well-being, Sustainability and Social Development*,
https://doi.org/10.1007/978-3-319-76696-6_10

Keywords Modernisation · Steam technology · Small firm · Class struggle ·
Safety · Nuisance

10.1   36 h Lugging Warm Stearin Around[1]

An odd gathering assembled on Monday, January 10, 1887 in one of the meeting
rooms of the second chamber of parliament. On one side of the table sat a reputable
commission of eight persons. Chairman was the former lawyer Herman Verniers
van der Loeff. As parliamentarian he stood out as a conservative liberal, vehemently
opposed to suffrage for workers, among other things. On the other side of the table
sat four women, uncomfortable, shy and poorly clothed. They were workers at the
Royal Factory of Wax Candles in Amsterdam.

The four women were being questioned by a parliamentary board of enquiry that
was investigating the well-being of workers in factories and workshops in the
Netherlands. In the course of earlier sessions it had already become clear that
Verniers van der Loeff played the leading role and that he did that with great zeal
and manifest expertise. In this case he had to mobilize all his resources to put the
women at ease and to get them to talk frankly about their situation. He succeeded
only in part. But a number of issues did surface. as the following fragments from the
transcript reveal[2]:

Chairman: … How long then do you work at a stretch?
Hendrika Kamphuizen: 36 hours
Chairman: Did that happen recently?
Henrika K.: About four weeks ago.
…
Chairman: What did you have to do during those 36 hours, tell us that. Walk around with
warm stearin?
Hendrika K.: Yes
Chairman: And transfer that to the trays?
Hendrika K.: Yes. in the machines.
Chairman: The machines. by this you mean the forms containing the candles?
Hendrika K.: Yes
Chairman: So you have to pour the vessels with stearin over into the forms and cut off the
candles. That is your work during the 36 hours. You were never able to sit?
Hendrika K.: No. sir.

---

[1] The introduction is based on: H.W. Lintsen et al., *Made in Holland: Een techniekgeschiedenis van
Nederland [1800–2000]* (Zutphen 2005), 155–157.

[2] *De arbeidsenquête van 1887: Een kwaad leven.* (Nijmegen 1981), Part 1 Amsterdam, p. 220–222.
The publication is a reprint of the labour enquiry of 1887 with an introduction by J. Giele.

Chairman: At night you were not given an hour to rest?
Hendrika K.: from 12 to 1.
Chairman: And during this time you could lie on the floor a bit?
Hendrika K.: Yes. I would look around for a soft plank.
…
Chairman: … Now is this good or bad for adolescent girls and women?
Hendrika K.: Almost nobody can take the night work. The next day most are ill and complain about pain in the legs, in the head, in the lower back…

The publication of the hearings and the report of the commission caused a wave of indignation. The public responded with complete disbelief. That such things were possible in the Netherlands. In hearing after hearing, one abuse followed the other: women who worked outrageously long hours, children that were beaten, workers subject to capricious bosses, low wages, unhealthy working conditions, dangerous machines and on and on.

Up to that time in the Netherlands, the debate on the factory system had been rather 'academic.' If authors (especially in academic circles) referred to concrete abuses, they usually cited reports from England, sometimes Belgium or another country. A very few were acquainted with the factory on the basis of actual observation. But the parliamentary enquiry of 1887 put a face on the Dutch factory system. About 150 people were interviewed. The 'bare' verbatim transcript of questions and answers made the introduction to factory labour almost a personal experience.

There is something remarkable about the parliamentary enquiries of 1887 and 1890. They loosened tongues about the factory system. But on closer reading it is obvious that more is at issue, as witness the following fragment from the enquiry, in which the home-working tailor Petrus Schröder is questioned:

Petrus S.: Then it happens – our trade is unpleasant on this point –that we have to work through the night, especially on Friday and Saturday nights; then all too often it happens that the children, no less than the father, if they are up to it, sleep for only an hour or work at one stretch until Sunday morning 10 o'clock…
Chairman: How old are the children of whom you now speak?
Petrus S.: They sometimes begin at age 9 or 10. Such children often have to help out between school time and until 11 or 12 o'clock in the evening…[3]

The description did not diverge very much from that given of the factory. It was even questionable whether craft labour as performed at home or in workshops was a better option than factory labour. Here too child labour, women's labour, accidents, unhealthy situations, occupational diseases, capriciousness, long working hours, monotony etc. were the rule. Craft work, as it was described, had suffered such conditions for centuries, but never had these abuses led to public indignation. Apparently, more was at issue in regard to the social debate on labour. It was not only a response to the emergence of the factory system, but also motivated by changed attitudes among the bourgeoisie to the well-being of the worker. The factory as a new phenomenon was a stimulus to think about the production factor

---

[3] *De arbeidsenquête van 1887: Een kwaad leven*, 220–222.

'labour' in general and to formulate new values and norms for quality of life. In other words, in addition to a structural change in the economy, one could also speak of a cultural revolution in society.

The factory was one of the most important technical-organisational innovations of the modern economy. It was characterised by a concentration of labour and capital (in the form of machines and apparatus). The modernisation of the economy was a broad phenomenon that affected agriculture and nutrition, infrastructure and transport, industry and the factory. Agriculture, nutrition, infrastructure and transport have been discussed in preceding chapters. In this chapter industry and the factory are in the limelight with a focus on coal as the energy source and the steam engine as prime mover. We start with the changes in industry but will certainly come back to the issue of labour. And the persistence of poverty will also draw our attention. Did the changes in industry contribute to the mitigation of poverty?

10.2 Finally an Industrial Nation and a Decline of Extreme Poverty[4]

The liberalisation of international trade since the 1840s had meant booming business for agriculture, but not for Dutch industry. Agricultural prosperity depended in the first place on the international competitive advantage of the Netherlands in the area of cattle husbandry, allowing trade in agricultural produce like butter, cheese and cattle to flourish. It was thus more attractive to invest in agriculture than in industry. In addition, domestic prices for foodstuffs rose to British levels and the cost of living increased.[5] That had far-reaching consequences for the domestic market and a series of economic sectors. The growth of the entire food processing industry, that counted for the bulk of industrial production, stagnated. The same was true of the construction, clothing and leather industries.

Finally, there were big problems in the 'colonial complex,' in which the Netherlands Trade Society (NHM) monopolised the most important flows of trade between the Dutch East Indies and the Netherlands.[6] It began with a crisis on Java at the beginning of the 1840s. Trade in calico all but collapsed and the NHM struggled with serious overcapacity. Even more problematic was the fact that the industrial sectors closely connected to the 'colonial complex' were not prepared for the liberalised international market. For years the NHM had protected the Dutch cotton, shipbuilding and maritime shipping sectors against foreign competition. It had moreover all but failed to invest in the modernisation of these sectors. Cotton, shipbuilding and maritime shipping found themselves adrift in heavy seas. The cotton

---

[4] For the first part of this section see: J.L van Zanden and A. van Riel, *Nederland 1780–1914: Staat, instituties en economische ontwikkeling* (Amsterdam 2000), 277–288.

[5] Van Zanden and van Riel, *Nederland 1780–1914*, 282.

[6] Van Zanden and van Riel, *Nederland 1780–1914*, 278–281.

Table 10.1 The distribution of the workforce across agriculture, industry and services (in percent)

	1850	1890	1910
Agriculture	40	36	30
Industry	31	32	34
Services	29	32	36
	100	**100**	**100**

Source: J.P. Smits, E. Horlings and J.L. van Zanden, *Dutch GNP and its components, 1800–1913* (Groningen 2000), Table C.1, 115–117

industry was the first to extricate itself from the legacy of the NHM. After the 1860s, the transition to factory production enabled cotton manufacturers to confront international competition. It took considerably longer for shipbuilding and maritime shipping. They regained an international presence only in the 1890s with the transition to steam and iron.

It seemed that it was only in the long term that Dutch industry began to profit from the liberalisation of economic life and the integration of domestic and international markets.[7] Its recrudescence began in the 1860s. Among the first signs were the establishment of bread and flour factories. The industrial renaissance was characterised by, among other things, specialisation – for example by a number of new agricultural industries like beetroot sugar fabrication and the margarine industry. The emergence of a shoe-making industry in the Langstraat (North Brabant) also fit this pattern. Dutch industry also became internationally more competitive. Wages lagged behind those of other countries, the cost of foreign coal had declined and the cost of imported raw materials and fabricated products had developed favourably. Modernisation proceeded across a broad front and industry began to assume a central place in economic development. The size of the industrial workforce began to exceed that of the agricultural workforce (Table 10.1). At the start of the twentieth century about an equal number of workers were employed in industry and in the service sector (each about 35% of the workforce).

Liberalisation and industrialisation exerted considerable influence on the development of extreme poverty in the Netherlands. The country was in a sorry state from the end of the 1840s up to the end of the 1860s (Graph 7.1).[8] Population growth and the growth of the gross domestic product declined in the 1840s and 1850s (Table 10.2). Even on a per capita basis, the GDP declined. To be sure, some branches of agriculture and trade flourished, but the profits accrued to only a small elite of farmers and merchants. The recession in industry and in other parts of the economy in connection with increasing inequality led to an increase in extreme

---

[7] Van Zanden and van Riel, *Nederland 1780–1914*, 283–287. Some branches of industry flourished thanks to chance circumstances. The printing industry, for example, fared well thanks to the rescinding of the newspaper seal (a tax measure) in 1871 that made newspapers considerably cheaper and increased the demand for them.

[8] See appendix 4.1 for a quantitative analysis of poverty in the nineteenth century.

Table 10.2 Contribution of agriculture, industry and services (in per cent of total) to the growth of the gross domestic product (GDP), 1840–1913

	1840–1850	1850–1860	1860–1870	1870–1880	1880–1895	1895–1913
Population growth	0.71	0.65	0.87	1.12	1.17	1.40
Growth GDP	1.11	0.59	2.38	1.74	2.10	2.37
Growth GDP per capita	0.40	−0.07	1.51	0.62	0.93	0.97
Contribution to growth of GDP						
Agriculture	34.3	−0.7	24.9	−6.6	11.6	2.6
Industry	8.1	−6.1	32.1	68.3	34.2	37.0
Services	57.6	106.8	43.0	38.3	54.3	60.4
	100	**100**	**100**	**100**	**100**	**100**

Source: J.L. van Zanden and A. van Riel, *Nederland 1780–1914. Staat. instituties en economische ontwikkeling* (Amsterdam 2000), table 4.2. 152 and table 8.1. 345

poverty. It is estimated that 30–40% of the Dutch population around 1860 lived under the poverty line. In some years that amounted to more than a million Netherlanders. The magnitude of extreme poverty would decline in the course of the 1860s. After 1870 that happened quite rapidly. At the beginning of the twentieth century about 6% of the Dutch population continued to be extremely poor.

From the 1870s on the Dutch population increased by more than 1% per annum and from the 1880s the GDP by as much as 2%. Industry, along with trade, was the chief engine of the economy (Table 10.2). This is not to say that that agriculture no longer played a role. Quite the contrary, agriculture, the food processing industry and the trade in food products were still the building block of most important complex in the Dutch economy.

10.3   Steam for Big and Small[9]

A crucial factor in the industrialisation process was the development of energy technology. From a technological perspective, energy is transformed into useful effects in the economy by the application of capital and labour. Labour's contribution is limited, namely by the amount of food that a worker or an animal can consume. For this reason a human can deliver continuously about 0.1 hp and a horse about 0.5 hp in the course of a day.[10] Mechanical prime movers break through these limits. With the technology of the time, windmills could deliver a maximum of 30 hp, but only

---

[9] See for this section: H.W. Lintsen, 'Een land met stoom', in: H.W. Lintsen et al. (eds.), *Geschiedenis van de techniek in Nederland. De wording van een moderne samenleving* (Zutphen 1995), volume VI, 191–216 and Lintsen et al., *Made in Holland*, 133–154.

[10] For the horsepower of different prime movers, see: Lintsen, 'Een land met stoom', notes 3–7 in chapter 7, 294–295.

Table 10.3 The average growth of labour productivity in industry. 1807–1913 (in percent)

Period	Average growth labour productivity per annum (%)
1807–1830	0.2
1830–1842	0.7
1842–1860	−0.6
1860–1890	4.8
1890–1913	1.3

Remark: The growth of labour productivity in the period 1860–1890 is above all concentrated in the years 1860–1875. After this it even declines a bit. See: Van Zanden and van Riel, *Nederland 1780–1914*, Graph 6.3 on p. 245
Source: J.P. Smits. 'The determinants of productivity growth in Dutch manufacturing, 1800–1913' (Paper presented at the workshop National Accounts, Utrecht 1992), Graph 1

Table 10.4 The number of prime movers in industry by type, 1850–1890

	±1850	±1860	±1880	±1890
Horse mills	1930	1710	910	570
Windmills	3050	3400	3120	1790
Water mills	470	500	250	160
Steam engines	290	820	2740	3930
Gas engines			10	20
Total	**5740**	**6430**	**7030**	**6470**

Remark: The figures are rounded off to the nearest ten
Source: H. Lintsen, 'Een land met stoom,' in: H. Lintsen et al. (eds.).*Geschiedenis van de techniek in Nederland. De wording van een moderne samenleving* (Zutphen 1995), part VI, table 7.1. 192

when the windmill was really big and there was a stiff breeze blowing. The average power of a windmill throughout the year was considerably lower, about 6 hp. Water mills in the Netherlands with their very modest hydraulic heads usually delivered no more than 10 hp. The production ceiling that the Netherlands was able to achieve with the classic technologies was once again shattered by steam technology. But first the country had to develop into a 'steam society,' that is to say into a society in which infrastructure, institutions, legislation, entrepreneurial behaviour and production processes were adapted to steam technology. That process was accomplished in the nineteenth century and in particular after 1850.

In the nineteenth century industrial labour productivity tripled. That increase was achieved almost entirely in the period 1860–1890 with an annual growth in labour productivity of 4.8% (Table 10.3). This growth was realised above all with steam engines and their associated production technologies. The number of steam engines increased rapidly after 1860 at the expense of wind, horse and water mills (Table 10.4). The transition to steam technology seemed an inevitable process. But for the entrepreneurs of the day it was anything but that.

Between 1860 and 1890 manufacturers were confronted with a variety of options for their production processes and with numerous uncertainties in decision-making. Prior to 1860 steam could not, under prevailing conditions, compete with wind and water. After 1890 there were still limited possibilities for the classic means of production. During the transitional phase, entrepreneurs had a choice of different possibly successful manufacturing strategies: continuing with the mill and cottage industry, combining mills and steam engines, adopting small steam engines or large-scale use of steam technology. The situation was made even more complex by a ceaseless stream of innovations. The changing prices of raw materials, new but unclear market expectations and changing social needs. The upshot was that entrepreneurs were not facing anything like a clear set of choices, but more often a confusing muddle. As much as possible, therefore, they adopted a course of gradualism as a strategy of coping with the possibilities and impossibilities of steam.

While the transition to steam followed a tortuous path, one could still speak of the inevitable advance of the steam engine. Three structural developments underlay this dynamic. In the first place, the running costs of steam engines continually declined, especially for small steam engines of up to about 20 hp. Steam engines became cheaper. The costs of repair and maintenance declined. Coal became cheaper, in part because of the elimination of excise taxes on coals in 1863. All told, running costs declined in the second half of the nineteenth century by roughly 50%.[11]

Further, steam engines were increasingly coupled to new and dedicated machinery, for example rollers instead of millstones in the grain mills, hydraulic presses instead of hammers in the oil mills, and churns instead of centrifuges in the dairy industry. It did not take long before the multiplication of production by means of a steam engine became both possible *and* necessary in order to compete with 'classic' means of production.

In the third place, opportunities for steam technology increased along with the size of markets. The entrepreneur opting for steam was in almost all cases 'doomed' to increased production and bigger markets. That not only applied to manufacturers with large steam engines, but also and especially to the entrepreneur with a small steam engine. Institutional barriers that hobbled large-scale production had to be eliminated, among other things urban autonomy, guilds and specific laws like the Milling Law. Various factors provided fertile soil for the upscaling of production. The construction of the railway network after 1860 and of tramways after 1880 provided access to local markets and laid the basis for national markets. The purchasing power of the population increased from the 1860s on. Producing for large and distant markets was a stimulus for steam, as the production of butter, margarine, paper, sugar and textiles showed.

Despite all this, the adoption of steam technology would not initially lead to the emergence of the factory with its large-scale production. Steam engines were first and foremost installed in small enterprises. The average power was low, around 13 hp. (Table 10.5). Only in the textiles and chemicals sectors did the average

---

[11] Lintsen, 'Een land met stoom', volume VI, table 7.7, 198.

Table 10.5 The number of steam engines in industry by sector. 1890

Sector	Number	% of total	Average power in hp
Food Processing	1308	33	10.6
Construction	467	12	15.9
Textiles	451	11	21.0
Ceramics	274	7	10.5
Shipbuilding/vehicles	265	7	3.4
Lumber	265	7	16.6
Chemicals	227	6	21.8
Metal	204	5	8.7
Misc.	464	12	
Total	**3925**	**100%**	**13.0**

Source: Lintsen, 'Een land met stoom', volume VI, table 7.6 and table on p. 270

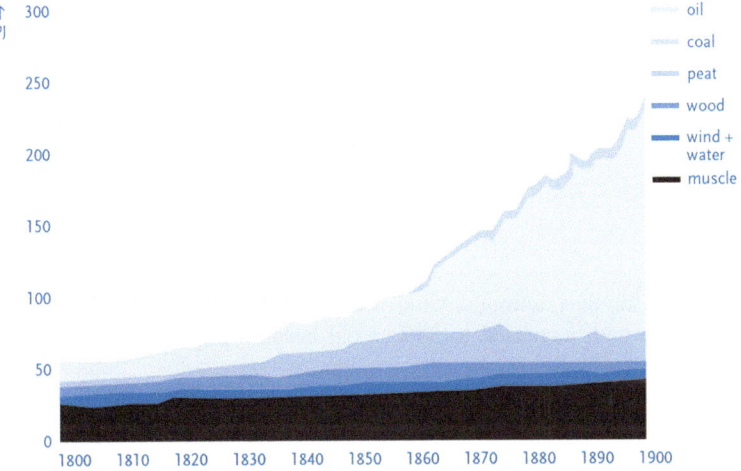

Graph 10.1 Consumption of different energy sources, 1800–1900
Source: B. Gales and H. Hölsgens, *Energy transitions in the Netherlands: Sustainability challenges in a historical and comparative perspective* (Groningen 2016), 11 and appendix 1

exceed 20 hp. These sectors had many large mechanised enterprises. The average size of industrial firms changed slowly. Around 1890 only 20% of the population worked in enterprises with 10 or more employees. Only after that time did middle and large-sized enterprises begin to appear. In 1909 as much as 40% of the workforce was employed in middle and large-sized firms. By the early twentieth century the factory began to dominate the industrial landscape of the Netherlands.

Energy consumption in the nineteenth century changed dramatically with the advent of steam technology. Total energy consumption in that century is estimated to have become five times as high, while the population increased by about two and a half times (Graph 10.1). Energy consumption increased at the same rate as the GDP (Graph 10.2). Around 1850, 84% of the energy was derived from the tradi-

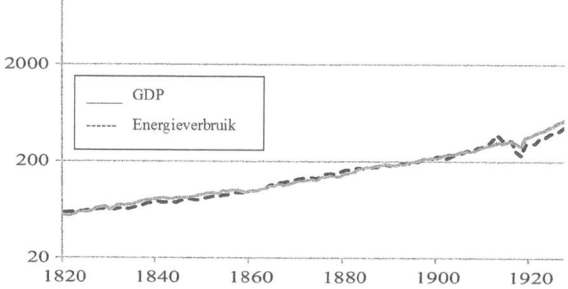

Graph 10.2 Economic growth and energy consumption, 1820–1920 (1860 = 100)
Source: B. Gales, A. Kander, P. Malanima and M. del Mar Rubio Varas, 'North versus South:
Energy transition and energy intensity in Europe over 200 years', *European Review of Economic
History* 11 (2007), no. 2, 219–253

Table 10.6 Energy consumption and greenhouse gas emissions in the Netherlands in 1910 and
2007 set off against the norms of the European policy agenda of 2007

	Situation 1900	Situation 2007	European Norm for 2020	Remark
Energy consumption	304,000 TJ	2,879,000 TJ	2,303,000 TJ	European policy agenda 2007: 80% of 1990
Greenhouse gas emissions	22,000 kton	178,000 kton	100,000 kton	European policy agenda 2007: 80% of 1990

tional sources: humans, animals, wind, water and turf. By the turn of the century
that percentage had decreased to 34%. Coal became far and away the dominant
source of energy.

On the eve of the First World War 88% of Dutch coal was being imported from
abroad. That made Dutch energy supplies vulnerable. Some were quite concerned
and advocated the exploitation of coal layers in the province of South Limburg (see
Chap. 15). Others trusted to international law, that made it difficult to proclaim
export restrictions and blockades against neutral countries in time of war. Such
countries could take it for granted that their trade flows would not dry up.

What were the consequences for the natural environment of this transition to an
economy based largely on fossil energy sources? Climate change and greenhouse
gas emissions were non-issues in those days. But even from a present-day perspective the transition was still hardly a problem. Energy consumption and the associated greenhouse gas emissions were low by present-day standards and lay far
beneath the norms that are now being pursued for the Netherlands (Table 10.6).
Matters were quite different, however, as far as safety was concerned and nuisances
from smells, soot and smoke.

10.4   Coal and Steam as Nuisance[12]

The increasing use of coal and steam had palpable consequences for the immediate surroundings. Safety was the first issue to attract government intervention. As early as 1824, King William I ruled that boilers should be inspected to prevent explosions. This decision was the basis of additional legislation, ultimately leading to the Steam Law of 1869.[13] In the meantime the government had delegated boiler inspection to the Steam Service (*Stoomwezen*). This service enjoyed an outstanding reputation and was held to be quite conscientious. Steam boilers had to be fitted with a number of devices such as two safety valves and an alarm system in case they ran out of water. Boiler manufacturers provided these devices as standard equipment. The regular inspections could not prevent the occurrence of occasional explosions. They were not frequent – as far as we can tell only four of them in the 1880s – but in some cases at the cost of human life.

The owner of a steam installation also had to deal with legislation that now falls under what we know as the Nuisance Law (*Hinderwet*). He always needed a permit and the surrounding residents could protest the installation of a steam engine. The

---

[12] See for this section: Lintsen, 'Stoom in ontwikkeling'.

[13] Nineteenth-century regulations concerning the safety of boilers included the following decisions and laws:

The 'Decision containing the specification of temporary safety measures on the use of steam engines' (*'Besluit houdende daarstelling van voorloopige veiligheidsmaatregelen bij het aanwenden van stoomwerktuigen'*, 6-5-1824, Staatsblad nr. 32).

The 'Decision concerning changes in the earlier legislated safety regulations on the use of steam engines' (*'Besluit houdende wijziging der vroeger voorgeschrevene veiligheids-maatregelen bij het aanwenden van stoomwerktuigen'*, 26-9-1833, Staatsblad nr. 58).

The 'Decision concerning specifications in regard to the inspection of the steam engines noted therein and the supervision of their use' (*'Besluit houdende bepalingen nopens het onderzoek der daarbij vermelde stoomwerktuigen en het toezigt op hun gebruik'*, 24-5-1855, Staatsblad nr. 40).

The 'Law regulating supervision of the use of steam apparatus' (*'Wet regelende het toezigt op het gebruik van stoomtoestellen'*, 28-5-1869, Staatsblad nr. 97).

The 'Decision to implement the law of May 28 1869, regulating the supervision of the use of steam apparatus' (*Besluit tot uitvoering der wet van 28 mei 1869, regelende het toezigt op het gebruik van stoomtoestellen*, 4-9-1869, Staatsblad nr. 154).

The 'Law concerning regulation of the supervision of the use of steam apparatus' (*'Wet houdende regeling van het toezicht op het gebruik van stoomtoestellen'*, 15-4-1896, Staatsblad nr. 69).

The 'Decision to implement the Steam Law (Law of 15 April 1896), (*'Besluit tot uitvoering der Stoomwet (wet van 15 april 1896)'*, 19-10-1896, Staatsblad nr. 163).

The 'Decision concerning the establishment of a protocol for civil servants, as stipulated in article 7 of the Steam Law (Law of 15 April 1896), (*Besluit tot vaststelling van eene instructie voor de ambtenaren, bedoeld in art. 7 der Stoomwet (wet van 15 april 1896)*, 14-1-1897, Staatsblad nr. 45).

In 1855 the government entrusted supervision to a separate agency, the Steam Service. Prior to that the government contracted the services of experts that performed inspections as an adjunct to their main occupations.

number of objections is unknown, but there were regular complaints.[14] An engineer sent to investigate conditions in the city of Zutphen heard from house painters working near a steam mill that 'windows painted two days ago were once again speckled with black particles.' He also reported that laundry hung out to dry appeared to be covered with black dust. Houses located downwind of the steam engine were unable to open their windows '…without breathing in choking fumes instead of fresh air.' Also, at some houses '…the gutters and courtyards were full of dust and everywhere in the neighbourhood the windows were apparently dirty and glazed with smoke.' Finally, the engineer examined the rainwater destined for consumption and found it to be 'completely black.'[15]

Nearby residents sometimes also had problems with the boiler and the steam engine themselves. A certain Mr. Wouters of Nijmegen was one of them and protested the construction of a steam grain mill in the shed next to his home. The shed and the dwelling were separated by a brick wall 'partly with a thickness of one brick and also a half brick.' To make matters worse, Wouter's box bed was positioned next to the shed, 'surrounded only by a half-brick thick wall and only 60 inches (=0,6 m – HL) distant from the sheathing of the boiler.' Wouters expected to sleep poorly. It was claimed that: 'Because of the situation of the boiler and chimney, so close to the box bed and separating wall, his house will become unbearably hot, which in addition to the noise and the vibration of the engine, the threat of fire and explosion of the boiler will make the inhabitation and in general the renting of a portion of the house impossible…'[16] The workshops equipped with steam engines were more often than not located in the narrow streets of the city centres, surrounded by houses in which people lived and worked and 'difficult to access in case of fire and dangerous for the entire neighbourhood.' But factories on the edges of towns could also be dangerous and a nuisance.

Complaints about emissions and other nuisances were only rarely considered sufficient cause to refuse a permit. The interests of industry were usually allowed to prevail. Still, the laws did have some effect on steam technology. Where the government considered the complaints grounded, conditions were imposed, for example increasing the height of a chimney, the construction of a free-standing foundation and the installation of spark arresters in the chimney. Tolerance for smells, soot and smoke would change in the twentieth century. That not only had to do with the increase of coal consumption. Changed attitudes toward smells, soot and smoke also played a role. The city-dweller of the nineteenth century was pretty case-hardened regarding smells and trash in the streets, canals and creeks. But,

---

[14] For the history of the Nuisance Law and its predecessors: J. Lintsen Sr., 'De werking van de hinderwet tijdens de industrialisatie van Nederland (1890–1910)', *Jaarboek voor de Geschiedenis van Bedrijf en Techniek JBGBT* 4(1987), 190–209.

[15] Report by J. van Ortt, hoofdingenieur, 27 juli 1863, no. 1561, Rijksarchief in Gelderland, G.S. Nijverheid inv.nr. 25.05112/2.

[16] Cited in: Lintsen, 'Stoom in ontwikkeling', 128.

increasingly, smells would be experienced as intolerable stenches and trash as waste and pollution.[17]

10.5   A Velvet Revolution[18]

The modernisation of the economy in England had aroused mixed feelings in the Netherlands. Many had looked with awe at the economic power that the country had developed. But many had also regarded the consolidation of a class society with trepidation and recoiled from the social tensions that that brought with it. Steam technology had been an important catalyst in bringing things to this pass. In England it had led to mechanisation, large-scale production and the emergence of the factory. With the factory system a new group of entrepreneurs had emerged that became the core of a bourgeois middle-class. Simultaneously, factory workers became a new group in the working class. In striking contrast to the luxury of the 'nouveau riche' stood the poverty of the factory proletariat. It created the fertile soil for an unprecedented class struggle.

Despite this, violent class conflicts would barely occur in the Netherlands in the nineteenth century and steam technology would be enthusiastically embraced. An important cause was the way in which steam technology in the Netherlands had been implemented in the first phase of industrialisation. Industrialisation proceeded across a broad front and spread throughout the entire country. A significant number of steam engines could be found in the food processing industry, the most important industrial sector and additionally in almost every branch of industry. One also found steam engines in both the coastal provinces and in the interior. To be sure, there was a visible concentration of steam engines in the provinces of South and North Holland and somewhat also along the large rivers, but large agglomerations of factories characteristic of industrialisation in regions like Lancashire, the Borinage, Lorraine, and the Ruhr region did not exist in the Netherlands.[19]

The Netherlands was unique in that the application of steam technology did not in the first instance lead to a factory system, but rather to the modernisation of workshops and small manufacturing enterprises. Craft production remained dominant. To be sure, labour relations in the large cities and elsewhere became more formal due to liberalisation and sharper competition, but these developments did not lead

---

[17] See for example: A. Corbin, *Pestdamp en bloesemgeur: Een geschiedenis van de reuk* (Nijmegen 1986). Also: Lintsen et al., *Made in Holland*, 56–58.

[18] See for this section among others: H.W. Lintsen, *Een revolutie naar eigen aard: Technische ontwikkeling en maatschappelijke verandering* (inaugural lecture Delft, 1990) and Lintsen et al., *Made in Holland*, 142–143.

[19] E. Nijhof and A. van den Berg, *Het menselijk kapitaal: Sociaal ondernemersbeleid in Nederland* (Amsterdam 2012), 45.

to a breakdown of the old social order.[20] The informal relations characterising economic life persisted. The owner of small workshops continued to work alongside his employees and was barely distinguishable from them. Above all, he did not become part of a new bourgeois elite.

It is remarkable how much attention was given after 1870 to appropriate technology for small businesses in circles of tradesmen, industrialists, engineers and politicians, despite the fact that they by no means distanced themselves from large-scale industry. A wealth of publications appeared on the use of small engines and machines. The zenith of this propaganda offensive was the exposition of craft machinery in 1907. According to some, modern technology made the rationalisation of small firms possible and thus contributed to the strengthening of the crafts. 'If mechanical power has truly come to serve the craftsman and to facilitate the development of his independence,' according to the director of the Trade School in the Hague in 1890, 'then we can foresee a new period of the flowering of small industry and prosperity for its practitioners.'[21] The dominant ideological currents shared the enthusiasm for steam technology and mechanisation. Liberals, confessionals and socialists ranged themselves solidly behind technical progress.[22]

Though the Netherlands was spared a disruptive class struggle, this did not mean that there were no heated debates on industrialisation. The parliamentary enquiries of 1887 and 1890 had put the labour issue squarely on the political agenda. Though there the question of poverty played an important role, the issue had assumed a different character. It had shifted from extreme poverty to the 'social question.'

Literature

Corbin, A. (1986). *Pestdamp en bloesemgeur: Een geschiedenis van de reuk.* Nijmegen: SUN.

Gales, B. et al. (2007). 'North versus South: Energy transitition and energy intensity in Europe over 200 years'. *European Review of Economic History*, 11(2), 219–253.

Giele, J. (1981). *De arbeidsenquete van 1887: Een kwaad leven. Part 1: Amsterdam.* Nijmegen: Link.

Hölsgens, H.N.M. (2016). *Energy transitions in the Netherlands: Sustainability challenges in a historical and comparative perspective.* Groningen: University of Groningen, SOM research school.

Lente, D. van (1998). *Techniek en ideologie: Opvattingen over de maatschappelijke betekenis van technische vernieuwingen in Nederland, 1850–1920.* Groningen: Wolters.

---

[20] J.L. van Zanden and A. van Riel point to the formalisation of the labour market in Amsterdam and in the province of Groningen and the region of the *Achterhoek*. Van Zanden and van Riel, *Nederland 1780–1914*, 305–307.

[21] Cited in D. van Lente, *Techniek en ideologie: Opvattingen over de maatschappelijke betekenis van technische vernieuwingen in Nederland, 1850–1920* (Groningen 1998).

[22] Van Lente, *Techniek en ideologie*, op. cit. The enthusiasm of liberals and socialists for technological progress is an international phenomenon. An almost total lack of opposition to technology is striking for the Netherlands.

Lintsen, H.W. (1990). *Een revolutie naar eigen aard: Technische ontwikkeling en maatschappelijke verandering* (Inaugural lecture). Technical University Delft.

Lintsen, H.W. (1993). 'Een land met stoom'. In H.W. Lintsen et al. (Eds.), *Geschiedenis van de techniek in Nederland: De wording van een moderne samenleving 1800–1890.* Zutphen: Walburg.

Lintsen, H.W. (2005). *Made in Holland: Een techniekgeschiedenis van Nederland [1800–2000].* Zutphen: Walburg Pers.

Lintsen Sr, J. (1987). 'De werking van de hinderwet tijdens de industrialisatie van Nederland (1890–1910)'. *Jaarboek voor de Geschiedenis van Bedrijf en Techniek JBGBT*, 4, 190–209.

Nijhof, E. and A. van den Berg (2012). *Het menselijk kapitaal: Sociaal ondernemersbeleid in Nederland.* Amsterdam: Boom.

Smits, J.P. (1992). 'The determinants of productivity growth in Dutch manufacturing, 1800–1913'. Utrecht: Paper presented at the workshop National Accounts.

Zanden, J.L. van and A. van Riel (2000). *Nederland 1780–1914: Staat, instituties en* economische ontwikkeling. Amsterdam: Balans.

Chapter 11
From Extreme Poverty to the Social Question. Well-being and Sustainability Around 1910

Harry Lintsen

Contents

Abstract The dynamics of the resource-production-consumption chains as analysed in the three previous chapters were interrelated with the dynamics of societal institutions. The strategies of professional groups and political parties eventuated in a corporatist state structure, with a societal midfield solidly anchored in a political system and in a state bureaucracy at local, provincial and national levels. New notions about extreme poverty and well-being acquired a legal framework around 1900, within which governments, entrepreneurs, citizens and workers could continue to work on the solution of different issues. The legal framework was also the origin of the welfare state as it would develop in the course of the twentieth century. The modernising economy was also characterized by the emergence of new professional groups and a modern knowledge infrastructure.

This is the context in which we must place the well-being monitor for 1910. Well-being and sustainability around 1910 are viewed from three perspectives. Judged by the societal agenda of 1850 much progress had been made, especially in regard to extreme poverty. But the societal agenda of 1910 shows that opinions about poverty have shifted and that in this period the qualities of the food supply, of public health facilities, of public housing and of work are the core issues. The

© The Author(s) 2018
H. Lintsen et al., *Well-being, Sustainability and Social Development*,
https://doi.org/10.1007/978-3-319-76696-6_11

present-day perspective on well-being differs little from that of 1910. But in regard to the issue of sustainability there are big differences.

Keywords Corporatism · Social question · Knowledge infrastructure · Monitor · Public health · Public housing · Labour

11.1 'The Material and Spiritual Side of the Social Question'

The general condition of the people in many respects compels us to be thankful. However, we should not ignore the fact that, both on the spiritual and material levels, a change in circumstances has come about that requires the government, more than hitherto, to give leadership and provide support … Not less does the material side of the social question continue to draw My attention.[1]

The Queen's speech to parliament, delivered by Queen Wilhelmina in 1901 from a text provided by the confessional government of prime minister Abraham Kuyper, outlined the new role that the national government had taken upon itself in the development of well-being. "More than hitherto' the national government decidedly saw a role for itself.

The Netherlands had just experienced two decades of prosperity. The growth indicators of the GDP were among the highest in the century. The magnitude of extreme poverty had shrunk to a historic low point. It was surely one of the reasons why Kuyper was satisfied with the 'general condition of the people.' The 'change' that was mentioned in the Queen's speech had to do with the Christian nature of the cabinet. Protestant Anti-Revolutionaries and Catholics had won the elections. Kuyper's cabinet took office after a period of three liberal cabinets.

What details did the Queen's speech provide for the 'spiritual and material' side of the social question? The 'material side' was a mixture of labour and poverty-related measures. The cabinet wanted to stop the emiseration of the vulnerable members of society by means of mandatory insurance against the consequences of illness, invalidity and old-age. They wanted to combat fraud with food supplies in order to protect the population against malnutrition. Mandatory insurance against industrial accidents served to protect vulnerable workers. New apprenticeship regulations sought to improve popular education. On the 'spiritual side' Kuyper's government wanted to uplift 'the moral character of public life' in particular to rein in 'addictions to games and drink.' And then of course there was the issue of special education that had already occupied politicians and the churches for years. The confessional parties wanted to have religiously-based schools subsidised by the government, in addition to the regular non-denominational public schools.

---

[1] Queen's address to parliament by Queen Wilhelmina, 17 September 1901. At www.troonredes.nl (consulted 26-9-2014).

Quite striking is the attention paid to the 'lesser welfare of the native population on Java.' The cabinet would take action on this issue and, among other things, closely monitor obedience to the regulations designed to protect native coolies.

But parliamentary opposition prevented the cabinet from making good on its ambitious social programme. To make matters worse, the big Railway Strike of 1903 seriously dented its image. Kuyper collided with striking railway workers and stevedores, whom he accused of 'criminal agitation.' Three laws were adopted – 'strangulation laws' according to the socialists – that prohibited civil servants and workers in key sectors (like the railways) from striking. This broke the strike. It was only in the area of colonial politics that the cabinet was able to cash in on its promises, particularly the introduction of the 'ethical policy' that aimed to promote the political and economic independence of the colonial population.

Poverty dominated the issue of well-being throughout the nineteenth century. A low quality of life was passed on through successive generations of the poor. Contemporaries acknowledged the seriousness of the issue, even as it took on a different guise in the course of the century. Around 1850 poverty was associated with extreme poverty. An important segment of the population had to make do without the minimal basic necessities like food, drink, clothing, warmth and housing. The struggle against poverty was initially aimed at providing for these minimal needs.

But in the course of time the issue became broader. Foreign reports about slums in the growing cities, about the miserable conditions in factories and about the exploitation of the proletariat on the one hand horrified the Dutch bourgeoisie and on the other hand incited fear of uprisings and revolution. When the first 'modern' strikes took place in 1869 – albeit on a modest scale – they stimulated a wide range of responses. One of them was the founding of the 'Committee for deliberating the Social Question' in 1870. The term 'Social Question' would become the dominant referent for a wide range of problems around poverty, extreme poverty, public health, public housing, popular culture and labour issues.[2]

At the time this also included qualitative aspects: the emphasis was no longer solely on the bare the survival of the poor, but rather on a life that was as healthy as possible; not only on sufficient food, but also on healthy food; not only on a roof over one's head, but just as much on a dwelling with hygienic facilities; not just on work, but also on skilled work under good working conditions. It was, finally, felt that the poor should not be obsequious, but above all 'civilized.'

After that things moved quickly. The Netherlands were faced with an increasing number of slums in rapidly growing cities. The modernisation of the economy made labour relations more distant and formal. The factory system began to penetrate into regions like Twente and North Brabant. The increase in the number of voters was also an important development. The constitutional reform of 1887 endowed many more citizens with voting rights. In 1880 12.3% of the male population could vote.

---

[2] See among others: J.L van Zanden and A. van Riel, *Nederland 1780–1914: Staat, instituties en economische ontwikkeling* (Amsterdam 2000), 314–317; E. Nijhof and A. van den Berg, *Het menselijk kapitaal: Sociaal ondernemersbeleid in Nederland* (Amsterdam 2012), 47–52.

In 1890 that had increased to 26.8%.[3] New social groups joined in the political process and hence acquired potential access to state power. Numerous organisations profiled themselves as protagonists of specific interests. Many claimed to be able to offer solutions to the social question. This issue would occupy a prominent place on the societal agenda around 1900.

In preceding chapters these shifts have been briefly noted. In this chapter we will summarize them and elaborate further. Who were the bearers of the new perspectives on poverty and quality of life? How and under what circumstances did they succeed in turning this into a political item? Why did this debate lead to an active role for a government that had formerly been passive in social matters?

The scope of extreme poverty declined in the nineteenth century. This development was embedded in trade-off processes. To what extent was the increase in quality of life paid for by the natural environment and the exploitation of natural capital? To what extent was depletion of resources elsewhere part of the picture? How sustainable was the new situation? The chapter closes with an evaluation of the situation at the outset of the twentieth century based on the monitor for well-being.

11.2 Building a New Corporatism: Citizens, Government, Entrepreneurs and Researchers

11.2.1 A New, Dynamic, Civil Society

After a period of near dormancy under King William I, the civil society came into its own in the second half of the nineteenth century. Farmers' organisations had organised themselves nationally. Workers first organised locally, then nationally. Associations of manufacturers followed in their wake. The rise of the labour movement led to the founding of the Netherlands Workers' Union in 1869 as the Dutch branch of the First International. The first modern political party was the Anti-Revolutionary party founded in 1879. The socialists followed suit with the Social-Democratic Federation in 1881, the liberals with the Liberal Union in 1886 and the Catholics with the Roman Catholic State Party in 1896. Then there were the professional organisations of physicians, engineers and economists and long-existing or newly founded associations with idealistic aims like the Society for the General Good of 1784 and the Peoples Federation against Alcohol Abuse of 1875.[4]

---

[3] Van Zanden and van Riel, *Nederland 1780–1914*, 314–315. Van Zanden and van Riel see the striving for the political and socio-cultural integration of new voters as one of the most important processes behind the rise of the social question, the struggle around special (religious) education, and the civilizing offensive after 1870. For figures, see table 7.1, 315.

[4] The People's Union against Alcohol Abuse (*Volksbond tegen Drankmisbruik*) was founded in 1875 under the name of Multapatior's Union against Intoxicating Spirits. In 1882 the name was changed to People's Union, Association against Alcohol Abuse, later referred to more briefly as People's Union against Alcohol Abuse.

Civil society certainly flourished, but did it also develop into a force able to influence the government in regard to the social question? Organisations could pursue their programs in three main ways: via societal action, via governmental organisations or via a political party. For workers, strikes were the weapon of choice. Up to the end of the 1880s the number of strikes was limited, with a peak of 33 strikes in 1872. After that the number of strikes would increase rapidly up to the First World War, with a peak of 400 strikes in 1913. Most strikes were spontaneous and barely affected government policy in any direct sense. The railway strike of 1903 was a watershed. After that, a large part of the workers' movement opted for a mode of unionization based on solid and nationally organised unions capable of being a force in Dutch politics.[5]

Professionals opted for another strategy. The example of the hygienists – a new generation of young and progressive physicians – shows how successful they could be.[6] The hygienists formed a solid national network, that was anchored in local medical associations and the national Society for the Promotion of Medicine (1849). They used this network to collect, exchange and tabulate statistics on diseases. They also created national standards for research protocols and the definition of diseases. This made the hygienists the most important suppliers of medical knowledge. Around 1870 their method of operation had been adopted by the government that in turn had assured itself of the cooperation of municipalities, citizens and physicians.

The hygienists also worked together with other groups on an hygienic programme that over the years came to include more than sanitary facilities. The quality of foodstuffs, public housing, working conditions and the establishment of new factories and workshops all came under their purview. An important role in this programme was played by the so-called health commissions, that had been set up in the 1850s and 1860s in a large number of municipalities and whose membership consisted of physicians, lawyers, scientists and engineers. A national breakthrough occurred in 1865 when a law was passed creating a State Medical Inspectorate. Almost all the hygienists became paid or unpaid staff of this State Inspectorate. In addition, every hygienist was member of a health commission or an association for promoting public health, often one they set up themselves. Sometimes they were members of city councils or became aldermen their cities. The medical community, and in particular the hygienists, thus acquired local influence on municipal health policy and national influence on emergent public health policy.

The third strategy – that of the political party – was initially developed by Abraham Kuyper.[7] On the one hand Kuyper worked on a number of civil organisa-

---

[5] S. van der Velden, *Stakingen in Nederland: Arbeidersstrijd 1830–1995* (second revised edition, Amsterdam 2009). For figures see the website of the International Institute for Social History, table with the number of strikes and shutouts per year, https://socialhistory.org/sites/default/files/docs/overzicht-aantallen-stakingen.pdf

[6] See for the below analysis: E. Houwaart, 'Medische statistiek', in: H.W. Lintsen et al. (eds.), *Geschiedenis van de techniek in Nederland: De wording van een moderne samenleving* (Zutphen 1993), volume II, 19–45.

[7] See for the role of Abraham Kuyper in Dutch politics: P. de Rooy, *Ons stipje op de waereldkaart: De politieke cultuur van modern Nederland* (Amsterdam 2014), 126–133.

tions to organize his protestant constituency and on the other on a political party that would maintain control over the civil initiatives and that would introduce protestant issues into parliament. Thus in 1869 he mobilised the existing Association for Christian National Schooling to start taking political action, in 1872 he was co-founder and founding chief editor of the journal *The Standard*, in 1878 he was instrumental in establishing the Free University in Amsterdam, and in 1886 he was the leader of the so-called *Doleantie*, a Calvinist current within the Dutch Reformed Church (*Nederlandse Hervormde Kerk*) that led to a schism and to the founding of the Reformed Churches in the Netherlands (*Gereformeerde Kerken in Nederland*). Finally he founded the Anti-Revolutionary Party on the basis of 'Our Program.'

It signified a revolution in political culture. Up to then, voters chose their representatives on the basis of their personal qualities. Ideally, parliamentarians were seated in parliament as independent persons and debated issues on the basis of their expertise and their interpretation of the general interest. A political party that defined the relevant political issues and stipulated who voters should choose as their representatives in parliament – as Kuyper was proposing – was anathema.

Other societal movements came to adopt Kuypers' strategy. This led at the outset of the twentieth century to a 'pillarised' social structure. Especially the confessionals and the socialists attempted to bring all the aspects of the lives of their constituencies – family, upbringing, education, habitation, work and leisure – within the sphere of influence of their own pillar. This consisted of a network of educational organisations, health care organisations, cooperative housing associations, sport associations, newspaper publishers, labour unions, farmers' organisations, employers' organisations and – in the case of a confessional pillar -churchly institutions.

Professional and party-political strategies led to the building of a new corporatism.[8] In the course of the nineteenth century the remnants of an old kind of corporatism consisting of guilds, marks and other kinds of corporatist organisation in autonomously functioning cities, had been dismantled. In its place a corporatist state structure emerged that was characterised by a societal midfield that had lodged itself firmly in a political system and a governmental bureaucracy at local, provincial and national levels.

11.2.2   The Birth of the Welfare State

Political debate in the second half of the nineteenth century consisted to an important degree of issue that would later become the 'social question' and in particular addressed the role that the government should play. Thorbecke had already seen to it that the constitution included poor relief as a governmental responsibility. He stated that 'a civilized nation … is duty bound to ensure as far as possible that its

---

[8] The thesis was proposed first by J. van Zanden and A. van Riel, see the section 'Collectieve actie en de opbouw van een nieuw corporatisme' in: Van Zanden and van Riel, *Nederland 1780–1914*, 322–329. Our analysis diverges somewhat, but the conclusion is the same.

members do not perish from want. Charity on the basis of church membership cannot do this...'[9] The Poor Law he submitted to parliament in 1851 was written in this spirit.

Opposition was fierce. According to the confessionals, the state had no business in this domain and the liberals regarded poor taxes as a thorn in their side. Hence the new poor law left everything as it was. A new attempt in 1901 to make poor relief a governmental responsibility encountered stiff resistance from the Anti-Revolutionaries. The draft law was 'a bold attempt at violating the freedom of the churches.' Ultimately, the Poor Law of 1912 would mark the tentative beginnings of more state influence – namely by municipalities – on poor relief.[10]

As we saw, the government had made a modest incursion into the domain of public health in the 1850s and 1860s. The mandatory establishment of health commissions by municipalities was rejected by parliament. It did, on the other hand, accept the founding of a State Medical Inspectorate. The national government made almost no effort to improve public hygiene. In this area, several municipalities with their health commissions took the lead, like Rotterdam, Amsterdam, The Hague and Utrecht.[11]

While a young generation of physicians (the hygienists) was active in the political domain of public health, a young generation of engineers was similarly active in the domain of labour.[12] The way had been paved by the social liberals, who were convinced that not everything could be left to the market. They demanded a greater role for the government in social issues. Parliamentarian Van Houten's 1874 Child Labour law was their first success.[13] The parliamentary enquiries of 1887 and 1890 forced a breakthrough in the debate and placed the issue of labour squarely on the political agenda. A new generation of engineers, educated in Delft, had appropriated this domain and enthusiastically supported the work of two of their fellow engineers in three liberal cabinets between 1890 and 1901.[14] The engineer P. van der Sleyden succeeded as minister in shepherding a Safety Law through parliament in 1895 and bringing the Public Nuisance Act of 1875 and the Labour Law of 1889 together in an elegant fashion. The Labour Inspectorate, dominated by engineers and charged

---

[9] M. van Leeuwen, 'Armenzorg, 1800–1912: Erfenis van de Republiek', in: J. van Gerwen and M. van Leeuwen (eds.), *Studies over zekerheidsarrangementen: Risico's, risicobestrijding en verzekeringen in Nederland vanaf de Middeleeuwen* (Amsterdam 1998), 284–286, quote on p. 284.

[10] M. van Leeuwen, 'Armenzorg, 1912–1965: Van centrum naar periferie', in: J. van Gerwen and M. van Leeuwen (eds.), *Studies over zekerheidsarrangementen: Risico's, risicobestrijding en verzekeringen in Nederland vanaf de Middeleeuwen* (Amsterdam 1998), 284–286. Citation from: Van Leeuwen, 'Armenzorg 1800–1912', 294.

[11] H. Buiter, *Riool, rails en asfalt: 80 jaar straatrumoer in vier Nederlandse steden* (Eindhoven 2005). See also: A. van der Woud, *Koninkrijk vol sloppen. achterbuurten en vuil in de negentiende eeuw* (Amsterdam 2010), 257–259.

[12] H.W. Lintsen, *Ingenieurs in Nederland in de negentiende eeuw: Een streven naar erkenning en macht* (Den Haag1980), 198–325.

[13] It prohibited children up to age 12 to work in factories. Notably, the prohibition did not apply to cottage industry, household services and agricultural labour.

[14] Lintsen, *Ingenieurs in Nederland in de negentiende eeuw*, 330–342.

with the implementation of the Labour Law, now also began to monitor adherence to the Safety Law. Furthermore, the inspectorate was empowered to nullify municipal permits granted under the Public Nuisance Act if the installation in question violated the Safety Law. After this victory, Cornelis Lely (the engineer with the plans for the closure of the Zuiderzee) managed, as minister, despite vociferous resistance, to get parliament to pass a draft Accident Law. The editor of the journal *De Ingenieur* wrote on this occasion:

> Lely has demonstrated once again that in social matters he is in the vanguard, straight through storms and heavy seas, never quitting his post, and in the process defying not only big industrialists but even more the objections advanced against his bill in the First Chamber ...[15]

The objections addressed above all the practical application of the law. Who would organise the accident insurance and bear the risks? Should that be left to the state or to civil society, in particular the employers? A compromise seemed to be the only workable solution.. Employers could have recourse to a Government Insurance Bank, but also to an insurance company. They established their own cooperative insurance company, *Centraal Beheer*. Nonetheless, the Government Insurance Bank was given a supervisory and policy role.

A similar question troubled the Housing Law of 1901 that aimed at improving public housing. This was ambivalent about whether government, private enterprise or citizens should take the initiative. The law enabled the government to subsidise certified housing associations and contractors who were exclusively committed to the interests of public housing.

By 1900 the social question had acquired a legal framework within which 'pillars,' governments, entrepreneurs, citizens and workers could proceed in solving different issues. In retrospect the legal framework can also be seen as the origins of the 'caring state' as that would develop in the twentieth century. After this period, well-being and quality of life acquired an entirely new dynamic.

11.2.3   The Emergence of a Modern Economy

From the 1840s on the Dutch economy underwent a radical transformation. It began with the liberalisation of foreign trade that cleared the way for the globalisation and accelerated integration of international markets. It encountered opposition from Dutch manufacturers. Various Chambers of Commerce and Factories called for protection for domestic industry because, 'every nation...must attempt to shield its trade and agriculture from being taken unawares and being destroyed by jealous neighbours.'[16] The government and parliament paid little heed to these appeals.

---

[15] 'Uit Ons Parlement', *De Ingenieur* 15(1900), 663. Quotation in: Lintsen, *Ingenieurs in Nederland in de negentiende eeuw*, 342.

[16] J.C.G. Schulte 'De Kamer van Koophandel en Fabrieken, 1842–1862' in: *De opkomst van Tilburg als industriestad: Anderhalve eeuw economische en sociale ontwikkeling*, 88–96. Zie ook:

They also succeeded in achieving an integration of domestic markets by means of legislation and the creation of new infrastructures. Modernisation of the economy subsequently proceeded along many fronts. Accelerated integration of the service sector came about with innovations on the money and capital markets. The production ceiling in agriculture was shattered thanks to specialisation and artificial fertiliser, while with the coming of artificial fertiliser the circular economy of the food chain changed into a linear one. Steam technology broke through the production ceiling in small firms and stimulated specialisation and the rise of large enterprises.

The relationships in the economy also changed because different groups tried to co-opt the modernisation process and exercise influence on the national government. Farmers' organisations had gone nationwide after 1850, founded cooperatives in the 1890s and since then worked closely with the government. Workers' associations were long a local phenomenon (with Amsterdam often in the vanguard) and organized by trade (with type-setters and diamond cutters as pioneers). Fifteen trade unions joined forces in 1906 and founded the Netherlands Association of Trade Unions. In 1909 several unions formed the Christian Trade Association and in 1912 the catholic union movement united in the Roman Catholic Workingman's Association.

Employers organisations followed a similar course. They began locally and by sector, especially in response to emerging workers' opposition. This was followed by national cooperation. The Association of Netherlands Employers of 1899 was the first important national employers' organisation. It emerged from worries within the group of textile factory owners in Twente with regard to the Accident Law that was then in preparation. Pillarisation of the employers' organisations mimicked that of the workers' unions. In 1907 there were 307 local and 35 national associations.[17] Factory owners did not react only in a defensive way to the workers' movement, socialism and the social question. Some, including Van Marken of the Delft yeast factory, the machine-builder Stork in Hengelo and the linen weaver Van Besouw in Goirle launched initiatives such as profit sharing, housing for workers and funds for ill, invalid and older workers.[18]

The modernising economy was also characterised by the rise of new professional groups. These included in particular mechanics, chemists, engineers, lawyers and physicians. They were the representatives of a modern knowledge infrastructure.

H.W. Lintsen and J. Korsten, *De veerkracht van de Brabantse economie: De Kamers van Koophandel en de kracht van netwerken [1840–2015]* (Hilversum 2007).

[17] B. Bouwens and J. Dankers, *Tussen concurrentie en concentratie: Belangenorganisaties, kartels, fusies en overnames* (Amsterdam 2012), 55.

[18] Nijhoff and van den Berg, *Het menselijk kapitaal*, 47–48.

11.2.4   The Foundations of a Modern Knowledge Infrastructure[19]

One of the most essential elements of the modern technological knowledge infrastructure was 'codified' knowledge. In craft-based manufacture, knowledge was mostly 'tacit,' that is to say embedded as non-reflexive routines in the heads and limbs of people. It was nearly impossible to explicate and could be acquired only through intensive practice. Knowledge was transferred from master to apprentice or by means of other modes of face-to-face contact. The development of knowledge occurred in practice, and communication took place via the spoken word and body language.

'Codified' knowledge was explicated knowledge, knowledge that was expressed in the written and printed word. Knowledge and communication were in principle no longer bound to one person and were hence de-contextualised in time and place. Books and articles could be consulted at any time. They could be preserved through time and consulted at different places. 'Codified' knowledge also expressed itself in more precise measuring and more systematic testing and experimentation. This knowledge was easier to share with others by means of printed books and journals. It could thus acquire a collective character, that is to say, become available to a professional community and the interested public.

The construction of a modern knowledge infrastructure in the second half of the nineteenth century took place to a great extent around 'codified' knowledge: new technological programs of study with their textbooks, new associations with their journals, new technical services with their technical literature and new laboratories with their scientific publications. This construction took place at all levels.

Technical education, for example, comprised the trade schools (the first dating from 1860), middle-level technical schools (like the Amsterdam School for Machinists, founded in 1878) and the Polytechnical School dating from 1864 (the successor to the Delft Academy for Engineers of 1842, that would in turn be transformed into the Technical High School, complete with the *ius promovendi*, in 1904), while the universities educated physicists and chemists. Many of the schools owed their existence to private initiative. The Polytechnical School and the universities were financed by the government.

Until 1890 only a modest number of students were enrolled in technical education. After 1890 their numbers increased rapidly. The trade schools, in particular, exhibited phenomenal rates of growth from about 1100 students in 1890 to more than 10,000 in 1915.[20] The Delft engineering school also profited from an increased interest in, among others, mechanical engineering, electrical engineering and chem-

---

[19] See for this section: M. Davids, H.W. Lintsen and A. van Rooij, *Innovatie en kennisinfrastructuur: Vele wegen naar vernieuwing* (Amsterdam 2013), 37–42; 52–64.

[20] P. Baggen and E. Homburg, 'Opkomst van een kennismaatschappij', in: J. Schot, H. Lintsen, A. Rip and A. Albert de la Bruhèze (eds.), *Techniek in Nederland in de Twintigste Eeuw* (Zutphen 2003), deel VII, 154.

ical technology. The interest was representative of an accelerated development of the modern knowledge infrastructure at the end of the nineteenth century. The first Dutch engineering firm was that of J. van Hasselt and De Koning, founded in 1881. In 1890, the national government established three new agricultural experimental stations, followed in 1913 by a State Industrial Service for Small Businesses. Between 1900 and 1910 the number of private and industrial laboratories increased remarkably.[21]

With its new modern knowledge infrastructure, the Netherlands was able to adopt new key foreign technologies in order to develop them further in domestic settings: synthetic chemistry on the basis of sulphuric acid and sodium carbonate; communications technology including telegraphy, telephony and telecommunication; electrical technology with the incandescent lamp and the electrical motor; organic chemistry on the basis of coal and mineral oil and finally mechanical engineering with steam technology, the internal combustion engine and the automobile.[22] In the course of the twentieth century, a number of these technologies would have a major impact on the issues of well-being and sustainability.

11.3 The Monitor of 1910: Well-being and Sustainability from Three Perspectives

How can we judge the endpoint of this period – around 1910 – in terms of well-being and sustainability? In this section we will shed light on this question from three perspectives. The perspective of 1850: to what extent was the societal agenda of the mid-nineteenth century realised? The perspective of 1910: which elements were praised by contemporary politicians and the public, what were the contemporary concerns and what was the agenda for the future? Finally, we look at the developments with present-day eyes. How do we judge the results of the period 1850–1910 by present-day norms? The monitor well-being 1910 provides a summary of the various perspectives (Table 11.1).

---

[21] J.J. Hutter, 'Nederlandse laboratoria 1860–1940, een kwantitatief overzicht', *Tijdschrift voor de Geschiedenis der Geneeskunde, Natuurwetenschapen, Wiskunde en Techniek* 9 (1986), nr.4, 153, graph 1.

[22] In the second half of the nineteenth century, scientific and technical developments in the Netherlands were mainly inspired by foreign examples. In some fields, however, the Netherlands played a leading role as in the field of dietetics with Gerrit Mulder (1802–1889) and in the application of 'proteins'. H.A.M. Snelders, *De geschiedenis van de scheikunde in Nederland: Van alchemie tot chemie en chemische industrie rond 1900* (Delft 1993), deel 1, 93–102.

Table 11.1 The monitor well-being of 1910 from the perspectives of 1850 and 1910 and the present-day perspective

Dashboard well-being 'here and now'						
Theme	Indicator	Unit	1910	Perspective 1850	Perspective 1910	Present day perspective
Population	Number inhabitants	million	5.9			
Material welfare and well-being						
Consumption, income	Consumptive expenditures per capita, constant prices	Index (1850=100)	200	+	−	−
	Income inequality, general	Gini coefficient 0–1	0.47	+	−	−
	Gender income inequality	% difference hourly wage M/F	?	+	−	−
Subjective well-being	Satisfaction with life	Score 0–10	?	?	?	?
Personal characteristics						
Health	Life expectancy	year	55	+	−	−
Nutrition	Height (military conscripts)	cm	173	+	−	−
Housing	Housing quality	% slums	60	+	−	−
	Public water supply	m³/capita	19	+	−	−
Physical safety	Victims of murder	number per 100.000 inhabitants.	0.4	+	+	+
Labour	Unemployment	% workforce.	3.3	+	+	+
Education	Level of education	years	5.8	+	o	−
Free time	Free time	hours per week.	?	+	−	−
Natural environment						
Biodiversity	MSA	% original biodiversity	54	+	+	−
Air quality	SO₂	kg SO₂/capita	25	+	+	+
	Greenhouse gas emissions	ton CO₂ /capita	3.8	+	+	+
Water quality	Public water supply	m³/capita	18.7	+	−	−
Institutional environment						
Trust	Generalised trust	% population with adequate trust	?	−	−	−
Political institutions	Democracy	Democracy-index 0–100	9.5	−	−	−

(continued)

Table 11.1 (continued)

Theme	Indicator	Unit	1910	Perspective 1850	Perspective 1910	Present day perspective
Dashboard well-being 'later'						
Natural Capital						
Energy	Energy consumption	TJ /capita	51	+	+	o
Non-fossil fuels	Gross domestic consumption	ton/capita	3.8	+	+	o
Biodiversity	MSA	% original biodiversity	54	+	+	−
Air quality	SO$_2$ emissions	kg SO$_2$/capita		+	+	+
	Greenhouse gas emissions	ton CO$_2$/capita	3.8	+	+	+
Water	Public water supply	m^3/capita	19	+	−	−
Economic Capital:						
Physical capital	Economic capital stock/capita	index (1850=100)	201	+	+	+
Financial capital	Grossnational debt	% gdp	71	+	+	−
Knowledge	Stock knowledge capital	Index (2010=100)	<0.5	+	o	−
Human Capital:						
Health	Life expectancy	years	55	+	−	−
Labour	Unemployment	% workforce	2.0	+	+	+
Educational level	Schooling	years	5.8	+	o	−
Social Capital:						
Trust	Generalised trust	% population with adequate trust	?	−	−	−
Political institutions	Democracy	democracy index 0–100	9.5	+	o	−

Theme	Indicator	Unit	1910	Perspective 1850	Perspective 1910	Present day perspective
Dashboard well-being 'elsewhere'						
Welfare						
Consumption, income	Development aid	% GDP	−	+	o	−
Natural capital						
Natural capital	Import of raw materials	ton/capita	3.3	+	+	−

(continued)

Table 11.1 (continued)

Legend

+	Not problematic or not problematized
−	Generally acknowledged as problematic
O	Under discussion: different opinions about the scale and nature of the problems
?	Unknown

Note: The signs − and O in the column of 1910 are the then important problematic themes. The column for 1850 shows with +, which agenda items of 1850 were realized in 1910 or were not problematized in 1850. The column of the contemporary perspective indicates with − which current themes would now be regarded as problematic. For the justification of the evaluation in the table, see the main text

11.3.1 Perspective 1850: An Enticing Paradise

What the monitor above all wants to express is the normative shift that occurred with respect to well-being and quality of life, in particular as regards poverty. From the perspective of 1850 – in terms of the political and societal agenda that then prevailed – the Netherlands had made great progress. The extent of extreme poverty had been radically reduced. Hunger had become a thing of the past. The Netherlander was healthier and lived longer. There was ample employment and the population was well-educated. Criminality was low. True, a large part of the population lived in slums and only a small part had access to clean drinking water, toilets and sewers. But public housing and public health were not issues of major concern to citizens in 1850 and were not on the political agenda. For the rest, the future of the Netherlands in 1910 looked quite auspicious through the eyes of 1850. Economic dynamics were such that the last remnants of extreme poverty would disappear. That alone was already a great accomplishment, but the expectations went further: the people would share in the increased level of well-being and would begin to enjoy a certain level of comfort and luxury.

In another respect too, considerable progress had been made according to the agenda of 1850. It is not to be found in the monitor because the issue of water management is not included. The rivers had for the most part been tamed thanks to normalisation and the digging of new river distributaries. Floods due to ice packs and ice jams had become pretty much a thing of the past.

In one respect the observer from 1850 would look with concern at the situation prevailing at the beginning of the twentieth century. It was the question whether democracy had developed positively with the considerable extension of suffrage. Increasing influence by all kinds of groups from among the people and the bourgeoisie had led to major tensions between rich and poor, between workers and capitalists, between Catholics and Protestants, and among religions, socialism and liberalism.

11.3.2 Perspective 1910: The New Agenda of the Turn of the Century

The monitor (Table 11.1) further indicates the contents of the new agenda on well-being around 1910. The question of poverty had by no means been solved. Aside from the fact that a hundred thousand Netherlanders continued to live in extreme poverty, public health remained abysmal. Child mortality was high. Adults still succumbed in great numbers to tuberculosis, typhus and other epidemic diseases. The common people lived in miserable circumstances. Hygiene in and around the dwellings and hygienic facilities in general gave much cause for concern. The labour situation in factories was very worrisome. The Netherlands faced the major challenge of finding a solution for the social question. Popular welfare, public health, public housing and labour all demanded quantitative and qualitative improvements.

The solution of the social question also demanded structural social change. Two unsolved issues remained at the top of the political agenda, namely universal suffrage (and with it the participation of all layers and groups of the population, including women, in political life) and the subsidising of special (i.e. religiously informed) education (and with it influence on the education and forming of the next generations).

11.3.3 Present-Day Perspective: Related and Deviant Values

From a present-day perspective the beginning of the twentieth century was a turbulent period in which economic prosperity and a belief in progress were accompanied with uncertainties, struggle and tensions. The monitor well-being (Table 11.1) indicates that from a present-day perspective we would emphatically agree with the importance of the social question. The new values that were then emerging about the various aspects of quality of life, are now completely accepted. They provided the foundations of a normative and legal framework that would be elaborated by successive generations in the twentieth century. But there are also two striking departures from the then current agenda: the perspective on natural capital and on the colonies.

The dominant vision on natural capital at the time was strictly utilitarian. Nature was to be dominated and made to serve humans. Land had to be won at the expense of water, wastelands reclaimed and subsoil resources exploited. Hunting animals that were a nuisance to agriculture was necessary and had in fact been supported by a bounty system since 1851. Neither civil society nor the political establishment breathed a word about the depletion of raw materials or about environmental damage due to the use of coal. Immediate neighbours could complain about the nuisance of smoke and soot on the basis of the Public Nuisance Act. But a discussion about the air as a collective good, the pollution of which threatened the quality of the environment, was still far in the future. We should of course realise that energy consumption at the time was considerably lower than at present and that levels of SO_2 emissions were far below present-day norms.

Equally unproblematized, was the change in material flows. In the fossil-fuel supply chains, domestic turf was replaced by foreign coal, a situation that made the Netherlands vulnerable on the point of its energy supply. Circular flows in the food chains were increasingly replaced by linear material flows. At the start of the supply chains farmers used ever more artificial fertiliser and imported cattle fodder. There was no discussion about what to do with human and animal faeces at the end of the production and consumption chain, once they lost their function as manure. Faeces, particularly those emanating from cattle husbandry, would become a big problem in the twentieth century.

The situation was a bit different in regard to encroachments on and damage to nature.[23] The urban bourgeoisie, with Amsterdam's elite in the vanguard, developed an increasing sensitivity to natural values. The confrontation with changes in the rapidly growing and industrialising cities will probably have played a role. The psychiatrist Frederik van Eeden, who with his colony Walden pursued a utopian society, repudiated cities '...that extend their filthy outskirts like diseased cell-tissues across the pure earth and that exhaust the life-force of humanity in a steady fire of luxury and sin...'[24] Foreign examples will also have inspired the bourgeoisie, such as with the founding of the Animal Protection Society in 1861. Art also provided inspiration, particularly through the '1880s movement' (of poets, painters, and authors) that etherealised nature and the landscape. Eventually the nature-loving bourgeoisie itself experienced the landscape in the course of day-trips by train, bike or automobile. In 1899 they founded the Bird Protection Society, in 1901 the Netherlands Natural-Historical Society and in 1905 the Association for the Preservation of Natural Monuments. Nature conservation was far from a political issue on anything like the scale of the social question. Nonetheless the movement scored some success, for example with the Bird Law of 1912. This prohibited the capture of migratory birds for consumption. But utilitarian arguments still prevailed. That said, it was not about 'utility' for agriculture or economy, but about 'utility' for humans.

The second point on which today's normative framework deviates from the 1910 perspective is the issue of colonial policy. In the nineteenth century, the colonies were regarded as sources of profit. In particular, gains from the Dutch East Indies comprised a substantial part of the Dutch state budget in the 1850s and 1860s. Average profits from the Cultivation System in those years amounted to between

---

[23] For the following see: J.L. van Zanden and S.W. Verstegen, *Groene geschiedenis van Nederland* (Utrecht 1993), 179–187; H.J. van der Windt, *En dan: wat is natuur nog in dit land? Natuur bescherming in Nederland 1880–1990* (Groningen 1995) and H. Rennes, 'Het Nederlandsche landschap in de twintigste eeuw', in: C. Boissevain, M. Bosboom and H. Rennes, *Typisch Hollands! De verandering van het Nederlandse landschap en de collectie Knecht-Drenth, 1900-heden* (Zutphen 2008), 59–65.

[24] Cited in: J. Bank and M. van Buuren, 'Utopisten en socialisten', in: *1900. Hoogtij van burgerlijke cultuur* (Den Haag 2000), volume 3, 449.

45% and 50% of the tax income of the national government.[25] Big projects like the construction of the railway network were financed with moneys from the colonies. On the other hand, there was the development in the colonies themselves of a monetary system, a land registry and infrastructures like the railways and irrigation systems. However these were intended primarily to help realise a financial surplus destined for the home country.

Dismantling of the Cultivation System took place in the second half of the 1860s. There had been repeated criticism of the Cultivation System from the left-liberal camp, but the chorus of criticism swelled on the publication of Multatuli's *Max Havelaar* in 1860. At the same time it became evident that private entrepreneurs had developed sufficient expertise and financial resources to assume the role of the Cultivation System. Over subsequent years private initiative turned out to perform better than the compulsory cultivation system.[26]

The liberalisation of the colonial economy had little effect on the standard of living of the farmers and coolies in the Dutch East Indies and on the widespread poverty, that considerably exceeded the extent of poverty in the Netherlands.[27] What in the Netherlands began to be called the 'social question,' became 'ethical politics' in reference to poverty in the colonies. The concept was first introduced by the liberal politician Pieter van Brooshooft in 1901 and was promptly included in the queen's speech for that year.[28] It referred to a 'debt of honour' that the Netherlands had to pay back or a 'moral calling' that the country had to fulfil. It was supposed to promote the independence of the Dutch East Indies. Initiatives were taken to promote public education, establish a public credit facility and to curb the privileges of native chiefs and rulers. Roads and irrigation works were further developed. The 'ethical politics' also served as legitimation for military expeditions to enforce changes. It did not, however, lead to the end of the colonial system.

From the perspective of 1910 the Netherlands faced a major challenge, namely the solution of the social question or in other words the improvement of quality of life in the broadest sense of the term. In addition, the tense political stalemates around education and suffrage demanded resolution. The struggle against the waters continued to be necessary, despite the fact that the threat of river floods seemed to have become a thing of the past.

From a present-day perspective we see two big issues looming in the background: the quality of natural capital (including the natural environment) and the colonial relationships.

---

[25] Van Zanden and van Riel, *Nederland 1780–1914*, tabel 5.1, 223. See especially the section 'Koloniale politiek en Batig Slot', 220–231.

[26] According to C. Fasseur, administrators in the Dutch East Indies had already started to reform the Cultivation System by the 1840s and 1850s. Van Zanden and van Riel, *Nederland 1780–1914*, 226.

[27] See for real wages in the Dutch East Indies in comparison with the Netherlands: table 4.6 Real wages of building labourers in selected countries, 1820s–2000s, in: J.L. van Zanden et al. (eds.), *How was life? Global well-being since* 1820 (OECD Publishing 2014), 81.

[28] E.B. Locher-Scholten, 'Brooshooft, Pieter (1845–1921)', in *Biografisch Woordenboek van Nederland*.   URL:http://resources.huygens.knaw.nl/bwn1880-2000/lemmata/bwn1/brooshooft [12-11-2013]. Consulted December 2016.

The biggest threat to well-being came from outside and was foreseen in the Netherlands by only a very few: the First World War. In many ways the war proved a turning point in Dutch history, not in the last place for the dance of well-being and sustainability.

Literature

Baggen, P. and E. Homburg (2003). 'Opkomst van een kennismaatschappij'. In J.W. Schot, H.W. Lintsen, A. Rip and A.A. Albert de la Bruhèze (Eds.), *Techniek in Nederland in de twintigste eeuw – Deel VII*. Zutphen: Walburg Pers.

Bank, J. and M. van Buuren (2000). 'Utopisten en socialisten'. In J. Bank and M. van Buuren (Eds.), *1900. Hoogtij van burgerlijke cultuur*. Den Haag: SDU uitgevers.

Bouwens, B. and J. Dankers (2010). *Tussen concurrentie en concentratie: Belangenorganisaties, kartels, fusies en overnames. Bedrijfsleven in Nederland in de twintigste eeuw*. Amsterdam: Boom.

Buiter, H. (2005). *Riool, rails en asfalt: 80 jaar straatrumoer in vier Nederlandse steden* (dissertation). Eindhoven: Technical University Eindhoven.

Davids, M., H.W. Lintsen, and A. van Rooij (2013). *Innovatie en kennisinfrastructuur: Vele wegen naar vernieuwing. Vol. 5. Bedrijfsleven in Nederland in de Twintigste Eeuw*. Amsterdam: Boom.

Hennes, R. (2008). 'Het Nederlandse landschap in de twintigste eeuw'. In C. Boissevain, M. Bosboom and H. Rennes (Eds.), *Typisch Hollands! De verandering van het Nederlandse landschap en de collectie Knecht-Drenth, 1900-heden*. Zutphen: Walburg Pers.

Houwaart, E. (1993). 'Medische statistiek'. In H.W. Lintsen et al. (Eds.) *Geschiedenis van de techniek in Nederland: De wording van een moderne samenleving 1800–1890*. Zutphen: Walburg.

Hutter, J. (1986). 'Nederlandse laboratoria, 1860–1940: Een kwantitatief overzicht'. *Tijdschrift voor de Geschiedenis der Geneeskunde, Natuurwetenschappen, Wiskunde en Techniek*, 9(54), 150–174.

Leeuwen, M. (1998a). 'Armenzorg, 1800–1912: Van centrum naar periferie'. In J. van Gerwen and M. van Leeuwen (Eds.), *Studies over zekerheidsarrangementen: Risico's, risicobestrijding en verzekeringen in Nederland vanaf de Middeleeuwen*. Amsterdam: Nederlands Economisch Historisch Archief.

Leeuwen, M. (1998b). 'Armenzorg, 1912–1965: Erfenis van de Republiek'. In J. van Gerwen and M. van Leeuwen (Eds.), *Studies over zekerheidsarrangementen: Risico's, risicobestrijding en verzekeringen in Nederland vanaf de Middeleeuwen*. Amsterdam: Nederlands Economisch Historisch Archief.

Lintsen, H.W. (1980). *Ingenieurs in Nederland in de negentiende eeuw: Een streven naar erkenning en macht*. Den Haag: Nijhoff.

Lintsen, H.W. and J. Korsten (2007). *De veerkracht van de Brabantse economie: De Kamers van Koophandel en de kracht van netwerken [1840–2015]*. Hilversum: Verloren.

Nijhof, E. and A. van den Berg (2012). *Het menselijk kapitaal: Sociaal ondernemersbeleid in Nederland*. Amsterdam: Boom.

Rooy, P. de (2014). *Ons stipje op de waereldkaart: De politieke cultuur van modern Nederland*. Amsterdam: Wereldbibliotheek.

Schulte, J. (1959). 'De Kamer van Koophandel en fabrieken, 1842–1862'. In H. van den Eerenbeemt and H. Schurink (Eds.), *De opkomst van Tilburg als industriestad: Anderhalve eeuw economische en sociale ontwikkeling*. Vught: Stichting tot Bevordering van de Studie der Sociale en Economische Geschiedenis.

Snelders, H.A.M. (1993). *De geschiedenis van de scheikunde in Nederland: Van alchemie tot chemie en chemische industrie rond 1900*. Delft: Delft University Press.

Velden, S. van der (2009). Stakingen in Nederland: Arbeidersstrijd, 1830–1995. Amsterdam: Amsterdam University Press.

Windt, H.J. van der (1995). *En dan: wat is natuur nog in dit land? Natuurbescherming in Nederland 1880–1990*. Groningen: Boom.

Woud, A. van der (2010). *Koninkrijk vol sloppen. achterbuurten en vuil in de negentiende eeuw*. Amsterdam: Bert Bakker.

Zanden, J.L. van and S.W. Verstegen (1993). *Groene geschiedenis van Nederland*. Houten: Het Spectrum.

Zanden, J.L. van and Riel, A. van (2000). *Nederland 1780–1914: Staat, instituties en economische ontwikkeling*. Amsterdam: Balans.

Zanden, J.L. van (2014). *How was life? Global well-being since 1820*. OECD.

Part II: New Problems 1910–1970

Chapter 12
The Situation Around 1910: A New Order

Frank Veraart

Contents

Abstract After tackling extreme poverty in the nineteenth century (see Chaps. 7, 8, 9, 10 and 11) the Netherlands faced new societal challenges around 1910: the food supply, public health care, public housing and labour issues. Water management also continued to be an important issue in well-being.

Pillarisation, universal suffrage and the emergence of political parties ensured a long period of political stability with a dominant role for the confessional parties.

Based on the monitors of 1910 and 1970, this chapter sketches the growth of material welfare, the improvement in personal characteristics and the investments in social, economic, and human capital. It also shows the emergent problems of natural capital, both within the Netherlands and in foreign countries. It also provides an overview of shifts in the three main categories of resources (organic, mineral, and fossil) and the associated supply chains. In this way the chapter forms the introduction to a more detailed analysis of trade-offs in the chains of agriculture and foods (Chap. 13), construction (Chap. 14) and energy (Chap. 15).

Keywords Monitor · Well-being · Suffrage · Pillarisation · Economic growth · Natural capital

This chapter is written by Frank Veraart with contributions by Fred Lambert and Jan-Pieter Smits.

12.1 1918 – The Counter-Revolutionary Breakthrough

> We are living in troubled times. The spirit of revolution is stalking our borders. As yet they
> remain closed to its pernicious entry, as yet our sober reason and a healthy Dutch calm
> deliberation prevail, that does not easily let itself be seduced into undertaking actions
> opposed to order and authority. Yet unrest can arise among our people. A certain agitation
> coupled with fear for things that may come.[1]

On the 12th of November 1918 an appeal went out to Dutch Catholics to be prepared. The previous day, the Social Democratic Workers' Party (SDAP) had rattled
the revolutionary sabres.

> Comrades, the mighty movement that is causing Europe to shake on its foundations is
> approaching the borders of our country and is urging Dutch workers too to fulfil their his-
> toric task… Do not let the moment pass, seize the power cast into your laps and do what you
> must and what you can.[2]

The SDAP leader P.J. Troelstra had treated his Rotterdam public to a fiery speech.
Troelstra called for an orderly and civilized seizure of power.

> Do not defile this great time with unworthy deeds, let it be said in times to come: the Dutch
> proletariat showed itself equal to its task, the Dutch proletarian revolution was the moment
> of glory in Dutch history.[3]

These words were followed by a 'lengthy and enthusiastic applause,' according to
the socialist newspaper *The People* (*Het Volk*).[4]

Sunday, November 10, 1918, the last day of the First World War, marked the start
of a week of great confusion and uncertainty. The German Emperor Wilhelm II fled
to the Netherlands. Hunger had Dutch cities in its grip and hungry soldiers were
becoming restless in their barracks. On Saturday morning, the mayor of Rotterdam,
A.R. Zimmerman, had discussed the proletarian seizure of power with leaders of the
SDAP. Troelstra called out the revolution on Monday. The government in The
Hague made an inventory of preventive measures and negotiated with the British
about fast shipments of food.

On November 12, Troelstra demanded the transfer of power in a long speech to
the Second Chamber of Parliament. The government did not comply. The non-
socialist papers announced emergency consultations of various societal organizations, but also the first food shipments and the fact that the Dutch fishing fleet had
put to sea. A few fearful days followed, but on the streets there was no sign of
anarchy or socialist revolution, only well-orchestrated demonstrations of appreciation for the royal family and the government. After several days Troelstra admitted
his 'mistake.' On November 18, the queen appeared, apparently spontaneously, at
one of the largest demonstrations on the Malieveld, the national mall, in The Hague.

---

[1] 'Katholieken..Paraat' in *De Tijd: godsdienstig-staatkundig dagblad*, 12-11-1918.

[2] 'Aan de Nederl. Arbeidersklasse' en 'Een Rede van Troelstra' in *Het Volk, dagblad voor de arbeiderspartij*, 12 November 1918.

[3] 'Een Rede van Troelstra' in *Het Volk, dagblad voor de arbeiderspartij*, 12-11-1918.

[4] 'Een Rede van Troelstra' in *Het Volk, dagblad voor de arbeiderspartij*, 12-11-1918.

The near-revolution had big political consequences. It reinforced the position of the monarchy and above all of the established confessional parties, i.e. those rooted in religious denominations.[5]

The revolution in the Netherlands was announced at a curious moment, at least so it seemed. Many of the original socialist ambitions to improve well-being had just been agreed to or had already been implemented since the turn of the century. The so-called 'social question' had become part and parcel of the political culture. Just about every politician was busy with the 'uplifting of the worker and the people.' Working conditions, public housing, public health and similar issues were all candidates for improvement. Nineteenth century initiatives arising in the 'societal midfield' had gained the support of the 'progressive liberals' (*sociaal liberalen*) and led among other things to the labour law (1889), the accident-law (1901) and the housing law (1901). These became the foundations of what would later be called the welfare state.

But at the time the question was what such a state would look like. The new laws still had to be applied in practice. Important issues like universal suffrage and the organization of education still had to be fought out. Extreme poverty had not yet been eradicated. Food supplies remained problematic. In short, at the outset of the twentieth century the quality of life of the poor and the workers was still – as we concluded earlier – the most important issue regarding well-being.

In many respects the First World War resulted in an about-face on the issue of well-being. The struggle for universal suffrage and education had been won. That created an entirely new political constellation and a new turn in political culture. The role of government changed fundamentally.

Since the nineteenth century male suffrage had been coupled to incomes, savings and diplomas. Between 1900 and 1913 the percentage of men allowed to vote increased from 46% to 67% (of men older than 23 years of age). The growth of the electorate benefitted especially the confessional and socialist parties. After 1900 the liberals began to lose power. By 1913 this led to a political impasse. None of the political power blocks represented a majority and none was prepared to give an inch in respect of their specific standpoint on universal suffrage.[6] Given the circumstances prevailing during the war – in which the Netherlands succeeded in remaining neutral – the parties were prepared to submit to a 'pacification' to put an end to this interminable dispute. In 1917 universal male suffrage was introduced, followed by women's suffrage in 1919. In the first elections under this new regime the confessionals, above all the Catholics, were the big winners. The liberals were the losers.

---

[5] F. Wielenga, *Nederland in de twintigste eeuw* (Amsterdam 2009), 73–77.

[6] Another issue was under discussion in those years, namely education. The political debate concerned mainly the question of the financing of 'special' (i.e. confessional) education with state funds. The debate became known as the 'school struggle.' The issue was resolved in a negotiated compromise between the confessional parties who wanted 'special' schools treated on an equal footing with non-denominational public schooling and liberal and socialist parties who wanted to introduce universal suffrage. The outcome of these negotiations became known as the 'pacification.' See: Wielenga, *Nederland in de twintigste eeuw*, 34–44.

Table 12.1 Total value of assets of 100 largest industrial companies as a percentage of GDP 1913–1990

	1913	1930	1950	1973	1990
United States of America	23	31	18	30	35
United Kingdom[a]	12	29	22	18	38
Germany	13	20	16	–	31
The Netherlands	20	38	62	88	77
The Netherlands Excluding Royal Dutch Shell	14	27	34	52	42

[a]Market share of stocks
Source: Van Zanden, *Een klein land in de 20e eeuw*, 1997

The socialist parties grew but it was not the political landslide that had been anticipated. This may have been the motivation for Troelstra's revolution.

The proclaimed proletarian revolution became a confessional counter-revolution. Much to the socialists' surprise and dismay, their proclaimed revolution forged a strong alliance among confessional parties. This coalition dominated the post-First World War political landscape.[7] The interbellum was the start of a period of political stability that persisted up to the mid-1960s. Catholics (30%), Protestants (25–30%), social democrats (20–25%) and liberals (10–20%) comprised the political spectrum. At a national level, the confessional parties colluded in their joint aversion to the 'red danger.' At a local level, especially in the bigger cities like Amsterdam, social democrats participated in government.

The First World War turned out to be a catalyst for the resolution of a number of thorny political issues. Wartime circumstances created an experimental setting to try out new approaches in direct state intervention in the economy. Due to Dutch neutrality and the maritime blockades, more emphasis began to be placed on national independence and autarchy. This translated into more intensive exploitation of the national territory. Investments increased, among others in the State Coal Mines. After extensive flooding in 1916 around the Zuiderzee, definite plans were made for closure and partial reclamation. The national government also saw itself compelled to intervene in the market for foodstuffs and other scarce commodities. Wartime conditions created closer relations of cooperation between government and business.

After the war there was a successful transition from a wartime to a peacetime economy as a new economic context for the development of well-being and sustainability emerged. As in many western countries in the twentieth century, modernisation of society was first and foremost understood as industrialisation (see Table 12.1). During the 'golden years' between 1923 and 1929, economic growth and labour productivity rose to great heights. Six companies contributed significantly: Shell, Philips, the State Mines (presently DSM), Hoogovens, AKU (presently AkzoNobel) and Unilever (a 1929 merger of the Margarine Union with the British Lever Brothers). It was during this period that the Netherlands acquired its characteristic dual economic structure with a top-six of big companies and a large number of smaller

---

[7] Wielenga, *Nederland in de twintigste eeuw*, 73–77.

enterprises. The share of the big companies in industrial employment increased threefold in the period between the wars.[8] They attached great importance to research and set up their own laboratories.

Forceful regulation of the economy by the government disappeared after the First World War, but cooperation among the firms continued. The forced wartime cooperation had led to the emergence of clusters in the chemicals and food-processing industries. After the war these were reinforced and assumed the form of cartels in various branches of industry. Despite the revival of international trade, the national perspective did not entirely drop off the radar. It remained a palpable factor in the production of food and in investments in basic industries.

The government continued to concern itself intensively with food supplies, public housing, education and other social issues. The political blocs had different opinions regarding the role of the government. Socialists preferred active governments with broad competencies in social-economic domains. The confessional parties preferred in principle a limited role for the national and local governments. They envisioned a corporatist model. This defined the state as a modern night-watchman, providing overall legal and financial frameworks. On this view, the execution of the societal agenda should be left to the 'societal midfield.' In this confessional perspective the 'midfield' consisted of specific organizations from within their own confessional community with in many cases close ties to politics and churchly authorities.

The Catholics succeeded in developing an institutional and governance system that penetrated deeply into the capillaries of Catholic life via countless associations. Catholics associated at work, in leisure time and in private life, read the same newspaper, went to the same kind of school, were members of a Catholic trade union, etc. The same was true of the socialists. The Protestant and liberal currents were less rigidly organized but here too many organizations and associations were founded on the basis of a specific shared ideology. This process of extremely fragmented and varied initiatives was the foundation for so-called 'pillarisation' and for the rise of a new political culture in which the pillarised 'midfield' interacted intensively with parliament, the government and the state apparatus. The different pillars each tried to shape societal development based on their particular perspective on the world. Authority and control were an important motivation. What influence did pillarisation have on the production of well-being? Did a specifically Dutch kind of welfare state come into being?

The 'midfield' was more inclusive than just the pillars. Even before the First World War it was densely populated by all kinds of professional groups like physicians and engineers. They were closely associated with the enactment of laws and would continue to participate in political debates on the social question. In time, new groups like spatial planners, psychologists and sociologists would also join in. Was the influence of these professionals in the pillarised Netherlands at all noticeable?

---

[8] Jan Luiten van Zanden, *Een klein land in de 20ᵉ eeuw, economische geschiedenis van Nederland, 1914–1995,* (Houten, 1997), 59; E. Bloemen, J. Kok and J.L. van Zanden, *De top 100 van industriële bedrijven in Nederland 1913–1990,* (Den Haag, 1993), 10.

The social movement around landscape and nature had a special role. As we saw earlier, around 1900 the countryside had a mesmerising effect on the bourgeois elite. They regarded the countryside as a still reasonably authentic idyll suffused with folklore and pastoral delights. Impressionist painters in artists' colonies like Laren, Katwijk and Plaswijk recorded this cherished landscape on canvas.

> The simple rural huts and farms, surrounded here and there by trees, are picturesquely spread out across the heathlike farmer's fields and marshes.[9]

A society like the Association for the Preservation of Natural Monuments (1905) wanted to conserve these special landscapes and their flora and fauna as in a collection. The General Netherlands Cyclists' Union (*ANWB*, 1883) stimulated the discovery of the idyllic countryside and 'nature' by bicycle, automobile or just walking. In these circles, country living was romanticized, it was a place of refuge and a perfect counterpart to modern industrializing city life.[10] Would these movements succeed in getting biodiversity, the environment and other aspects of sustainability on the political agenda?

Various movements active prior to 1914 had survived the First World War. In addition, the war had created entirely new frameworks for well-being and sustainability. These form the starting point for an analysis of the period 1910–1970. We will start with a comparison of the monitor of well-being for 1910 with that of 1970 and of the material flows in both years.

12.2 Well-being 'Here and Now': A Life longer and Happier, 1910 Versus 1970

Dutch population grew from 5.9 million in 1910 to 13.0 million in 1970. Real per capita consumptive expenditures increased by a factor of 1.75, while income inequality declined (Table 12.2). This meant that in this period extreme poverty was nearly banished. At the outset of this period some 8.5% of the population lived below the poverty line. In absolute terms, this still represented a sizable number of 500,000 people.

Living conditions for the urban poor were difficult. According to socialist critics the cities were:

> modern, ugly and repulsive heaps of humanity, (…) cancerous growths on the beautiful earth, ruined by smoke and stench and degenerate people (…)[11]

---

[9] J.S. Göbel, *De Plasmolen: Gids voor de bezoekers van den Plasmolen en omgeving*, (1910). The colonies even drew groups of American painters who came to paint what was in their eyes a 'Dutch Utopia,' the as yet unspoilt Dutch countryside.

[10] A. van der Woud, *De nieuwe mens: De culturele revolutie in Nederland rond 1900* (Amsterdam 2015), 217–25.

[11] The socialist Frederik van Eeden cited in Jan Bank and Maarten van Buuren (eds.), *1900, Hoogtij van Burgerlijke Cultuur*, vol. 3, Nederlandse Cultuur in Europese Context (Den Haag 2000), 449.

Table 12.2 Dashboard well-being 'here and now,' 1910–1970

Theme	Indicator	Unit	±1910	±1970	Corresponding CBS methods
Population		million inhabitants	5.9	13.0	
Material welfare and well-being					
Consumption, income	Consumptive expenditures per capita, constant prices	index (1850=100)	200	340	⇧
	Income inequality, general	Gini coëffic. 0–1	0.47	0.36	⇧
	Gendered income inequality	% difference in hourly wage M/W	?	29%	?
Subjective well-being	Satisfaction with life	score 0–10	?	7.4	?
Personal characteristics					
Health	Life expectancy	years	55	75	⇧
Nutrition	Height	cm	173	182	⇧
Housing	Housing quality	% slums	60	6	⇧
	Public water supply	m³/capita	19	109	⇧
Physical safety	Murder victims	number per 100.000 inhabitants	0.4	0.7	⬇
Labour	Unemployment	% workforce	2.0	1.6	⇧
Education	Level of Education	year	5.8	9.0	⇧
Free time	Free time	in hours/week	?	47.9	?
Natural environment					
Biodiversity	MSA	% original biodiversity	54	66	⇧
Air quality	SO_2	kg SO_2/ capita	4.6	21.0	⬇
	Greenhouse gas emissions	ton CO_2/capita	3.8	10.1	⬇
Water Quality	Public water supply	m³/capita	18.7	109.0	⇧
Institutional environment					
Trust	Generalised trust	% population with adequate trust	?	?	?
Political institutions	Democracy	democracy-index 0–100	9.5	39.0	⇧

Legend

⇧	Positive development
⬇	Negative development
⟺	Not positive/not negative
?	Unknown or irrelevant

Source: See note 23 of Chap. 2

The poorest members of society lived in hovels, in cellars fronting on cramped and crowded alleyways. It was hardly surprising that politicians aimed to improve these living conditions. Of the entire Dutch population a bit more than 38% lived in towns with more than 20,000 inhabitants, and 22% in the three largest cities, (Amsterdam 10%, Rotterdam 7% and The Hague 5%).

The majority of Dutch people in 1910 lived in the countryside. Daily life for the smallholders and agricultural workers on the sand grounds had hardly changed. Working conditions were still unremittingly heavy. Added to this was the huge social problem of agricultural unemployment around 1900. Many day-labourers abandoned the crop-farming regions and the grasslands for the cities. The unemployed were put to work in reclamations, forestry, but also in 'unproductive labour like breaking up rocks, unwinding rope and similar things that have been invented to maintain an illusion of industry.'[12] Living conditions and housing were miserable. For example, in Emmen, in the province of Drenthe, 80% of the dwellings had only one-room.

Between 1910 and 1970 the Dutch economic landscape totally changed. The share of agricultural labour fell from 28% to 7%. Industrialisation transformed Dutch society. By 1970, 28% of the workforce was employed in industries, 31% in international services and 35% in the building trades and other services.[13] The difference in income between rich and poor diminished and most likely also the pay-differential between men and women. Regarding the latter, exact figures for 1910 are lacking. Women's share of the labour force remained nearly constant between 1910 (23%) and 1970 (26%). We can estimate that at the end of the nineteenth century women earned half or two-thirds of what men earned.[14] Opinions differed greatly on the subject of female labour and its remuneration. In 1924, for example, laws on legal competence were enacted, preventing women from working as civil servants. These were retracted in 1956. In the early 1950s the average gross hourly wage of women amounted to 56% of what men earned. In 1958 European agreements led to regulations enforcing equal pay. By 1960 the average hourly wage of women had risen to 60% and by 1970 to 70% of the gross hourly wage of men.[15]

On average, material welfare increased, but were the Dutch more satisfied with their lives? Here too figures for 1910 are lacking, as are time-series data for the period. But we can assume that the picture must have been quite erratic in view of the two world wars and a big economic recession. Increased satisfaction might have been expected after the Second World War thanks to economic growth and the augmentation of the welfare state. We shall return to this.

The nutritional situation in 1970 relative to 1910 was much improved with more and better food. Conditions during the First World War had drawn attention to the

---

[12] Citation from the *Landbouwkundig Tijdschrift* in: Auke van der Woud, *De nieuwe mens*, 221.

[13] J.L. van Zanden and R.T. Griffiths, *Economische geschiedenis van Nederland in de 20ᵉ eeuw*, (Utrecht, 1989), 27.

[14] H. Pott-Buter and K. Tijdens (eds.), *Vrouwen, leven en werk in de twintigste eeuw* (Amsterdam 1998), 180.

[15] Pott-Buter and Tijdens (1998), 183.

production and distribution of food. During the war and the depression of the 1930s, food production acquired a national character in which a balance had to be sought between continuity of production and acceptable prices. For the unemployed and those on welfare the depression was a time of 'hunger and misery.' Nightmares became reality during the 'hunger winter' at the end of the Second World War when access to and distribution of food became critical. In this period, food security was one of the crucial aspects of well-being. Improvements in nutrition as well as in medical technology became manifest in increases in average height and age.

Housing conditions also improved greatly. The number of occupants per dwelling decreased from 4.45 to 3.42. The number of dwellings with fewer than three rooms declined from 68% to 8%. Most homes in 1970 had facilities like a toilet (86%), running water (99%) and electricity (99%). The number of years of schooling increased. Leisure time and travel increased. In 1952 44% of the population left home for vacation, of which 7% to foreign destinations. By 1969 this number had increased to 65% of which 28% to foreign destinations.[16]

The development of the institutional environment was a final indicator that pointed to improvement in well-being. The reform of the voting system increased democratic participation. Compulsory voting ensured turnouts of about 95%, of which only a small minority (7–10%) submitted invalid votes. In this way, a large part of the population participated in the political process. The district system was replaced by a system of equal representation. This stimulated the formation of national political parties along ideological lines. These parties were the political instruments of the pillarised social institutions.

Developments between 1910 and 1970 were not necessarily linear. The period was characterised by external vulnerabilities: two world wars and the Great Depression of the 1930s. Unemployment in both 1910 and 1970 amounted to about 4% of the workforce. In the 1930s it climbed to above 17%, and for the male workforce to about 20%.[17] Unemployment during the depression left deep scars in the collective memory. Not for nothing did the populace demand of the post-war government, that it devotes a good deal of attention to the creation of jobs.

Data on housing shortages in 1910 and 1970 also fail to provide an adequate picture of the problematic nature of well-being in the intervening decades. At 1% this seemed quite low for 1910, even though many lived in miserable conditions. Around 1970 the housing shortage stood at 5%. But in this case too the problem actually reached its zenith between the two sample years. In the latter 1940s, the Netherlands experienced a housing shortage reaching nearly 20%. Public opinion surveys revealed that combatting the housing shortage in the 1950s and early 1960s was considered the most important issue for the government to work on.[18]

---

[16] Nederlands Instituut voor de Publieke Opinie (NIPO), *Zo zijn wij: De eerste vijfentwintig jaar NIPO-onderzoek* (Amsterdam 1970), 40–41.

[17] B. Lodder, *Twee eeuwen beroepsbevolking* (Den Haag 2010).

[18] Nederlands Instituut voor de Publieke Opinie, *Zo zijn wij: De eerste vijfentwintig jaar NIPO-onderzoek*, 112.

There are two areas for which the monitor in this period shows a problem. First of all in the area of physical safety. The number of murders per hundred thousand inhabitants was 0.4 in 1910 versus 0.71 in 1970. Up to 1965, it rose marginally. After 1965 the number of murders increased, particularly in the big cities. With greater frequency, men between the ages of 14 and 65 were the victims. The increase seems to be associated with new forms of criminality.[19]

A second, more inclusive problem, was the change in the natural environment. Biodiversity (in terms of MSA) declined, especially due to scale increases in agriculture. Industrial water and air pollution assumed serious proportions. These issues were recognized by some contemporaries. Heated debates were carried on about the effects of carbon dioxide on the climate, with dire predictions both of a new ice age and of global warming.[20] It was seen as one of the many, but not the most important, environmental problems. It was the local problems of air and water pollution from which the first groups emerged that critically began to question the 'price of progress.' In these years, modernisation became ever more 'contested.'

12.3   Well-being 'Later': Materials and Energy for a Better Future, 1910 Versus 1970

The substantial improvements in quality of life in 1970 relative to 1910 had been achieved by the investments of successive generations of Netherlanders in economic, human and social capital (Table 12.3). These resources were essential for the development of well-being 'later.' The foundation of social capital – namely political participation and trust in the political institutions – had been laid during and just after the First World War with universal suffrage and pillarisation. For the next half century this ensured reasonably stable social relations. This came to an abrupt end in the 1960s. Increasing secularisation and individualism brought new opinions and a rearrangement of the 'societal midfield.'

As far as economic and human capital are concerned, we can only conclude that the period 1910–1970 shows an extremely positive development. Economic capital grew impressively. Capital-intensity (quantity of capital per worker) rose remarkably

---

[19] P. Nieuwbeerta and I. Deerenberg, 'Trends in Moord en Doodslag, 1911–2002,' *Bevolkingstrends*, Centraal Bureau voor de Statistiek, 1st quarter (2005): 56–63.

[20] Around 1970 less than 10% of the scientific meteorological and oceanographic investigations into the effects of CO_2 emissions made reference to global warming or cooling. In 1971 S.I. Rasool and S. Schneider published an article on the rising levels of sulphate aerosols in the atmosphere that could lead to global cooling and a new ice age. *Time Magazine and Newsweek* reported these findings in respectively 1974 and 1975. In the course of the 1970s about 60% of the relevant scientific studies predicted global warming, about 10% global cooling and 40% made no predictions about temperature change. D. Nuccitelli, *Climatology versus Pseudoscience, Exposing the Failed Predictions of Global Warming Skeptics* (Santa Barbara 2015), 19–22.; Th. C. Peterson, W. M. Connolley, and J. Flerck, 'The myth of the 1970s global cooling scientific consensus', *Bulletin of the American Meteorological Society* 89, no. 9 (2008): 1325–37.

Table 12.3 Dashboard well-being 'later': 1910–1970

Theme	Indicator	Unit	±1910	±1970	Corresponding CBS methods
Natural capital					
Energy	Energy consumption	TJ / capita	0.05	0.16	⬇
Non-fossil subsoil assets	Gross domestic consumption	ton/capita	3.8	9.4	⬇
Biodiversity	MSA	% original biodiversity	54	66	⬆
Air quality	SO_2 emissions	kg SO_2/capita	4.6	21.0	⬇
	Greenhouse gas emissions	ton CO_2/ capita	3.8	10.1	⬇
Water	Public water supply	m³/capita	19	109	⬆
Economic capital					
Physical capital	Economic capital stock /capita	index (1850=100)	141	518	⬆
Financial capital	Gross national debt	% GDP	71	48	⬇
Knowledge	Stock knowledge capital	index: 2010=100	< 0.5	30	⬆
Human capital					
Health	Life expectancy	years	55	75	⬆
Labour	Unemployment	% workforce	2.0	1.6	⬆
Level of education	Schooling	years	5.8	9.0	⬆
Social capital					
Trust	Generalised trust	% population with sufficient trust	?	?	⬆
Political institutions	Democracy	democracy index 0–100	9.5	39.0	⬆

Legend

⬆	Positive development
⬇	Negative development
⬌	Not positive/not negative
?	Unknown or irrelevant

Source: See note 23 of Chap. 2

during the Great Depression. In order to save on labour costs, entrepreneurs increasingly invested in labour-saving technologies. The increase of this form of capital accelerated after the Second World War and this set the stage for an extremely high rate of growth in Dutch productivity, certainly by international standards. A better educated population was an important investment in human capital. Laws on compulsory education that came into force after 1917 ensured an increase in the number of secondary school students and an increase in the length of schooling. Education become more accessible and achievable for all layers of the population.

Around 1910 the Dutch economy was in a period of transition. It was a mix of traditional and new sectors. Services (51%), industry – excluding foods and luxury foods – 24%, and agriculture (25%) shaped Dutch economic activity. The considerable

agricultural interests were coupled in the agrarian sector (16%) and the food processing and luxury foods sector (9%). In this respect, the economy still exhibited a very
traditional structure. New activities were visible in the contributions of a diverse set of
industrial sectors. These constituted another quarter of the Dutch economy. Textiles,
construction, and raw metals were the most important sectors. Commercial services,
among which trade, transport, and retailing, constituted the biggest piece of the Dutch
economic pie in 1913.

In the years after the First World War, the Netherlands in many respects invested
in a modern economy. Big companies like Shell, the State Mines, and Philips pioneered in chemicals and electrical engineering – the leading sectors of the second
industrial revolution. Steel, the new trend-setting material, was produced on a large
scale as was the electric motor as the new source of motive power. A modern knowledge infrastructure emerged with the industrial laboratory as an icon. The government continued to invest in and elaborate the number of state and semi-state
laboratories. An important step was the founding in 1932 of TNO, the Central
Organization for Applied Scientific Research. Additionally, other knowledge institutes came into being, like the provincial Economic-Technological Institutes (ETIs).
The government also invested in vocational training. In the course of time, for
example, technical education was provided at four different levels: lower, middle,
higher and university-level programs of study.

In this period the role of the state increased substantially. It was responsible for
maintaining the constitutional state and busy passing social and economic legislation that would eventually lay the foundations for a future welfare state. This was
evident in the state budget. Using income from taxes and other sources these expenditures increased after the First World War from several tens to nearly one hundred
euros per inhabitant.

The government had always invested in water management and infrastructure.
'Large works' characterized this period. 1918 marked the commencement of the
project to close and partially reclaim the Zuiderzee, a project that was completed in
1968 with the drainage of Flevoland. The great flood of 1953 initiated the Delta
Works, a rigorous improvement of and shortening of the coastline. From the 1930s
the government invested in a national road plan, shaped around a new network of
limited access highways. At regional and local levels new basic services like running water, electricity, and municipal sewers and gas were implemented. The efforts
of provinces and municipalities resulted in a high degree of coverage throughout the
country.

The improved well-being also had a price. The Netherlands had to surrender part
of its natural capital. The landscape changed considerably. In 1910 the countryside
was still a kaleidoscope of different landscapes. A large variety of small-scale farmland alternated with large tracts of frequently open land. Every region was characterized by its own crops, cattle, and farming techniques. Agricultural statistics in
1910 distinguished 83 different 'farming regions.'[21] These were increasingly

---

[21] J. Bieleman, *Boeren in Nederland, Geschiedenis van de Landbouw, 1500–2000* (Amsterdam
2008), 36–41; Van der Woud, *De Nieuwe Mens*, 185–90.

replaced after the Second World War by large-scale farming and urban expansions. This had an adverse effect on biodiversity.

Time series data for energy consumption and CO_2 emissions both show that up to the mid-twentieth century there was a modest increase in pressure on the environment. A period of strong growth set in after the Second World War, certainly by comparison with other western countries. After 1950 the consumption of materials and energy increased dramatically under the impact of economic growth and increased consumption. Mass production led to price reduction per unit product and increased purchasing power to mass consumption. The refrigerator, the radio, the electric iron and other consumer goods had already been introduced before the war. Developments like the Marshall Plan and the orientation to America, the introduction of the free Saturday and the emergence of large department stores contributed to the advent of a consumer society. There was a new attitude toward disposal and reuse. The quantity of household garbage per capita increased substantially.

12.4   Well-being 'Elsewhere': From Colonial to Global Trade, 1910 Versus 1970

In consequence of two world wars and the Great Depression of the 1930s the Dutch economy became more autarchic in the first half of the twentieth century. In addition the economic slow-downs during the war and depression years reduced the need for energy and therefore also demands on the natural capital of other countries. Energy supply was above all solved domestically by developing the coal mines in the province of Limburg. Exceptions were materials that were scarce or non-existent in the Netherlands. These included building materials like wood and metals, and specific raw materials like cotton for the textile industry and plant oils, and coffee and cacao beans for the food processing industry.

Up until the Second World War the Dutch East Indies, present-day Indonesia, was an important factor in Dutch foreign relations. The Dutch East Indies were regarded as an inseparable part of the kingdom. Like other colonial powers the Netherlands had also developed an ever more effective dual policy since the turn of the century, a policy that sought to invest in the land *and* the population of the colony. In the context of the latter, administrators sought to promote the development of well-being among the colonial population as well, by establishing schools, libraries, and stimulating cooperatives and savings banks.[22]

Investments in the land included above all exploration, making hinterlands and resources accessible, exploitation and other forms of modernisation. Firms like the petroleum company Shell, the Batavian Petroleum Company (*Bataafsche Petroleum Maatschappij*) and the mining company Billiton Inc. played an important role in the exploitation of the natural resources of the colonies. The main offices, from which

---

[22] H. Baudet and I.J. Brugmans, eds., *Balans van beleid: terugblik op de laatste halve eeuw van Nederlandsch-Indië* (Assen 1984), 35–65.

Table 12.4 Dashboard well-being 'elsewhere,' 1910–1970

Theme	Indicator	Unit	±1910	±1970	Corresponding CBS methods
Material Welfare					
Consumption, income	Development aid	% gdp	–	0.6	⬆
Natural capital					
Natural capital	Import of raw materials	ton/capita	3.3	8.6	⬇

Legend

⬆	Positive development
⬇	Negative development
⬌	Not positive/not negative
?	Unknown or irrelevant

Source: See note 23 of Chap. 2

trade was coordinated, were located in the Netherlands. In the domain of agricultural products there were also close ties with companies located in the Netherlands. These flows of trade created the foundation for industries in refining and in producing consumer goods using raw materials from the Orient. After Indonesian independence many of these relations continued to exist, although the companies involved also broadened their horizons.

Decolonisation had far-reaching consequences for the diplomatic and political relationships with foreign countries. Administratively, the Netherlands became 'smaller' and had to realign itself in its foreign relations. To a greater extent than previously this became an orientation to the role of the Netherlands in a West-European context. In this way the Netherlands once again became strongly embedded in the world economy.

The monitor well-being 'elsewhere' reveals an ambiguous situation (see Table 12.4). The relationship with the poor countries was called 'development aid.' In this context the Netherlands helped implement some of the technical aid programs set up by the United Nations after the Second World War. After 1960 bilateral aid programs were started in the areas of education, agriculture and health care in the poorest countries. Experience and knowledge gained in the colonies were dusted off, applied here once again and further developed. The higher goals were fighting hunger and poverty and the improvement of living conditions in these regions.

Import of goods from abroad reached previously unheard of levels. Industrial growth and well-being in the Netherlands in the period 1950–1970 were nourished to a large extent with imported natural resources. Rotterdam became the largest European petrochemical complex. Dutch harbours became the points of transhipment in streams of agrarian products and metal ores. Portions were processed here or sent on to the European hinterland. Rotterdam acquired the appropriate name of 'Europoort.'

12.5   Natural Capital and Material Flows 1913 Versus 1970

> Get on out there, gentlemen Delft mining students (…) If we are not yet become a nation of sluggards, then extracting money from mining must also be an important item on our national agenda; the equal of trade and shipping, agriculture and traditional industry (…). If we be only granted 'time of life,' then will we see the Golden Century return in palpable form, then will our descendants, thanks to burrowing ever deeper into the core of the earth, experience a time that they can give their children lumps of gold and diamonds to play with (…) We have been asleep; long asleep, but the nap is over.[23]

Mining engineer G.P. Rouffaer made no secret of his optimistic expectations about the future of the Netherlands thanks to mining activities. The occasion was the appointment of two full professors in mining technology at the Delft Institute of Technology and the increased political interest in domestic geology. A State Agency for the Exploration of Subsoil Assets (*Rijksdienst voor de Opsporing van Delfstoffen*) was established in 1903. This agency was set up to promote and coordinate the exploration of subsoil assets. Politicians and engineers were convinced that through intensive exploitation of natural capital it would be possible to develop well-being and thus launch a new 'Golden Century.' This conviction turns out to contain a lot of truth, if we compare the use of natural capital and its derivative material flows in 1913 with those of 1970. The increase in well-being, as we could see in the monitor, was accompanied by an enormous input of energy and matter. In this section we provide an overview of the changes in the material flows.

The availability of raw materials (domestic production and import) increased spectacularly in terms of weight (Table 12.5). Between 1913 and 1970, three times more bio-raw materials (including agricultural products and wood) were used, ten times more mineral subsoil assets (sand, gravel, ores etc.), and six times more fossil subsoil assets (among others coal and petroleum). Use of materials also increased in a relative sense, that is per capita. The use of bio-raw materials grew from somewhat more than 2400 kg/per capita to about 3300 kg/per capita, mineral subsoil assets from 1300 kg to 6100 kg per capita and fossil subsoil assets from 2800 to 8000 kg per capita. The growth of mineral and fossil subsoil assets is statistically the most impressive. In the top-ten list of material resources by weight for 1970, they occupy the first seven places (over against the first two places in 1913, see Table 12.6). Mineral oil (68,920 kton) and natural gas (25,650 kton) are on top in a class by themselves, followed by gravel, sand, rocks, coal and clay. The Dutch delta in 1970 was the scene of energetic construction in pursuit of a comfortable and safe existence and that cost a lot of energy.

There was a remarkable shift to processed products (Table 12.7). Raw materials and subsoil assets were ever more seldom directly consumed, but instead served as inputs to industry. That was already the case for mineral subsoil assets, but by 1970 it also held for bio- and fossil raw materials. Imports also increasingly consisted, in both an absolute and relative sense, of processed products, a development mirrored in Dutch exports as well (Table 12.8). These shifts are typical of production and

---

[23] G.P. Rouffaer, in het Tijdschrift van het Koninklijk Nederlandsch Aardrijkskundig Genootschap, Tweede Serie, deel XXIII, 1906, p. 1034–1042, geciteerd in Patricia E. Faasse, *De ontdekking van de ondergrond, anderhalve eeuw toegepast geowetenschappelijk onderzoek in Nederland* (Utrecht 2002).

Table 12.5 Raw materials in the Netherlands 1913–1970 in kton and ton/capita

	1913	1970	Ratio 1910:1970
Bio raw materials:			
Gross Available (kton)	14,740 kton	42,400 kton	1: 2.9
Bio/capita (ton/cap.)	2.4 ton/cap.	3.3 ton/cap.	1: 1.4
% import	22%	22%	1: 1.0
% export	14%	9%	1: 0.6
Mineral subsoil assets:			
Gross Available (kton)	8,040 kton	80,120 kton	1: 10.0
Mineral/capita	1.3 ton/cap.	6.1 ton/cap.	1: 4.7
% import	34%	38%	1: 1.1
% export	7%	8%	1: 1.1
Fossil subsoil assets:			
Gross Available (kton)	17,430 kton	104,890 kton	1: 6.1
Fossil/capita	2.8 ton/cap.	8.0 ton/cap.	1: 2.9
% import	80%	69%	1: 0.9
% export	33%	22%	1: 0.7
Total raw materials:			
Gross Available (kton)	40,210 kton	227,400 kton	1: 5.7
Raw materials/capita	6.8 ton/cap.	17.5 ton/cap.	1: 2.7
% import	50%	49%	1: 1.0
% export	21%	16%	1: 0.8

Remark: Gross available = domestic production + imports
Source: F. Lambert, *Massastromen in Nederland. In de jaren 1850, 1913, 1970, 2010* (researchrapport Technische Universiteit Eindhoven, oktober 2016)

Table 12.6 Ten most massive raw materials (in kton) and the proportion imported (in procent), 1913–1970

		1913	1970	1913	1970
		Raw material (kton)	Raw material (kton)	Import (%)	Import (%)
Bio Raw Materials	Total, of which	14,740	42,400	22	22
	Milk	3,210	8,330	1	0
	Cereals	2,830	6,180	68	78
	Potatoes	2,650	5,040	2	0
	Sugar beets	1,710		3	
Mineral Subsoil Resources	Total, of which	8,040	80.120	33	38
	Clay	3,450	8,470	2	10
	Sand	1,750	22,980	9	5
	Gravel	1,200	27,170	67	47
	Stone	1,060	10,390	100	68

(continued)

Table 12.6 (continued)

		1913	1970	1913	1970
		Raw material (kton)	Raw material (kton)	Import (%)	Import (%)
Fossil Subsoil Resources	Total, of which	17,430	104,890	81	69
	Coal	15,610	9,160	88	53
	Turf	1,600		6	
	Mineral oil		68,920		97
	Natural gas		25,650		0
	Total	**40,007**	**227,400**	**50**	**49**
		35,085	197,040	45	44

Source: F. Lambert, *Massastromen in Nederland. In de jaren 1850, 1913, 1970, 2010* (researchrapport Technische Universiteit Eindhoven, oktober 2016)

Table 12.7 The portion of raw materials and subsoil resources industrially processed (in percent), 1913–1970

	1913	1970
	Industrial processing of raw materials and subsoil resources	Industrial processing of raw materials and subsoil resources
Bio-raw materials	51%	69%
Mineral subsoil resources	60%	60%
Fossil subsoil resources	7%	62%

Remark 1: Industry, excluding mineral and fossil subsoil asset extraction
Remark 2: Mineral resources such as clay and sand directly used in building activities (in for example dikes) are not treated as processed materials
Source: F. Lambert, *Massastromen in Nederland. In de jaren 1850, 1913, 1970, 2010* (researchrapport Technische Universiteit Eindhoven, oktober 2016)

Table 12.8 Import of processed products in kton and percent, 1913–1970

	1913	1970
Total import	26,000 kton	155,920 kton
Import processed products	6,060 kton	44,640 kton
Import processed products (% of Total)	23%	30%
Total export	11,420 kton	95,500 kton
Export processed products	3,110 kton	62,020 kton
Export processed products (% of Total)	27%	65%

Source: F. Lambert, *Massastromen in Nederland. In de jaren 1850. 1910, 1970, 2010* (researchrapport Technische Universiteit Eindhoven, oktober 2016)

Table 12.9 Ten most sizeable processed products (in kton) and the portions imported and exported (in percent), 1913–1970

		1913	1970	1913	1970	1913	1970
		Processed products (kton)	Processed products (kton)	Import (%)	Import (%)	Export (%)	Export (%)
Food Processing Industry	Flour	750		30		10	
	Bread	860		0		0	
	Processed fodder	1260	13,020	8	25	13	6
Wood-Industry	Woodworking	1450		61		0	
Mineral Processing Industry	Utility ceramics (e.g. bricks)	3100	7960	10	4	0	7
	Concrete products	580	11,600	0	5	0	4
	Concrete and asphalt		26,000		0		0
	Cement		6930		33		2
Chemical Industry	Cokes	1330		64		50	
	Artificial fertilizer	880		81		30	
	Petroleum products		75,330		16		56
	Anorganic chemicals		7250		60		20
Metal- and Machinery Industry	Metal products	950		86		4	
	Machine building	610		20		6	
	Raw steel		5330		5		3
	Steel mill products		8020		41		35
	Iron slag		5320		66		1

Source: F. Lambert, *Massastromen in Nederland. In de jaren 1850, 1913, 1970, 2010* (researchrapport Technische Universiteit Eindhoven, oktober 2016)

consumption chains that grow longer and more complex. That development too, as we shall see, made its mark on well-being and sustainability in the period 1910–1970.

Shifts in material flows moreover reflected the development of a new type of economy. An important shift is that in the fossil subsoil assets. Turf disappeared from the top ten most massive raw materials and coal fell from the first place in 1913 to the sixth in 1970. They were replaced by mineral oil and natural gas (Table 12.5). This also marked the end of cokes as a coal product from the list of top ten processed commodities, while the category of petrochemical products crowned the new top ten list (Table 12.9).

In building materials, wood and woodworking lost their dominant role in construction, bricks maintained their position, but new processed building materials were introduced: cement, concrete and concrete products. In the metalworking and machinery industry steel replaced iron as the material of choice for metal products and machinery.

The shift in processed products in the area of bio raw materials is also interesting. In 1913, flour, bread and grain fodder were on the top ten list of processed bio products. Flour and Bread were absent in the 1970 list. Cattle fodder remained in the top-ten of most sizeable processed products (Table 12.9). It was mainly utilized in meat producing cattle breading. This is the origin of important issues in the period after 1970 such as the manure-surplus and the bio-industry.

Autarchy was an important issue in the Netherlands in the period 1910–1917. During the First World War, the country was faced with the shortage of two essential raw materials: grain and coal. In 1913 the Dutch imported 68% of their grain and 88% of their coal (Table 12.6). And yet it appears that subsequently the situation did not improve. In 1970 grain imports were even more voluminous (with an import percentage of 78%). And the Netherlands was almost wholly dependent on foreign sources for mineral oil (with an import percentage of 97%). That said, it was fully independent as far as its natural gas was concerned (0% import). The striving for autarchy had consequences for the exploitation of natural capital, the environment and the landscape. It is an important question for the following chapters.

In 1913, the Netherlands faced a big challenge. As the monitor 1910–1970 shows, there was a lot of work to be done in the area of well-being, from a present-day no less than a contemporary perspective. Extreme poverty persisted. Public health and public housing were important issues. The satisfaction of primary needs left much to be desired, both in a quantitative and qualitative sense. At the same time the population continued to grow and to inhabit an unsafe delta.

To solve these issues, the Netherlands looked to a new way of exploiting natural capital, 'here' in the Netherlands and 'elsewhere.' The changes in material flows point to this. But that brought new problems to the fore. The monitor for 1910–1970 reveals that these were situated in the domain of the natural environment and natural capital. The most important question is thus: in what way is the dynamism in the material flows related to changes in well-being and sustainability? To make this analysis we take as our points of departure the bio-materials (agriculture and foods), mineral substances (construction and building materials) and fossil substances (energy and plastics).

Literature

Bank J. and M. van Buuren (2000). *1900. Hoogtij van burgerlijke cultuur*. Den Haag: SDU uitgevers.

Baudet, H. and I.J. Brugmans (1984). *Balans van beleid, terugblik op de laatste halve eeuw van Nederlandsch-Indië*. Assen: van Gorcum.

Bieleman, J. (2008). *Boeren in Nederland: Geschiedenis van de landbouw 1500–2000*. Amsterdam: Uitgeverij Boom.

Bloemen, E., J. Kok and J.L. Zanden van (1993). *De top 100 van industriele bedrijven in Nederland 1930–1990*. Den Haag: Adviesraad voor het Wetenschaps- en Techniekbeleid.

Faasse, P.E. (2002). *De ontdekking van de ondergrond, anderhalve eeuw toegepast geowetenschappelijk onderzoek in Nederland*. Utrecht: Nederlands Instituut voor Toegepaste Wetenschap.

Göbel, J.S. (1910). *De Plasmolen: Gids voor de bezoekers van den Plasmolen en omgeving*.

Lodder, B. (2010). *Twee Eeuwen Beroepsbevolking*. Den Haag: Centraal Bureau voor de Statistiek.

Pott-Buter, H. and K. Tijdens (1998). *Vrouwen, leven en werk in de twintigste eeuw*. Amsterdam: Amsterdam University Press.

Nederlands Instituut voor de Publieke Opinie (NIPO) (1970). *Zo zijn wij: De eerste vijfentwintig jaar NIPO-onderzoek*. Amsterdam: Elsevier.

Nieuwbeerta, P. and I. Deerenberg (2005). *Trends in moord en doodslag, 1911–2002*. Den Haag: Centraal Bureau voor de Statistiek.

Nucitelli, D. (2015). *Climatology versus pseudoscience, exposing the failed predictions of global warming skeptic*. Oxford: Praeger.

Peterson T.C., W. M. Connolley and J. Flerck (2008). 'The myth of the 1970s global cooling scientific consensus'. *Bulletin of the American Meteorological Society,* 89(9), 1325–37.

Wielenga, F. (2009). *Nederland in de twintigste eeuw*. Amsterdam: Boom.

Woud, A. van der (2015). *De nieuwe mens: De culturele revolutie in Nederland rond 1900*. Amsterdam: Bert Bakker.

Zanden, J.L. van and R.T. Griffiths (1989). *Economische geschiedenis van Nederland in de 20ᵉ eeuw*. Utrecht: Het Spectrum.

Zanden, J. L. van (1997). *Een klein land in de twintigste eeuw: Economische geschiedenis van Nederland 1914–1995*. Houten: Spectrum.

Chapter 13
Agriculture and Nutrition: The End of Hunger

Frank Veraart

Contents

Abstract A robust supply of healthy food was the challenge in the domain of agriculture and foods in the twentieth century. Despite the agrarian successes of the nineteenth century (see Chap. 8), two world wars and the Great Depression had rendered food supply a persistent core element of government policy. Investments in agriculture like reclamations and land re-allocation transformed the landscape. Cooperation among the government, knowledge institutes and industry promoted mechanisation of agriculture. The use of artificial fertilisers and crop protection substances became widespread. Mixed farms transformed into specialized enterprises. The supply chains of agricultural products became longer and more complex.

In the food processing industry too innovations led to long international supply chains and new processing methods. New relationships between producers and consumers were the result. Consumer had to be convinced of the quality of food products by means of government quality control and informational campaigns.

The new production chains were a major contributor to the degradation of the natural landscape and the reduction of biodiversity, both domestically and internationally. This culminated in growing social unrest and by 1970 in a more critical view of developments in agriculture and the food processing industry. This was the prelude to measures in the area of sustainable agriculture and food production (see Chap. 18).

Keywords Agriculture · Food supply · Reclamations · Land re-allocation · Rationalisation · Specialisation · Longer supply chains · Biodiversity · Food quality · Consumption patterns

13.1 Ode to Winter

'I hear the bitter winter calling
The snow keeps falling, falling, falling
We're singing gaily young and old
We're getting now so nice and cold
The cold will make us strong and hearty
My father's job has just departed.
Food's becoming very dear
The rent is due all through the year.
We're very far from being merry,
My youngest sis has dysentery.
That gives us one less mouth to feed,
How good is God! How great his deeds.'

This satirical poem, *Ode to Winter,* was read to the Second Chamber of parliament in December 1903 by the SDAP (Social-Democratic Workers' Party) parliamentarian K. ter Laan in the budget debate during deliberations on primary education.[1] In the debate, Ter Laan pressed for extra finances for food and clothing for children. The verse, according to Ter Laan, had been reprinted in Belgian, French, German and Austrian newspapers to underscore the abysmal living conditions in the Netherlands. Amsterdam city councillors responded irritably and even saw sacrilege in the verse. The political squabble caused by the poem illustrated that around the turn of the century food was an important and timely issue.

Food production and nutrition would remain on the political agenda throughout the twentieth century. Part of the problem was the distribution of food. As in the poem, the poor were still living from hand to mouth, though as a rule the Netherlands had more than enough food. The situation had improved since the nineteenth century, but what additional measures were called for? Food security was a major issue. Two world wars had demonstrated that the food system was vulnerable in extreme situations. What policies had to be developed to banish hunger permanently? The distribution of food and food security belong to the classic core of well-being and sustainability and go back to the nineteenth century and earlier. In large part they determined the quality of life of the Dutch population. These problems would be largely solved in the twentieth century but in their place new sustainability issue

---

[1] Staten-Generaal *Handelingen Tweede Kamer* 1903–1904 15 December 1903, p. 937.

arose. The food problem shifted to issues of quality and healthy nutrition. Additionally problems with the environment and the landscape came to the fore.

We shall use three case studies to illustrate the changing character of sustainability: (1) reclamation and land consolidation, (2) crop protection and artificial fertiliser and (3) food quality: the so-called 'disk of five'. We close this section with an overview of dynamics in the food supply chain between 1910 and 1970.

13.2 The Transformation of the Landscape

Well into the 1970s food security was the subtext of innovations and policy in the domain of agriculture and nutrition. In 1985, Th. M. Bakker of the Institute for Agricultural Economics was able to conclude in his dissertation that the problem of food security in the Netherlands had been solved. Using mathematical models he demonstrated that:

> ... were we in the future, for whatever reason, unexpectedly to be confronted with a lengthy disruption of the supply of food and food constituents from abroad, then we would not in the short term ... have to fear for a famine.[2]

Given that the means of production remained intact and provided that an annual energy input of 31.8 MJ would be available for producing artificial fertiliser, it would be possible on the basis of a diet composed of 'cereals...and a sliver of pork' to achieve a consumption of 2350 kcal per capita per day.[3] A century after the agrarian crisis the food supply seemed secure, even in times of war when the Netherlands would have to make do without imports of agricultural products.

---

[2] T. M. Bakker, *Eten van eigen bodem: Een modelstudie* (Den Haag 1985), 14.

[3] Bakker, *Eten van eigen bodem,* 11. In Bakker's dissertation different theoretical scenarios are quantitatively assessed. The question at issue is whether the territory of the Netherlands and the associated agricultural produce could supply enough food in the event of a complete moratorium on international imports. These scenarios are based on the transformation of agricultural production with the feeding of the domestic population as primary goal. In this way four scenarios are worked out in the dissertation. In the table the most important features:

Scenario	Land use	Energy equivalent 1 = 31,8 MJ	Input – output energy production	Input-output energy chain	Population size	Menu
1 – Minimum	50%	1	26:100	61:100	13.8	One-sided
2 – Maximum	100%	5.5	52:100	162:100	30.0	One-sided
3 – Responsible	76%	1.7	–	–	16.4–17.9	Responsible
4 – Habitual	100%	2.3	–	–	13.8	Unchanged

This was possible thanks to radical changes in agriculture and land-use. From the time of the agrarian crisis in the nineteenth century, there had been increasing cooperation between the government and various organizations active in agriculture. Agriculture was not to be protected by import duties but rather stimulated. Specialisation and intensification would allow Dutch agriculture to claim a new position in the free international market propagated by the government. The first sign of the new cooperative spirit was the formation in 1893, by the Minister of Agriculture, of the Netherlands Agricultural Committee (NLC), a national coalition of the provincial Agricultural Boards. The NLC emphasized the importance of improving agricultural technology. In 1898 a Department of Agriculture was set up within the Ministry of Internal Affairs, which enabled structured consultations between the government and the organised agricultural sector to take place.[4] Agricultural research and education was further professionalised. Between 1904 and 1918 the National Agricultural School (established in 1876) acquired a fully academic status.[5]

One of the most visible agricultural developments was the expansion and modification of agricultural land by means of draining polders, reclaiming land and land consolidation. Yields were increased by introducing crop improvement, artificial fertiliser and chemical agents to protect crops.[6] Yields per hectare of potatoes, sugar beets, grains and other crops grew steadily (Graph 13.1). Labour productivity in agriculture grew apace. Farming acquired an industrial and entrepreneurial character. The supply of food increased and dependency on food from abroad declined. We start with reclamation and land consolidation.

The Dutch landscape underwent a transformation in the twentieth century. Large regions of moorland disappeared to become farming land and forests. Agricultural plots and grasslands were in turn swallowed up by urban expansion. Researchers at the Alterra institute at the Wageningen University calculated that in the year 2000 only 7% of the moors were still located where they had been in 1900. For deciduous forests the figure was 25% and for pine forests 36%. In the twentieth century only large parcels of grassland and urban areas remained untouched (Fig. 13.1).[7] Open natural terrain declined from more than 6000 km² to less than 2000 km² in the 1970s. The greatest decline occurred in the interbellum and immediately after the Second World War (Table 13.1).

From the outset of the century, aficionados of nature and their associations kept a sharp eye on encroachments into natural terrain. After the ornithological associations founded in the final decade of the nineteenth century, 1901 saw the founding

---

[4] Piet de Rooy, *Ons stipje op de waereldkaart: De politieke cultuur van modern Nederland* (Amsterdam 2014), 205.

[5] J. van der Haar, *De geschiedenis van de Landbouwuniversiteit, (Deel I) van school naar hogeschool, 1873–1945* (Wageningen 1993).

[6] Jan Bieleman, *Boeren in Nederland: Geschiedenis van de landbouw, 1500–2000* (Amsterdam 2008), 463.

[7] W.C. Knol, H. Kramer and H. Gijsbertse, *Historisch grondgebruik Nederland: Een landelijke reconstructie van het grondgebruik rond 1900* (Wageningen 2004).

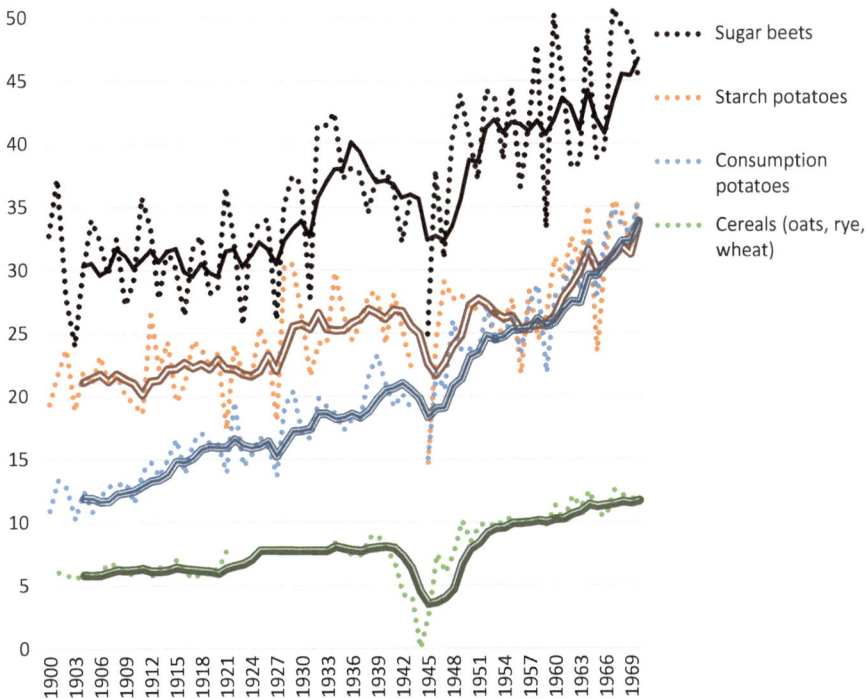

Graph 13.1 Yields of potatoes, sugar beets and cereals 1900–1970 (tons/ha)
Source: CBS – *Landbouw vanaf 1851* (*Agriculture since 1851*)

of the more generic Netherlands Natural History Association and 1905 the founding of the Association for the Preservation of Natural Monuments. Initiators of these associations were representatives of the societal elite who either via networks or professional functions were closely associated with landscape development. The preservation of 'rare' and 'curious pieces' of plants and animals belonged to the core values of the movement. The associations established nature preserves and agitated for legal protection for rare species.[8] After the economically inspired laws on hunting (1852), fisheries (1857) and useful animals (1880), the associations scored a success in the Bird Law of 1912, that protected birds living in the wild. In addition the associations exerted themselves in favour of the Forest Laws of 1917 and of a law on Scenic Beauty that was passed in 1928.[9] These laws laid a legal foundation under their ambitions. But what was the relationship between the value

---

[8] H. van der Windt, 'De totstandkoming van 'de natuurbescherming' in Nederland,' *Tijdschrift voor Geschiedenis* 107(3) (1994), 485–507.; H. van der Windt, *En dan: Wat is natuur nog in dit land?, Natuurbescherming 1880–1990* (Amsterdam 1995).; J. Cramer, *De groene golf: Geschiedenis en toekomst van de Nederlandse milieubeweging* (Utrecht, 1989).

[9] M. Coesèl, J. Schaminée, and L. van Duuren, *De natuur als bondgenoot: De wereld van Heimans en Thijsse in historisch perspectief* (Zeist 2007), 208–213.

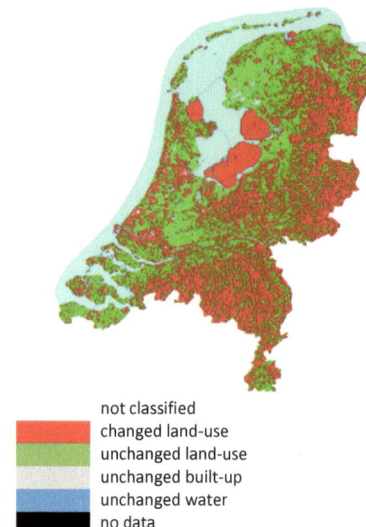

	1900	1970	2000
Forest	10.8	10.3	10.4
Moorland	13.9	1.2	1.1
Grassland	42.0	50.8	47.0
Farmland	27.9	29.3	27.4
Built up areas	1.9	6.8	12.3
Sand/dune	1.8	1.3	1.4
Miscellaneous	1.6	0.3	0.4

Table: percentual change in land-use 1900, 1970 and 2000 (excluding water).

Map: Change in land-use between 1900 and 2000

not classified
changed land-use
unchanged land-use
unchanged built-up
unchanged water
no data

Fig. 13.1 Changes in land-use 1900–2000
Source: H. Kramer and W.C. Knol, *Historisch grondgebruik Nederland: Grondgebruik rond 1970 in 500 meter grids* (Wageningen 2003); W.C. Knol, H. Kramer, and H. Gijsbertse, *Historisch grondgebruik Nederland, Een Landelijke Reconstructie van het grondgebruik rond 1900* (Wageningen 2004). The table shows the percentual figures of land-use. In the studies just cited data is provided in km². However there is an enormous discrepancy in measured surface area between 1970 and 2000. In the measured surface areas for 1900, 1970 and 2000 the largest discrepancies are to be found in the category 'water,' which is why it is left out of the table

Table 13.1 Land-use in the Netherlands in km²

	Agrarian		Forest		Open natural terrain	
		+/– in %		+/– in %		+/– in %
1900	21,160		2520		6240	
1905	21,270	0.5	2570	1.9	5840	−6.8
1910	21,540	1.3	2600	1.2	5530	−5.6
1915	22,010	2.1	2570	−1.2	5110	−8.2
1920	22,180	0.8	2480	−3.6	4930	−3.7
1925	22,530	1.6	2490	0.4	4480	−10.0
1930	22,950	1.8	2540	2.0	3910	−14.6
1935	23,350	1.7	2550	0.4	3570	−9.5
1940	23,240	−0.5	2580	1.2	3580	0.3
1945	21,680	−7.2	2410	−7.1	3270	−9.5
1950	25,050	13.5	2420	0.4	2730	−19.8
1955	25,330	1.1	2450	1.2	2370	−15.2
1960	25,520	0.7	2680	8.6	2360	−0.4
1965	25,690	0.7	2890	7.3	1900	−24.2
1970	25,520	−0.7	2980	3.0	1990	4.5
1975	25,160	−1.4	3080	3.2	1690	−17.8

Source: CBS – Bodemgebruik per provincie vanaf 1900

of nature and the utility of reclamation and land consolidation? How were the different values weighed in practice?

13.2.1 Reclamations

At the end of the nineteenth century, reclaiming the moors was above all a matter of planting new forest. The use of artificial fertiliser also made it possible to transform the heath meadows into grazing pasture or crop fields. Reclamation was carried out by private firms - except for state-owned lands for which an agency called State Forest Management (*Staatsbosbeheer*) was called into being in 1899. Two firms played a pivotal role: the Heidemij (the *Nederlandse Heidemaatschappij* founded in 1888) and the Grondmij (the *Grondverbeterings- en Ontginningsmaatschappij* founded in 1915). They advised on the transformation into forest, crop fields or pasture; proposed improvements in drainage and water-level management; and executed the plans.

Pressure increased on available agricultural acreage during the First World War. The government emphatically encouraged reclamations (and new polders) in order to increase food production. After 1920 reclamations became far less profitable. Profitability was partly dependent on the ability to export agricultural produce. The weakened purchasing power of German consumers as a result of the hyperinflation of the early 1920s as well as increasing competition from Danish, Australian and New Zealand products in the British market caused falling prices.[10] The situation only worsened during the economic crisis.

In cooperation with the agricultural organizations, the government exerted itself to the utmost to protect agriculture and the many small family farms. This so-called 'green front' in which the government and the agricultural organizations cooperated closely, aimed to guarantee a domestic and foreign market for agricultural produce. Agricultural policy continued to pursue the expansion of agrarian acreage. This seemed counterintuitive given overproduction and the low prices of farm produce. But the increase of acreage aimed both to decrease dependency on foreign agricultural produce, especially cereals, as well as to strengthen the position of small farmers and at the same time provide work for the legions of the unemployed.[11] In the first half of the twentieth century on the sand grounds in the east and south of the Netherlands, farms increased their acreage by more than 33% thanks to reclamation (Table 13.2).[12]

---

[10] Bieleman, *Boeren in Nederland*, 209.

[11] De Rooy, *Ons stipje op de waereldkaart*, 210.; P.H.M. Thissen, 'Van Heide tot boerenland en bos, Regionale verscheidenheid in heideontginningslandschappen, 1850–1940,' in M. de Harde and H. van Triest, *Jonge landschappen, 1800–1940* (Utrecht 1994), 21–37.

[12] Bieleman, *Boeren in Nederland*, 408.

Year	Area of Wastelands in ha
1875	766,000
1890	713,000
1900	591,000
1910	543,000
1920	482,000
1930	378,000
1940	270,000
1945	260,000
1955	205,000
1965	235,000
1975	225,000

Table 13.2 Reclamation of 'wastelands', 1875–1975

Source: N.H. Lier, *Een Bont Patroon, Vijfendertig Jaar Cultuurtechniek* (Wageningen 1981), 14

Initially, nature lovers were enthusiastic about reforestation, but this began to wane as the pace of heath reclamation for agriculture increased.[13] Though the State Forest Management (*Staatsbosbeheer*) and the municipalities tended to spare parcels that were scenically attractive, this was not enough for the Society for the Preservation of Natural Monuments. The society bought scattered parcels of heath in the provinces of Brabant, Overijssel and Drenthe, as well as in the Veluwe region in order to preserve regional variety. In 1929 the society set up the Heath Fund, based on public donations. Its first purchase was the Dwingelerveld in Drenthe, a 37 km^2 region of slightly accidented heathland dotted with fens and bogs. These purchases were controversial, as the 1924 acquisition of the De Campina region near Boxtel in Brabant demonstrated; in order to prevent local disturbances, this purchase was conducted in secret.[14]

In 1932, Nature Monuments, the ANWB (cyclists' and motorists' association) and other nature preservation associations, set up the Contact Commission for Nature and Landscape Protection, with the aim of representing the interests of nature and the agrarian landscape to the government. These associations agitated against the structural changes in agriculture, like reclamation and land consolidation.[15] These preservationists, however, had a difficult time propagating the interests of nature preservation, certainly when toward the end of the 1930s the persisting economic crisis claimed increasing attention for the problem of unemployment.

'There is a danger that our generation, in order to temporarily mitigate some of the suffering of unemployment, will forever ruin much of beauty in our country to the detriment of all that come after us,'

---

[13] Van der Windt, *En dan: Wat is natuur nog in dit land?,* 103.

[14] Thissen, *Van heide tot boerenland en bos*, 30–31.

[15] J. Dekker, *Dynamiek in de Nederlandse natuurbescherming* (Utrecht 2002), 26–28.

as the engineer J. Loeff, from the Gooi region near Utrecht, put it in 1939.[16] His remark was incited by the plans for poldering and reclamation proposed by the government inspector for work projects, engineer J. Th. Westhoff in a report to the minister of social affairs, C.P.M. Romme. The report occasioned a heated debate between representatives of agriculture and preservationists about the use and function of natural regions. It got much publicity in the national and regional papers and journals, including professional journals.

Preservationists saw Westhoff's plans as a serious threat with dire consequences for nature. What would remain after reclamation would be 'little parks.'[17] The prominent socialist politician Henri Polak climbed the barricades against the destruction of what he called 'treasures of beauty.'[18] Frisian preservationists associated with It Fryske Gea emphasized the consequences of the disappearance of nature for the local economy. On the other hand, the Royal Netherlands Agricultural Committee was a staunch advocate of drastic expansion of agricultural acreage. Under the banner of 'The Netherlands, dare to live,' chairman H.D. Louwens articulated his criticism in local and regional newspapers:

> … Natural beauty can be enjoyed only with a clear conscience. When natural beauty has to be maintained at the cost of the pleasure in labour of the unemployed and of lost life-chances for young farmers then it can no longer be experienced in innocent joy, because then it is burdened by a dark cloud of human suffering. True joy, including the joy of free nature, can never be egotistical.[19]

Nature was a luxury product, subservient to economy and welfare.

13.2.2 Land Consolidation

After the Second World War the transformation of the landscape acquired an extra dimension due to land consolidation projects. The government's agricultural policy took a new turn. From efforts to maintain employment for the many small farmers and farm labourers, policy turned to mechanisation, rationalisation and upscaling of farm work. Sicco Mansholt, minister of Agriculture between 1945 and 1958, shaped this so-called 'structure policy' that was supported by new laws and by intensive cooperation between agricultural organisations and government agencies. The Agriculture Foundation, founded in 1945 by the three farmers' organisations and farm workers' unions, consulted monthly with the Minister of Agriculture and Food Supply. With the founding of the Agricultural Board (*Landbouwschap*) in 1954 this

---

[16] 'Utrechtse en Hollandse plassen bedreigd' in *Utrechts volksblad : sociaal-democratisch dagblad*, 02–05-1939.

[17] 'Natura' in *Het Vaderland: Staat- en letterkundig nieuwsblad*, 12-10-1939.

[18] 'N.V.V. en Plan-Westhoff' in *Het volksdagblad: dagblad voor Nederland*, 6-2-1939.

[19] H.D. Louwes 'Nederland durf te leven' in onder andere *Nieuwsblad van het Noorden* (3-1-1939) *Leeuwarder nieuwsblad* (3-1-1939) *De Graafschap Bode* (4-1-1939) , *Nieuwsblad van Friesland* (4-1-1939), *Leeuwarder courant* (4-1-1939), *Zaans volksblad* (4-1-1939) *Utrechts volksblad* (4-1-1939) en *De Tijd* (4-1-1939).

corporatist consultation assumed the form of a 'national agrarian parliament.'[20] Government policy aimed at the continued strengthening of the international competitive position of Dutch agriculture. A stable market for agricultural produce was created thanks to guaranteed national and later European selling prices.

Agricultural organisations like farmers' co-ops devoted themselves to improvements in the supply chain. They invested in the supply and shipment of raw materials and products, via their own channels like slaughterhouses, dairies, egg hatcheries and vegetable auctions. In the post-war years most farms shifted from mixed to specialized production. Farms became links in longer food supply chains. The meat-poultry supply chain, for example, became lengthened with 'multipliers' who supplied 'breeders' with fertilized eggs, who supplied 'fatteners' with chicks, who delivered chickens to the slaughterhouses.[21]

Government policy aimed at structural improvements in productivity per farm. Policies were supported by new ideas and approaches taken from United States Marshall Plan technical support programs, research at Wageningen Agricultural University, and national agricultural schools. Education and advice via Government Agricultural Consultants directly and indirectly stimulated the mechanisation and rationalisation of Dutch farms.[22] One of the more ambitious and controversial aspects of the plans was a policy aimed at scale increases.[23] With larger parcels of land, agricultural efficiency could be further increased.

This made land consolidation one of the core aims of the structural policy. Due to purchases and divisions in the past, farmland was often fragmented into small parcels divided by fences, paths, hedgerows, wood banks, ditches, brooks etc. The aim of land consolidation was to unite the fragmented pieces into larger wholes and to optimize access, soil quality, drainage and water supply in the service of efficient production. Land consolidation implied a physical intervention, in addition to an economic and legal trajectory of land exchange and value-compensation. In land consolidation areas, the land was levelled, ditches moved, hedgerows and wood banks cut down and accessibility improved with new roads.

The first modest - and voluntary - land consolidation projects date from the nineteenth century. After the turn of the century the Royal Netherlands Agricultural Committee embraced the idea. After the First World War the national government began to formulate a policy for land consolidation, building among others on land consolidation laws in which procedures and financial arrangements, including gov-

---

[20] De Rooy, *Ons stipje op de waereldkaart*, 211–22.

[21] H. Veldman, E.van Royen and F. Veraart, *Een machtige schakel in de Nederlandse land- en tuinbouw: De geschiedenis van Cebeco-Handelsraad, 1899–1999* (Rotterdam 1999), 177–192.

[22] J. Grin, J. Rotmans, and J. W. Schot, *Transitions to sustainable development: New directions in the study of long term transformative change* (New York 2009), 287–88.

[23] Especially groups of small farmers feared the plans for upscaling. Mansholt denied that upscaling was a goal in itself, but he kept insisting on upscaling via land consolidation. In any case, later policy measures and the centralisation of marketing would later on lead to agricultural upscaling. See: Grin, Rotmans, and Schot, *Transitions to sustainable development,* 285–290.

ernment funding, were laid down.[24] The unification of various commissions concerned with reclamation, drainage and land consolidation, led in 1938 to the founding of the Central Commission for Cultivation Technology (CCC). The Heidemij (a reclamation firm) played an important role in consultation and execution.[25]

The various agricultural and governmental agencies also consulted with nature conservation organisations, represented by the Contact Commission for Nature and Landscape Protection. In 1948 exchanges among the different parties led to the founding of the Consultation Commission for Nature and Landscape Protection the aim of which was to manage local conflicts around reclamations. The consultation platform, that in the end met only 13 times, had a difficult time finding a niche in the rapidly changing and ever more tightly organised agrarian world. After the creation of the Agricultural Board in 1954, the consultations came to a halt for the time being.[26]

By then, the structural transformation of agriculture had shifted into a higher gear. In 1954 a new land consolidation law came into effect.[27] The procedures were further simplified by equalizing the legal status of land renters and owners. In the voting on the consolidation plans, abstention was now to count as a vote *for* a new landscape plan. A special organisation, the Cultivation Technology Service, developed the plans in the name of the CCC and looked after the financing of the projects. Execution of the plans fell to the reclamation firms Heidemij, Grontmij and other engineering bureaus.[28] These measures aimed to speed up the process of land consolidation in order to meet the challenge of declining global prices for agricultural produce.[29]

The approach bore fruit. The large number of applications required a certain prioritization. That resulted in 1958 in a Multi-annual Plan for Land Consolidation. Plans were brought together in so-called 'regional improvement projects'. These

---

[24] S. van den Bergh, *Verdeeld land: De geschiedenis van de ruilverkaveling in Nederland vanuit een lokaal perspectief, 1890–1985* (Groningen 2004), 42–45. The first law dated from 1924. A new land consolidation act in 1938 increased the options for those wishing to take initiatives and granted new subsidies. This law proved inadequate for larger land consolidation projects. Temporary legislation filled the breach, as in the Land Consolidation Law Walcheren of 1947, initiated by the inundations during the war, and the Law on Land Consolidation of Emergency Zones after the 1953 flood disaster.

[25] In 1935 the State Commission for Drainage, the Advisory Commission for Reclamation of Wastelands and the Central Commission for Land Consolidation were consolidated into the Central Advisory Commission for Cultivation Technology (*Centrale Cultuurtechnische Adviescommissie*). In 1938 this was renamed the Central Commission for Cultivation Technology. Van den Bergh, *Verdeeld Land,* 45.

[26] Dekker, *Dynamiek in de Nederlandse natuurbescherming,* 84–85; J. Dekker, 'De dynamische opstelling van het landbouwschap ten aanzien van het milieu, 1948–1972,' in *Jaarboek voor Ecologische Geschiedenis,* (Eekhout 2008).

[27] H. Buiter and J. Korsten, *Land in aanleg: De dienst Landelijk Gebied en de inrichting van het platteland* (Zutphen 2006), 19–47.

[28] Buiter and Korsten, *Land in aanleg,* 60–61.

[29] Van den Bergh, *Verdeeld land,* 46–51.

projects included more than just land consolidation. Extension services and schooling in the areas of technical knowledge and entrepreneurship were set up to support
the modernisation of farming. A Development and Renewal Fund was called into
being in 1963 to help small farmers who wanted to stop. In this way policy makers
wanted not only to 'improve' the physical structure, but the social structure as
well.[30]

The agencies, firms and governments involved in land consolidation were primarily interested in promoting agriculture and the economy. Initially, there was
broad political and social support. In the 1960s, landscape consolidation projects
had increasingly to be coordinated with urban and infrastructural developments.
This created more room for recreational areas and other urban claims on the landscape. On paper the interests of nature and landscape were incorporated into the
new land consolidation legislation that came into force in 1954. Around 1950 nature
preservation comprised 1% of the budget. This grew to 6% around 1970.[31]

There were incidental controversies. In 1964, for example, farmers levelled a
number of wood banks in a land consolidation region in Twente before the plans had
been approved. This unilateral action caused considerable consternation within
nature conservation organizations and led to questions in parliament. Despite the
commotion, the farmers were not sanctioned. Within the Cultivation Technical
Service, initiatives were taken behind closed doors to discuss and align the different
social interests. In this context, H.P. Gorter, director of the Society for Natural
Monuments, was appointed to the Central Committee for Cultivation Technology in
1968 as representative of the nature conservation organizations.

Social unrest around land consolidation projects increased toward the end of the
1960s. The vote for a new land consolidation plan in the town of Tubbergen in the
Twente region, for example, elicited violent protest. The protest was aimed particularly at the voting procedure. In the first vote the plans had been adopted with 27
votes in favour and 12 against. The votes of the 2938 entitled voters who didn't
show up counted as being in favour of the plans. This, in accordance to the rules that
were in effect – rules once adopted to simplify land consolidation. This outcome
was followed by riots in which the mayor's house was set on fire and the police were
called out. Opposition in Tubbergen was broad-based. Unrest was caused particularly by the undemocratic nature of the land consolidation procedures. Besides providing an occasion to revise the procedures, the unrest also created a window of
opportunity for bringing non-agrarian interests to centre stage. This inspired the
restructuring of government agencies around a broader vision of land planning,
incorporating the interests of agriculture, nature and landscape planning.[32]

The challenge for the period after 1970 was a more thorough integration of
nature values and recreation into land planning. By 1985 this resulted in a land-
planning law that put an end to land consolidation informed exclusively by an agrarian perspective. But into the 1980s, the agricultural lobby continued to wield

---

[30] Van den Bergh, *Verdeeld land,* 52; Bieleman, *Boeren in Nederland,* 467–473.

[31] J.L. van Zanden and S.W. Verstegen, *Groene geschiedenis van Nederland* (Houten 1993), 83.

[32] Buiter and Korsten, *Land in aanleg,* 77–79.

Table 13.3 Changes in surface area of different landscape types with variegated ecotopes

	1920	1950	1976	1988
Sandbanks/mudflats	24,000	26,000	12,800	12,000
Marsh (excl. forest)	36,000	26,500		21,000
Sand drifts	12,700	11,000	4300	1800
Living high moors	33,000			160
Excavated and drained high moors	23,400			9800
Heathland	377,000	108,000	79,000	35,800
Forest	248,000	242,000	309,000	328,700

Source: R.J. Bink et al., *Toestand van de Natuur 2* (Wageningen 1994), 93

significant influence. Between 1924 and 1985, in 452 projects, 1,490,520 hectares of land was newly parcelled. Land consolidation projects were pending for another 448,630 hectares. Over the course of 60 years, about two million hectares of agricultural acreage had been reorganized or were about to be so.[33] That amounted to some 59% of the surface area of the Netherlands.

The interventions in the landscape due to reclamations and land consolidation contributed to the decline of diversity in flora and fauna. Landscape types like marshes, mud flats and living peat moors as well as landscape elements like wood banks disappeared (Table 13.3). Specific ecotopes became smaller and more widely separated. Changes in water management, in part carried out to make it easier to work with agricultural machines, led to desiccation in many places.

The changes caused some species to disappear, while others flourished. Research into the historical development of biodiversity has demonstrated that, especially in the second half of the twentieth century, humans negatively influenced Dutch biodiversity.[34] Reclamations and land consolidation made the landscape more monotonous. Among animals, it was especially the reptiles, amphibians and butterflies that were victimized by the changes in the landscape. Ecotopes disappeared, became smaller and more fragmented.[35] The decline was strongest in the riverine zones, sand grounds and the hilly country.[36] Possibly not by chance the regions that were most affected by the physical interventions (see Fig. 13.1).

---

[33] J. Bieleman, Landbouw, in *Techniek in Nederland in de twintigste eeuw* (Zutphen 1993), 63.; Bieleman, *Boeren in Nederland*, 468.

[34] J. Noordijk et al., De Nederlandse biodiversiteit, *Nederlandse Fauna*, 10, 339.

[35] Noordijk et al., *De Nederlandse biodiversiteit*, 339–354.

[36] J.A. Weinreich and C.J.M. Musters, *Toestand van de natuur: Veranderingen in de Nederlandse natuur* (Den Haag 1989), 225.

13.3   The Modern Farm and the Environment

The transformation of the landscape was only one of the aspects of the agrarian transition. Another aspect was the modernisation and rationalisation of the farm itself. The government took a leading role in this process. It programmed research, put much effort into extension services and supported education. The knowledge infrastructure that emerged would later become known as the OVO-triangle (for the Dutch words *Onderzoek* – Research –, *Voorlichting* – Extension –, and *Onderwijs*, – Education). It occupied itself among other things with mechanisation, artificial fertiliser and crop protection. The Netherlands developed the highest agricultural productivity in the world and also used the most artificial fertiliser and crop protection substances (insecticides and herbicides) per hectare. These radical innovations contributed substantially to the solution of the food problem, but at the cost of initiating new ecological calamities. What were the motive forces behind these developments? How did the Dutch deal with the new sustainability challenges?

13.3.1   Artificial Fertiliser

Artificial fertiliser together with an increase in the number of cattle sounded the death knell of the pre-modern cycle of fertiliser production. The increase of livestock in the last decades of the nineteenth century put an end to the fertiliser shortage. It broke open the conventional fertiliser cycle. Heath meadows became largely superfluous for the making of fertiliser.[37] The collection of human faeces in cities also became irrelevant, a development that was also applauded and encouraged in hygienist circles. The fertiliser cycle between the city and the countryside came to an end. The collection of faeces in barrels was replaced by hygienic sewer systems.[38]

The use of artificial fertiliser shows a steady growth since the beginning of the twentieth century, with the exception of the two war periods (Graph 13.2). The zenith was reached in the 1980s. More than 700 kilotons of artificial fertiliser were used in agriculture in this period, equivalent to about 350 kg per hectare.[39] Before the Second World War, the growth was associated with a number of factors.

The knowledge infrastructure as it developed from the end of the nineteenth century stimulated the use of artificial fertiliser. Government agricultural advisors shared experiences in the area of fertilization at the state agricultural experimental stations (*Rijkslandbouwproefstations*) with individual farmers and farmers' study

---

[37] M.T. Knibbe (2000), 'Feed, fertilisers, and agricultural productivity in the Netherlands, 1880–1930,' *Agricultural History* 74(1): 39–57.

[38] This was reinforced by public health and hygienic initiatives. H. van Zon, *Een zeer onfrisse geschiedenis. Studies over niet-industriële vervuiling in Nederland, 1850–1920* (Groningen 1986), 147–157.

[39] E. Homburg, *Groeien door kunstmest, DSM Agro, 1909–2004* (Hilversum 2004), 94.

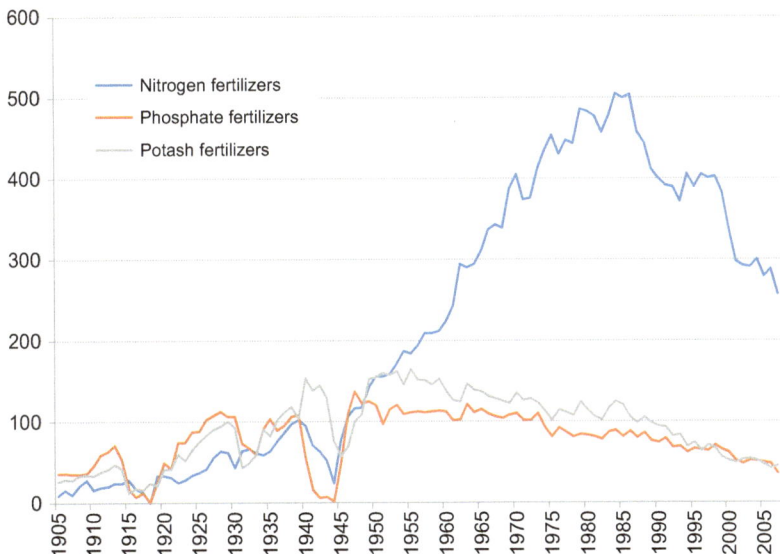

Graph 13.2 Use of artificial fertiliser in the Netherlands 1905–2007, in kilotons
Source: CBS t/m 1998: Tweehonderd jaar statistiek in tijdreeksen, from 1999: Land- en tuinbouw-cijfers 2008 (LEI)

groups. Analysis of and knowledge about fertilisers was concentrated in 1915 at the experimental stations in Maastricht and Groningen. On the basis of this specialization the stations were in a better position to develop procedures and guidelines for the application of artificial fertiliser and to combat fraud. In addition, traders and importers, keen to market the new product, set up their own advisory bureaus for artificial fertiliser.[40]

Farmers' cooperatives also played an important role. By means of collective purchasing they were able to break the power of the German trade cartels. In this manner the Central Bureau of the Netherlands Agricultural Committee (CB), founded in 1899, developed into the biggest importer of artificial fertilisers (Graph 13.3).[41] Membership of such purchasing associations was popular. In 1920 more than half the farmers were members of a cooperative.[42]

Production and trade in artificial fertiliser were important factors in the increased use of artificial fertiliser and acquired an ever-more national character. On the eve of the First World War, artificial fertiliser was produced in the Netherlands in six superphosphate factories and a number of smaller factories producing the nitrogen-based fertiliser ammonium sulphate, mostly prepared from a waste product of cokes

[40] Homburg, *Groeien door kunstmest,* 29–30.

[41] Veldman, Van Royen and Veraart, *Een machtige schakel in de Nederlandse land- en tuinbouw,* 22–56..

[42] Bieleman, *Boeren in Nederland* 287.

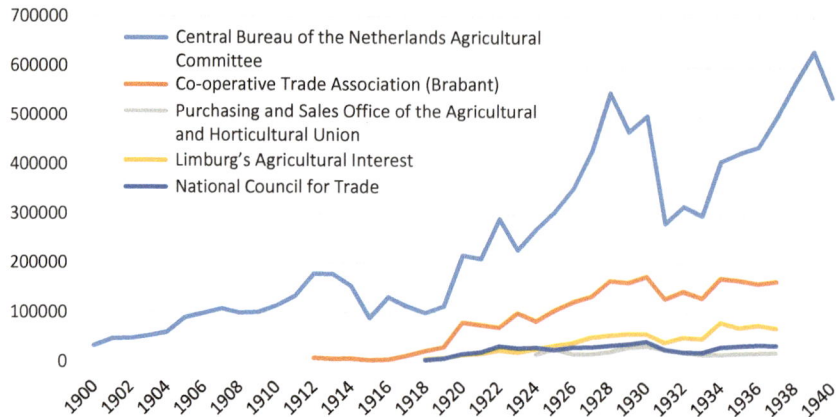

Graph 13.3 Trade in artificial fertilisers by Cooperative Associations 1900–1940
Source: H. Veldman, E. van Royen en F. Veraart. *Een machtige schakel in de Nederlandse land- en tuinbouw, de geschiedenis van Cebeco-Handelsraad, 1899–1999* (Rotterdam, 1999), 54

production.[43] The First World War interrupted the supply especially because raw materials like ammonia and saltpeter were also important ingredients in the production of explosives. Inadequate supplies forced the Dutch superphosphate factories into mergers. During the war, the government compelled merchants, co-operatives and importers to work together in acquisition associations. In the course of the war and immediately thereafter they pursued a more independent course with respect to foreign suppliers and raw materials. In December 1919, following the example of various municipal gas factories, the State Mines started producing ammonium sulphate at the cokes factory of the Emma mine. This was followed in 1923 by the production of this fertiliser at the cokes factories of the Hoogovens iron and steel plant in IJmuiden.

The invention of the Haber-Bosch process to synthesize ammonia, patented in 1910 and first implemented in 1913 by the German chemical giant BASF, was an important technical innovation. The First World War facilitated its rapid diffusion.[44] In 1929, using this process to synthesize ammonia, the Company for the Exploitation of Cokes Oven Gas (MEKOG) – a joint venture of Hoogovens and Shell – started the production of artificial fertiliser. This was followed in 1930 by the State Mines

---

[43] The merger bore the name United Chemical Factories (*Vereenigde Chemische Fabrieken (VCF)*. The Amsterdam Superphosphate Factory (ASF) also belonged to this combination but continued its acitivities under its own name. In 1948 the combination changed its name to Albatros.

[44] The Haber-Bosch process was developed by two Germans: the academic Fritz Haber and the engineer Carl Bosch. Their process produced ammonia by combining hydrogen from methane gas with nitrogen from the air. In 1913 the Badische Analin und Soda Fabrik (BASF) set up the first large-scale ammonia production facility based on this process. Between 1916 and 1923, via obscure routes, French, Italian, German and Swiss chemists succeeded in developing alternatives to the Haber-Bosch process. These alternatives were marketed on the basis of licenses. See Homburg, *Groeien door kunstmest* 17.

Table 13.4 Share of different nitrogen fertilisers in Dutch agriculture, fertilizing years 1923/1924–1928/1929

	1922/1923	1924/1925	1926/1927	1928/1929
Chile Saltpeter	83.3	57.6	32.6	32.1
Ammonium Sulphate	14.7	39.9	57.5	41.7
Calcium Nitrate	1.6	1.4	5.9	17.3
Calcium Cyanamide	0.4	1.1	1.0	3.1
Other	–	–	3.0	5.8

Source: E. Homburg, *Groeien door kunstmest, DSM Agro 1929–2004* (Hilversum, 2004), 43, table 2.2

with its Nitrogen Binding Company (SBB). That same year in the town of Sluiskil, another artificial fertiliser factory under the name of Compagnie Néerlandaise de l'Azote (CNA) opened its doors. Its production of 45,000 tons of artificial fertiliser per annum made it for a brief time the largest factory of its kind in Europe.[45] Between 1921 and 1927 no fewer than 28 factories for the production of synthetic ammonia and 31 nitrogen binding factories sprung up all over Europe.

Price drops in the 1920s as a result of overproduction in the Dutch and European artificial fertiliser industry was the final factor in the increasing use of this commodity. Prices dropped dramatically. Other fertilisers like those derived from urban wastes and faeces disappeared from the market.[46] At the same time, prices for agricultural produce were also depressed. Farmers responded by increasing production and using more fertilisers.

Well into the 1920s, Dutch agriculture still used mostly imported fertilisers like Chile salpeter. But this import was gradually abandoned and Dutch industry began to dominate artificial fertiliser production (Table 13.4). The sector looked for solutions to the price drops. The State Mines secured a significant national market in 1930 by signing contracts with the agricultural co-operatives Central Bureau (CB) the Co-operative Trade Association (CHV) and the Limburg's Agricultural Interest (LB). After 1927, the European nitrogen industry formed a cartel with agreements about limiting production and dividing the market. In 1930, the *Convention de l'Industrie de l'Azote* (CAI) allowed the State Mines and MEKOG as part of the so-called 'Holland group' to divide the domestic market between themselves. CNA at Sluiskil would produce exclusively for the foreign market. The Central Nitrogen Sales Office (CSV), set up in 1934, centralized the sale of nitrogen-based fertilisers within the Netherlands. In time the cartel succeeded in stabilising prices. But during

---

[45] Homburg, *Groeien door kunstmest,* 94.; E. Homburg, 'Van carbo- naar petrochemie, 1910–1940', in J.W.Schot et al. (ed.) *Techniek in Nederland in de twintigste eeuw, deel II: Delfstoffen, Energie en Chemie* (Zutphen 2000) 332–357.

[46] The Company for Ammonia Manufacture from Faecal Matter (1890–1902) and the First Netherlands Factory for Chemical Products (1902–1915) produced ammonia salts and ammonium sulphate from the then extant Liernur system of sewerage. Homburg, *Groeien door kunstmest,* 39; Van Zon, *Een zeer onfrisse geschiedenis,* 147.

Table 13.5 Use of nitrogenous artificial fertilisers in selected European countries, 1913–1951 (in kg/hectare)

	1913	1925	1926	1927	1928	1938/39	1950/51
Netherlands	7.1	16.5	20.5	21.6	26.3	37.2	71.0
Belgium	16.0	19.7	18.7	19.8	19.3	–	–
Germany	7.2	10.0	11.2	13.6	13.9	23.7	26.3
France	0.2	3.1	3.0	3.4	4.0	9.4	16.5
England	2.3	2.0	1.8	2.1	4.0	4.8	17.4

Source: E. Homburg, *Groeien door kunstmest, DSM Agro 1929–2004* (Hilversum, 2004), 44, table 2.3; 'Hogere opbrengsten van tuinbouw en veeteelt door doelmatige stikstofbemesting' in *Nieuws van Staatsmijnen*, 15, July 1955, 3

the 10 year period between 1922 and 1932 the price of ammonium sulfate declined from 18 guilders to about 5 guilders per 100 kilograms.[47]

Firms were glad to contribute to the accumulation of knowledge. Using their own test fields, they studied the effects of fertilisation and communicated this to a broader public via advertisements, consultancy and exhibitions. The Nitrogen Binding Division of the State Mines tested their own fertilisers in various test fields of the Government Agricultural Advisors. This clarified the superiority of the new fertiliser developed by the State Mines – Chalk-ammonium saltpeter (KAS) over sulphuric acid ammonia. These results were touted in the advertisements of the Agricultural Bureau of the State Mines – set up specifically with this aim in mind. After 1935 cooperation within the Dutch nitrogen industry intensified. The Agricultural Bureau acquired an ever-more general character and in 1948 became part of the Central Nitrogen Sales Office (CSV).[48]

During the interbellum, Dutch use of artificial nitrogenous fertilisers surpassed that of the surrounding countries (Table 13.5). Dutch use grew from 70 kg per hectare in 1950 to 100 kg around 1960 and almost 200 kg per hectare in the early 1970s (see Table 13.6). Dairy farming in particular exhibited the strongest post-war growth in use of nitrogenous artificial fertilisers. Of the 430,000 tons deposited on the land in 1975, more than 300,000 tons were used to fertilize grazing pastures.

Increased use of artificial fertiliser was realised thanks to the emergence of close cooperation among firms, research institutes and farmers generously supported by the government. These exertions increased both the productivity per hectare as well as the productivity of agrarian labour. The effects were amplified by plant breeding aimed at developing varieties adapted to the rapid absorption of nutrients.[49]

This spectacular development was celebrated as a victory in the struggle to solve the food question. In the late 1960s artificial fertiliser surpassed animal fertilisers in the category of nitrogenous fertilisers (see Table 13.6). In the early 1970s experts at the State Consultancy for Soil and Fertilisation Issues at Wageningen and the Institute for Soil Fertility at Haren first expressed concern about possible

---

[47] Homburg, *Groeien door kunstmest.*

[48] Homburg, *Groeien door kunstmest,* 85–90.

[49] Bieleman, *Boeren in Nederland,* 285.

Table 13.6 Animal manure production and fertilizer use, 1935–1995

	1930	1935	1940	1945	1950	1955	1960	1965	1970	1975	1980	1985	1990	1995
Application of artificial fertiliser in kg/ha farmland														
Nitrogen (N)	19	28	41	36	70	80	96	137	177	210	241	251	206	208
Phosphate (P_2O_5)	47	45	24	23	54	48	48	51	49	44	42	44	38	32
Potassium Oxide (K_2O)	41	36	66	26	69	72	59	61	56	54	62	62	49	35
Manure production cows, pigs, poultry in kg/ha farmland														
Nitrogen (N)	.	.	.	.	91	.	110	122	163	196	240	280[a]	287	292
Phosphate (P_2O_5)	.	.	.	.	49	.	58	64	78	92	114	126[a]	110	107
Potassium Oxide (K_2O)	.	.	.	.	102	.	122	133	183	221	265	310[a]	305	322

[a]Data 1984

Source: CBS – Historische Reeksen Landbouw 1899–1999

over-fertilisation using conventional and artificial fertilisers. The priorities for environmental policy in 1972 first mentioned problems with fertiliser surpluses. The solution proposed by the KVP (Catholic Peoples Party) minister of Public Health and Environmental Hygiene was to advocate the planting of crops, like green maize, that supported a high fertilisation intensity. The relevant agricultural experts saw the increased interest in these crops as a '...positive development to prevent possible soil pollution by fertilisers.'[50] The solution was a textbook example of a corporatist arrangement involving the political establishment and agricultural organizations. Ranks were closed in the 'green front' and on the sand grounds the cornfields shot out of the ground like mushrooms. For the time being fertilisation hardly encountered resistance. That was certainly not the case with the use of crop protection substances.

13.3.2   Crop Protection

The use of chemical crop protection substances incited the first post-war criticism of the agricultural modernisation process. The spectacular development of chemical crop protection substances much resembled the introduction of artificial fertilisers. Here too, public-private cooperation within the network of agricultural organisations, government and private industry played a key role.

The first experiments with the chemical control of mould infections, especially in the fruit and flower bulb sectors, date from the 1930s. After the Second World War the introduction of dichloro-diphenyl-trichloroethane (DDT) initiated a revolution in the use of chemical insecticides. Successes during the war with combatting typhus (lice) and malaria (mosquitos) stimulated further research. The substances generated much enthusiasm. Broad international appreciation was expressed in the Nobel prize accorded to the Swiss chemist Paul Müller for developing DDT.

> As lovers of the potato plant, exotic insects like the Colorado beetle constitute a threat to 'our daily meal.

as the cinema newsreel Polygoon put it in 1947.[51] Chemical insecticides like DDT were a formidable weapon against this enemy. In 1950, following up on the success of DDT, Shell developed a number of chlorinated hydrocarbon compounds like Aldrin, Dieldrin, Eldrin and Telodrin as alternative and even more poisonous insecticides. In the 1950s there was much optimism about the use of insecticides (and also fungicides and herbicides) and the possibilities offered by these chemical applications.[52]

---

[50] F. Bloemendaal, *Het mestmoeras* (Den Haag 1995), 12–13.

[51] In 1947 and 1948 the cinema newsreel Polygoon Journal featured items on the fight against the Colorado Beetle. See website Beeld en Geluid, Open Beelden: www.openbeelden.nl/media/670176/Strijd_tegen_de_coloradokever

[52] J. Bieleman, 'Gewasbescherming, dieren en gewassen in een veranderende landbouw', in J.W. Schot, H.W. Lintsen, A. Rip and A.A. Albert de la Bruhèze (eds.), *Techniek in Nederland in de twintigste eeuw, deel 3: Landbouw, voeding* (Zutphen 2000), 210.

The modernisation of agriculture in the 1950s and 1960s was in the first place inspired by the drive for greater production and by the development of a professional and efficiency-based approach. Chemical crop protection substances helped to save labour. Mixed farms made way for specialized enterprises. In the market-garden and crop-farming sectors the number of different crops declined. This period also saw the emergence of agricultural contractors that assisted farmers with the use of specialized machinery. Plant breeding technologies aimed at suppressing leaf growth in order to promote richer ears. While this combination of developments delivered higher yields, it also caused new problems. Fewer leaves allowed more light to reach the soil and weeds proliferated. With monocultures diseases and plagues had a better chance of spreading. Shared use of (specialized) machinery increased the chance of infestation. International trade became subject to so-called 'phyto-sanitary' regulation, which specified that export product had to be a 100% free of diseases, moulds and insects. All these developments together promoted the increased use of plant protection substances.[53]

From the very beginning, the use of chemical insecticides, herbicides and fungicides had had its detractors. As early as 1949 the director of the Wageningen Plant Disease Service had warned about excessive use of chemical plant protection substances. One of the first visible problems was that pesticides also targeted useful insects like bees or the natural enemies of pests. A second effect, especially evident in the struggle against moulds, was the development of rapid tolerance to the chemical substances, thanks to mutations in the organism. These effects stimulated a search for other means of pest, plant and mould control, varying from plant breeding for resistance, substances that were absorbed by the plant and specific plant protection substances per crop and per disease.[54]

Beyond this reorientation to the use of chemical insecticides, herbicides and fungicides within the agrarian cluster, there was also increasing criticism from outside. In her 1962 book *Silent Spring,* the American author Rachel Carson described the effects of chemical insecticides, especially DDT, on the entire ecosystem. The Dutch translation, *Dode Lente,* followed a year later. It provided the foundation for a critique of modern agriculture and food supply chains from a broader environmental perspective. Spokespersons were biologists, ecologists, nature preservationists and critics of industrialisation, i.e. parties outside the powerful public-private agrarian network.

Massive bird mortality in Drenthe (1966) and Zeeland (1968) inspired a chorus of criticism of crop protection substances. The livers of the dead birds contained high concentrations of the pesticide Dieldrin.[55] From 1969 on, legal measures were

[53] Bieleman, *Gewasbescherming, dieren en gewassen in een veranderende landbouw*, 222–223.

[54] Bieleman, *Gewasbescherming, dieren en gewassen in een veranderende landbouw*, 213–221.

[55] J.W. Copius Peereboom, *Chemie, mens en milieu: Schadelijke stoffen in milieu en voeding, een studie over chemische milieuverontreiniging* (Assen 1976.), 126–127. Birds of prey like the hawk and sparrow-hawk recovered after limits were imposed on the use of crop protection substances. In J. van Zoest, ed., *Biodiversiteit*, (Utrecht 1998), 133.

taken against pest control substances containing mercury. This was followed in 1972 by a ban on DDT. Despite this, the Netherlands remained the global front-runner in the use of pest control substances well into the 1980s. The road taken, in which land and labour productivity were the guiding principles in agricultural inno-vation, proved hard to leave.[56]

The professor of animal systematics and geography, K.H. Voous of the Free University in Amsterdam had described developments in 1970 in terms of the 'sor-cerer's apprentice, who could initiate a process, but who could not manage the side-effects of the much desired increase in agricultural yields. Humanity has poisoned nature. Can it *itself* remain healthy?'[57]

After the physical deterioration of nature caused by transformations of the land-scape, a qualitative deterioration set in as modernisation and the use of fertilisers and chemicals caused desiccation, over-fertilisation and environmental pollution. In addition to the direct effects of pest control substances noted above, fertilisers also caused changes in vegetation. The spread of bush grass in the dunes, common rush in peat marshes and purple moor-grass in the heathlands were the consequences. The appearance of these species changed local ecosystems of mushrooms, insects, reptiles and birds.[58] The loss of biodiversity in the Netherlands since 1950 was caused for about 30% by landscape transformations and for 60% by acidification and lowering of groundwater levels as part of the modernisation of agriculture.[59]

Nature and the environment appeared for the time-being to have footed the bill for solving the problem of food security. New challenges for sustainable develop-ment emerged from the agricultural transitions that delivered food and well-being. Pesticides inspired the first measures to make agricultural developments more sus-tainable, to inoculate them with a concern for the integral quality of life, including concern for ecological aspects and natural resources. This was the social agenda that came to fruition after 1970.

13.4   Foods and Nutritional Patterns

Since the end of the nineteenth century, concern about food shortages has been accompanied by concern about food quality. The supply chain of production, distri-bution, and preparation of food changed dramatically in the twentieth century. The distance between locations of production and of consumption increased. The trans-formation of the food supply chains impinged in different ways on aspects of sus-tainability, not only in the Netherlands but also elsewhere. Thinking in terms of a

---

[56] Bieleman, 'Gewasbescherming, dieren en gewassen in een veranderende landbouw, 222.

[57] K.H. Voous, *Natuur, milieu en mens* (Kampen 1970), 47.

[58] Noordijk, *De Nederlandse biodiversiteit*, 339–354.

[59] M.P. van Veen et al., Halting biodiversity loss in the Netherlands: Evaluation of progress, *Netherlands Environmental Assessment Agency* (2010), 7.

food supply chain became ever more current in the course of the twentieth century. Three facets claim our attention.

The hygienic concerns of the nineteenth century were extended into the food supply chain. Food quality and hygiene came to play a central role in the activities of the food processing industry, users' organisations and government. New patterns emerged in the food trade, in distribution and in processing in the kitchen. Monitoring was instituted to protect honest trade and educational campaigns were launched to inform consumers about food preparation and healthy nutrition.

In the second place, the industrialisation of the food processing sector led to lengthening and diversification in production chains. This shifted the effects of changing nutritional patterns to locations across the border. International trade in foodstuffs and industrial processing into foods were responsible for new foods in Dutch households and influenced economic and ecological systems elsewhere. This was associated with a chemical-analytical perspective on agricultural and fisheries products as raw materials for foods. Foods were now defined in terms of a basic or bulk product (in the trade jargon: a commodity) and in terms of proteins, fats, and vitamins. The interchangeability of raw materials transformed the food supply chains into a network of material flows. The flows of raw materials were reconstituted in factories into new compound foods, varying from margarine to soups and instant dinners.

A third facet was an increasing concern about healthy nutrition. This commenced with research into the one-sided nutritional patterns of the poor. After the Second World War a new problem arose. The formerly sober menu had been replaced by one laden with a surfeit of foods. Availability was no longer a problem, but rather the right choice from what was on offer. In the context of the different and distinct interests of private companies, consumers and governments, new informational strategies emerged in combination with efforts to combat the new welfare diseases that emerged in the wake of new nutritional patterns.

13.4.1 Food Quality: Commodities Law, Trademarks and the Modern Housewife

At the turn of the century, concern about food quality emerged within two distinct social settings. In 1905 a handbook for research into food quality, the *Codex Alimentarius*, was published and quickly became the bible for the professional-scientific approach of the internationally oriented hygienist movement. In addition, in the agrarian sector cooperation among science, private industry and government had produced a system of agricultural schools and experimental stations. Improvements in methods of chemical analysis that increased and objectified insights into the composition of foods, had made both developments possible.[60]

[60]A.H. van Otterloo (2000), 'Voeding,' in J.W. Schot et al. (eds.), *Techniek in Nederland in de twintigste eeuw, deel 3: Landbouw, voeding*, 254–256.

National regulation of food quality bogged down around the turn of the century in disputes about measurement methods, standards and finances among tradesmen, producers, scientists and civil servants. However, local arrangements emerged in Rotterdam (1893), Leiden (1901), Dordrecht (1909) and Drenthe (1916). The national government did develop quality standards for, for example, meat (1902) and butter (1904) mainly with an eye to international trade. During the First World War the necessity for a generic arrangement became more evident. The number of surrogates, fakes and products treated with dyes and unacceptable preservatives increased. In view of these developments, in 1919 a nationwide Food and Drugs Act came into effect based on the *Codex Alimentarius*.[61]

The Food and Drugs Act counted among its supporters the bona fide tradesmen and producers. They regarded the new law as a means to protect their own trade against fraudulent operators. In addition they began to employ trademarks and advertisements in order to draw attention to their industrially processed foods and their qualities. Thanks to packaging they were also able to present their products in a handy and hygienic fashion.

From the 1920s on, stiff competition became endemic in the distribution channels for food. Industrial producers like the margarine producer Jurgens acquired shares in the grocery chains of De Gruyter and Albert Heijn. Albert Heijn in turn invested in their own factories for cakes and candies. In addition various purchasing cooperatives emerged, like Enkabé (1929), De Spar (1932) and the wholesaler's group Schuitema (1934).[62]

These developments in the production and distribution of foods coincided with ideas on the modernisation and rationalisation of the household, with the Dutch Association of Housewives (NVvH) as most important spokesperson. The NVvH had been founded in 1912 as a union of mostly middle-class housewives. By way of courses and educational materials on nutrition, cooking and hygiene, the NVvH tried to tackle the 'servant question.' Servants were becoming ever more scarce and expensive and the association wanted to make 'housewives more conscious of the task they were obliged to fulfil.'[63] The hygienically packaged foods and the first composite food products like stock cubes, baking powders and pudding powders merged seamlessly with the ambitions of the NVvH and the modernisation of the household.

Companies focused their attention on the modern housewife. Brands like Maggi, Calvé, Honig, Van Nelle and Verkade became trusted names in Dutch kitchens. The bond between producers and modern households were forged by means of trading stamps, coupons, sales and competitions:

> **The old-fashioned housewife** had recourse to her own home-made store of fat: beef fat, lard or a mixture of the two, provided by the annual slaughter in November. **For the modern housewife**, who is no longer wont to keep such stocks in the cellar or the cupboard, it

---

[61] Van Otterloo, 'Voeding', 258–261; H. W. Lintsen (ed.), *Tachtig Jaar TNO*, (Delft 2012), 108–109.

[62] Van Otterloo, 'Voeding', 264.

[63] A. H. van Otterloo, *Eten en eetlust in Nederland, 1840–1990* (Amsterdam 1990), 157.

is of importance that she can have access to fat that is just as nutritious, just as tasty and just as economical, that can repeatedly be used in small quantities and in a fresh condition. **This fat she finds in Delfrite** (bold text in the original citation)[64]

The brand names of the food processing industry became the shining examples of food quality, hygiene and the modern household.

After the Second World War the modernisation of the household got a new impulse from changes in the trade and distribution of domestic articles. During the war and as an extension of Dutch agriculture and fisheries a deepfreeze industry had developed with the German armies as an important customer. After the war, agrarian organisations agitated for a civilian expansion of this deepfreeze supply chain. Deepfreeze companies succeeded in wheedling the necessary scarce funds from Sicco Mansholt, Minister of Agriculture, in order to invest in deep-freezers. These companies and food producers like Unilever developed new forms of food conservation, processing and preparation. Vegetables, fish, meat and complete meals were offered as new deepfreeze product. The foodstuff supply chain branched and made detours that ran via industry to the cooling and deepfreeze facilities in stores and cold storage facilities. In 1956 only 3% of the Dutch households had a refrigerator, but by the early 1960s the low temperature supply chain had been completed with the massive purchase of refrigerators and freezers by households. In 1962 almost 20% of the households had a refrigerator and by 1972, 88%.[65]

In 1948 the first self-service store opened its doors, to be followed in the early 1950s by large grocery chains like De Gruyter and Albert Heijn. Regulations regarding the concentration of shops that dated from the 1930s and that were intended to protect grocers, butchers and greengrocers and their specific skills initially limited the spread of self-service stores. But after a recommendation by the Social-Economic Council pointing to the rich variety of packaged products, the regulations were modified in the early 1960s. This made room for new kinds of stores like self-service stores and supermarkets with a more elaborate assortment and lower prices.[66] In the kitchens the foods were prepared in entirely new ways.

13.4.2 *International Food Supply Chains*

The international trade in foods long remained limited to products that could be easily conserved, including grain, butter, cheese and pigs. At the end of the nineteenth century technical innovations like refrigeration made it possible to transport perishable foods over ever greater distances. British industrialists developed 'cold chains' that guaranteed the supply of meat products. With the aid of refrigerator ships, Great

---

[64] Collection brochures Johannes van Dam, cited in A. H van Otterloo, *Eten en eetlust in Nederland, 1840–1990*, 166. Bold in original text.

[65] B. Sluijter, *Kijken is grijpen, zelfbedieningswinkels, technische dynamiek en boodschappen doen in Nederland na 1945* (Eindhoven 2007), 148–64.

[66] Sluijter, *Kijken Is Grijpen*, 133–40; Van Otterloo, 'Voeding', 284–95.

Britain imported meat from remote regions like New Zealand, Australia and Argentina. In this last country around 1900 British industrialists had invested in about 16,000 km of railways that connected the harbour of Buenos Aires with its hinterland. In the name of welfare the Argentinian army launched violent campaigns to 'clean out' the native population of the pampas. Australian Aboriginals too were driven from fertile lands to the benefit of international trade.[67] The removal of the 'savages' was doubtless legitimated to everyone's satisfaction at the time, but it nonetheless shows the brutal and radical way in which the expansion of, in this case the British, food supply chain was effected.

Where Great Britain was often the endpoint of the supply chain, the position of the Netherlands was different. It was more like a node of supply chains. It exported pork, eggs, milk and margarine – much of it to Great Britain and Germany – and imported raw materials like cattle fodder and vegetable oils.[68]

The network shaped by Dutch margarine producers illustrates the influence of such supply chains outside the Netherlands. In 1870, in the town of Oss, two families by the name of Jurgens and Van den Bergh had set up margarine factories to produce artificial butter using animal fats, a by-product of the local slaughterhouses. Both entrepreneurs profited from the growing demand from England. In 1911, as the price of animal fats rose, both firms succeeded in obtaining a license on a German process whereby liquid vegetable oils and animal fats could be hardened by means of hydrogenation. The process opened new markets for raw materials. Competitors in this search for the cheapest raw materials were the soap manufacturers.

One of the new possibilities was the use of whale-oil. Anton Jurgens and the British soap manufacturer Lever Brothers toyed with the idea of establishing their own whaling fleet. This proved a bridge too far. Instead, in 1913 a whale pool was set up, a cooperative venture among Dutch, British and Austrian soap and margarine manufacturers to purchase whale-oil in common from the Norwegian whalers.[69] These kinds of cooperative efforts formed the foundation for the later merger between the Dutch-German-Austrian *Margarine Unie* (1927) and the British Lever Brothers to form Unilever in 1929. One of the reasons for the merger was control over raw materials. As buyer, Unilever dominated trade in whale oil. This put the company in a position to pressure the Norwegian government for favourable selling conditions for margarine and soap.[70]

The big demand for oils and fats by the soap and margarine industries was largely responsible for the resurgence of modern whaling. British and Norwegian fleets

---

[67] P. Högselius, A. Kaijser, and E. van der Vleuten, *Europe's Infrastructure Transition: Economy, War, Nature* (New York 2016), 120–121.

[68] A. A. Albert de la Bruheze and A. H. van Otterloo, The Milky Way: Infrastructures and the Shaping of Milk Chains, *History and Technology* 20(3) (2004): 249–70.

[69] F.J.M van de Ven, *Anton Jurgens Hzn, 1867–1945, Europees ondernemer, bouwer van een wereldconcern* (Zwolle 2006), 137–140.

[70] P. Thonstad and E. Storli, Big business and small states: Unilever and Norway in the interwar years, *Economic History Review* 66(1) (2013): 109–31.

dominated the whaling industry. They developed factory ships and new harpooning techniques. These made it possible to hunt for a wide variety of whales in Antarctic waters.[71] But what had initially appeared to be an endless supply of whales soon began to diminish. By the 1930s, negotiations on international agreements were underway. However, the International Whaling Commission (IWC) mandated with a coordinating role, had great difficulty maintaining headway in the face of the diverse geopolitical interests. Governments regularly leaned heavily on the IWC or unilaterally terminated agreements. Given continuing demand it appeared impossible to prevent decimation of the whale population.[72] By the 1970s populations of most species of whale had been reduced to a mere 10% of their 1946 numbers. Whaling became one of the first symbols of the emerging nature and environmental movement as embodied in organisations like the World Wide Fund for Nature (WWF) and Greenpeace.[73]

In addition to whaling, Unilever and its predecessors invested in plantations and oil factories in Africa, South America and the Dutch East Indies. The investments aimed to guarantee the supply of raw materials and to decrease dependency on middlemen. These concerns also moved them to invest in their own fleets. After the 1929 merger, Unilever consolidated the joint African possessions of tens of thousands of hectares of plantations, oil factories and trading posts under the flag of the United Africa Company (UAC), a trading firm at a discreet distance from the parent firm Unilever. The UAC functioned as a trading post for Unilever products and as supplier of raw materials. It exploited plantations for vegetable oil and fruit and ran its own merchant fleet.[74] The company also invested outside of Africa in the Solomon Islands, Malaysia, and the Dutch East Indies. Two thirds of the plantations cultivated palm oil, raw material for the margarine and soap industries.[75] After difficult years during the depression and the war, the UAC developed into an important component of the Unilever empire. As priority buyer from the UAC, Unilever had a dominant position in the international trade in vegetable oils and fats. After the Second World War, 10% of the multinational's profits came from the African trade.[76] The plantations also proved important to the firm for the supply of raw materials in the long term.[77]

---

[71] Factory ships made it possible to undertake longer campaigns to the richer fishing grounds around Antarctica. New harpoon techniques also made it possible to hunt whales that sunk after being killed. J. R Bruijn en J. C.A. Schokkenbroek, *De laatste traan: Walvisvangst met de Willem Barentsz, 1946–1964* (Zutphen 2012), 14–19.

[72] Bruijn en Schokkenbroek, *De laatste traan*, 251–56.

[73] A. Kalland, Management by totemization: Whale symbolism in the anti-whaling campaign,' *Arctic* 46(2) (1993): 124–33. D. Toke, Epistemic Communications and Environmental Groups, *Politics* 19, no. 2 (1999): 97–102.

[74] D.K. Fieldhouse, *Merchant capital and economic decolonization: The United Africa Company, 1929–1987.* (Oxford 1995), 176–225.

[75] Fieldhouse, *Merchant capital and economic decolonization*, 450.; G. Jones, *Renewing Unilever: Transformation and tradition* (New York 2005), 197–203.

[76] W.J. Reader, *Vijftig jaar Unilever, 1930–1980* (London 1980), 82.

[77] Jones, *Renewing Unilever,* 197–201.

The cultivation of market crops like palm oil, coffee, tea and cacao had various consequences for well-being and sustainability. Worries were expressed about erosion, droughts and the fluctuation of water levels. But forests were mainly seen as serving industries and trade.[78] At the time, the loss of biodiversity was not an issue. Jungle was replaced by plantations with a monoculture. Differentiated agricultural systems were traded in for one-sided plantations. In Southeast Asia the survival of the Orangutan was threatened by palm oil production. We also see a transformation of local economies in Asia and Africa. Unilever bought out the self-sufficient farmers, who subsequently had to survive as labourers. Where local food production no longer provided adequate nutrition, recourse was taken to more expensive import of food. In the period of decolonisation after the Second World War the economic structures also remained virtually intact and market crops contributed to the increasing inequality between the West on the one hand and Africa and Asia on the other.

The example of the growth of the Dutch margarine industry shows how flows of raw materials and sustainability issues were transported beyond the national borders. The production of food was dissociated from the places where raw materials were produced. 'Blue Band' and 'Zeeuws Meisje' – popular Dutch margarine brands - were made of whale oil and palm oil. International trade networks in raw materials made it possible for Dutch consumers to pay 'not a penny too much' for their margarine. The costs with respect to economic and ecological sustainability were paid in Africa and Asia. The growth of Dutch margarine production was made possible by exploiting natural capital elsewhere; in Antarctic waters and in the tropical plantations.

13.4.3   Healthy Nutrition and Excess

Around 1890 food costs still made up about half the household budget. In Amsterdam in 1935 this had declined to 22%. The poorest segment of the population spent 35–37% of the family budget on food and the wealthiest less than 10%.[79] In the first half of the twentieth century, diets became more varied, but the menu of the underclass continued to consist mostly of bread, milk, porridge and potatoes, once in while supplemented by a bit of meat.

Around the turn of the century socially concerned physicians began to investigate the diets of workers. They concluded that these exhibited little variety and suffered from a dearth of calories and nutrients.[80] Food shortages during the First World War and the consequences for public health led to the founding of the Netherlands

---

[78] D.A. Zoethout, *De plant in nijverheid en handel*, (Amsterdam 1914), 84–87.

[79] For 1850 data see van Otterloo, 'Voeding', 282; data 1935 in 'Inkomsten, Uitgaven, Verbruik En Physiologische Waarde van Voeding, uit Verschillende Kringen Der Bevolking Gedurende de Periode 1 Maart 1934–28 Februari 1935,' Statistische Mededeelingen (Amsterdam: Bureau van Statistiek der gemeente Amsterdam, 1935), 45.

[80] Van Otterloo, 'Voeding', 241.

Institute for Popular Nutrition in 1919. The instigator, the physician dr. E.C. van Leersum, enjoyed the support of food processing companies and local government, including the city of Amsterdam. The Institute began to investigate the relationship between nutrition and health. In 1922 it was integrated into the University of Amsterdam.

The economic crisis and the mounting threat of war in the 1930s moved the government to take action. It was involved in the founding of the Commission for Household and Family Education (1934) that disseminated nutritional information via exhibitions, classes and demonstrations. It also supported the Foundation for Household Education in the Countryside (1935) that provided information about cooking, hygiene and health in cooperation with the Dutch Association of Housewives (NVvH) and cooking teachers. The TNO research institute - for the most part funded by the state – was augmented in May 1940, as German armies invaded the country, with the addition of a Central Institute for Nutritional Research (CVO). At the same time a Nutritional Board was founded as a sub-commission of the Health Board.[81]

The focus of food research shifted from quantity to quality. In the interbellum, for example, a basis was laid for knowledge about vitamins. New insights revealed that vitamins were not only medicines against specific diseases, but also essential for a balanced nutrition. Via the Nutritional Board and TNO these insights rapidly penetrated into private industry and the societal 'midfield.' In 1927, for example, the food industry succeeded in fabricating a tasteless vitamin concentrate from cod liver oil that could be added to margarine. The importance of vitamins was exploited in advertisements to prize products containing such supplements. Knowledge of food quality was transmitted via informational and educational activities like cooking lessons, cookbooks and magazines.

Supplements also entailed risks. In 1960 some 100,000 people in the Netherlands suffered from an allergic reaction. After an investigation lasting several weeks it transpired that the 'blister disease' was caused by the emulsifier ME18, that had been added as an anti-spatter substance to the margarine brand *Planta*. The additive had been approved by the State Institute for Public Health after animal tests. The so-called *Planta affair* was widely publicized in the press. After Unilever had paid damages, but without admitting guilt, the affair was publicly framed as a mistake by a big and complex company. The affair did have an effect on food research: it put the safety aspects of industrially processed foods on the agenda. TNO pumped more resources into research in this field. Unilever became one of the most important clients. Next to specific research contracts it also funded half of the collective research done by the TNO Food Organization.[82]

After the Second World War, increasing well-being manifested itself in the menu of the Dutch household. The consumption of meat, sugars, edible fats, and luxury products like nuts, peanuts and chocolate articles increased, while grain and potatoes

---

[81] Van Otterloo, 'Voeding', 266–268.; H. W. Lintsen *(eds.), Tachtig jaar TNO* (Delft 2013), 109.

[82] Lintsen, *Tachtig Jaar TNO,* 121–122; Van Otterloo, 'Voeding', 288.

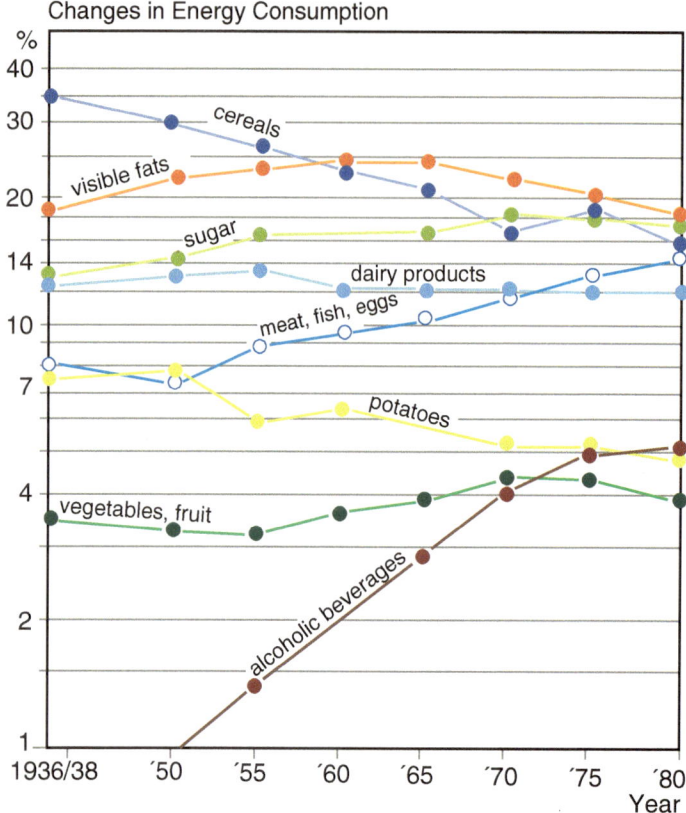

Graph 13.4 Percentual contribution of groups of foods to the average energy consumption of the human body in the Netherlands in the period 1936–1980
Source: J.F. de Wijn and W.A. van Staveren, *De Voeding van Elke Dag* (Utrecht 1986), 166

constituted an ever smaller part of daily energy input (Graph 13.4).[83] The new nutritional habits were encouraged in magazines. On their trips abroad, Dutch travellers encountered other food cultures. And countrymen returning from the Dutch East Indies and foreign guest-workers brought new dishes with them to the Netherlands. How could the country cope with this superabundance? In 1953 the Extension Service for Nutrition, modelled on an American example, introduced the so-called 'disk of five,' a slogan that rhymes mnemonically in Dutch: *schijf van vijf*. The disk described the basic nutritional groups that ought to compose a healthy daily diet: grain products, vegetables and fruit, dairy products, meat and fish and finally fats. This nutritional advice emphasized variety and quality. The consumption of vitamins, minerals and animal protein was to be encouraged.[84]

The food industry was only too willing to contribute to an ever more varied menu. Under the banner of the modernisation of cooking it introduced pre-prepared,

---

[83] J.F. de Wijn and W.A. van Staveren, *De voeding van elke dag* (Utrecht 1986), 166.

[84] Wijn and Staveren, *De voeding van elke dag.*; Lintsen, *Tachtig jaar TNO,* 117.

deep frozen and composite foods. In addition they attempted to change eating patterns by introducing new 'eating moments.' Such innovations often proved more difficult than imagined, but gradually new 'eating moments' for snacks, in-betweens, and appetizers penetrated into the daily lives of most Netherlanders. New foods like French fries, croquettes and sorbet became common. Old foodstuffs acquired new identities like 'cheese in the fist' (lump of cheese without bread) and 'Hollandse Nieuwe' (literally 'Holland's New,' referring to salted herring caught in the current season).[85] Food became part of a lifestyle. Its primary function, the preservation of the human body, retreated ever further into the background.

13.5 New Food Chains, New Problems

For a long time the food supply chain had a rather simple structure. For a large number of products it ran from the farmer via the market to the plate. Often producer and consumer were one and the same: the farmer who ate his own produce and the city dweller with his own vegetable patch. In part, producers marketed their own products, like butter and cheese. For another part, products came from afar, like cereals and coffee. The food industry was limited to simple processing: grinding grain, baking bread, extracting oil etc.

Starting at the end of the nineteenth century, the food supply chain got more and more complex. First, it was evident that the chain was getting longer, that is to say, characterized by an increase in the number of links between producer and consumer and in the geographic distance between links. Second, chains were becoming more differentiated. Links split up into a variety of processing activities. Finally, the food chains began to assume the character of networks. They were coupled to organisations that did not belong to the primary production and consumption process.

In the changing food chains new dynamics emerged around the problem of food supply. This aspect of well-being had long remained limited to food distribution and food security. This remained relevant in the twentieth century, in particular due to the two world wars in which the Netherlands was confronted with hunger and scarcity. In addition, food quality moved higher up the societal agenda. Problems of excess and changing nutritional habits came to the fore. New also was the problematic issue of sustainability, in particular the landscape and biodiversity. Who addressed these problems?

At the start of the supply chain a dense network evolved that was composed of farmers, agricultural organizations, suppliers (among other things of machinery and artificial fertiliser), traders, banks (especially the Farmers' Lending Banks), research institutes and the government. The government took a leading role, partly inspired by the war situations, the Great Depression of the 1930s and the reconstruction after the Second World War. It invested in a flourishing agricultural sector that contributed to economic growth and international trade. This ambition had to dovetail with

---

[85] Van Otterloo, ed., 'Voeding,' 283; H. W. Lintsen, *Made in Holland, Een techniekgeschiedenis van Nederland 1800–2000* (Zutphen 2005), 49–50; Lintsen, *Tachtig jaar TNO*, 115.

the ambition to secure a dependable food supply for the Dutch population. The Second World War was a watershed for this policy. Until 1940 the numerous small farmers had been the focus of attention and reclamations and new polders had to contribute to conserving jobs and food security. After 1945 the government and the agricultural network devoted all their attention to upscaling, land consolidation and the rationalisation of large farms in the context of an emergent European Union. Opposition to this policy was limited and had little effect. Conservation organisations had a marginal position in the network. Other agricultural variants like the biological-dynamic approach survived only in the societal periphery.

Another network, separate from the agricultural network, emerged in the middle of the food supply chain. It was based on the increasing interchangeability of raw materials for foodstuffs, the separation of raw materials into valuable elements, the reconstitution of materials in new products, the improvement of transport and growing international flows of commerce. The position of firms like Unilever changed from being a link in the chain to being a node in an international network. In this way, the production of food in the Netherlands became intimately linked to new sustainability issues abroad. Governments and international organisations negotiated some practices like whaling, but with limited success. Other issues like the social and ecological effects of large-scale plantations were utterly ignored. That happened only in the course of the 1970s.

Finally, new networks developed in the middle and at the end of the food supply chain. They consisted of food processing firms (with a dominant position for Unilever) grocery stores and purchasing organisations (with a growing role for self-service chains like Albert Heijn), government agencies (like the Nutrition Council), research institutes (for example, the TNO Nutrition Organization) and consumer organisations (originally the Dutch Association of Housewives, later also the Consumers Union). Government bodies, research institutes and consumer organisations were the spokespersons for the issue of food quality. The food industry and food retailers jumped on this bandwagon. Around 1900 the emphasis shifted from the caloric content of foods to hygiene and perishability. In the 1930s interest increased in healthy foodstuffs like vitamins. With increasing prosperity after the Second World War organisations slowly but surely began to express concerns about the new feeding habits. The recommended diet in the 1950s - two bread meals and a main meal according to the 'disc of five' – was threatened by an enormous range of foodstuffs that could be consumed at any moment of the day.

After 1970 the latent agricultural problems would lead to widespread protests. They presaged the demise of the agricultural network that had come to full flower in this period.

Literature

Albert de la Bruheze, A.A. and A.H. van Otterloo (2004). 'The Milky Way: Infrastructures and the shaping of milk chains'. *History and Technology*, 20(3), 249–269.
Bakker, T.M. (1985). *Eten van eigen bodem: Een modelstudie*. Den Haag: Landbouwinstituut.

Bergh, S. van den (2004). *Verdeeld land: De geschiedenis van de ruilverkaveling in Nederland vanuit een lokaal perspectief, 1890–1985*. Groningen: Nederlands Agronomisch Historisch Instituut.

Bieleman, J. (1993) 'Landbouw'. In H.W. Lintsen et al. (Ed.), *Geschiedenis van de techniek in Nederland. De wording van een moderne samenleving 1800–1890*. Zutphen: Walburg.

Bielieman, J. (2000). 'Gewasbescherming, dieren en gewassen in een veranderende landbouw'. In J.W. Schot, H.W. Lintsen, A. Rip and A.A. Albert de la Bruhèze (eds.), *Techniek in Nederland in de twintigste eeuw, deel 3: Landbouw, voeding*. Zutphen: Walburg Pers.

Bieleman, J. (2008). *Boeren in Nederland: Geschiedenis van de landbouw 1500–2000*. Amsterdam: Uitgeverij Boom.

Bloemendaal, F. (1995). *Het mestmoeras*. Den Haag: SDU Uitgevers.

Bruin, J.R. and J.C.A. Schokkenbroek (2012). *De laatste traan: Walvisvangst met de Willem Barentsz, 1946–1964*. Zutphen: Walburg Pers.

Buiter, H. and J. Korsten (2006). *Land in aanleg: De dienst Landelijk Gebied en de inrichting van het platteland*. Zutphen: Walburg Pers.

Centraal Bureau voor de Statistiek (1935). *Inkomsten, uitgaven, verbruik en physiologische waarde van voeding, uit verschillende kringen der bevolking gedurende de periode 1 maart 1934–28 februari 1935*. Statistische Mededeelingen CBS, Amsterdam.

Coesèl, M., J. Schaminée and L. Van Duuren (2007). *De natuur als bondgenoot: De wereld van Heimans en Thijsse in historisch perspectief*. Zeist: KNNV Uitgeverij.

Copius Peereboom, J.W. (1976). *Chemie, mens en milieu: Schadelijke stoffen in milieu en voeding, een studie over chemische milieuverontreiniging*. Assen: Van Gorcum.

Cramer, J. (1989). *De groene golf: Geschiedenis en toekomst van de Nederlandse milieubeweging*. Utrecht: Van Arkel.

Dekker, J. (2002). *Dynamiek in de Nederlandse natuurbescherming*. Utrecht: Universiteit van Utrecht.

Dekker, J. (2008). 'De dynamische opstelling van het landbouwschap ten aanzien van het milieu, 1948–1972'. In M.J. Kraker and H.J. van der Windt (eds.), *Jaarboek voor ecologische geschiedenis 2008: Klimaat en atmosfeer in beweging*. Eekhout: Academia Press.

Fieldhouse, D.K. (1995). *Merchant capital and economic decolonization: The United Africa Company, 1929–1987*. Oxford: Clarendon Press.

Grin, J., J. Rotmans, and J. W. Schot (2009). *Transitions to sustainable development: New directions in the study of long term transformative change*. New York: Routledge.

Haar, J. van der (1993). *De geschiedenis van de Landbouwuniversiteit: Van school naar hogeschool, 1873–1945*. Wageningen: Landbouwuniversiteit Wageningen.

Högselius, P., A. Kaijser and E. van der Vleuten (2016). *Europe's infrastructure transition: Economy, war, nature*. New York: Palgrave Macmillan.

Homburg, E. (2000). 'Van carbo- naar petrochemie, 1910–1940'. In J. Schot, H.W. Lintsen, A. Rip and A.A. Albert de la Bruhèze (Eds.), *Techniek in Nederland in de twintigste eeuw: deel II: Delfstoffen, Energie en Chemie*. Zutphen: Walburg Pers.

Homburg, E. (2004). *Groeien door kunstmest: DSM agro, 1929–2004*. Hilversum: Uitgeverij Verloren.

Jones, G. (2005). *Renewing Unilever: Transformation and tradition*. New York: Oxford University Press.

Kalland, A. (1993). 'Management by totemization: Whale symbolism in the anti-whaling campaign'. *Arctic*, 46(2), 124–133.

Knibbe, M.T. (2000). 'Feed, fertilisers, and agricultural productivity in the Netherlands, 1880–1930'. *Agricultural History*, 74(1), 39–57.

Knol W.C., H. Kramer and H. Gijsbertse (2004). *Historisch grondgebruik Nederland: Een landelijke reconstructie van het grondgebruik rond 1900*. Wageningen: Alterra.

Lintsen, H.W. (2005). *Made in Holland: Een techniekgeschiedenis van Nederland [1800–2000]*. Zutphen: Walburg Pers.

Lintsen, H.W. (eds.). (2013). *Tachtig jaar TNO*. Delft: TNO.

Noordijk, J. et al. (2010). De Nederlandse biodiversiteit: Flora en fauna van Nederland. *Nederlandse Fauna*, 10, 1–512.

Otterloo, A.H. (1990). *Eten en eetlust in Nederland, 1840–1990.* Amsterdam: Prometheus.

Otterloo, A.H. (2000). 'Voeding'. In J.W. Schot, H.W. Lintsen, A. Rip and A.A. Albert de la Bruhèze (eds.), *Techniek in Nederland in de twintigste eeuw, deel 3: Landbouw, voeding.* Zutphen: Walburg Pers.

Reader, W.J. (1980). *Vijftig jaar Unilever, 1930–1980.* London: Heinemann.

Rooy, P. de (2014). *Ons stipje op de waereldkaart: De politieke cultuur van modern Nederland.* Amsterdam: Wereldbibliotheek.

Sluijter, B. (2007). *Kijken is grijpen: Zelfbedieningswinkels, technische dynamiek en boodschappen doen in Nederland na 1945.* Eindhoven: Technische Universiteit Eindhoven.

Staten-Generaal (1903). *Handelingen Tweede Kamer* 1903–1904, 15 December 1903.

Thissen, P.M.H. (1994). 'Van heide tot boerenland en bos: Regionale verscheidenheid in heideontginningslandschappen, 1850–1940'. In M. de Harde and H. van Triest (eds.), *Jonge landschappen, 1800–1940: Het recente verleden in de aanbieding.* Utrecht: Matrijs.

Thonstad, P. and E. Storli (2013). *Big business and small states: Unilever and Norway in the interwar years.* Economic History Review 66(1), 109–133.

Toke, D. (1999). Epistemic communications and environmental groups. *Politics,* 19(2), 97–102.

Veen, M.P. van (2010). *Halting biodiversity loss: Evaluation of Progress.* Netherlands Environmental Assessment Agency.

Veldman, H.E., E. van Royen and F. Veraart (1999). *Een machtige schakel in de Nederlandse land- en tuinbouw. De geschiedenis van Cebeco-handelsraad, 1899–1999.* Rotterdam: Cebeco groep.

Ven, F.J.M. van de (2006). *Anton Jurgens Hzn,1867–1945: Europees ondernemer, bouwer van een wereldconcern.* Zwolle: Waanders.

Voous, K.H. (1970). *Natuur, milieu en mens.* Kampen: J.H. Kok.

Weinreich, J.A. and C.J.M. Musters (1989). *Toestand van de natuur: Veranderingen in de Nederlandse natuur.* Den Haag: SDU Uitgeverij.

Wijn, J.F. de and W.A. van Staveren (1986). *De voeding van elke dag.* Utrecht: Springer Media.

Windt, H. van der (1994). 'De Totstandkoming van 'de Natuurbescherming' in Nederland'. *Tijdschrift voor Geschiedenis,* 107(3), 485–507.

Windt, H. van der (1995). *En dan, wat is natuur nog in dit land? Natuurbescherming 1880–1990.* Amsterdam: Boom.

Zanden, J.L. van and Verstegen, S.W. (1993). *Groene geschiedenis van Nederland.* Houten: Het Spectrum.

Zoest, J. van (1998). *Biodiversiteit.* Utrecht: KNNV.

Zoethout, D.A. (1914). *De plant in nijverheid en handel.* Amsterdam: Elsevier.

Zon, H. van (1986). *Een zeer onfrisse geschiedenis. Studies over niet-industriële vervuiling in Nederland, 1850–1920.* Groningen: proefschrift

Chapter 14
Building Materials and Construction: Constructing a Quality of Life

Frank Veraart

Contents

Abstract Catastrophes and new societal ambitions energized the huge construction effort undertaken between 1910 and 1970. The floods of 1917 and 1953 led to enormous investments in coastal defences. The government also undertook major investments in the construction of roadways and other infrastructural works. New building codes, damage incurred during the Second World War and population growth incited new housing construction on a colossal scale. Demand for building materials grew apace.

The need for wood and mineral subsoil resources transformed nature and landscapes in the Netherlands and at foreign sites. Dutch forestry practices were rationalised. Imports from the Baltic regions by and large met the Dutch demand for wood. But the creation of monocultures and production forests in these regions reduced local biodiversity. Gravel and marl were mined above all in the province of Limburg. That led to tensions with local stakeholders. Gravel extraction transformed the floodplains of the Meuse into a lake landscape. It led directly to the Excavation Law, the first environmental law in the area of land-use. After 1970, regulations concerning land-use and new landscape values would regularly inspire conflicts in the national supply of building materials (see Chap. 19).

This chapter is written by Frank Veraart with contributions by Harry Lintsen.

© The Author(s) 2018
H. Lintsen et al., *Well-being, Sustainability and Social Development*,
https://doi.org/10.1007/978-3-319-76696-6_14

Keywords Construction · Housing construction · Coastal defences · Infrastructure
· Building materials · Wood · Iron · Minerals · Cement · Concrete · Gravel ·
Limburg · Land-use

14.1   Could the Flood Disaster Have Been Prevented?

1953. The night of January 31st to February 1st. A northwest wind roared across the
North Sea at hurricane force. The storm had already lasted for hours. An enormous
mass of water was being driven at great speed in the direction of the southwest
Netherlands. Around midnight it was supposed to be low tide, but the waters had not
retreated. Quite the contrary, the waters continued to rise and looked ominous; it
was black and full of driftwood. The next high-tide was a super high-tide.

> (That afternoon...) you saw masses of water charging from afar over the Oosterscheldt. It
> was magnificent, magnificent! We stood there thrilled by the tremendous power of nature.

as the soldier Jo Leune and his girlfriend Suus Priem from Stavenisse later recalled.[1]
The majority of the population was not really worried. Many people were visiting
or at a party, as was normal for Saturday night. The floods of 1906 and 1916 survived only as the vague memory of an older generation. Toward the end of the war
Walcheren had been flooded. But that had been done on purpose to support the
Allied invasion of the continent. In this night it was certainly exceptionally heavy
weather, but here and there measures had been taken and flood-boards erected. That
would suffice.

Several hours after midnight the Dutch delta was transformed into an unimaginable pandemonium. In some places sirens had wailed or church bells chimed. In
many places there had been no warning at all. The dikes burst at more than 150
places. Around 140,000 hectares of land were flooded. 1835 people lost their lives.
Thousands of head of cattle were drowned. Nearly 50,000 buildings were destroyed
or damaged. The economic damage was estimated at about 700 million euros (at
current prices equivalent to more than 10 billion euros).

God had shown his hand, according to the deeply pious part of the inhabitants of
Zeeland. In the book *In the Grip of the Water Wolf*, published in response to the
disaster, it was written:

> God saw the work of man...and they did what was evil in His eyes. They loved money and
> power...and violence ruled in the streets... Then He used His right and he chastised us, so
> that the whole of humanity should learn that love is more than violence.[2]

---

[1] Cited and translated from: K. Slager, *De ramp: Een reconstructie: 200 ooggetuigen over de
watersnood van 1953* (Goes 1992), 15.

[2] C. Baardman, *In de greep van de waterwolf* (Den Haag, 1953), 171.

Others emphasized the mystical power of nature. Nature surprised humans time and again. Humans might think they had conquered nature with their modern technologies. But that was a misconception.

But we must ask: Is this perception accurate? Had the Netherlands no inkling of these kinds of extraordinary natural phenomena in 1953? Was there no knowledge of the consequences of the unusual circumstances that had presented themselves? A day after the disaster of February 1st, 1953 an emotional engineer wrote:

> Natural disasters! But our civil engineers already knew beforehand where these 'natural disasters' would strike. They warned in reports and expert advice about the dikes that were too low and too weak…[3]

He pointed to research undertaken by the Research Unit of the Estuaries, Tidal Rivers and Coasts of the Dutch Agency for Public Works and Water Management, *Rijkswaterstaat.* The research unit was set up in 1929. During the 20 years prior to the flood disaster an average of 50 persons per year worked on the problem of the delta region, that is, about 1000 person-years in total. Old conceptions made way for new theories and traditional rules of thumb were abandoned. Endless measurements and soundings were taken and registered. New instruments and new approaches like the application of statistical methods and factor analysis of storm surges were introduced.[4] The investigations arrived at a clear conclusion. The delta region was a kind of time-bomb and vulnerable to big threats. Draconic measures would be needed to turn the tide.

But even with all this knowledge the disaster could not have been prevented. New approaches take time to penetrate into the domain of policy and politics. During the Second World War the Netherlands was occupied with other matters. Designing and constructing a new coastline would require gigantic investments, investments that were unthinkable from a social perspective during the Depression and post-war reconstruction.

'Calamities produce policy' is an oft-cited saying in civil engineering circles. After the disaster things moved quickly, among other things because the plans were already prepared. The disaster took place on February 1 1953. On the 21st of that month the Delta Commission was sworn in. Between 1953 and 1955 the commission published five fundamental and radical reports. In 1955 it proposed a Delta Law that was ratified by parliament in 1958. By then the *Rijkswaterstaat* had already been long at work.

The vulnerable delta is one of the fundamental aspects of well-being and sustainability in the Netherlands. The well-being monitor of Statistics Netherlands (CBS) lacks an indicator to reference this issue. This is hardly surprising because the set of indicators we use in this study conforms to international guidelines. A specifically Dutch issue like the flood-prone situation cannot be coupled to one of these

---

[3] R. Verloren van Themaat, 'Watersnood 1953', *De Ingenieur* 65(1953), nr. 6, A 56.

[4] P.J. Wemelsfelder, 'Wetmatigheden in het optreden van stormvloeden', *De Ingenieur* (1939), B, nr. 31 and J. van Veen, *Te verwachten stormvloeden op de benedenrivieren* (internal report, Rijkswaterstaat, directie Benedenrivieren, afdeling Studiedienst, 1939).

indicators. But for the Netherlands it is an ancient and crucial issue. This was the case in the twentieth century and it will remain the case for future generations.

14.2 Well-being, Vulnerability and Liveability

In addition to the vulnerable delta, poverty was one of the biggest determinants of well-being and sustainability in the past. Despite the positive developments at the end of the nineteenth century, the differences among living conditions of the various segments of the population were still very large. The unemployed and low-paid workers lived in hovels, while industrialists, bankers and other well-off prominent citizens fitted their opulent dwellings with modern delights like bathrooms, electric illumination and telephones. Public housing demanded significant improvement. In addition, population growth also demanded a continual growth in welfare. This in turn required growth in business and in transport, energy and communications infrastructures.

The further elaboration of the infrastructures, the defences against flooding and the construction of liveable housing for the poor dominated the enormous construction agenda of the twentieth century. This effort had far-reaching consequences for the quality of life and for natural capital in the Netherlands and abroad. The effects on natural capital had two dimensions, a spatial and a material. Construction of housing, roads, utility buildings and urban layout had implications for the use of space. In the course of the twentieth century it proved necessary to reflect ever more profoundly on the division and utilisation of the country's limited surface area. It was also impossible to build without a supply of materials. Millions of kilos of wood, bricks, concrete and steel found their way to construction sites. Between 1910 and 1970 building materials were among the biggest material flows in the Dutch economy (see Tables 12.5 and 12.8).

In this section we briefly describe the water management and infrastructural developments. To an important degree these were extensions of nineteenth century policy. Subsequently we shift our focus to public housing construction for the poor and to urban development. After that the supply side of the enormous building effort demands our attention. Where did the raw materials come from? What influences did this have on changes in the environment and the landscape?

14.2.1 Safe Behind the Dikes

The closure and reclamation of the Zuiderzee (1919–1969) and the Delta Works (1956–1998) were the calling cards of Dutch civil engineering. After the reclamation of the Haarlemmermeer using the world's biggest steam engine, plans were made, starting in the mid-nineteenth century, for the next step: the reclamation of an entire sea. The first proposals still seemed like fantasies, but in 1886 liberal circles founded a Zuiderzee Association that began to develop concrete plans. The young engineer Cornelis Lely authored the various technical reports published by the association. In 1891 Lely left the association to become Minister of Public Works, Trade and Industry. During his first and second terms as minister nothing came of the Zuiderzee plans. In 1913 he accepted a third term, but only on condition that now a decision would be taken in favour of closure and reclamation.

The outbreak of the First World War again threatened the plans. Though the Netherlands succeeded in remaining neutral, the war in the surrounding countries nonetheless caused material and food shortages. The storm surge and floods of 1916 that affected the region of the Zuiderzee caused local food shortages. This not only put the food situation squarely on the agenda but also provided an important impulse for the Zuiderzee plans.

Together with the Delta Works these were the dizzying heights of Dutch hydraulic engineering in the twentieth century. The works confirmed the image of the Netherlands as a nation of dike-builders. Dike-builders who with perseverance and investments had wrested land from the sea and who, after completing the Delta Works, could resist the sea as well.

14.2.2   New Roads and the Unification of the Netherlands

At the turn of the century the Netherlands was a patchwork of loosely connected regions. During the twentieth century these were stitched together by various infrastructures, that reduced time needed for travelling, communications and hauling goods. Important infrastructural works included the construction of canals, the improvement of the rivers and the building of road, energy and communications networks.[5] In the twentieth century the Netherlands was a colossal 'work in progress.'[6] The country was literally smothered in networks of conduits, pipes, rails, asphalt and electromagnetic waves. Every bit of the Netherlands, wet or dry, was eventually connected up with a larger whole. Links among households, farms and factories were established throughout the country. Intentionally or not, the material infrastructures thus became an important medium of the 'unification of the Netherlands,' the socio-spatial integration of regions and communities within the national territory.[7]

They also contributed to economic growth and the growth of well-being. Networks were built up and broken up again or partially lost their function. The

---

[5] Erik B.A. van der Vleuten, 'De materiële eenwording van Nederland,' in *Techniek in Nederland in de twintigste eeuw - Techniek en modernisering, balans van de twintigste eeuw*, by Johan W. Schot et al., VII (Zutphen: Walburg Pers, 2003), 42–73.

[6] A. van der Woud, 'Stad en land: Werk in uitvoering,' in D. Fokkema and F. Grijzenhout, *Nederlandse Cultuur in Europese Context* (Den Haag 2004).

[7] H. Knippenberg and B. de Pater, *De eenwording van Nederland: Schaalvergroting en integratie sinds 1800* (Nijmegen 1988).

nineteenth century telegraph network lost some of its functions to the telephone network as it unfolded into a national and international network in the twentieth century. The local gas networks suffered competition from the electricity networks, that were ultimately organized at a provincial level. The railway trunk network branched out after 1880 with connections to numerous 'secondary' local rail networks and inter-local tramways. There were also 'tertiary' municipal tramway networks. Around 1930 there were about 6500 km of railways and tramways. Large portions were later torn up.

The network that made the largest claims on space and spatial planning was the road and highway network. The ANWB, founded in 1883 as an organisation for cyclists, developed into one of the most important advocates of improvements to the road infrastructure. In 1920 this organisation, together with the Royal Institute of Engineers (KIvI), organised the First Netherlands Road Congress. The congress accelerated the road improvement program that was already underway. Additional consultations between the government and interest groups led to a national road plan by 1924, a plan that was ratified by parliament in 1928.[8]

In the 1930s ideas about roads exclusively for motorised traffic gained currency. The national road plan, especially in the western part of the country, gradually became a plan of limited access highways. The ANWB, the KIvI and other organisations saw the highway as an alternative and successor to the railways. Construction of the first highway between The Hague and Utrecht began in 1934 and took 5 years. Other sections followed.[9] After 1958, bowing to popular pressure, the Minister of Transport and Public Works increased the pace of highway construction. Between 1950 and 1970 the total length of the highways grew from about 100 to a 1000 km. In 1970 all the roads in the Netherlands taken together had a collective length of 77,000 kilometres.[10]

The investments in communications, energy and road infrastructure contributed to the spread of industries throughout the country. The textile factories in Brabant and Twente were heavily dependent on their connections to the maritime ports in the western part of the country. Coal from the Limburg mines was able to fuel the Dutch economy only after this region had been opened up by the canalisation of the river Meuse and railroad connections. The development of infrastructures fostered industrialisation throughout the Netherlands.

Building of infrastructures and industrialisation contributed significantly to improving the quality of life. That had been demonstrated unequivocally in the nineteenth century. At the beginning of the twentieth century there was still a yawning gap between rich and poor. Part of the elite was of the opinion that the poor should have more of a share in the increase in well-being. This focused attention on another aspect of the building agenda, the construction of public housing.

---

[8] G. Mom and R. Filarski, *Van transport naar mobiliteit: De mobiliteitsexplosie (1895–2005)* (Zutphen 2008), 173–201.

[9] Mom and Filarski, *Van transport naar mobiliteit*, 197.

[10] Mom and Filarski, *Van transport naar mobiliteit*, 314–317.

14.3   Living and Quality of Life in the City

> We cannot say in the abstract: this or that is a good habitation; that depends entirely on the situation in a given municipality. When one hears, for example, that people live in a house made of turf, then this sounds very miserable to a city-dweller, but I have seen such turf houses that I would far prefer to some rooms in one of the big apartment buildings in the big cities, where one lives between stone walls, but where one loses all freedom…where outside of one's room there is no single domain in which one is one's own boss. In a dwelling on the heath, even if it is made of turf, one is just as well protected from cold or heat; one is one's own boss and is not bothered by neighbours, piano-players and such.[11]

The quality of dwellings, argued the parliamentarian for the Christian Historical Union (CHU), squire A.F. de Savornin Lohman, in 1901 in the Second Chamber deliberations on the new housing bill, was subjective. Reports on housing conditions in the Netherlands had produced a clear image of the hovels and miserable workers' dwellings. But opinions diverged on what was to count as good housing.

Inspired by and based on foreign examples, Dutch Hygienist groups in the societal midfield had developed ideas about daylight and ventilation and about sanitary, cooking and sleeping facilities. At the end of the nineteenth century they themselves set an example with the construction of workers' housing. At the political level the right approach remained up for grabs. Liberal and confessional politicians feared that the demands made on housing would infringe the rights of landlords and that housing would become unaffordable for the poor. The social-liberal cabinet led by N.G. Pierson resolved the debate by introducing the Housing Law of 1901.

14.3.1   The Housing Law and Housing Construction

The Housing Law aimed at the elimination of hovels and the improvement of the quality of dwellings. Starting with the census of 1899 information on housing had been gathered with each census. This revealed that more than half the dwellings had only one or two rooms, a statistic that provided an estimate of the proportion of hovels in the total housing supply. The housing statistics of various cities over a number of years showed the effect of the Housing Law. In a half century the number of one and two room dwellings declined sharply (Table 14.1).

Eliminating the hovels was work for the municipalities. Municipal inspectors made an inventory of the many workers' homes in order to be able to advise on the housing situation. The residents of these homes regarded the members of the commission with mixed feelings. For them, a declaration that their homes were uninhabitable meant eviction or higher rent. The prominent socialist member of Amsterdam's city council, F.M. Wibaut, member of the Public Health Commission from 1907 to 1914, noted a typical reaction:

---

[11] Handelingen Tweede kamer 1900–1901, 'Wettelijke bepalingen betreffende de volkshuisvesting', 60ste Vergadering −15 Maart 1901, 1218.

Table 14.1 Development of the number of rooms per dwelling as a percentage of the total number of dwellings in different cities

		Number of rooms per dwelling (in %)				
		1	2	3	4	5 and more
Amsterdam	*1900*	19.1	25.7	22.4	23.6	9.2
	1930	10.2	8.9	17.5	26.4	37.1
	1947	0.6	3.1	15.2	32.5	48.5
	1956	0.6	3.2	15.3	31.6	49.4
Rotterdam	*1900*	20.1	41.1	16.8	14.3	7.7
	1930	5.3	12.8	15.7	21.7	44.5
	1947	0.3	3.4	8.9	27.5	59.9
	1956	0.3	3.0	10.2	31.5	55.0
Groningen	*1900*	34.8	31.7	12.0	11.2	10.4
	1930	6.8	16.4	16.6	16.4	43.8
	1956	1.1	5.3	11.5	18.3	63.9
Nijmegen	*1900*	15.0	32.3	16.8	16.1	19.7
	1930	1.9	6.7	10.6	13.6	67.2
	1956	0.0	1.9	6.1	13.8	78.2
Tilburg	*1900*	2.5	37.5	31.9	19.8	8.3
	1930	0.7	4.5	12.2	22.1	60.5
	1956	0.0	1.1	9.7	10.5	78.7
Maastricht	*1900*	31.2	32.6	12.4	11.4	12.5
	1930	7.8	18.2	17.5	15.9	40.6
	1956	0.5	3.0	7.0	11.1	78.4
Enschede	*1900*	7.5	38.3	23.6	20.0	10.5
	1930	1.6	4.0	6.1	10.8	77.5
	1956	0.0	0.4	5.1	11.1	83.4
Emmen	*1900*	80.0	13.2	3.6	2.5	0.7
	1930	31.6	25.9	15.0	10.0	17.5
	1956	3.1	11.2	14.8	16.2	54.7

Data from ´Uitkomsten der Woningstatistiek´ behorende bij Volkstellingen 1899/´Woningtelling en Gezinsstatistiek' behorende bij Volkstelling 1930/'Volkstelling annex woningtelling' 1947/'Algemene Woningtelling' 1956
Source: NIWI- KNAW/CBS (www.volkstelling.nl)

> Gosh, sir, just let us live here. I'm sixty-four. Was born here in this cellar. My husband is sixty-six and for him it's alright too. Never a doctor. If we want fresh air we have the street. Brought up twelve children here. Figuring out sleeping places was a real puzzle. But it went fine... It was a good family. All of them married. Two dead, all of them healthy.[12]

In their reactions the residents reflected the arguments of the commission members about health, light and fresh air. The focus on living conditions meant that decisions were made about and for the lowest classes. The supervisors and elite champions of workers' housing were mocked and were known among the hovel-dwellers paradoxically enough as 'poor-people-rejectors.'[13]

---

[12] F.M. Wibaut, *Levensbouw, memoires*, Querido, Amsterdam, 1936, cited in Egbert Ottens, *'Ik moet naar een kleinere woning omzien want mijn gezin wordt te groot', 125 Jaar Sociale Woningbouw in Amsterdam* (Amsterdam 1985), 22.

[13] Ottens, *'Ik moet naar een kleinere woning omzien want mijn gezin wordt te groot'.*

The Housing Law compelled municipalities to establish a department of building and dwelling inspection and to formulate building codes. Construction plans that satisfied legal demands were financially supported by the national government. The law took account of local differences. The law's implementation took a long time, among other things due to the time it took to formulate local building codes and the guarantees against misuse of state financing, the lengthy debates on the proper rents for subsidized dwellings, the discussion about the role and recognition of building societies and the delaying tactics of powerful landlords.

In 1904 the national government published the first 'guidelines' that provided a point of departure for state financing. In subsequent years ministerial letters and new guidelines in 1927 ensured harmonisation of the building codes.[14] This enabled the national government to be specific about the quality of workers' housing. Alcoves were forbidden and separate rooms for cooking, sleeping and toilet facilities became mandatory. Every dwelling had to have an outside space. The toilet in the dwelling was not allowed to open into the kitchen. Bedrooms had to have a window.[15] The admission of air and light became the new adage, with the typical Dutch *'doorzonwoning'* - a house equipped with large glass panes front and back so as to admit sunlight both morning and afternoon - as the characteristic outcome. The minimum surface area of a dwelling was originally fixed at 30 m². In 1920 this was upgraded to 40 m². The average size of Dutch dwellings in that year was 47 m².[16] The national average dwelling size increased starting in the 1950s to 109 m² in 1970. The surface area differed according to the type of housing; the surface area of non-subsidized dwellings grew especially rapidly in the 1960s (Table 14.2).[17]

Subsidies were granted to recognized housing corporations. In order to streamline discussions among municipalities, the national government, and housing corporations, a Housing Council was established in 1913 in which the certified housing corporations united themselves.[18] During the First World War, housing construction ground to a halt. Construction workers and materials were scarce. The government eased the restrictions on loans for housing construction and framed a Housing Shortage Law. This mandated municipalities to support housing corporations or take housing construction into their own hands. Many municipalities adopted the latter course and set up their own housing agency or housing firm. In 1917 munici-

---

[14] N. de Vreeze, (Ed.), *65 miljoen woningen: 100 jaar Woningwet en wooncultuur in Nederland* (Rotterdam 2001), 105–111.

[15] J. Huisman et al., *Honderd jaar wonen in Nederland, 1900–2000* (Rotterdam 2000), 16.

[16] Vreeze, *65 miljoen woningen*, 107; Huisman et al., *Honderd jaar wonen in Nederland, 1900–2000*, p. 25. For more detailed analyses of the development of dwelling size in the city of Groningen see: A.M.L. Diepen, 'Spatial Aspects of Housing,' in *Green Households? Domestic Consumers, Environment and Sustainability*, ed. Klaas Jan Noorman and Ton Schoot Uiterkamp (London 1998), 105–7.

[17] L.A. Bruggeman, 'Kwalitatieve woningdocumentatie (KWD) 1948–1970: Enkele kwaliteitsaspecten van de nieuwbouw van woningen over de afgelopen 30 Jaar,' kwalitatieve woningdocumentatie (Zoetermeer 1981), Table V.

[18] Wouter Beekers, *Het bewoonbare land: Geschiedenis van de volkshuisvestingsbeweging in Nederland* (Amsterdam 2012), 107–33.

Table 14.2 Development of dwelling size in the Netherlands: average surface area in m^2

	Housing law housing			Unsubsidized housing			Total new housing[a]		
	One-family homes	Multi-family homes	Average	One-family homes	Multi-family Homes	Average	One-family homes	Multi-family homes	Average
1948	56	50	55						
1950	56	46	53						
1952	54	44	49						
1954	55	47	51						
1956	58	50	54						
1958	58	50	54						
1960	57	49	53						
1962	57	51	54	70	56	66	62	52	58
1964	60	52	57	68	59	66	63	54	60
1966	67	60	63	75	66	73	69	60	66
1968	66	59	63	85	62	82	72	58	68
1970	69	59	65	89	57	85	75	55	69
1972	70	58	66	90	56	88	78	54	72
1974	67	52	63	95	66	93	78	52	72
1976	67	53	64	100	60	98	80	52	74

The average surface area is the sum of the surface areas of the kitchen, living room, and bedrooms
[a]Housing Law, Unsubsidized, and Premium dwellings
Source: Bruggeman, L.A. "Kwalitatieve Woningdocumentatie (KWD) 1948–1970, Enkele Kwaliteitsaspecten van de Nieuwbouw van Woningen over de Afgelopen 30 Jaar." (Zoetermeer 1981), tabel V

palities also acquired the right to appropriate housing corporations.[19] Despite the long start-up phase and difficult circumstances, between 1901 and 1918 the housing corporations built, with government support, more than 50,000 dwellings. The number of corporations rose from 400 in 1914 to 1300 in 1920.[20] This made corporative building the most important force in housing construction in this period.

After the First World War the national government took stock of its position, reduced subsidies and coupled them to slum clearance. Housing corporations and municipalities saw this about-face as a death blow to the Housing Law and protested fiercely. Exerting themselves, they plumbed alternative sources of financing and in this way managed to triple the production of new homes between 1918 and 1940.[21] In the total housing production the share of corporative housing construction declined to 10 or 20%.

The production of new housing by housing corporations and private parties grew from about 20,000 dwellings per year around 1900 to 30,000 just before the out-

---

[19] Beekers, *Het bewoonbare land*, 135–142.

[20] P. Ekkers, *Van volkshuisvesting naar woonbeleid* (Den Haag 2006), 63.

[21] Beekers, *Het bewoonbare land*, 172.

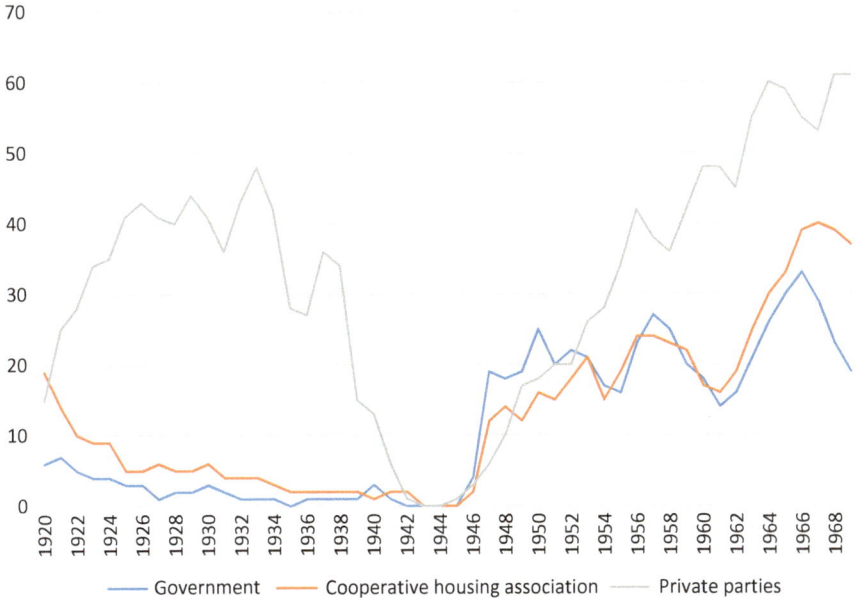

Graph 14.1 Housing production by the government, housing construction corporations and private parties 1920–1970
Source: CBS – Historie bouwnijverheid vanaf 1899

break of the First World War. In the interbellum housing production increased further to about 50,000 dwellings around 1930, after which it declined to a bit more than 30,000 per year. After the Second World War the Ministry of Public Housing and Reconstruction assumed responsibility for housing construction. The number of new dwellings per year rose from 50,000 per year in the early 1950s to more than 150,000 in the early 1970s. (See Graph 14.1)

These building efforts brought the average number of occupants per dwelling down from five around 1900 to four on the eve of the Second World War. After the war the number of occupants per dwelling climbed again until another decline set in during the 1960s. At the end of the 1970s the figure was three occupants per dwelling.

Throughout the twentieth century there was a shortage of dwellings in relation to the number of households. In 1900 the shortage amounted to 100,000 dwellings for 1.1 million households (9% housing shortage). Housing construction reduced this shortage up to the First World War, but due to production cutbacks and an increase in the number of households the shortage again increased to more than 130,000 dwellings in the early 1920s (8% housing shortage). After that there was a decline until the Second World War. War damage, a moratorium on construction during the war, shortages of materials and the post-war increase in the number of households forced the housing shortage up to a half million dwellings for 2,6 million households (20% housing shortage) at the end of the 1940s. For the second time in the twentieth century public housing became a core priority of the national state. The

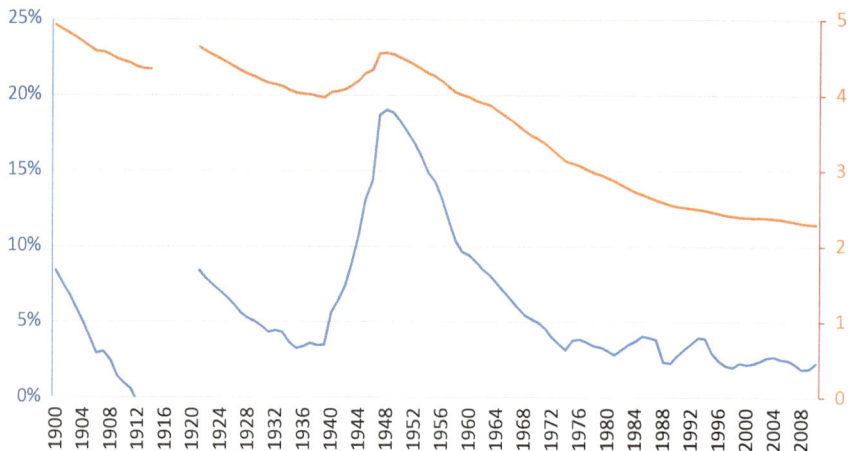

Graph 14.2 Percentage of households without a dwelling and average number of occupants per dwelling in the Netherlands, 1900–2010
Source: CBS - processing of data regarding number of dwellings, households and inhabitants by author

enormous housing production of the 1950s and 1960s once again reduced the housing shortage to about 150,000 dwellings (4% housing shortage). With these building efforts the government succeeded in limiting the housing shortage in the post-war years. But even after the 1970s some 3% of the households still did not have its own dwelling (Graph 14.2).

The focus of attention of building contractors, housing corporations and the government gradually changed. From slum clearance, attention shifted to fighting the housing shortage. The Housing Law explicitly defined the minimum requirements for a Dutch dwelling. The massive share of one and two-room dwellings had assuredly been reduced but without any palpable increase in the size of dwellings. With the stipulation of minimal requirements housing shortage became the new name for a shortage of quality dwellings. Housing shortage shaped the building activities of post-war ministers of reconstruction and public housing. In view of the unheard-of production of more than 123,000 dwellings in 1969, the Minister of Public Housing and Spatial Planning, W.F. Schut, felt able to announce the end of the housing shortage.[22] It seemed that the worst of it was indeed over. However, the end of the housing shortage was announced time and again simply because it persisted, albeit at a lower level than before.

With the shift of attention from the quality of dwellings to the struggle against the housing shortage, more effort was put into the production of new dwellings. Elimination of the hovels and slum clearance became secondary issues. New housing construction was situated in the urban peripheries, satellite towns, and so-called 'growth cores.' The fixation on new construction meant that in the 1960s portions of the old cities were still deteriorating into slums. An investigation by the Commission

---

[22] 'Opheffing van woningnood schuift steeds verder op' in *De Waarheid,* 14 januari 1969, 1 en 4.

Table 14.3 Number of rooms per dwelling 1899–1971

	Dwellings	With 1 room	With 2 rooms	With 3 rooms	With 4 rooms	With 5 rooms	With 6 or more rooms
	x1000	%	%	%	%	%	%
1899	1091	28.3	30.7	17.6	15.3	15.3	8.1
1909	1237	18.7	29.2	20.0	21.3	21.3	10.8
1920	1366	.	.	.	.	.	.
1930	1824	7.4	14.5	16.3	18.3	18.2	25.3
1947	2117	2.2	8.2	11.4	17.8	24.0	36.4
1956	2546	0.9	4.5	10.4	17.9	28.4	37.9
1971	3873	0.2	1.5	7.9	16.6	38.0	35.9

Source: CBS – Historie bouwnijverheid vanaf 1899

for Slum Clearance and Redevelopment of the Ministry of Public Housing revealed that in 1957 approximately 6% of the housing supply, or 145,450 dwellings, consisted of hovels. The provinces of North and South Holland topped the list with respectively 30,000 and 29,650 hovels. The housing problem was thus far from solved, but it assumed an entirely different aspect in the 1960s. In the statistics the improvements in housing quality became increasingly visible in the growth of the number of rooms per dwelling (see Table 14.3).

The pauperization of the cities was intensified by the growing suburbanisation of the 1960s. By means of redevelopment and 'city formation' according to modernist principles, city governments tried to maintain their inner cities as economically viable centres.[23] Growing welfare was responsible for the flight of families with increased purchasing power from the cities. New sustainability issues around residency came to the fore, because, where were these people to live? With this, the housing question and urban development after the Second World War acquired an emphatically spatial dimension.

14.3.2   Expanding Cities 1900–1970

The growth of the cities was governed by two different spatial dynamics. In the first place it was about urban density: how many inhabitants are there per square kilometre? This was closely related to the housing developments just described. A second spatial aspect concerns the expansion of cities and their infrastructures into the countryside. How much land was used for the city and for the infrastructures that connected the cities to each other? Both these spatial aspects were consequential for deliberations about how to use the territory of the Netherlands, for the food supply, industrial development, nature and residence patterns.

---

[23] H. De Liagre Böhl, *Steden in de steigers: Stadsvernieuwing in Nederland 1970–1990* (Amsterdam 2012), 17.

Table 14.4 Ten cities with the highest population density on January 1, 1850, 1900, 1950 and 2000 (inhabitants per km²)

| 1850 | | 1900 | | 1950 | | 2000 | |
× 1000		× 1000		× 1000		× 1000	
Leiden	21.4	Amsterdam	22.0	The Hague	8.5	The Hague	6.5
Amsterdam	13.5	Maastricht	14.4	Utrecht	7.6	Voorburg	6.2
Rotterdam	11.9	Vlaardingen	9.2	Leiden	7.2	Leiden	5.3
Maastricht	10.6	Schoonhoven	9.1	Haarlem	6.3	Haarlem	5.0
Schoonhoven	5.1	The Hague	8.5	Voorburg	5.7	Maassluis	4.4
Breda	4.2	Breda	7.5	Amsterdam	5.3	Amsterdam	4.4
Vlaardingen	4.2	Haarlam	7.1	Groningen	5.3	Cappelle aan den IJssel	4.4
Gouda	4.2	Leiden	6.6	Urk	4.9	Gouda	4.2
Haarlem	4.1	Enschede	6.4	Rotterdam	4.4	Schiedam	4.1
Delft	3.5	Rotterdam	6.1	Bussum	4.4	Bussum	3.8

Remark: The population density depends on the surface area of the municipality. Due to the redrawing of municipal boundaries in the second half of the twentieth century the population density of a number of cities declined. If the population density is expressed as the number of inhabitants in a radius of 25 km, The Hague, Rotterdam and Amsterdam are the cities with the highest population density

Source: P. Ekamper, R. van der Erf and N. van der Gaag, *Bevolkingsatlas van Nederland, Demografische Ontwikkeling van 1850 tot Heden*, (Den Haag 2003), 43–49

Limited housing space and a limited surface area due to city walls and moats made nineteenth century cities very densely populated. This situation was basically the same in 1900 (Table 14.4).[24] After changes to the Fortress Law in 1874, the utility and necessity for city walls had disappeared and urban expansion became possible. The national government forced the cities to formulate detailed expansion plans as a condition for acquiring the formerly military zones. This marked the starting point of a more structured approach to urban development, one shaped by the city government in consultation with experts and interested parties. The new urban space on and around the old walls was devoted to the elaboration of traffic and transport possibilities and new utilities like gasworks and electrical power plants. Many railway stations came to be located in the late nineteenth century expansions of the city, just outside the old defensive works.[25]

The new international ideas about light, air and space, previously applied to the design of dwellings, also seized the imaginations of urban planners. Dutch urban planners were influenced by German handbooks. Around the turn of the century international congresses in London (1906), Vienna (1908), Düsseldorf (1913) and Ghent (1913) contributed to the spread of ideas and knowledge amongst urban planners.[26]

---

[24] P. Ekamper, R. van der Erf and N. van der Gaag, *Bevolkingsatlas van Nederland: Demografische ontwikkeling van 1850 tot heden* (Den Haag 2003), 42.

[25] L. de Klerk, *De modernisering van de stad 1850–1914: De opkomst van planmatige stadsontwikkeling in Nederland* (Rotterdam 2008), 198–227.

[26] De Klerk, *De modernisering van de stad 1850–1914: De opkomst van planmatige stadsontwikkeling in Nederland*, 274–289.

The first nineteenth-century city parks like Rotterdam's Maaspark (1853) and Amsterdam's Vondelpark (1865) had become green enclaves surrounded by densely built up neighbourhoods. In the eyes of the improvers of housing conditions these parks were too small. Toward the end of the century they developed a new vision for parks: in addition to offering a touch of nature, they should also provide space for sports and games.[27] Renowned architects, naturalists and politicians like J.Kruseman, H.P. Berlage, J. van Hasselt, D. Hudig, C.A. den Tex en Jac.P Thijsse testified as members of Amsterdam's Housing Council in 1909 that

> 'parks, green areas and sport facilities are an absolute necessity for every single neighbourhood. Where they have once been established their zoning may never be changed into that of building plots again.'[28]

In 1908 the Housing Law was amended to allow a zoning specification of 'park'; by 1921 it had become embedded in the law. With the parks, nature was brought into the city. In 1908 Rotterdam's city government pronounced that parks served to compensate the working population for 'the lack of rural, dust-free pedestrian routes and attractive spots.' A year later the director of Rotterdam's municipal public works, G.J. Jongh, presented his plans for parks in the Blijdorppolder and around the Kralingen Lake. The latter consisted of a terrain of 315 hectares the plans for which included not only a park and woods, but also athletic fields, ponds and playgrounds as well as plots for villas and allotment gardens. The cost was estimated at 7.7 million guilders, but according to the director it would result in 'a kind of natural monument, that would be of lasting value to future generations.'[29]

Around the turn of the century, architects and urban planners in Germany and Great Britain developed ideas for so-called garden cities. In these types of urban expansions there was more room for greenery and private gardens. Socially conscious housing corporations embraced the garden city concept and implemented it in the Netherlands. In the more generous use of residential space they saw possibilities for elevating the working masses. With the aid of architects like H.P. Berlage (Amsterdam and Utrecht), C.J. van Eesteren (Amsterdam), M.J. Granpré Moliere (Rotterdam) and W.M. Dudok (The Hague) housing corporations realized garden city neighbourhoods, villages and districts on the outskirts of the large cities. Even middle-sized cities like Hilversum and Apeldoorn were developed according to these concepts. In the expanding industrial cities big firms established new garden towns, like the garden village 't Lansink developed by the Stork company in Hengelo or garden village Heyplaat developed by the Rotterdam Drydock Company as well as the Philipsdorp and Drentsdorp, built by the Philips company in Eindhoven. In the mining region of the (Catholic) province of Limburg the project to curtail the 'socialist threat' and a Catholic parochial structure went hand in hand with the creation of a village structure in the mining colonies, developed by the Catholic housing construction corporations.[30]

---

[27] K. Bosma et al., *Bouwen in Nederland, 600–2000* (Zwolle 2007), 586.

[28] J. Kruseman (secretary), *Rapport over de Amsterdamsche parken en plantsoenen*, commission consisting of H.P. Berlage, J. van Hasselt, D. Hudig, J.Kruseman, C.A. den Tex and Jac.P Thijsse, Amsterdamse Woningraad, 1909, cited in Klerk, *De Modernisering van de Stad 1850–1914*, 219.

[29] Bosma et al., *Bouwen in Nederland, 600–2000*, 595–97.

[30] W. Rutten, 'Een archipel van koloniën: Wonen in de mijnstreek,' in A. Knotter, *Mijnwerkers in Limburg: Een sociale geschiedenis* (Nijmegen 2013), 432.

The point of departure for both the creation of new city parks as well as the garden city concept was to bring the joys of nature closer to the urban population. Nature was seen as the counterpart of the urban and as the purveyor of healthy air, rest and space for relaxation. But the paradoxical result of these developments was that the spatial footprint per urban household increased, with the consequence that the cities laid an ever-increasing claim on space, a development that proceeded at the cost of the landscapes surrounding the cities.

During the interbellum, urban expansion plans were aimed not only at the city, but more and more pointedly at the surroundings as well. The national road plan designed by engineers in the 1930s was the occasion for the first spatial reconnaissance of the Netherlands as a whole. Urban planners developed ideas for a National Plan 'that indicates the land-use of the Dutch territory and that aims to promote the harmonious development of the surface of our country along pre-determined lines' in the words of the urban planner F. Bakker-Schut in 1937.[31] During the German occupation these centralistic spatial planning ideas acquired an institutional grounding in the State Service for the National Plan, established in 1941. The service was mandated to develop national plans for different facets of the ordering of the national territory. This included planned spaces for nature, traffic, built-up areas, agriculture and industry. By way of preparation, the service made lists of areas of special scientific or landscape significance and compelled municipalities to report land transactions and activities in those areas.

This centralist approach came under fire from lower levels of government that wanted to retain their municipal development plans and their regional structure plans. After the war, ideas about central planning and the rule of expertise acquired wider currency. In the economic sphere this was visible in the founding of the Central Planning Office in 1945. National plans in the domain of spatial planning first appeared in the form of a Building Plan and an Industrialization Memorandum in 1949. According to this memorandum the best chances for economic recovery were to be found in the large cities in the west and in the industrial regions of Twente, Brabant and South Limburg. The accompanying Building Plan described how the still scarce building materials should be distributed among these regions.[32]

In 1956, the spatial planners at the State Service for the National Plan assessed the effects of existing policy in a report entitled *The West...and the rest of the Netherlands*. It described the exodus from Zeeland, Limburg and the northern provinces. The study provided the impetus for a first government memorandum on spatial planning, that appeared in 1961. The document aimed at 'a balanced development of our country as a whole.' The plan anticipated a redistribution of economic activities, thanks to which the boundaries between 'the west' and 'the rest of the Netherlands' would become more vague.

The policy of regional economic stimulation proved to be a success, not least thanks to economic developments in the 1960s. In the middle of that decade the

---

[31] F. Bakker Schut cited in H. van der Cammen and L. de Klerk, *Ruimtelijke ordening: Van grachtengordel tot vinex-wijk* (Utrecht 2003), 157.

[32] Cammen and de Klerk, *Ruimtelijke ordening*, 207.

CBS calculated that with this kind of growth in the economy and in welfare the population would number about 20 million by the year 2000. The population of the 'Randstad' (the horseshoe-shaped urban cluster in the western part of the country) would grow from 5.5 to 9 million inhabitants. These predictions caused much consternation and precipitated a flood of regional and national plans. In response to the different ideas, the government published its second spatial planning memorandum in 1966, with concern for a broad quality of life.

> 'Our country has entered a phase in which more than in the past the government and society must take to heart a concern for the maintenance of a good living environment.'

The spatial plans opted for so-called 'bundled de-concentration,' an effort to spread suburbanisation to smaller centres in urban zones around the big cities. The metropolitan district became the new spatial unit for planning.[33] The plan reserved a remarkable amount of space for nature and recreational landscapes. The spatial planners expected that increasing leisure time would lead to 'a recognition of the creative possibilities of the natural environment.' In these spatial plans nature thus acquired a new, specifically recreational function.[34]

The implementation of the memorandum was institutionally supported by the Law on Spatial Planning enacted in 1965. This law defined the frameworks for the different planning levels of municipality, provinces and national state. Following the subsidiarity principle - decentralize what can be decentralized - the centre of gravity of planning came to rest with the municipalities. With their zoning plans they shaped the local space. The provincial regional plans harmonized the supra-municipal functions of the region. National plans described zones for housing, working, agriculture, recreation and nature. In addition the law provided for harmonisation of procedures for permissions and protests at all levels.[35]

This spatial policy was in part responsible for steering Dutch urban development away from the kind of dense metropolitan spatiality characteristic of London or Paris. In the Netherlands population growth was spread out over many cities that were closely tied together by roads and railroads.

The urban density in the centres of the old cities stood at about 150–200 dwellings per hectare around 1970. Post-war outskirts had densities of 75 dwellings per hectare and in smaller settlements the housing density stood at about 25–30 dwellings per hectare.[36] Urban and infrastructural development claimed ever more space in the course of the twentieth century. Between 1900 and 1970 this increased from 2 to 6% of the national surface area.

The planning of space had originated around the planning of national infrastructures like roads and canals. Subsequently attention shifted to a fair distribution of economic development throughout the country. In the 1960s the idea took hold that space was becoming a scarce commodity. It became necessary to think about how

---

[33] Cammen and de Klerk, *Ruimtelijke ordening,* 208–215.

[34] H. Meyer, *De staat van de delta: Waterwerken, stadsontwikkeling en natievorming in Nederland* (Nijmegen: Van Tilt 2016), 132.

[35] Cammen and Klerk, *Ruimtelijke ordening,* 177–78.

[36] Verstedelijkingsnota 1976.

to apportion space among the various needs and desires of society. In spatial terms, improvements in housing quality and in material prosperity tended to be made at the cost of heath meadows and sand drifts. These fell prey to urbanization, agriculture and forestry (Figure and Table 13.1).

14.4   The Demand for Building Supplies and Construction Materials

The construction of infrastructures and housing had not only a spatial, but also a material dimension. Building activities were responsible for an enormous increase in the use of building materials. Where did these enormous quantities of building materials come from? Which societal organisations shaped the procurement and production of the building materials that made housing construction and urban development possible? How did the procurement and production of these materials influence changes in natural capital domestically and abroad?

Clay, sand, gravel, bulk ceramics (including bricks and roof tiles), stone and wood were the dominant materials in 1913. In 1970 construction materials were also part of the sizeable material flows. Concrete, concrete products and cement had relegated the role of clay and clay products to a second place. Together with raw metals and chemical products they formed a broad spectrum of material flows within the industrialised society of the Netherlands.

The use of building materials and activities in the sphere of public housing exhibited a parallel upward development (Graph 14.3). Though building materials were not used exclusively for housing construction, the large scale of this activity shaped the demand. In addition to the increasing use of classic building materials like wood and bricks, the use of concrete and later steel became ever more prominent. These developments were the result of the application of new building methods and changing use of materials. An increasing proportion of construction work - especially after the Second World War - was carried out in concrete, composed of cement, sand and gravel.

Only a portion of the necessary building materials were produced in the Netherlands. The domestic production of building materials consisted around 1900 above all of clay products like bricks for buildings and streets, and roof tiles. During the 1920s this was supplemented by domestic production of cement at the marl pits near Maastricht and of steel at the Hoogovens in IJmuiden. With the increasing use of concrete, domestic production of gravel also increased sharply. The extraction of clay, marl and gravel had, above all, local implications for the immediate surroundings. This was not the case with imported building materials, like wood, iron and a portion of the cement. These influenced natural capitals elsewhere. The First World War made the dependence on these materials painfully clear. This had diverse consequences for construction in the Netherlands and the development of the domestic building materials industry.

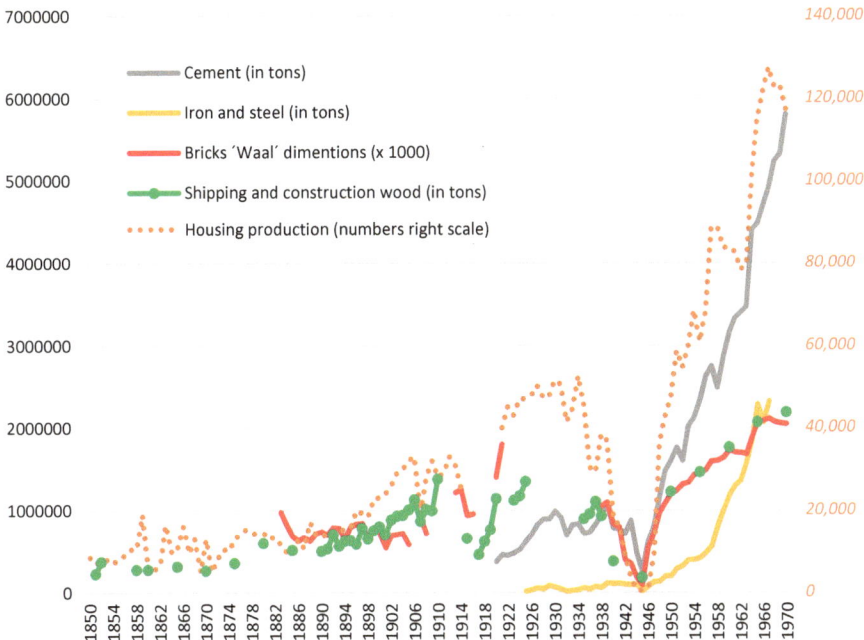

Graph 14.3 Utilization of wood, bricks, cement, iron and housing production in the Netherlands, 1850–1970

Source: wood – trade statistics for various years; G.B. Janssen, *Baksteenfabrikage in Nederland, 1850–1920* (Zutphen 1987); J.A. de Jonge, *De Industrialisatie in Nederland Tussen 1850 en 1914* (Nijmegen 1976), 497–501; Joh. de Vries, *Hoogovens IJmuiden, 1918–1968, Ontstaan En Groei van een Basisindustrie* (Amsterdam 1968); J.C.A. Everwijn, *Beschrijving van Handel en Nijverheid in Nederland* ('s Gravenhage 1912)

14.4.1 Wood Production and Resources from Foreign Forests

Prior to the First World War the Dutch building sector was heavily dependent on imports. Resources like wood, cement and iron were found elsewhere. What was the origin of these raw materials and how were they processed in the Netherlands?

Among the most common traditional building materials was wood. But for a long time the Netherlands had little (remaining) forest. Forest acreage in 1900 amounted to a bit less than 268,000 hectares, about 8–9% of the country's total surface area. This was 0.05 hectare of forest per inhabitant. This made the Netherlands, together with Great Britain (3.9% forest cover and 0.03 ha/inhabitant) and Denmark (6.3% forest cover and 0.10 ha/inhabitant) one of the least forested countries of Europe.[37]

Since the mid-nineteenth century forest acreage was on the increase again in connection with projects to reclaim the wastelands; the sand drifts and heathlands

---

[37] Data CBS – statistics land use and A.G. Malcom, *De houthandel van Nederland* (Rotterdam 1930), 13.

(see previous chapter). Private and public organizations like the Heidemij (1888) and the State Forest Service (*Staatsbosbeheer* 1899) coordinated the planting of new forest. At its founding the State Forest Service had assumed responsibility for 2000 hectares of forest and more than 10,000 hectares of so-called wastelands. Making these lands profitable was in fact the main purpose of this organisation.[38] Nature conservationists and the ANWB tourist organisation also welcomed the cultivation of new forest by the state. Together with a number of foresters they advocated the planting of a natural forest with a variety of hardwoods and pines. Their plea fell on receptive ears and this resulted in a program to plant not only fir trees on the sand grounds, but also Douglas spars, the Corsican pine and various kinds of larches. The second director of the State Forest Service, E.D. van Dissel, appointed in 1902, himself anchored concern for 'natural' re-forestation in the organization's DNA. From 1907 until his retirement in 1937, in addition to his leadership of the Forest Service, he was also a board member of the Association for Natural Monuments. This meant that from very early on the executive of the Forest Service had an opening to nature and to natural forests.

The chief activity, however, was still the creation of forests for wood production and the improvement of the soil. Only at extremely unique locations did the service abstain from planting new forest in order to preserve special natural landscapes.[39]

In 1910 a bit more than half of the Dutch forests consisted of various species of softwoods. For the rest they were composed of 20% oak forest, 8% other hardwoods, 5% reeds and wicker, and 13% other coppice wood. The most important building applications were in carpentry, railroad ties and beams for shoring up mineshafts.[40] Domestic wood production was limited in relation to the demand for wood, but not unimportant. This became clear during the First World War as imports of wood and coal were throttled. Wood prices soared and people took to the woods in search of alternative fuels. Associations devoted to nature, heritage and tourism applied to the government to take measures to protect the forests and natural beauty. In 1917 the government submitted an emergency law to appease the persistent complaints about increasing logging activities and illicit forays to gather firewood.[41] The law prohibited the chopping of wood in forests of unusual natural beauty, in immature forests and where damage to the wood supply might be expected. The State Forest Service was mandated to issue permits. In 1922 the emergency law was transformed into a permanent one. This was intended to protect forests against plagues, diseases and fire, as well as illegal logging.[42]

---

[38] M. Coesèl, J. Schaminée and L. van Duuren (2007), *De natuur als bondgenoot: De wereld van Heimans en Thijsse in historisch perspectief* (Zeist 2007), 164–67.

[39] J. Buis and J.P. Verkaik, *Staatsbosbeheer 100 jaar: Werken aan groen Nederland*, ed. Fred Dijs (Utrecht 1999), 11–49.

[40] J.C.A. Everwijn, *Beschrijving van handel en nijverheid in Nederland* (Den Haag 1912), 208.

[41] There had been proposals as early as 1908 and 1912 for a forestry law, especially with a view to preventing plant diseases, plagues and fire. The associated violations of private property had prevented adoption.

[42] Buis, Verkaik and Dijs, *Staatsbosbeheer 100 jaar,* 51–69.

During the Second World War the Dutch wood supply suffered considerable attrition. The German occupiers forcibly requisitioned more than 4.7 million cubic metres of wood from the Dutch forests. During the so-called 'Hunger Winter' of 1944–45, another million cubic metres of firewood disappeared. The annual pre-war wood production of 550,000 cubic metres more than doubled during the war years. In order to satisfy an acute post-war demand for mining wood, permission was granted by the Allied authorities after the war to chop wood in German forests near the border.[43] The appraisal of the exact war damage took until 1950. It then transpired that 55,000 hectares (13%) of the Dutch forest acreage had been lost during the war.[44]

After the war, rational approaches gained the upper hand in forestry, just as they had in the broader agricultural sector. Foresters saw it as their obligation to create new forest as quickly as possible in order to satisfy the demand for wood. This was an opening for agrarian, rational and business-economic approaches. Increasing the profitability of forest production was now informed by academic research, as it had long been in the traditional wood producing countries.[45] It led to an increase of production forests characterized by monocultures, a lack of biodiversity and vulnerability to diseases.

In the end, Dutch wood consumption relied heavily on foreign imports. This was a long-standing tradition. Since the late middle ages wood had been imported via the big rivers and across the sea. By the end of the nineteenth century the famous *Höllander Holzfloss* (Dutch wood-raft) trade with its enormous wood rafts floating down the Rhine, had come to an end. Around 1900 only 1% of the Dutch wood supply (about 10,000 tons) came to Dutch wood vendors via this route.[46] The decline was an example of ecological over-exploitation causing the demise of a commodity flow. The pirating of the forests around Württemberg on the shores of the high Rhine was stimulated by the increasing demand from the river's lower reaches, in the industrialising Ruhr and the export to the Netherlands.[47] Nineteenth century deforestation in Germany led to a focus on *Nachhaltigkeit* (sustainability) in forest management. There was extensive debate on how this might be achieved. Foresters and wood merchants saw sustainability above all in terms of optimization of the production cycle.[48]

In the twentieth century, Dutch wood imports came primarily from the Baltic countries. For the Netherlands, Russia had been the most important source before the First World War; thereafter Sweden and Finland became the main suppliers.

---

[43] Up to 1948 the State Forestry Service had been able to recover 550,000 cubic metres of wood from German forests. This was about the annual growth of the Dutch forests and equivalent to 8 years of timber cutting in the 1930s.

[44] Buis, Verkaik and Dijs, *Staatsbosbeheer 100 jaar*, 91–94.

[45] Buis, Verkaik and Dijs, *Staatsbosbeheer 100 jaar*, 100–112.

[46] R. Stenvert, *Kerkkappen in Nederland, 1800–1970* (Zwolle 2013), 52–53.

[47] J. Buis, *Holland houtland: Een geschiedenis van het Nederlandse bos* (Amsterdam 1993), 139–40.

[48] J. Radkau, *Holz - Wie ein Naturstoff geschichte schreibt* (München 2012), 229–42.

Table 14.5 Import of wood according to country of origin 1905, 1930, 1962 en 1970 (in kton)

	1905	1930			1962		1970	
	Timber	Timber	Mining wood	Poles	Wood, shaped or simply worked	Poles, pilings, posts & other wood in the rough	Wood, shaped or simply worked	Poles, pilings, posts & other wood in the rough
Russia/USSR	1,060,944	388,739	29,288	23,132	313,873		108,464	
Latvia		128,715		4063				
Estonia		22,745		158				
Finland		316,952	28,302	427	639,317	15,832	326,691	1557
Sweden	302,496	222,733	2500	10,002	918,055	8082	659,312	19,394
Norway	70,050			1293	7760	2411	7516	1334
Poland		74,393	19,091	7788	28,756		20,606	
Germany	195,694	128,924	85,534	74,572				
GDR					92,762	6964	120,630	37,992
DDR					19,980		20,840	
Belgium	73,524	12,702	3830	11,050	45,828	92,243	27,225	110,500
France			246	242	90,983	917	60,755	313
Austria	13,264				61,529	359	46,143	179
Czechoslovakia			34	384	143,459	76	72,348	72
USA	313,456				19,256		19,172	
Canada					29,588		3400	
Total (world)	2,209,049	1,463,515	168,895	134,288	2,599,020	127,149	1,740,521	172,822

Source: CBS – Jaarstatistiek Handel 1905, 1930 (www.historisch.cbs.nl) en UN Comtrade Database (comtrade.un.org)

(Table 14.5). In the late 1920s the Netherlands was the most important export market for the Finnish wood industry after Great Britain.[49] In this period the wood trade constituted 90% of Finnish exports. Under the influence of scientific research and Darwinist conceptions, German forestry practices were further developed in Finland. Soil Forestry was optimized thanks to botanic knowledge and new methods of production. Dense monoculture forests were the ideal in which the Finnish forestry authorities and the industrial foresters found common cause. In this context, 'sustainable forestry' was redefined as the maximisation of a theoretical growth potential of a particular soil type. The approach consisted of razing contiguous parcels of forest and replanting. Other kinds of forestry like selective felling were discouraged by regulations.

After the Second World War the demand for wood increased even more. In Finland this resulted in further intensification of forest production. By means of

---

[49] Malcom, *De houthandel van Nederland*, 21.

draining high moors, fertilizing the forests and improving access to the forests for men and machines, the wood yield per hectare was further increased.[50]

In this way, the growing demand for wood from the Netherlands, among others, did not lead to deforestation in the wood-producing countries. But it did have serious repercussions in these regions. The industrialisation of forestry in the twentieth century led to extended forests in monoculture. The original forest was transformed into much younger production forests. This had far-reaching consequences for the decline of local biodiversity, not only because of the monoculture in these woods, but also because of the disappearance of a variegated landscape.[51]

The domestically produced wood and the overwhelming share of imported wood was further processed in the Netherlands in sawmills and carpentry shops and in the furniture industry. In 1913 Zaandam was the largest wood import harbour on the European continent. Fifty percent of the Dutch wood import was unloaded here. The supply of wood changed as a result of the foreign industrialisation of forestry. Wooden beams - roughly sawn trunks - were gradually replaced by already sawn wood. For the exporting firms this meant more value added to the export product. In addition, ships could be more efficiently loaded with sawn wood. Waste products like sawdust, pulp and cellulose became raw materials for the paper industry that emerged in the forestry countries.[52]

14.4.2   Steel and Cement: Vacillating Support for Basic Industries

In 1898 the Society for the Promotion of Factory and Trade Industry presented plans for a national steel-foundry. A blast furnace facility would stimulate the development of Dutch shipyards, machine builders and construction firms.[53] In 1913 the board of the Society for Industry (*Maatschappij van Nijverheid*) appointed a commission to explore the expansion of cement production.[54] Dutch neutrality and the troubled supply of materials and resources during the First World War provided a new context for both plans. More than ever before, the Netherlands was thrown back upon its own resources. What would happen to trade after the war? Would not the combatant nations first look to their own needs when it came to raw materials? In an exploration of the post-war situation, the Delft professor I.P. de Vooys definitely saw chances for a leading position for the Netherlands if rapid measures could be taken

[50] J. Kotilainen and T. Rytteri, 'Transformation of forest policy in Finland since the nineteenth century,' *Journal of Historical Geography* 37 (2011): 429–39.

[51] T. Myllyntaus and T. Mattila, 'Decline or increase? The stading Timber stock in Finland 1800–1997,' *Ecological Economics* 41 (2002): 271–88.

[52] P.J. Middelhoven, *Hout en trouw: De geschiedenis van een familiebedrijf de wed. stadlander en middelhoven H = houthandel te Zaandam* (Zaandijk 1975), 55.

[53] J. de Vries, *Hoogovens IJmuiden, 1918–1968* (Amsterdam 1968), 23–29.

[54] A. Heerding, *Cement in Nederland* (IJmuiden 1971), 72–75.

to invest in industrial and infrastructural activities. But he had doubts about the 'general feeling regarding the big technical-economic, and thus above all national, problems.' He referred to the Zuiderzee Works, the electrification of the country, the founding of a national cement industry, steel plants and rolling mills, exploitation of subsoil assets, the industrialisation of the Dutch East Indies and the expansion of domestic road and waterway networks.[55] De Vooys was enthusiastic about a Dutch steel foundry. The state was obliged to support this project financially:

> We have started to exploit our coal fields… But for years now the lack of blast furnaces has kept just as many engineers busy as the reclamation of the Zuiderzee… Both stimulate the creative urge to provide for the future by undertaking a large enterprise.[56]

Prohibitions on export of iron and steel products to the Netherlands by both England and Germany placed the plans for a Dutch blast furnace facility in an entirely new light. Under the leadership of the chairman for the state commission for the distribution of iron and steel, the engineer-entrepreneur H.J.E. Wenckebach, plans were developed during the war for the founding of an iron and steel plant. In the plans for the facility Wenckebach not only included the production of iron and steel, but also the exploitation of other material flows like the production of cokes gas and cement using blast-furnace slag. By early 1917 the plans were ripe for financial negotiations with firms and banks. The latter particularly insisted on financial participation by the state. After tough negotiations and parliamentary debates, the Government Gazette of July 1918 was able to announce state financing for the blast furnace facility.[57] A lobbying effort by the metalworking industry had been able to convince the banks and the government to risk an investment of 11.3 million euros (160 million euros at current value) for the blast furnace facility. The government guaranteed 30% of the investment.

In 1919 the first harbours were excavated at the site of the future blast furnaces in IJmuiden. In 1920 the site was connected to the Haarlem-Uitgeest railway line and construction was started on the first blast furnace. On January 22, 1924 the first blast furnace was ignited and the production of Dutch raw iron and steel could begin.

Considerably more effort was needed to establish a national cement industry. There had been some modest attempts to mine marl and establish a cement factory.[58] In 1919 the government installed a commission to study the chalk and cement problem. This commission charged the Government Service for Exploration of Subsoil Resources (*Rijksopsporingsdienst van Delfstoffen*) with investigating the commercial feasibility of mining chalk in Limburg.[59] In 1921 the commission published a provisional report recommending a state-run cement factory. But it was critical of

---

[55] A. Heerding, *Cement in Nederland* (IJmuiden 1971), 72–75; I.P. de Vooys, 'De economische taak van den ingenieur na den oorlog' in *De Ingenieur* 32 (1917), nr 33, 597–611, (quote p. 600).

[56] I. P. de Vooys, 'Het hoogovenplan' in *De Ingenieur* 33 (1918), nr 14, 243–244.

[57] J.J. Dankers and J. Verheul, *Hoogovens 1945–1993: Van staalbedrijf tot twee-metalenconcern* (Den Haag 1993), 19–32.; Heerding, *Cement in Nederland*, 85.

[58] Heerding, *Cement in Nederland*, 60–69; A. Nieste, *Van mergel tot cement: 70 jaar ENCI, 1926–1996* (Maastricht 1996), 17–25.

[59] Heerding, *Cement in Nederland*, 71–93.

the quality of the raw materials in the Limburg subsoil, despite successful experiments during the war.[60] During the economic recession of the early 1920s the possibility of financing by the government also receded from view. The debate had a welcome side-effect. In response to the proposed exploration, foreign cement vendors lowered their prices in the Netherlands, aiming to undermine the founding of a domestic cement factory.

During the interbellum, Belgium developed into an important cement-exporting country, with a share of world trade amounting to 17–26%. In the Netherlands, cement cartels and suppliers from Belgium, Germany, England, France, Switzerland and Austria and even Poland fought for a share of the Dutch market. In 1925, during this battle for a market share, a Belgian-Swiss consortium applied for permission to set up the First Netherlands Cement Industry (ENCI) at the Sint Pietersberg in Maastricht.

The establishment of the cement factory and the mining of marl in an open pit was not uncontested. The Natural History Fellowship in Limburg protested vehemently. The polemic issued in a petition against a permit for the ENCI, a situation that brought the question before parliament. In subsequent negotiations the factory reached agreements with the city and the local planning board, but the Ministry of Education, Arts and Sciences kept making additional demands. This led to revision and modification of the permit, such that a number of historically valuable ruins were protected and a wooded hillside of the Sint Pietersberg was retained in order to partly hide the unsightly excavation. The permit was finally issued in 1925. In January 1927 construction on the factory commenced and in 1928 the first cement oven was fired up.[61]

With the founding of the ENCI the Netherlands became less dependent on the import of cement. The locally produced cement was used to make cement mortar, but chiefly for concrete. After the Second World War concrete became a favourite building material. Between 1958 and 1970, concrete production rose from 460,000 to 6.6 million cubic metres. Relative to the total national surface area, Dutch use of concrete in this period exceeded that of Belgium and Germany. In 1970 53% of the concrete was used for housing construction, 36% in utility construction, 8% in hydraulic engineering and 3% for road construction.[62]

The increasing demand for concrete had consequences for the Dutch landscape at those locations where marl and so-called additive materials like gravel and sand were found. In addition, ENCI was among the industries with the highest level of carbon dioxide emissions. That was also the case for the Hoogovens. In the final decades of the twentieth century this point became an element in discussions about sustainability. The transformations in the landscape had become an item on the societal agenda much earlier.

[60] Quote from 1924 by Prof. J.A. Van der Kloes, founder of Dutch cement testing, cited in Heerding, *Cement in Nederland*, 65.

[61] Heerding, *Cement in Nederland*, 91–105; Nieste, *Van mergel tot cement*, 28–29.

[62] A.A. van der Vlist, *Tussen cement, zand en grind... en beton: 50 jaar betonmortelindustrie in Nederland, 1948–1998* (Driebergen 1998).

14.4.3 Gravel Pits and Marl Quarries, What Is Lost?

The discussion about the ENCI pit flared up again in 1948 after new proposals to expand the excavations. Once again local administrators and the company stood on opposite sides of the fence. Municipal, provincial and national advisements by the State Service for the National Plan, founded in 1941, hindered further expansion of the pit. In 1946 this agency had pronounced the Sint Pietersberg to be a protected nature zone. Local cultural and conservationist interests appeared to be blocking the expansion.

As an alternative for ENCI's expansion plans, the State Service for the National Plan developed the 'Valley Plan.' This was a proposal to excavate portions of the Margraten Plateau near the town of Sibbe, nine kilometres northeast of the ENCI pit. ENCI had also studied this location during the war, but had concluded that such a relocation was not economically feasible. In 1948 new negotiations about the extension of the existing permits took place among ENCI, provincial authorities and representatives from the Ministries of Reconstruction and of Education, Culture and Science. After long and difficult negotiations, the 'Valley Plan' was dismissed. What followed was a new expanded 60 year concession at the existing location with an annual ceiling of 600,000 tons. Permission was also given to excavate the wall created earlier between the Meuse and the pit so that the factory could expand further. By way of compensation, terrain to the west of the pit would be spared and there would be an investigation into the cultural history of those marl caves that were to succumb to the excavation. Finally a plan would be developed concerning the condition of the pit after marl mining had been terminated.[63]

The discussions about the conditions for granting a permit were unusual for the time. The duration and acerbic quality of the debate were probably due to the proximity of the city of Maastricht and the historical role of the Sint Pietersberg. In addition, the deliberations were shaped by chance occurrences and local, sometimes even individual interests. The controversy around the excavations was above all a local problem. The huge demand for cement made it into a national interest.

The intensive building efforts relating to post-war reconstruction and the struggle against the housing shortage brought the procurement of building materials and their coordination more forcefully into the national political arena. This became clear in relation to the delving of gravel - next to cement a second important raw material for making concrete. In the flat countryside and close to rivers, gravel pits quickly became lakes, resulting in loss of land and effects on water management. Prior to the turn of the century the use of the land was the business of the owners. From the mid-nineteenth century on, gravel mining in the rivers was controlled by the need for permits and by rules policed by the *Rijkswaterstaat*. In the early twen-

---

[63] Nieste, *Van mergel tot cement*, 45–64.

tieth century new hydrological insights placed ever more restrictions on gravel mining in the rivers themselves. The gravel industry shifted its activities to the floodplains and to sites in close proximity to the rivers.[64]

In 1950, in view of the enormous demand for materials, the Minister of Reconstruction and Public Housing concluded that the existing regulations for sand and gravel extraction were inadequate. He proposed to make the State Service for the National Plan responsible for coordinating the winning of sand and gravel. In the course of parliamentary debates, a few parliamentarians pointed out that decisions about sand and gravel extraction addressed not only a material question, but that recreational and natural values had to be taken into consideration as well. But the plans were above all economically motivated: in the words of the responsible minister, 'as concerns this sand winning, its costs (may) not be so high as to force the land prices for public housing to rise to intolerable levels'.[65]

In 1951, in response to the proposed changes, the Minister of Transport and Public Works appointed an inter-ministerial commission to investigate the question of sand and gravel extraction.[66] In February 1953, only a few days after the flood disaster in the southwest Netherlands, the Communist newspaper *De Waarheid* reported on clay and sand pits in the river area near Culemborg:

> Agricultural land is being exported with permission from the government. While in the southwest of our country, thousands are fighting for every centimetre of land that can be salvaged from the sea, while in the reclamation of the Zuiderzee 17,000 to 18,000 guilders are spent in order to wrest a hectare of land from the sea, at Culemborg 78 hectares of floodplains have been sold in order to be excavated. This is an inestimable loss for our country. This concerns excellent river clay, one of the most fertile soil types that we have.[67]

The owner of this land had sold it to an excavating company that exported to Belgium. But everything had been done quite legally. National, provincial and regional authorities were stalemated. The *Waarheid's* true interest was the tenant farmers who were being driven from their fields by landowners and dealers in sand in what they called a 'capitalist chaos.' The paper pinned its hopes on 'engineering circles' who would propose a national sand mining plan in which sand would be mined at one central location. Opponents of these plans pointed to the differences in

---

[64] H. van Heiningen, *Diepers en Delvers: Geschiedenis van de zand en grindbaggeraars* (Zutphen 1991), 259–322.

[65] Minister cited by parliamentarian Ten Hagen in *Handelingen der Tweede Kamer* 1950–1951, 30ste Vergadering, 7 December 1950, 812.

[66] *Handelingen der Tweede Kamer* 1951–1952, Kamerstuk Tweede Kamer 1950–1951 kamerstuknummer 1900 IX A ondernummer 15, Memorie van Antwoord, Rijksbegroting voor het dienstjaar 1951 (Wederopbouw en Volkshuisvesting), 18–19; *Handelingen der Tweede Kamer*, Algemene Beschouwingen, Bijlage A Tweede Kamer, Rijksbegroting van het dienstjaar 1951, 1900 IX B (Verkeer en Waterstaat), V 202, 50.

[67] 'Met toestemming van regering worden cultuurgronden geëxporteerd' in *De Waarheid*, Februari 14, 1953, 3; 'Kostbare kleilaag afgegraven omdat Belgen zand willen' in *De Waarheid*, May 16, 1953, 4.

sand qualities and the special properties of river sand and gravel. According to the paper such a decision 'could best be left to the engineers. They have no 'mushy' plans in their head.'[68]

The reports of the excavations at Culemborg led to parliamentary questions being put to the Minister of Transport and Public Works about possibly putting a stop to the gravel mining there. In view of the existing legal framework, the minister regarded this as impossible. He added that new statutes were in the making.[69] After this, discussions about excavations flared up repeatedly in response to new planned mining announcements and rumours.

After gravel mining had been relocated to the floodplains, the province of Limburg had become the most important supplier of gravel. In 1938, in order to acquire a say in the mining activities, the province formulated an 'ordinance against the infringement of natural beauty by excavating or digging in the ground.' Halfway through the 1950s the provincial authorities modified the ordinance to require refilling of the gravel pits. The gravel companies rejected this as an impossible demand due to the lack of sufficient material with which to refill the pits. The province threatened to take the matter to court and with that the debate once again moved to parliament.

In 1958 parliament asked the Minister of Agriculture to formulate a standpoint regarding the loss of agricultural land to gravel mining along the Meuse. The minister determined that gravel mining was undertaken haphazardly and that due to combined sand mining much more land was dug up than anticipated. In view of the scarcity of arable land he ordered an investigation into the economic rationality of gravel mining at a national level. At the same time he expressed a preference for refilling the gravel pits, with the costs being borne by the gravel vendors. According to the minister this would imply a reduction of profits made by gravel miners by 10–30% and a price hike for concrete of 1–3%.[70] The Catholic Peoples' Party (KVP) parliamentarian W.J. Droesen welcomed this idea, though he remained critical of the strictly economic approach to the gravel mining:

> I assume that gravel mining, from a national economic point of view, is a necessity, but that doesn't give the nation the right to totally mutilate the Limburg countryside for the future. The modern excavating and dredging machines snatch about 50 hectares of superior acreage from the shores of the broad river Meuse each year, creating enormous disorderly holes up to and soon around the village centres. In order to prevent the destruction of an old and beautiful landscape, the costs of refilling should not be weighed on an analytical balance, but the Netherlands should say: Beloved Limburg, we need your gravel for our construc-

[68] 'Ingenieurskringen voor nationaal zandwinningsplan' in *De Waarheid*, May 28, 1953, 3.

[69] *Aanhangsel tot het Verslag van de Handelingen der Tweede Kamer*, Deel III, Zitting 1952–1953, nr. 3043, 3069.

[70] *Verslag van de Handelingen der Tweede Kamer*, Zitting 1957–1958, nr. 4900, 'Rijksbegroting voor het Dienstjaar 1958, Hoofdstuk XI, Landbouw Visserij en Voedselvoorziening, voorlopig verslag', nr 11, 20; *Verslag van de Handelingen der Tweede Kamer*, Zitting 1957–1958, nr. 4900 'Rijksbegroting voor het Dienstjaar 1958, Hoofdstuk XI, Landbouw Visserij en Voedselvoorziening, Memorie van antwoord', 37.

tions, but we want to repair your damage, even if that means that concrete will become a few percent more expensive.[71]

Fearing a possible stagnation of the gravel supply, the Ministry of Reconstruction and Public Housing initiated mediation between the province and the gravel dredgers. These negotiations led to a levy on gravel which could be used to fill up the pits. The levy of 0.20 euro per ton of gravel and 0.11 euro per ton of sand was enacted in August 1958. The agreement also compelled the gravel dredgers to hand over the empty pits to the province. Using the funds from the levies that were collected in a so-called gravel fund, the province would then take responsibility for filling up the pits and restoring them to agricultural use.[72]

The provincial authorities settled on mine rock as suitable fill material. The first trials were undertaken after talks with the State Mines in 1959.[73] These proved successful and in the 1960s hundreds of tons of mine rock were used to fill up the gravel pits. In 1968 more than 4.5 million tons of mine rock was processed, enough to restore 475 hectares of land. In that year the contract was upgraded to an annual bulk of 3.5 million tons. This was partly made possible by excavating the mine rock hill at the Maurits State Mine, which incidentally made it possible to expand the chemical plants located on the mine site. With this mass of stones another 200 hectares of gravel pits could be refilled.[74]

In 1961, after more than a decade of negotiations by various political interests in inter-ministerial and provincial commissions, no less than six ministers submitted a proposal to parliament for a Law on Subsoil Excavations (*Ontgrondingswet*).[75] Parliament insisted on strong ties with the Ministry of Public Housing. Parliamentarians also advocated harmonisation with the law on spatial planning that was then in preparation.[76] The new law was published on October 27th 1965. It harmonised the different provincial regulations. As part of an application for a digging concession, the sand and gravel mining industry was now obliged to submit plans for re-use, accompanied by financial underpinnings. The execution of the permits became a provincial task and was to be aligned with the provincial regional

---

[71] *Verslag van de Handelingen der Tweede Kamer*, Zitting 1957–1958, 29 session 11 December 1957, 3437.

[72] Province of Limburg, 'De afronding van het grindfonds', appendix to Letter Provinciale State van Limburg, 4 oktober 2005, onderwerp: Afronding ontgrindingsfonds (via web: http://portal.prvlimburg.nl/psonline/ accessed 16-8-2012); H.B. Kramer, 'De grintwinning In Midden-Limburg: Een evaluatie van alternatieve winningsgebieden,' *Tijdschrift Voor Economische En Sociale Geografie*. 58, nr mei/juni (1967), 113–125.

[73] Soil mechanics reports by T.H.Huizinga in 1942 had proven the suitability of these materials for dikes and banks.

[74] 'Afgraving steenberg Maurits levert materiaal voor vulling van 200 ha grindgaten in Midden-Limburg' in *Nieuws van de Staatsmijnen*, Oktober 25 1968, 8.

[75] *Handelingen der Tweede Kamer*, Zitting 1960–1961, 'Regeling omtrent ontgrondingen (Ontgrondingwet), Memorie van Toelichting', nr 6338.

[76] The law on spatial planning entered into force in June 1961.

plans.[77] The incorporation of the proposals into provincial regulation took another few years. While discussions and provincial planning continued, the Law on Subsoil Excavations silently came into force on September 1, 1971.[78]

The nature of the discussion gradually changed. Acquisition of materials, water management and agrarian interests had motivated the law, but the creation of recreational areas now also became a growing consideration.[79] This became visible at the presentation of the regional plan for gravel mining locations by the Limburg provincial authorities in 1969. In the ensuing discussion new actors came to the fore. At the behest of municipalities along the Meuse River, H.B. Kramer, a geographer at the Catholic University of Nijmegen, had executed an economic cost-benefit analysis of the provincial proposals. On the basis of these - for that time modern - calculations, he was able to propose an economically viable alternative in which the gravel pits along the Meuse would be transformed into an 'international' centre for aquatic sports. This could produce new economic activity, certainly desirable in view of the immanent closure of the mines. Kramer concluded that filling in the gravel pits and restoration of the agrarian function had become less relevant in view of agricultural developments in a European context. The cost-benefit analysis revealed the economic advantages of continued gravel mining along the Meuse and the transformation of the region into a zone for aquatic sports with 'international allure.'[80]

The provincial council was receptive to these arguments and consulted the tourist organisation ANWB and the Royal Dutch Yachting Association (KNWV). In 1971 these organisations published a voluminous joint report exploring the possibilities for aquatic sports in the gravel pits. According to both organisations there were good chances for this region that in the past had only very limited possibilities for waterborne recreation. The proximity of Germany was regarded in this connection as an extra plus. The gravel pits as new nature zones with unique economic possibilities would support the economic development of the region.[81]

After 1970 contracts for gravel mining in Panheel were signed by the gravel mining companies and the province. The agreements also included a twenty-year moratorium on gravel mining outside of the planned zone.[82] The 1970s were the high water mark for gravel dredging in Limburg. The activities were not without consequences. Gravel hunger transformed the landscape around Roermond and at many places along the Meuse.

---

[77] The figures are mentioned in: *Tweede Kamer zitting 1963—1964* no. 6338, ´Memorie van Antwoord, Regelen omtrent ontgrondingen (Ontgrondingenwet), Wijziging van de onteigeningswet: Wijziging en aanvulling van de Rivierenwet´, Submitted 20 april 1964, 2.

[78] *Handelingen der Tweede Kamer*, Zitting 1973–1974, ´Aanhangsel tot het verslag van de Handelingen der Tweede Kamer´, Vragen nr 840, 1681; P. Ike, D*e planning van ontgrondingen* (Groningen 2000), 31–41.

[79] *Handelingen der Eerste Kamer*, Zitting 1965–1966, ´Regeling omtrent ontgrondingen (Ontgrondingwet), etc.´ 3ᵉ vergadering 5 oktober 1965, 23.

[80] Kramer, 'De grintwinning in Midden-Limburg'.

[81] ANWB en KNWV, 'Grintgaten in Limburg' (Den Haag 1971).

[82] Kramer, 'De grintwinning in Midden-Limburg'.

14.5 The State, Construction and Well-being

Since time immemorial, construction consisted of short supply and command chains among purveyors of raw materials, builders and clients. Building materials were transported directly to the construction site, as in the case of sand for raising the ground level, or processed and then transported to the construction site, as in the case of clay and bricks. In a geographic sense there was often a long supply chain. Many construction materials, for example wood and stone, had to be imported. For the rest, the construction site was a node for the supply of various materials, including wood, mortars and stone. It also resembled a network because the construction site gathered a variety of trades like carpenters, bricklayers and excavation workers who processed these materials.

In the twentieth century the supply chains for mineral substances (construction and building materials) became more and more complex, just like those of bio-materials (agriculture and foods) and fossil substances (energy and plastics). New chains were added, among others those of steel, with new links like the processing of steel in metal products, machines and constructions. Supply chains thus became longer and more differentiated. New actors concerned themselves with the supply chains. The networks around specific links became more ramified. The fundamental dynamic of the supply chains was the increasing size of the enormous flows of materials. This caused tensions and problems in the chain and its environment.

With respect to the issue of well-being and sustainability the role of the state in the construction supply chain was crucial. From the beginning of this period the state was a new and dominant actor. The Housing Law of 1904 gave it directive powers over the construction of public housing. In subsequent decades the state defined the quality of dwellings for the poor and on that basis financially supported housing construction. Together with housing corporations and municipalities it achieved a significant improvement in public housing construction, one of the big issues in quality of life in this period.

The influence of the state extended to other supply chains as well. The First World War had made it clear what the international embargo on raw materials like wood, cement and steel meant for the Netherlands. The same confrontation followed in the Second World War. During both wars the state did its best to allocate flows of materials. Post-war reconstruction also demanded government steering in order to prevent a descent into chaos.

During peacetime, the state made efforts to diminish the vulnerability of material flows, among other things by creating the prerequisites for a more independent position vis a vis foreign suppliers. After the First World War it co-financed the iron and steel plant Hoogovens, which reduced Dutch dependency on foreign suppliers to 5% of the total market. Dependency on foreign suppliers of natural stone (in particular marl for making cement) decreased: imports declined from 100% in 1913 to 68% in 1970. Dependency on foreign gravel also decreased: imports declined from 67% in 1913 to 47% in 1970 (Table 12.8). It should be noted that the latter developments were the result of initiatives by private entrepreneurs. Nonetheless here too the state played an important role. It intervened in local conflicts around the extraction of raw materials in order to guarantee a stable supply. The Excavation Law of 1971 became the legal framework within which the state, the provinces and other parties would solve conflicts in the future.

The local conflicts and state interventions had to do with the interests of raw materials extraction, agriculture, recreation and landscape. In the 1920s landscape conservation became an issue in the marl excavations in Maastricht. Later, during gravel dredging, priorities shifted from agrarian interests to that of recreation and landscape conservation. Limburg became the territory in which national and local interests were weighed in the balance. Landscapes were sacrificed for the material needs of housing, utility buildings, roads and the Delta Works. Zeeland was wrested from the sea while Roermond changed into a centre for aquatic recreation, a win-win situation for safety and the economy.

Ultimately the state was also confronted with the consequences of the construction agenda in this period. The expanding cities and the building of infrastructures infringed increasingly on the available space. That provided fertile soil for conflicts. It was the first time that politicians and national policy makers had the feeling that, in an incoherent fashion, the Netherlands was becoming 'full' of buildings and constructions. Residential quality, the human environment, economic capital and natural capital had to weighed in claims on space. The Law on Spatial Planning of 1965 created the legal framework in which discussions on well-being and sustainability were to be carried on.

Literature

Baardman, C (1953). *In de greep van de waterwolf.* Den Haag: Voorhoeve.

Beekers, W. (2012). *Het bewoonbare land: Geschiedenis van de volkshuisvestingsbeweging in Nederland.* Amsterdam: Boom.

Bosma, K., A. Mekking, K. Ottenheym and A. van der Woud (2007). *Bouwen in Nederland, 600–2000.* Zwolle: Waanders.

Bruggeman, L.A. (1981). *Kwalitatieve woningdocumentatie (KWD) 1948–1970: Enkele kwaliteitsaspecten van de nieuwbouw van woningen over de afgelopen 30 jaar.* Zoetermeer: Ministerie voor Volkshuisvesting Ruimtelijke Ordening.

Buis, J. (1993). *Holland houtland: Een geschiedenis van het Nederlandse bos.* Amsterdam: Prometheus.

Buis, J., J.P. Verkaik and F. Dijs (1999). *Staatsbosbeheer 100 jaar: Werken aan groen Nederland.* Utrecht: Matrijs.

Cammen, H. van der and L. de Klerk (2003). *Ruimtelijke ordening: Van grachtengordel tot vinexwijk.* Utrecht: Het Spectrum.

Coesèl, M., J. Schaminée and L. Van Duuren (2007). De natuur als bondgenoot: De wereld van Heimans en Thijsse in historisch perspectief. Zeist: KNNV Uitgeverij.

Dankers, J.J. and J. Verheul (1993). *Hoogovens 1945–1993: Van staalbedrijf tot twee-metalenconcern.* Den Haag: Sdu Uitgeverij.

Diepen, A.M.L (1998). 'Spatial aspects of housing': In K.J. Noorman and T. Schoot Uiterkamp (Eds.), *Green households? Domestic consumers, environment and sustainability.* London: Earthscan Publications.

Ekamper, P., R. van der Erf and N. van der Gaag (2003), *Bevolkingsatlas van Nederland: Demografische ontwikkeling van 1850 tot heden.* Den Haag: Nederlands Interdisciplinair Demografisch Instituut.

Ekkers, P. (2006). *Van volkshuisvesting naar woonbeleid.* Den Haag: Sdu Uitgeverij.

Everwijn, J.C.A. (1912). *Beschrijving van handel en nijverheid in Nederland.* Den Haag: NV Boekhandel.

Heerding, A. (1971). *Cement in Nederland.* IJmuiden: Cementfabriek IJmuiden.

Heiningen, H. van (1991). *Diepers en delvers: Geschiedenis van de zand en grindbaggeraars.* Zutphen: Walburg pers.

Huisman, J. et al. (2000). *Honderd jaar wonen in Nederland, 1900–2000.* Rotterdam: Uitgeverij 010.

Ike, P. (2000). *De planning van ontgrondingen.* Groningen: Geo Pers.

Klerk, L. de (2008). *De modernisering van de stad 1850–1914: De opkomst van planmatige stadsontwikkeling in Nederland.* Rotterdam: NAI Uitgevers.

Knippenberg, H. and B. de Pater (1988). *De eenwording van Nederland: Schaalvergroting en integratie sinds 1800.* Nijmegen: Sun.

Kotilainen, J. and T. Rytteri (2011). 'Transformation of forest policy in Finland since the 19th century'. *Journal of Historical Geography,* 37, 429–39.

Kramer, H.B. (1967). 'De grintwinning in Midden-Limburg: Een evaluatie van alternatieve winningsgebieden'. *Tijdschrift voor Economische en Sociale Geografie,* 58, nr mei/juni, 113–125.

Liagre Böhl, H. de (2012). *Steden in de steigers: Stadsvernieuwing in Nederland 1970–1990.* Amsterdam: Bert Bakker.

Malcom, A.G. (1930). *De houthandel van Nederland.* Rotterdam.

Meyer, H. (2016). *De staat van de Delta: Waterwerken, stadsontwikkeling en natievorming in Nederland.* Nijmegen: Van Tilt.

Middelhoven, P.J. (1975). *Hout en trouw: De geschiedenis van een familiebedrijf, de Wed. Stadlander en Middelhoven houthandel te Zaandam.* Zaandijk: Klaas Woudt Uitgever.

Mom, G. and R. Filarski (2008). *Van transport naar mobiliteit: De mobiliteitsexplosie (1895–2005).* Zutphen: Walburg Pers.

Myllyntaus, T. and T. Mattila (2002). 'Decline or increase? The standing timber stock in Finland 1800-1997'. *Ecological Economics,* 41, 271–88.

Nieste, A. (1996). *Van mergel tot cement: 70 jaar ENCI, 1926–1996.* Maastricht: Stichting Historische Reeks Maastricht.

Ottens, E. (1985). *'Ik moet naar een kleinere woning omzien want mijn gezin wordt te groot',* 125 jaar Sociale Woningbouw in Amsterdam. Amsterdam: Gemeentelijke Dienst Volkshuisvesting.

Radkau, J. (2012). *Holz - Wie ein Naturstoff geschichte schreibt.* München: Oekom verlag.

Rutten, W. (2013). 'Een archipel van koloniën: Wonen in de mijnstreek'. In A. Knotter (Ed.), *Mijnwerkers in Limburg: Een sociale geschiedenis.* Nijmegen: Vantilt.

Slager, K. (1992). *De ramp: Een reconstructie. 200 ooggetuigen over de watersnood van 1953.* Goes: De Koperen Tuin.

Stenvert, R. (2013). *Kerkkappen in Nederland, 1800–1970.* Zwolle: WBooks.

Vleuten, E.B.A. van der (2003). 'De materiële eenwording van Nederland'. In J.W. Schot, H.W. Lintsen, A. Rip and A.A. Albert de la Bruhèze (Eds.), *Techniek in Nederland in de twintigste eeuw - Techniek en modernisering, balans van de twintigste eeuw.* Zutphen: Walburg Pers.

Veen, J. van. (1939). *Te verwachten stormvloeden op de benedenrivieren.* Den Haag: Rijkswaterstaat, directie Benedenrivieren, afdeling Studiedienst, internal report.

Verloren van Themaat, R. (1953). 'Watersnood 1953', *De Ingenieur* 65, nr. 6, A 56.

Vlist, A.A. van der (1998). *Tussen cement, zand en grind… en beton: 50 jaar betonmorte-lindustrie in Nederland, 1948–1998*. Driebergen: Vereniging van Ondernemingen van Betonmortelfabrikanten in Nederland.

Vreeze, N. de (2001). *65 miljoen woningen: 100 jaar Woningwet en wooncultuur in Nederland.* Rotterdam: Uitgeverij 010.

Vooys, I.P. de (1917). ´De economische taak van den ingenieur na den oorlog´, *De Ingenieur* 32, nr 33, 597–611.

Vooys, I.P. de (1918). ´Het hoogovenplan´, *De Ingenieur* 33, nr 14, 243–244.

Vries, J. de (1968). *Hoogovens IJmuiden, 1918–1968: Ontstaan en groei van een basisindustrie.* Amsterdam: Koninklijke Nederlandsche Hoogovens en Staalfabrieken NV.

Wemelsfelder, P.J. (1939). 'Wetmatigheden in het optreden van stormvloeden', *De Ingenieur*, B, nr.31.

Woud, A. van der (2004). 'Stad en land: Werk in uitvoering'. In D. Fokkema and F. Grijzenhout (Eds.), *Nederlandse cultuur in Europese context*. Den Haag: Sdu Uitgeverij.

Chapter 15
Energy and Plastics: Toward a Fossil Land of Milk and Honey

Frank Veraart, Rick Hölsgens, and Ben Gales

Contents

Abstract Two energy transitions characterised the period 1910–1970: the rise and fall of a national mining industry and the shift from coal to oil and natural gas. Domestic coal made the Netherlands less dependent on foreign supplies. World wars and economic crises long inspired a lifestyle based on low energy consumption.

An energy-intensive lifestyle emerged after the 1960s with the import of cheap oil and the discovery of natural gas in Groningen. The discovery also led to the attraction of energy-intensive industries, to the massive use of natural gas in greenhouse farming and to a national gas grid for households.

Oil and gas also laid the basis for the production, processing and use of plastics. These became the symbol of modernity and of the rise of the consumer society. As packaging material and raw material for cheap consumer goods they also initiated the waste society and formed the iconic example of the linear economy. The products of this linear economy ended up *en masse* on the rapidly growing waste heaps.

Increasing energy consumption in industry and households caused local air pollution. The first investigations of and policy measures in the area of air pollution

© The Author(s) 2018

H. Lintsen et al., *Well-being, Sustainability and Social Development*,
https://doi.org/10.1007/978-3-319-76696-6_15

were initiated from the viewpoint of public health. Pollution mobilised local resistance against the excesses of modernisation. Local environmental groups were the cradle of a broad societal concern about the environment, ecology and climate change in the following decades (see Chaps. 17, 18, 19, 20 and 21).

Keywords Energy supply · Coal · Oil · Natural gas · Plastics · Waste · Linear economy · Air pollution · Environmental movement

15.1 Working on a National Energy Supply

15.1.1 Vulnerable Energy Supplies and Public Welfare

'A country that does not know how to use its natural sources of wealth proves that it does not deserve them,' according to the second-chamber parliamentarian from the province of Limburg, W.H. Nolens, speaking in December 1897.

> We all know that societal income, the material component of national well-being, comes into being through the combination of natural givens with labour, capital and entrepreneurial ambition, that the size of this income will depend on the quantity of each component and from the more or less correct proportions with which they are brought into contact with one another...[1]

At the end of the nineteenth century, subsoil core samples had demonstrated mineable layers of coal under significant portions of South Limburg. Despite various requests for mining concessions and the construction of a railway line to the south, a flourishing coal-mining industry remained a distant vision. Nolens characterised coal mining as vital to:

> the national interest...the economic independence of our country...A measure by the German Government to temporarily prohibit the export of coal would be disastrous for our country, even now.[2]

---

[1] W.H. Nolens in Handelingen Tweede Kamer, 35ste vergadering, 22-12-1897, 675, italics in original.

[2] W.H. Nolens in Handelingen Tweede Kamer, 35ste vergadering, 22-12-1897, 675, italics in original, Nolens referred to an article entitled ´Een Nationaal Belang,´ *De Ingenieur* 4(1889), no. 25. Nolens' remarks suggest that policy makers above all feared the Netherland's dependency on Germany. Around the turn of the century German coal increased its market share in the Netherlands and the latter became ever more dependent on the Ruhr region. But public opinion also referred to other suppliers. See, for example, H.C. van der Houven van Oordt and G. Vissering, *Economische beteekenis van afsluiting en drooglegging der Zuiderzee*, (Leiden 1901), 250: 'Should war with England ever break out again, then the State Mines will be of decisive importance.' The Boer Wars in South Africa made such a conflict conceivable. An even greater threat was perceived in 'that huge spider' the United States.

But it must be noted that the Netherlands also had some advantages. Because of British competition the Rhineland-Westphalian coal syndicate kept prices on the Dutch market lower than in their home market.

His call to modify the Mining Laws were heard by the government, that subsequently made efforts to promote the development of domestic mining.

From the perspective of well-being, Nolens' argument was telling. First of all, he made an explicit connection between national well-being and the exploitation of natural capital. In the nineteenth century, in imitation of England, the Netherlands had invested in steam engines and railways. This was accompanied by a transition in energy supply. Coal increasingly supplemented Dutch demands for energy. In 1870 the Dutch energy supply still consisted for 62% of classic sources: turf and wood, wind and water and the muscle power of humans and animals. Around 1900 this share had declined to 34%.[3]

The gradual energy transition to coal made the Netherlands completely dependent on foreign supplies, which made the country vulnerable. This was Nolen's second point. He regarded the development of a domestic coal supply as the necessary precondition for a stable and sustainable development of the economy and of well-being.

The theme of foreign dependency recurred regularly throughout the twentieth century. Between 1870 and 1970 the Netherlands passed through two energy transitions. The first was the transition from traditional forms of energy to coal (1870–1910), the second the transition to oil and natural gas (1950–1970). The first transition was already anticipated by the end of the nineteenth century, but almost no one could imagine that after a bit more than 50 years domestic coal production would be traded in for oil and natural gas. From a present-day perspective the energetic values of coal, oil and natural gas and an increasing energy consumption seem almost autonomous drivers for this transformation process.[4] The truth lay elsewhere.

The new fuels that shaped the modernisation of society were scarce goods throughout most of the twentieth century. The two world wars and their aftermaths led to rationing. Foreign conflicts meant fluctuations in price and delivery. This absolute and relative scarcity was a continuous spur to improvements in energy consumption. In this chapter we shall first of all consider the energy question. How dependent was the Netherlands and how did the country develop in the twentieth century? From being a net importer of energy around 1900, by 1970 the Netherlands had changed into a net exporter of energy, thanks to the discovery of natural gas in the Groningen subsoil. What effects did this have on the Dutch economy and energy consumption?

Coal, oil and natural gas varied in composition. The raw materials had in many cases to be processed before they could be used. The purification and refining of fossil assets were the basis for the development of chemical complexes. These com-

---

[3] H.N.M. Hölsgens, *Energy transition in the Netherlands: Sustainable challenges in a historical and comparative perspective* (Groningen 2016), 20.

[4] The energetic or combustion values are expressed in units of produced energy (joule) per kilogram. For methane - the major fraction of natural gas - this is between 50–55 MJ/kg, for petrol (gasoline) this is 44–47 MJ/kg, for diesel this is about 45 MJ/kg, anthracite coal is 27 MJ/kg, lignite coal about 15 MJ/kg, and wood and dry turf also about 15 MJ/kg.

plexes produced not only fuels, but other products as well. After the Second World War, plastics became an important sector with companies like DSM and Shell deploying a great deal of innovative prowess. We discuss this sector in the second part of the chapter.

We conclude with the ecological downsides of the fossil transition. What effects did production and consumption have on air, water and soil? How did these negative effects gradually find their way onto the political agenda? How did these evolve into the origins of the present-day sustainability challenges? We focus in particular on air pollution.

15.1.2 Domestic Coal Production

Parliamentarian Nolens' declamations have to be viewed in the light of the slow and difficult development of coal mining in the Netherlands. On the eve of the twentieth century coal was mined in only two small mines, the Domaniale and the Neuprick Mines in Kerkrade. From 1850 on, geological explorations confirmed the presence of coal over a larger area in South Limburg. Around 1860 and especially after 1876, 15 concessions were granted on the basis of French mining laws dating from 1810. In this period of liberal dominance, many entrepreneurs succeeded in staking a claim. These concessions were above all a matter of speculation and the creation of reserves by foreign mining companies. In 1891 the Minister of Public Works, Trade and Industry put a stop to this by recalling the concessions and granting the Company for the Exploitation of Limburg Coal Mines a concession for the Oranje-Nassau Mine. From the very beginning – with the digging of the shafts – the enterprise struggled with technical problems and accidents. Policy makers became increasingly irritated with the delay. In 1898 new concessions were granted for the exploitation of the Willem, Sophia and Laura en Vereeniging mines.

In early 1901 Minister Lely proposed establishing state-owned mines in Limburg. Instead of a limited area, the government reserved the entire territory known to contain coal in its subsoil at that point in time. This was, internationally speaking, a unique move. In addition to economic and nationalist motivations, social considerations played a role. A state enterprise, according to the Catholic parties, could prevent labour conflicts and the disruption of traditional society. The debate about the State Mines took place during the 'coal crisis.' The price of German coal had risen steeply. Subsequently Great Britain imposed duties on its coal exports. Given these circumstances the public and parliament concurred in Lely's proposition.

In 1903 construction was started on the Wilhelmina State Mine and in 1906 the Emma State Mine opened its gates, followed by the Hendrik State Mine in 1911. At the same time various private mines started producing: Oranje Nassau (1899, with a second mine in 1904), Willem Sophia (1902) and Laura (1907).[5] In barely a decade

---

[5] Ad Knotter (eds.), *Limburg Kolenland, over de Geschiedenis van de Limburgse Kolenmijnbouw* (Zwolle 2015), 66–67.

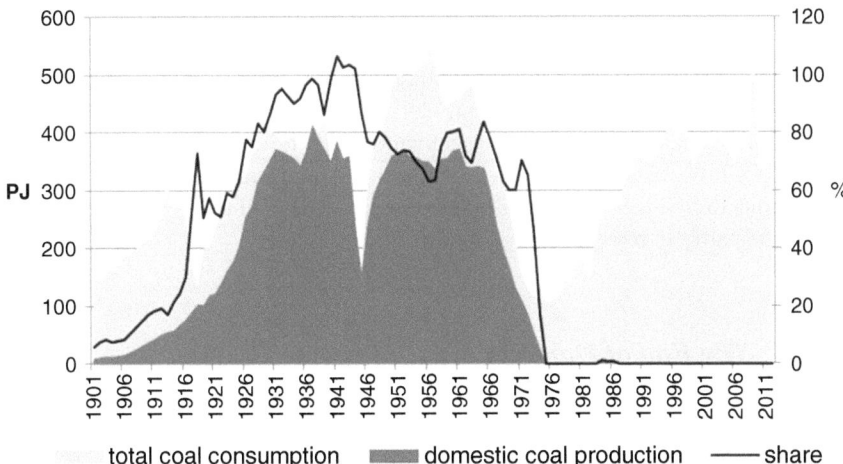

Graph 15.1 Consumption and domestic production of coal in petajoules (left scale) and share of domestic production in total consumption in percent (right scale)
Source: B. Gales and H. Hölsgens in: H.N.M. Hölsgens, Energy transition in the Netherlands: Sustainable challenges in a historical and comparative perspective (Groningen 2016), 207–217

a serious mining industry had come into existence thanks to much hard work both by private parties and the state.

The domestic production of coal increased steadily after 1910. Nonetheless the spectre of foreign dependency came to haunt the country during the First World War. Coal shortages followed on the abrupt cessation of foreign imports. Only after difficult negotiations and in exchange for other important 'war commodities' like food, was the Netherlands again able to import very modest quantities of coal. In the final years of the war this led to rationing and the closing of factories.

It took until the mid-1920s before coal consumption again reached pre-war levels. In the meantime, however, domestic production was meeting 60% of the demand. In 1937 the Dutch mines even supplied about 90% of the domestic demand. In the first years of the Second World War, coal production remained stable, but after 1942 war circumstances began to play a role and production declined. After the war the ideal of self-sufficiency was quietly laid to rest.[6] Imports met an increasingly large part of the demand for coal and this caused no worries. Up to the announcement of the Limburg mine closures in the mid-1960s, the state mines were responsible for delivering 60–80% of the domestic demand for coal (Graph 15.1).[7]

Self-sufficiency and foreign dependence became increasingly unimportant factors. But the preoccupation with a responsible use of subsoil resources remained undiminished. Policy makers feared that the mining companies, in pursuit of quick profits, would first concentrate on the richer layers and thereby make the extraction

---

[6] H.N.M. Hölsgens, 'Resource vulnerability and energy transitions in the Netherlands since the mid-nineteenth century' in *Energy Policy*, Forthcoming.

[7] Hölsgens, *Energy transition in the Netherlands*, 42–43.

of other layers impossible. By also taking on the thinner layers from the outset, costs would rise by 10–15% but in the end the entire supply would be utilized. This became an important point in the debate on the nationalisation of the private mines that took place after 1946. In the event, nationalisation was not pursued but a 'thin layers policy' became the norm. A less than optimal level of productivity became the price for less waste.[8] The tradition of responsible use was continued in the 1970s in decisions to first tackle the smaller natural gas fields and to keep the big field at Slochteren partly in reserve.

15.1.3   The End of Domestic Coal

Coal still played an important role in the post-war energy supply and thus occupied a central place in Dutch society. The decision, in 1952, to open a new State Mine (Beatrix) in Vlodrop attested to that. The previous year, the Netherlands had become part of the European Coal and Steel Community (ECSC), a cooperative trade and production agreement signed by France, Germany, Italy, Belgium, Luxembourg, and the Netherlands. This cooperative venture had developed along the transnational cartels of the powerful iron and steel industries of especially France, Germany, Belgium and Luxembourg. From a Dutch perspective, the emergence of West-European cooperation in the ECSC was a complex juxtaposition of national interests aimed at the rapid recovery of export markets, at monetary integration at the behest of the USA and of opening up markets for coal and steel that were important for reconstruction and industrialisation.[9]

With this European cooperation, domestic mining activities found themselves in a new international playing field. The Suez crisis of 1956, and particularly the blockage of the canal, moved energy to the top of the European agenda. Reports by the ECSC high authority demonstrated that West European energy demands could not be met with coal mining. Increasing demand would have to be met by importing coal and oil. Nuclear power also offered hope. In 1958, with the founding of the European Economic Community (EEC), cooperation on the European market inten-

---

[8] Staatscommissie ingesteld bij Koninklijk Besluit van 26 November 1946 No. 1, Den Haag 1948, 25–26, 33.

[9] The cooperation among France, Germany, Italy, Belgium, Luxembourg and the Netherlands in the ECSC had diverse origins. Important roots were the formation of economic cooperation between Belgium and Luxembourg in the 1920s, the economic cooperation in the framework of the Benelux since 1944, as well as post-war international cooperative arrangements like the United Nations Economic Commission for Europe (UNECE) and the Organisation for European Economic Cooperation (OEEC). More on this in W. Kaiser and J. Schot, *Writing the rules for Europe: Experts, cartels and international organisations*, (London 2014), 212–16. Concerning the national and macro-economic considerations of the Dutch government in the realisation of the ECSC, see: J. Luiten van Zanden and R.T. Griffith, *Economische geschiedenis van Nederland in de 20e eeuw*, (Utrecht 1989), 247–54.

sified. At the same time Euratom started a program to satisfy increasing energy demands with nuclear energy.

These developments turned Dutch coal mining into a plaything in a world of covert national subsidy programs in the surrounding countries.[10] Moreover, oil began to compete with coal. Mounting imports of foreign coal and oil eroded profitability. Process improvements in the steel industry meant a much lower demand for cokes, leading to overproduction. To add insult to injury, the discovery of natural gas in Groningen cast an entirely new light on the national importance of coal mining.

Mining engineers pointed out that Dutch coal mines were among the most modern in Western Europe. But the government had doubts about the size of the reserves and the profitability of mining. They placed these doubts over against the alternative of national exploitation of natural gas and a choice for the latter became inevitable. In 1962 the first decisions for the gradual phasing out of the mines were taken. In 1965 the Minister of Economic Affairs, Joop den Uyl, announced the definitive closing of the mines in a speech in the municipal theatre in Heerlen, headquarters of the State Mines. Nine years later, colliers employed by the Oranje Nassau mine brought op the last of the Dutch coal.[11]

In 1958 more than half the energy consumption was still provided by coal, while the share of oil was about 40%. By 1964 the figures were the inverse.[12] Statistics on energy consumption show the fuel transitions in the 1960s across different sectors (Table 15.1). About half of the energy was consumed by electricity production and industrial applications (both 24%). These sectors consumed mainly oil and coal. In industry, coal demand declined only gradually, but its consumption was dwarfed by the growing use of oil and natural gas from the late 1960s onward. Transportation was largely oil-driven and accounted for about 10% of the energy demand. Households and small businesses were the main consumers of coal (41% in 1962). In the late 1960s these small-scale users also transferred to oil.

With this shift in energy consumption, dependency on foreign sources again increased. And, as a result of the increasing use of oil and the founding of OPEC in 1960 – the cartel of oil-exporting countries – Dutch energy supplies became even more vulnerable.[13] Just as at the outset of the twentieth century, the Netherlands was now also in large measure dependent on foreign suppliers. But now too it had the option of partially compensating its dependency. By the mid-1960s the enormous

---

[10] On the development of mining and the national subsidy programs of the participants in the ECSC see B. Breij, *De mijnen gingen open, de mijnen gingen dicht* (Alphen aan den Rijn 1991). 125–128 and 147–165.

[11] The history of the mine closures is more complicated than can be described here. How the situation evolved and which arguments were used by different groups of engineers and politicians is described in detail in B. Gales, 'Delfstoffen.' in J.W. Schot et al. (eds.) *Techniek in Nederland in de twintigste eeuw - Delfstoffen, energie, chemie* (Zutphen 2000), 61–65. and C.E.P.M. Raedts, *De opkomst, de ontwikkeling en de neergang van de steenkolenmijnbouw in Limburg* (Assen 1974), 198–201.

[12] Hölsgens, *Energy transition in the Netherlands,* 31–32.

[13] Calculations concerning the vulnerability of the Netherlands with respect to energy supplies and different energy sources can be found in Hölsgens, *Energy transition in the Netherlands,* 19–59.

Table 15.1 Energy consumption by fuel and sector 1958–1974 (in PJ)

		1958	1962	1964	1966	1968	1970	1972	1974
Electricty production	Coal	152	145	137	131	133	73	34	32
	Oil		117	230	303	202	514	243	170
	Gas		7	9	13	40	133	245	312
Industry	Coal		110	105	92	95	81	76	80
	Oil		161	182	198	234	168	293	353
	Gas		5	7	42	132	266	434	499
Transport	Coal		2						
	Oil		118			185		250	252
	Gas					2		3	3
Other	Coal	309	238	202	130	93	46	19	9
(Incl. Households)	Oil		219	363	440	318	574	348	214
	Gas		10	20	63	166	306	518	607
All	Coal	460	495	444	353	322	201	129	121
	Oil		614	775	941	939	1255	1135	989
	Gas		22	36	117	340	706	1200	1420
Grand Totals		460	1131	1255	1412	1601	2162	2464	2531
Electicity production		152	268	376	447	375	721	522	515
Industry			277	294	332	461	515	803	932
Transport			120			188		253	255
Other (incl. houseoulds)		309	466	585	634	577	925	886	829
Total		460	1131	1255	1412	1601	2162	2464	2531

Sources: CBS Statistical Yearbooks 1967, 1971, 1973, 1975

gas field in Groningen had turned the Netherlands into a net exporter of energy. Worries about scarcity, at least for the coming decades, were a thing of the past.

15.1.4 A Warm House

Concerns about energy scarcity, foreign dependency and costs had governed the use of fossil resources up to the end of the 1960s. Frugality was woven deeply into the fabric of Dutch society, and extended to the use of fuels. In firms, innovations for energy-saving and for increased production often went hand in hand. For example, the brick and glass processing industries introduced new ovens so that progressively less fuel would be needed per unit product.[14] The effects of such investments could be seen in the relationship between energy consumption and investments in machines and apparatus. Over the long term these figures exhibit a continual decline, with the exception of short periods before and after the First World War and at the end of the 1960s.[15]

---

[14] G.B. Janssen, *Baksteenfabrikage in Nederland, 1850–1920* (Zutphen 1987), 276–78.; E.J.G van Royen and H. Buiter, *Grofkeramische industrie* (Zeist 1994).; B.A. van Veen, *Glas- en glasbewerkingsindustrie* (Zeist 1994).

[15] Hölsgens, *Energy transition in the Netherlands*, 81–83.

In the household, frugal use of fuels was encouraged by the building codes. The national government was able to influence the construction of public housing via various guidelines. These so-called 'suggestions' were inspired by the ideology of simplicity and sobriety. Guidelines in the area of heating dated from the 1920s and assumed a stove in the kitchen and a hearth in the living room. These had to provide warmth for the entire home. Compared to surrounding countries Dutch heating facilities were extremely frugal. From time to time Netherlanders suffered from cold *en masse*.[16]

During the Second World War, architects developed new concepts in housing construction. Among other things this included heating the home by means of central heating or block heating, which were held to improve the comfort of the home as well as its compatibility with the future.[17] But these ideas were at loggerheads with the stringent post-war restrictions imposed by the government on scarce materials and means. The government was able to control the production of materials via its system of distribution and to control the specification of housing standards via its systems of financing. It was even able to regulate the configuration of neighbourhoods by means of its ground-price policy.[18]

Given the limited financial means, the government tried to augment housing production by implementing a radical standardisation of floor plans for dwellings. In this way the government consciously or unconsciously also shaped the level of comfort and heating of the home. These politics had a favourable effect on the low expenditures for domestic fuels and on the energy rationing around and during the Second World War. The fear of future energy shortages caused these measures to remain in force until well into the 1950s.

It seemed that the occupants themselves were not overly concerned about the cold in the house. Dissatisfaction seemed limited to experts, who kept insisting on structural improvements to facilities in the home. But households plotted their own course in improving domestic comfort. Supplementary heating by means of mobile electric or petroleum stoves supplied the desired comfort. A survey conducted in Utrecht in 1948 revealed that more than 20% of the households kept the cold at bay in this fashion. A nationwide survey in 1957 revealed that 27% of the Dutch households used an electric and 6% a petroleum stove as supplementary heating. In 1964, 61% of the households owned an electric stove and only 24% of the households had no form of supplementary heating.[19]

The guided heating policy seemed to be effective on the supply side. Modest heating facilities limited the demand for domestic solid fuels. But portable stoves compensated the lack of comfort in the homes. From a financial point of view this

---

[16] B. Gales, 'Gemütlich Am Ofen?', *Zentrum Für Niederlande-Studien - Jahrbuch*, 2009, 91–112.

[17] Gales, 'Gemütlich Am Ofen?', 100.

[18] K. Schuyt and E. Taverne, *1950: Welvaart in zwart wit* (Den Haag 2000), 204–5.

[19] P. van Overbeeke, *Kachels, geisers en fornuizen: Keuzeprocessen en energieverbruik in Nederlandse huishoudens, 1920–1975* (Hilversum 2001), 179.

Table 15.2 Different types of heating in percent, 1947–1998

	1947	1957	1966	1968	1970	1972	1974	1978	1981	1998
Coal stove/hearth	97	84	58	} 61	43	26	17	1	0	0
Oil Stove	0	8	19					2	1	0
Gas stove/hearth	0	3	11	21	31	39	42	40	34	11
Individual central heating	} 3	5	8	11	16	24	30	47	55	78
Block/neighborhood/district heating			4	7	10	11	11	10	10	11

Source: Peter van Overbeeke, *Kachels, Geisers en Fornuizen*, (Eindhoven 2001)272

did not seem to be the ideal solution. Electric stoves cost 200 cents per 100 mega-joules (about 28 kwh) of heat.[20] The detour of using electricity to produce warmth also entailed extra energy costs. The energetic efficiency of coal-fired electricity plants in the Netherlands in the late 1950s hovered around 28%.[21] In exchange for comfort and ease, households were prepared to suffer higher energy consumption and higher costs.

In comparison with surrounding countries stoves long remained commonplace. In Sweden, by 1961, 99% of the new residences had central heating. With only 5%, the Netherlands struck a discordant note among other West European nations.[22] Around 1960, architects and societal organisations became ever more insistent in addressing the question of heating. 'Good Living' (*Goed Wonen*), a foundation for promoting a modern lifestyle, typified the bedroom as a polar region. In its view it was necessary to put an immediate end to this dire situation.[23] The transition to natural gas marked the turning point. From the early 1960s on the number of new housing units in which central heating was installed grew explosively. After 1965 almost all new flats were equipped with central heating; around 1970 this was also the case for one-family homes. In that year only 26% of the dwellings in the Netherlands had central heating; by the early 1980s this was about 50% and only around the turn of the millennium had homes without central or collective heating become an exception (see Table 15.2). The guided energy policy reverberated long after its inception.

---

[20] Other options were bottled gas (180 cents), municipal gas (105 cents) and petroleum (55 cents) for the same quantity of heat. See van Overbeeke, *Kachels, geisers en fornuizen*, 180.

[21] This meant that most of the energy in coal was lost in friction, heat and transformation losses. J.H. de Boer, 'Primaire en secundaire energiebronnen,' in *Bevolkingsgroei en energie-Verbruik* (symposium at the University of Amsterdam held in the summer of 1957, Assen: Van Gorcum & Comp NV, 1958), 76. Internationally in the 1950s and 60s the efficiency varied between 30–40%. See V. Smil, *Energy in world history* (Oxford 1994), 171–75.

[22] Central heating in homes in West European countries in 1961: Switzerland (90%), Denmark (76%), Belgium (70%), France (44%) and West Germany (24%). Van Overbeeke, *Kachels, geisers en fornuizen*, 165.

[23] Gales, 'Gemütlich Am Ofen?', 107.

15.1.5 Natural Gas to Spare

The first drilling for natural gas was in 1959, but it took more than 4 years before the different parties were able to reach agreement on an acceptable way to exploit this new natural resource. Since the interbellum, exploration for and exploitation of oil and gas had been the monopoly of the Dutch Petroleum Oil Company (*Nederlandse Aardolie Maatschappij, NAM*). The NAM was a joint venture of the oil companies Shell and Esso. Earlier gas finds had been entirely financed and distributed by the State Gas Company (SGB).

But given the gigantic size of the Groningen gas field, this construction became untenable. Against this background, negotiations commenced, about building an infrastructure of pipelines, the sale of gas domestically and abroad, and the division of the income. The role of the government was a complicating factor for Shell and Esso. Foreign governments of oil-producing countries demanded a big cut in sales incomes from the oil companies. Were the Dutch government to make similar demands with respect to the Groningen gas, this would compromise Shell's and Esso's negotiating position elsewhere. The Minister of Economic Affairs, J.W. de Pous, had to weigh the considerable investments of these companies in the Netherlands, like the oil refineries in Rotterdam's Botlek area, against the possibility of extraction via a state enterprise.

In the course of the negotiations, the government proposed that its other energy enterprise, the State Mines (nowadays DSM) become a partner in the extraction and distribution of the Groningen natural gas. In 1962, after nationalistic sentiments within the government had been quelled, the minister presented a 'Memorandum regarding the Natural Gas'. This led to the founding of the Gas Union (*Gasunie*) in 1963, a joint venture in which DSM participated for 40%, Shell for 25%, Esso for 25% and the state for 10%. Thanks to this division of shares as well as corporate taxes, about 70% of the income accrued to the state.[24]

In the course of the negotiations, Esso and Shell developed plans for selling natural gas to households and small businesses. The oil companies calculated that this could be the most profitable market. The gas could be coupled to existing municipal gas networks by means of a national distribution network, which would make it immediately competitive with coals and fuel oil. Households were to utilize the gas for space heating, cooking and warm water. Additionally, plans were made to develop a so-called premium market for applications in the chemical, metal and ceramic industries. Private users and firms consuming large quantities of gas could bargain for significant discounts.[25]

---

[24] G.P.J. Verbong, 'Energie,' in J.W. Schot et al. (eds.), *Techniek in Nederland in de twintigste eeuw - Delfstoffen, energie, chemie*, (Zutphen 2000), 206–10., W. Kielich, *Ondergronds rijk: 25 jaar gasunie en aardgas* (Groningen 1988), 41–45., A. Correlje, *Hollands welvaren: De geschiedenis van een Nederlandse bodemschat* (Hilversum 1998), 27.

[25] Correlje, *Hollands welvaren*, 21–32.

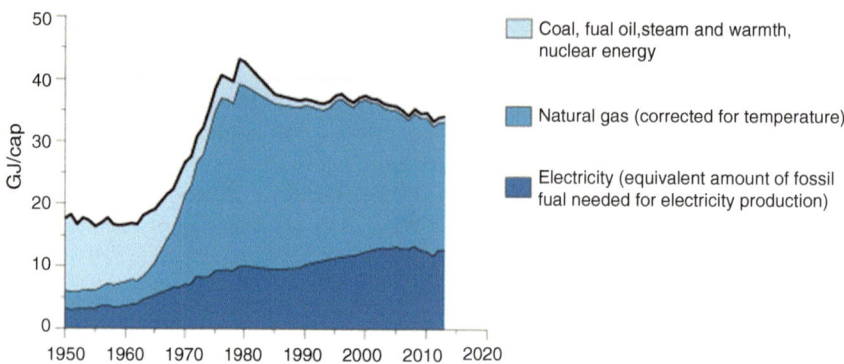

Graph 15.2 Domestic energy consumption per inhabitant 1950–2012
Source: CBS/okt14, www.clo.nl/nl003617

In 1963 the Gasunie laid down the first high-pressure pipelines that were to deliver the gas from Slochteren to the municipal gas companies. Within 5 years almost all onshore Dutch municipalities were connected to the natural gas network. In the cities and towns gas companies were busy modifying and installing new heating and cooking appliances. In the early 1970s households became one of the largest consumers of natural gas (see Table 15.1 and Graph 15.2).

By the mid-sixties expectations about nuclear energy were driving the sale of gas to new heights. It was self-evident that nuclear energy was destined to become the energy source of the future and to replace fossil fuels. That would be the end of natural gas profits. After 1966 natural gas was increasingly used to fuel conventional electric power plants. Low gas prices were employed to seduce industries with large energy requirements into investing in the Netherlands. In this way a consortium consisting of the Billiton Company, Alusuisse and Hoogovens founded an aluminium smelting facility, Aluminium Delfzijl (Aldel), in the province of Groningen. The government agreed to a consignment of 13 billion cubic metres of natural gas, enough to produce 60,000 tons of aluminium in 40 years. The aluminium ore was provided by the Billiton Company and came from bauxite mines in Surinam.[26]

Other energy-intensive sectors also profited from the declining energy prices due to the availability of natural gas. Natural gas, for example, utterly transformed the greenhouse farming industry. Greenhouses were heated with natural gas and also supplied with carbon dioxide (CO_2). The cultivation of vegetables in commercial gardening under glass increased between 1960 and 1970 from 4200 to 5300 hectares, the cultivation of cut flowers from 500 to 1700 hectares. In addition there was a yield increase thanks to the addition of CO_2. The yield of tomatoes, for example, increased from 10 to 15 kilograms per square meter.[27] Along such devious paths,

---

[26] H. Beukema, *Lichtmetaal op zware grond: 40 jaar Aldel* (Delfzijl 2006).

[27] J. Bieleman, *Boeren in Nederland: Geschiedenis van de landbouw, 1500–2000* (Amsterdam 2008), 553–560.

Groningen's natural gas had fundamental consequences for energy use and the industrial structure of the Netherlands.

Natural gas, oil and coal were not only used to make fuels. They were also raw materials for countless chemical products and materials. A new type of material was the synthetic materials, especially plastics.[28] Prior to the Second World War there had been a number of basic inventions in this area, especially in Germany and the United States. After the war, plastics technology created a revolution. A 1949 a Dutch treatise on the state of the art in plastics opened with these sentences:

> Next to atomic physics and radar there is probably no domain that has so exercised the public imagination as that of plastics. Throughout the world new materials have come into use and new applications for these materials have been found, thanks to which new industries have come into being or existing industries have expanded beyond recognition...[29]

In the post-war Netherlands, plastics were the fastest growing industrial sector. At the same time it very quickly won the heart of the welfare society. It offered the masses ease, comfort and pleasure. At the same time, plastics were from the very first controversial materials. They were associated with poor quality, consumerism and environmental problems. The history of plastics is a sublime illustration of the tension between well-being and sustainability. As a part of the fossil material flows, plastics will get special attention in the next section and in the next part.[30]

15.2   Wellfare with Plastics

15.2.1   The Netherlands as Leader

Plastics are built up from long chains of mostly carbon atoms, with molecules that are 1000 to 100,000 times bigger than molecules of substances like water and sugar. During manufacture they can be moulded or are fluid and at the final stage they assume a more or less permanent form. In the course of moulding or casting, i.e. plastic shaping, these synthetic materials acquire their material function.

---

[28] In the past, elaborate factory complexes had developed around fossil subsoil resources. An early example was the rise of the synthetic dye industry in Germany after 1870, an entirely new branch of industry that settled among other places in the neighbourhood of coal basins. In the Netherlands, small-scale chemical industry flourished around the production of city-gas in municipal gasworks. In Limburg a large-scale chemical complex developed under the control of the State Mines once coal deposits there began to be exploited.

[29] R. van de Kasteele, *Het kunststoffengebied: Chemie, grondstoffen en toepassingen* (Amsterdam 1949), 13.

[30] See for the following: H.W. Lintsen, M. Hollestelle and R. Hölsgens, *The Plastics Revolution. How the Netherlands became a global player in plastics* (Eindhoven 2017), the prologue and part I. Furthermore: E.M.L. Bervoets en F.C.A. Veraart, 'Bezinning, ordening en afstemming 1940–1970', in: J.W. Schot et al., *Techniek in Nederland in de twintigste eeuw* (Zutphen 2003) Part VI, 214–239.

Bakelite – invented around 1907 by the Belgian Leo Baekeland – was the first synthetic material derived from a fossil raw material and substances derived therefrom (in this case coal and its derivatives).[31] The three plastics that as bulk products would dominate the market after the Second World War had been developed in the 1920s and 1930s in large foreign chemical enterprises: polystyrene at the American Dow Chemical Company and the German IG Farben, polyvinyl chloride (PVC) at IG Farben, and polyethylene at the British ICI. The most important plastic fibre, nylon, was invented at the American chemical firm DuPont.

Prior to the Second World War, Dutch firms played next to no role in plastics technology. In 1923 Philips became one of the companies that produced and processed Bakelite. It was used to make speakers, radio cabinets and insulating base plates for x-ray apparatus, but also for lightbulb fittings, plugs and switches. A few firms specialized in Bakelite moulding.[32] After the war the Netherlands lagged far behind the USA, England and Germany. Despite this, within 20 years it would succeed in becoming one of the front-runners in plastics technology.

An important reason was the presence of raw materials for plastics production. DSM was perched on top of the coal in the Limburg subsoil. Shell had access to oil and refineries in Pernis. Pernis and Europoort developed into the largest storage and trans-shipment location for crude oil in Europe. That attracted foreign firms. The plastics industry thus represented a classically Dutch type of industry, namely the *trafiek*, a processing and value-adding industry based on flows of trade.

A second reason was that the Netherlands was able to appropriate plastics technology very quickly. Shell, DSM and AKU (General Artificial Silk Union) invested large sums in research and development and built world-class competencies in the production and processing of plastics. TNO, the largest public research organisation

---

[31] As early as the mid-nineteenth century, materials that could be molded were the object of extensive experimentation. Various materials like paper-maché (based on paper and glue) and vulcanised natural rubber (based on natural rubber and sulphur) were not (and are not) counted as synthetic materials. Parkesine, named after its inventor Alexander Parkes (1813–1890), was the first substance that later acquired that predicate, albeit somewhat half-heartedly. Some speak of a half-synthetic substance because cotton was the basic material. The cotton was treated with a mixture of nitric and sulphuric acid and subsequently mixed with vegetable oil and organic solvents. This produced a kneadable dough that could be cast, formed, cut and painted to produce a variety of products like medallions, billiard balls, buttons and letter openers. But the material was brittle, breakable and flammable. An important improvement was the addition of camphor, that made the material strong and flexible. The American John Wesley Hyatt (1837–1920) succeeded at the end of the nineteenth century in producing this material (that now is called celluloid) and its derivative products at an industrial scale.

[32] With a bit of good will we can also count the artificial horn and artificial silk industry as part of the plastics sector. But these were in fact half-synthetic plastics (to speak in contemporary terms) because the artificial horn was prepared from casein, a by-product of the dairy industry and the artificial silk from cellulose derived from wood and cotton. The International Artificial Horn Industry (IKI) for example, belonged to the plastics sector. ENKA (the First Netherlands Artificial Silk Factory in Arnhem, founded in 1912) was the first producer of artificial silk in the Netherlands. Other artificial silk factories followed. In 1928 the ENKA took over one of those firms, the Holland Artificial Silk Industry and merged in 1929 with the German *Vereinigte Glanzstoff Fabriken AG* to form the Algemene Kunstzijde Unie NV (AKU).

in the Netherlands, founded the Plastics Institute TNO, that developed into the national centre of expertise in plastics.

In the 1960s the sector grew at an extraordinary rate. The Netherlands expanded its plastics production capacity so quickly that by the mid-1970s it ranked among the top producers worldwide.[33] Domestic consumption rose from 1.7 kg per inhabitant in 1950, to 9.1 kg in 1960, to 35 kg in 1971.[34] The most important market for plastics in the 1970s was construction (30%). Products included water conduits, gutters, rain pipes, bathroom fixtures, and sinks. A second big category was packaging (23%) such as bags, bottles, crates and shrink-wraps. Then there were several smaller market segments among which were transport (6%) especially applications in cars; households (5%) with utilitarian objects, decorations and toys. In addition there were numerous other markets and application domains, like electronics, machine-building, furniture, paints and medicines.

15.2.2   The Plastics Revolution

The career of plastics after the Second World War can justifiably be characterised as a revolution. Where plastics were still marginal materials just before the Second World War, only 30 years later they had become one of the most prominent materials in the Netherlands. The revolution was caused by, among other things, the extremely low oil prices at the time, a raw materials cost-factor that immediately made plastics price-competitive with classical materials like wood, metal, cotton and wool. Another important factor was the great variety of plastics – each with its own functionality.[35] What also made plastics so attractive was the promise of

---

[33] Production of plastics per capita in various countries in 1963 and 1975 (kg).

	Production (kg/cap) 1963	Estimated production (kg/cap) 1975
West-Germany	24.3	78
United States	20.6	45
England	13.7	?
Italy	12.4	?
Netherlands	11.2	106
France	10.5	41

[34] Initially this was significantly less than the average American, the largest user of plastics. The difference between America and the Netherlands (and other Western European countries) diminished in the course of the 1950s. Around 1970 Germany would become the largest user with 62 kg per year per capita.

[35] F. Van der Most, F.V. Homburg, E. Hooghoff and A. Van Selm, 'Nieuwe synthetische producten: plastics en wasmiddelen na de Tweede Wereldoorlog', in J.W. Schot, H.W. Lintsen, A. Rip and A.A. Albert de la Bruhèze (eds.), *Techniek in Nederland in de Twintigste Eeuw – deel II* (Zutphen 2000), 364.

mass-production, the possibility of imitation and the allure of modernity. Growing welfare did the rest.

Crockery, statues, medals, chains, dolls and a cornucopia of consumer articles could be produced cheaply and *en masse*. These kinds of products were formerly made of wood, leather, glass, earthenware, metal or ivory. And components of durable products like cupboards, tables and chairs could be made of plastic or treated with synthetic paints. Dresses, shirts, socks and other clothing made of tricot weaves could also be fabricated with synthetic fibres. Many products had formerly only been affordable for the middle and upper class. Now they came within the reach of the working class. And with plastics, modern design also entered the home. Middle and upper-class consumers could flaunt their good taste with radio cases of satiny black laminates in streamlined forms, elegant tables with shiny plastic surfaces and a chrome-plated frame or cast plastic clocks of eccentric design.

Originally, plastic had a bad reputation: a plastic product was cheap junk and of poor quality. This improved in the course of time. In the 1960s a plastic product stood for 'strong, hygienic, washable, lightweight, attractively coloured and...a pleasant design.'[36] Plastic played an essential role in the rise and the shaping of consumer society.

15.2.3   Symbol of the Linear Economy

Low production costs also made plastics eminently suitable as packaging material. The light weight, the cheap raw materials and the ease with which it could be shaped, increasingly caused plastics to be seen as the ideal material for disposable packaging.[37] This, together with the rise of the self-service store, resulted in increasing use of pre-packaged food products and beverages. Experts in the packaging industry stated in 1957: 'The most important development in cast plastics insofar as it relates to packaging is the acceptance of the idea that packaging is made to be thrown away.'[38]

Recycling of rags, rubber, glass and paper declined and more and more material was disposed of as waste or burned. In earlier times the Netherlander was careful of his possessions and was constantly refurbishing and re-using them. Now there were goods that one threw away after using them briefly and only once. The first disposable packaging was made of paper, glass or tinned steel, materials that were increas-

---

[36] G. Staal, 'Het wonder, het wantrouwen en de weerstand', in M. Boot, A. Von Graevenitz, H. Overduin and G. Staal (eds.), *De eerste plastic eeuw: Kunststoffen in het dagelijks leven* (Den Haag 1981), 22.

[37] G. Hawkins, 'Made to be wasted: PET and topologies of disposability', in J. Gabrys, G. Hawkins and M. Michael (eds.), *Accumulation: The material politics of plastic* (London 2013), 49–67; J.L. Meikle, 'Materia Nova: Plastics and Design in the U.S., 1925–1935', in S. Mossman and T. Morris (eds.), *The development of plastics* (Cambridge 1994).

[38] In Hawkins 'Made to be Wasted', 5 cited in Lintsen, Hollestelle and Hölsgens, *The Plastics Revolution*, part I.

Table 15.3 Household waste in the Netherlands in 1972

	Veg.Fruit & Garden (GFT)	Paper	Plastics	Glass	Ferro	Textile
Specific gravity: kg/m³	300	120	50	300	400	250
in million kg	1,534	800	156	348	100	66
in million m³	5.1	6.7	3.1	1.2	0.2	0.3
per household						
in kg	116	60	12	26	7	5
in liters	385	503	234	87	19	20

Processed data of CBS – Composition of Household Waste, 1940–2011

ingly replaced by plastic. The consumer society also became a waste society. That was a new phenomenon and plastic waste became its symbol.

Until the early 1970s, the debate on plastic waste remained restricted to those processing the waste. Between 1950 and 1970 the volume of household waste per inhabitant doubled. In 1972 the four main waste categories in weight and volume were vegetable, fruit and garden waste (GFT), paper, glass and plastics. Despite their as yet brief presence in Dutch households, in 1972 about 12 kg of plastic was thrown out per household. That amounted to a volume of about 234 litres (see Table 15.3).

It became more and more difficult to turn the garbage into compost. The increasing glut of materials like glass, paper and plastics that decomposed only with difficulty, if at all, forced waste processors to switch to new methods. The new composition of municipal wastes increased their caloric value, causing many municipalities to consider investments in waste incinerators.[39] This appeared to solve the spatial aspect of the growing waste burden. The problem went up in smoke, at least partly. A fine solution, so it seemed, but certainly one that fits in with a linear production and consumption chain. With the advent of plastics the circular economy became ever more distant.

15.3   Dark Clouds Gather Above Well-being and the Human Environment

15.3.1   The 'Super Pipe'

On September 26, 1965, the under-minister of social affairs and public health ceremonially rammed the first of 196 piles into the ground that would serve as a foundation for the 'Super Pipe,' a 213 meter-high smokestack at the Shell refinery complex at Pernis. The smokestack was completed in 1968 and was part of the fight against air pollution. Twenty-five processing units were connected to the smokestack that

---

[39] W. van Dieren, *Een grondige zaak: 50 jaar vuilafvoermaatschappij VAM, 1929–1979* (Amsterdam 1979), 131–36.

could dispose of 1000 m^3 of waste gasses per second. Not without a touch of pride, Shell refinery engineers announced in 1969 that since the smokestack had come on line,

> it has become necessary to install more sensitive apparatus in the Shell monitoring stations in order to measure sulphur dioxide concentrations [...] We have tried to follow the cloud with a mobile laboratory. Despite all our efforts it has proved impossible to define the location where the cloud touches the ground. Nothing remains of the sulphur-dioxide concentrations after the cloud has travelled several kilometres in the higher air layers.[40]

Air pollution problems seemed to have dissolved, literally and figuratively. It might have seemed surprising that an under-minister of public health should have performed the start-up ceremony at the Shell refinery. From a present-day perspective the 'Super Pipe' is certainly in no way a solution for air pollution. The presence of the under-minister shows that at that time air pollution was seen as a public health issue.

Urban liveability and environmental problems belonged to the most important political and societal topics at the end of the 1960s and the beginning of the 1970s. Local groups of experts called attention to the issue. They mobilised local residents and journalists. New political currents politicised the environmental issues.[41]

This was the context in which environmental legislation around surface water (1970) and air pollution (1972) was passed. This raises various questions. How serious was air pollution? How did different groups throughout the twentieth century look upon air pollution and what conclusions and interventions followed? We shall attempt to answer these questions in the following section. We start with the last question.

---

[40] Reports on the 'Super Pipe' in: ´Tegen de luchtvervuiling: Reuzentoren voor Shell in Pernis´ in *De Telegraaf* 28 februari 1964; ´Eerste paal voor Superpijp van Shell´ in *Gereformeerd Gezinsblad* 27 september 1965 and citation from Henk Thonen 'Vuilstort'der Staatmijnen ook nog onder raffinaderij wolken´ in *Limburgs Dagblad,* 11 oktober 1969. Furthermore F.V. Homburg, A. Selm, and P.F.G. Vincken, 'Industrialisatie en industrie-complexen: De chemische industrie tussen overheid, technologie en markt', in J.W. Schot, H.W. Lintsen, A. Rip and A.A. Albert de la Bruhèze (eds.), *Techniek in Nederland in de twintigste eeuw – deel II* (Zutphen 2000),. 376–401.

[41] Top 5 problems according to survey of Dutch voters

	1967	1971	1972	1977
Unemployment	**29**	3	11	**57**
Housing shortage	21	**28**	15	6
Environment	1	22	**18**	4
Political Problems	11	4	12	3
Income and Prices	8	7	15	7

From Aarts 1989 Cited in H.T. Siraa, A.J. van der Valk, and W.L. Wissink, *Met het oog op de omgeving: Het Ministerie van Volkshuisvesting, Ruimtelijke Ordening en Milieubeheer, 1965–1995* (Den Haag, 1995), 234.

15.3.2 Polluted Air as a Nuisance

For centuries local authorities issued regulations to limit water and air pollution seen as a nuisance and a threat to health. These rules were aimed at perceptible pollution, i.e. pollution that you could see and smell. In 1896, the various regulations were nationally integrated into the form of the Nuisance Law. This law was intended to prevent nuisance and dangers caused by industrial activities, for example explosions, poisoning of water and polluted air.[42] Stench, smoke, soot and ash were the directly perceptible components of air pollution. The measures prescribed in the Nuisance Law aimed at these components, such as zoning for smelly enterprises like slaughterhouses, tanneries and steam-powered factories.

In 1905, the minister of Agriculture, Industry and Trade contacted the executive board of the Society for Industry (*Maatschappij van Nijverheid*) in order to discuss possible new articles in the Nuisance Law 'against smoking factory chimneys.' Inspired by German and British examples, the industrialists advocated the promotion of 'economic steam production,' whereby thanks to total combustion less soot was produced. The consultation led neither to the subsidized stokers' courses advocated by the industrialists nor to new rules. On the minister's advice the Society for Industry founded the Association for the Promotion of Smoke-free Stoking. This 'in order to combat smoke formation by factory chimneys on our own power.'[43] In 1910 the association counted 54 members responsible for 244 boilers. In that year the association merged with the Netherlands Association against Water Pollution to form the Netherlands Association against Water, Soil and Air Pollution. The members of the association included industrialists, but also public health inspectors and representatives of public works and municipal utilities. The association emphasized that it was not 'hostile' or 'less sympathetic' to the interests of agriculture and industry. The universal interest in clean air, water and soil had to be achieved by mutual consultation and enlightenment.[44]

Water pollution was the dominant topic in the association's journal, *Water, Soil, Air*. The sporadic articles on air pollution were largely accounts of British and German research into the effects of coal combustion. Smoke was seen as a danger to health chiefly in closed spaces and in cases of direct inhalation of soot clouds. Municipal health services investigated the effluents of specific factories after complaints by nearby residents. Incidental investigations were carried on in cooperation with research institutes like agricultural experimental stations and the Central

---

[42] T. J. Dijkstra, Het bezwaar: De beleving van leefomgevingshinder in de periode 1870–2000 in de Friese havenstad Harlingen (Groningen 2006), 54–58.; E.M.T. Beenakkers, *Aandacht van de overheid voor bodembescherming: Sinds wanneer?* (Den Haag 1991), 7–15.; Siraa, van der Valk, and Wissink, *Met het oog op de omgeving,* 230–233.

[43] J. de Kuijser, 'Mededeelingen betreffende de vereeniging tot bevordering van rookvrij stoken' in *Water Bodem, Lucht,* vol 1. 1910–1911, p. 7–10

[44] Redactie, 'De Oprichting Der Vereeniging´ in *Water, Bodem Lucht,* 1(1910–1911), 19–20; see also H. van der Windt, 'De Totstandkoming van ´de Natuurbescherming´ in Nederland,' *Tijdschrift Voor Geschiedenis* 107(3) (1994): 485–507.

Laboratory for Public Health, founded in 1909. The solution was found in optimizing the combustion of coal and diluting the smoke concentration. Given the right chimney-height the latter could be achieved and this, most critics agreed, solved the problems.[45]

15.3.3   A National Monitoring Network for Air Pollution

In December 1930 sixty-three people died in a deadly fog in the Meuse Valley town of Engis in Wallonia (Belgium). Immediately after the disaster, it was speculated that poisonous chemicals were to blame. Subsequent investigation showed the disaster was the result of serious air pollution. This conclusion caused hardly a ripple in the Netherlands, despite the participation of the Leiden physician, Willem Storm van Leeuwen, in the investigation. Air pollution – as the reports of the Association against Water, Soil and Air Pollution attest – was a foreign problem especially in the industrialised areas of Germany, England and Belgium. In those regions, for the time being, no measures were being taken to curb pollution.[46]

In medical circles there was, however, concern about air pollution and consequences for labour and public health. In the Netherlands, the inspector of public health, M.J.N. Schuursma, spoke out about the dangers of air pollution.

> Air pollution due to smoke, exhaust gases and condensation nuclei are prevalent where there are large concentrations of industry and can produce irritation and danger. As industrialisation increases it is desirable to pay attention to this matter.'[47]

In 1948, the city of Rotterdam established a commission for Soil, Water and Air. In the 1950s this body investigated excessive quantities of fluoride and sulphur-dioxide in the region's atmosphere caused by artificial fertiliser factories, electrical power plants and oil refineries. On the basis of data from 18 air-quality measurement stations, the commission was able for the first time in the Netherlands to chart the geographical dispersion of pollution.

London's 'Great Smog' of December 1952 gave an extra impulse to the public health aspects and encouraged a more integral approach to pollution. Medical research concluded that about 4000 people died as a result of the smog. In Great Britain the event led to the making of new laws on air pollution that came into force in 1956. The legislation focused not only on industry, but also on the use of coal in domestic stoves and cooking stoves. In the Netherlands too, investigations focused

---

[45] Redactie, 'Luchtverontreininging door eene loodaschbranderij' in *Water Bodem, Lucht*, 5(1915), 89–91; J. De Kuijser, 'Luchtverontreininging en hinderwet aangelegenheden', in *Water Bodem, Lucht*, 6(1916), 45–56 and E. Buijsman, *Er zij een meetnet...: Een geillustreerde geschiedenis van het luchtmeetnet van het RIV(M)* (Bilthoven 2003), 18–19.

[46] B. Nemery, P.H.M. Hoet, A. Nemmar, 'The Muese Valley fog of 1930: an air pollution disaster', in *The Lancet*, Vol. 357, (2001); Redactie, 'Mist in het Maasdal' in *Water, Bodem, Lucht* 21(1931), 8–9; E. Buijsman, *'De Moordende Mist: De Ramp in Maasvallei bij Luik in 1930* (Houten 2010).

[47] Buijsman, *Er zij een meetnet*, 20–21.

more explicitly on regions rather than on specific industrial plants. Rotterdam was a prime example of a region with a high concentration of industry and was the first municipality to focus attention on the regional situation. In 1953 TNO followed in Rotterdam's footsteps, investigating pollution in the environs of the Hoogovens. In 1956 the National Institute for Public Health (RIV) started investigations into air pollution around the State Mines in Limburg and in Amsterdam's harbour area.[48]

The measurements showed an extremely high level of air pollution. Scientists began to use ever more forceful terminology. In 1958, prof. W.F.J.M Krul, chairman of TNO's Institute for Health Technology, stated that the future habitability of the 'Randstad Holland will depend in large measure on whether the technical and economic problems of air pollution will admit of a solution.'[49] But according to the minister of Social Affairs and Public Health:

> ... the air pollution is [...] not so serious, that it endangers public health...According to the National Weather Service (KNMI) the Netherlands does not have to fear for inversions such as those that in combination with a number of other factors cause London "smog".[50]

But in January 1959 and December 1962 Rotterdam suffered 'winter smog' under the influence of the polluted air. This added fuel to the fire in the debate among scientists, politicians and policy makers, a debate that now found its way to the pages of the public media. The research focused on the health effects of air pollution. The Rotterdam Commission for Soil, Water and Air registered a slight increase in mortality rates, hospitalisations and sick-leave, but did not consider the situation alarming. It concluded that air pollution still 'played an insignificant role compared to sickness and mortality associated with traffic accidents, poor nutritional habits, substandard housing and spoiled food.' And other research that focused especially on the increase in lung cancer concluded that explanations should be sought in smoking, rather than in polluted air.[51] Public review of the various investigations expressed increasing concern about air pollution. The unease was not quelled, despite the soothing conclusions. Protest groups like the Society against Air Pollution in and around the New Waterway (1963) and the Committee for the Habitability of the Waterway Region (1968) became the protagonists of the dissatisfaction. Municipal politicians, scientists and local experts took the initiative. The local groups continued to demand attention for liveability and environmental issues.

---

[48] E. Buijsman, *Een geannoteerd overzicht van publicaties over chemische samenstelling van lucht en neerslag in Nederland* (Houten 2011), 10.

[49] Cited in J.W.Tesch 'Volksgezondheidsaspecten van luchtverontreiniging' in *Water Bodem, Lucht*, 50(1960), 44.

[50] 'Verslag van het mondelinge overleg betreffende afdeling VII (Volksgezondheid)', *Rijksbegroting voor het dienstjaar 1960, Handelingen van de Tweede Kamer*, zitting 1959–1960, Kamerstuk Tweede Kamer 1959–1960 kamerstuknummer 5700 XII ondernummer 19, p. 17–18.

[51] 'Uit Rotterdams rapport blijkt: Sterke luchtverontreiniging werkt ziekte in de hand- Lichte toename sterfgevallen tijdens zware mist', in *De Waarheid*, 25 april 1984, p. 3; 'Dr. Bierstekers verrast in proefschrift: Verontreinigde lucht zo gevaarlijk nog niet.' in *De Tijd, dagblad voor Nederland*, 30–6-1966, p.5.

In 1963 the government moved to appoint a Council on Air Pollution. A year later it produced a draft law on air pollution. The law was ratified in 1970 and came into force a year later. It coupled the air pollution of emissions to specific limiting values. The limit value was defined as that point where emissions 'can negatively affect human health or create a nuisance for humans, or cause damage to animals, plants or goods.'[52] The law was a so-called framework law. General Administrative Ordinances were to fill in the further details.

The new regulative structure was thus impossible without knowledge of air quality. In 1965 the ministry of Social Affairs and Public Health consulted with experts in the Inspectorate of Environmental Hygiene, Royal Dutch Meteorological Institute (KNMI), TNO and RIV (National Institute for Public Health). These talks resulted in a number of investigations and a design for a national air quality monitoring network. The RIV was assigned the task of realising a completely automated measurement network. By 1967, in cooperation with the KNMI, TNO and Philips, it had a test network up and running in Twente. Automation placed new demands on the measurement system and the processing of the data. The first automated piece of apparatus was a sulphur-dioxide monitor designed in part by Philips. While experts realised that other substances were also relevant for determining the level of air pollution, sulphur-dioxide levels thus became the benchmark for air quality. The first fully automated measurement network, using so called 'sniffing poles' to monitor air quality, was set up in 1969 in Rotterdam's harbour region at the behest of local health services and the Rijnmond Regional Authority (*Openbaar Lichaam Rijnmond*). Provinces and municipalities got involved in the plans for a national monitoring network. In November 1970 the under-minister of public health granted approval. In 1975 a national air-quality monitoring network was completed.[53]

15.3.4   The Overture to New Sustainability Problems

During the 1960s for the first time an overall picture of the national diffusion of air pollution became available. How serious was air pollution? The measurements focused on two important components: sulphur-dioxide (SO_2) and so-called 'black smoke,' dust particles smaller than four microns. Both substances were related to the combustion of coal, the customary fuel in industry and households. The concentration of sulphur-dioxide was measured by means of sampling and chemical analysis, that of black smoke by means of filtration techniques.

The first investigations focused on problems around specific factories and on extreme situations such as the smog episodes in the winters of 1952 and 1962. The

[52] Cited in and translated from: G.H. Dinkelman, *Verzuring en broeikaseffect: De wisselwerking tussen problemen en oplossingen in het Nederlandse luchtverontreinigingsbeleid (1970–1994)* (Amsterdam 1995), 36.

[53] Buijsman, *Er zij een meetnet.*, 13–75; E. Buijsman, 'Van mosterdgas naar zwaveldioxide: Over de oorsprong van de eerste zwaveldioxidemonitor,' *Studium* 3(8) (2015),159–62.

Table 15.4 Extreme air pollution in Europe, 1952 and 1962

	Sulphur Dioxide (SO_2) in $\mu g/m^3$	Smoke (dust) in $\mu g/m^3$
1952		
London (8 Dec)	1800–2000	1200–1500
1962		
Rotterdam & Surr. (5 Dec)	1040–1610	160–530
Paris (7 Dec)	840–1260	600–800
Ruhr (6 Dec)	2300–4000	–
London (5 Dec)	1400–4650	1700–4550
Norm 2016		
Daily average (max 3 days)	125 (Limit value)	
Hourly average (24x p.j.)	350 (Limit value)	
Hourly average (3 hours)	500 (Alarm value)	

Sources: E. Buijsman, *Er Zij een Meetnet..., Een Geillustreerde Geschiedenis van het Luchtmeetnet van het RIV(M)* (Bilthoven 2003), 38 and F.M.W. de Jong and P.J.C.M. Jansen, *Luchtnormen Geordend*, (Bilthoven 2010)

measurements showed that the sulphur-dioxide levels at Rotterdam in 1962 began to approach those of the Big Smog in London 10 years earlier (Table 15.4).

Systematic measurements dating from the 1960s revealed high concentrations of sulphur dioxide and 'black smoke.' The concentrations of both substances declined after the mid-1960s (Graph 15.3). The transition from coal to natural gas had beneficial side-effects. The switch liberated the Netherlands from the strongly polluting coal fumes in the cities and the surrounding industrial parks. The heyday of choking air pollution seemed to be definitely on the way out by the end of the 1960s.

The difference in rates of decline between sulphur dioxide and black smoke were the first indication of other developments (Graphs 15.3 and 15.4).[54] Measurements in Utrecht showed that motorised traffic contributed strongly to the presence of dust particles (Graph 15.5).[55] More research was needed. The need for more data on the composition of the air and its consequences for health shaped a new agenda for research institutes and policy makers. Local interests groups developed into environmental organisations. The new organisations in the 'societal midfield' developed new strategies to influence public opinion and government. While during the first decades after the war, water and air pollution were still dismissed as inevitable collateral consequences of welfare, around 1970 some were concluding that the price of this kind of welfare was far too high. New demands on the quality of air, water and soil became the overture to the sustainability problems of the following decades.

---

[54] Buijsman, *Een geannoteerd overzicht van publicaties over chemische samenstelling van lucht en neerslag in Nederland*; E. Buijsman, *Meten waar mensen zijn: De ontwikkeling van stedelijke luchtkwaliteit in Nederland* (Houten 2010).

[55] E. Buijsman, *Stof in Nederland: Een reconstructie van historische metingen in Nederland* (Houten 2010).

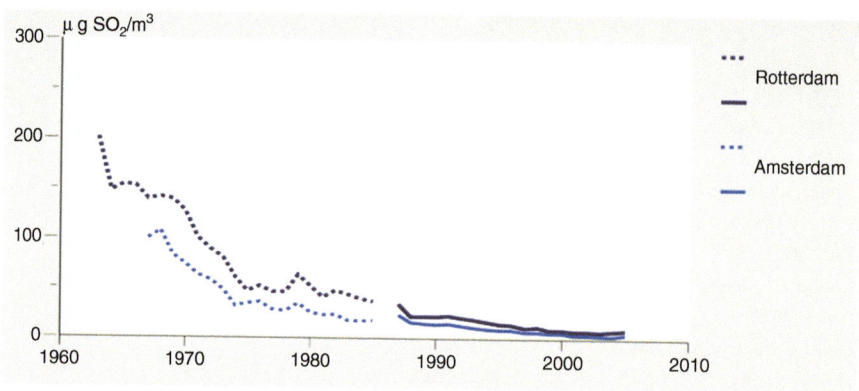

Graph 15.3 Yearly average sulphur dioxide concentrations in Amsterdam and Rotterdam, 1965–2005
Source: E. Buijsman, *Meten Waar Mensen Zijn, de Ontwikkeling van Stedelijke Luchtkwaliteit in Nederland* (Houten 2010), 14–17

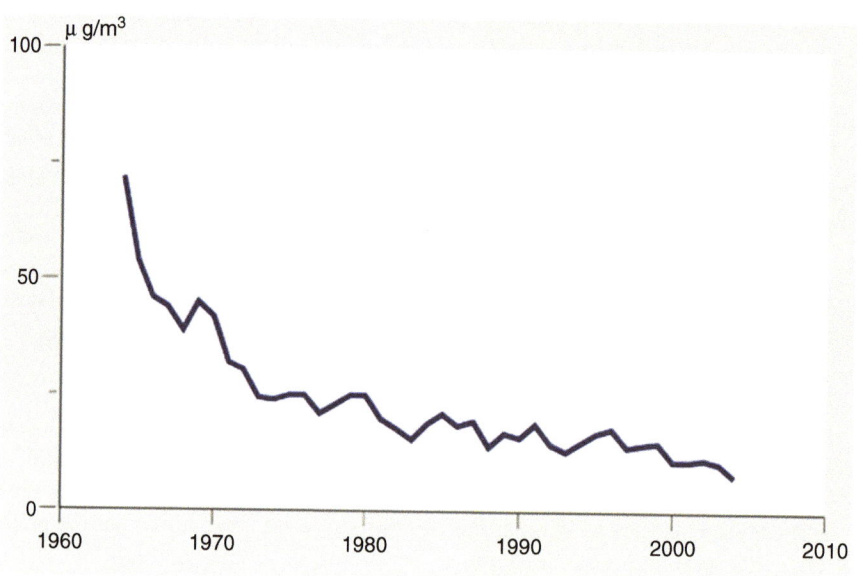

Graph 15.4 Concentration black smoke, 1963–2005
Source: E. Buijsman, *Meten Waar Mensen Zijn, de Ontwikkeling van Stedelijke Luchtkwaliteit in Nederland* (Houten 2010), 14–17

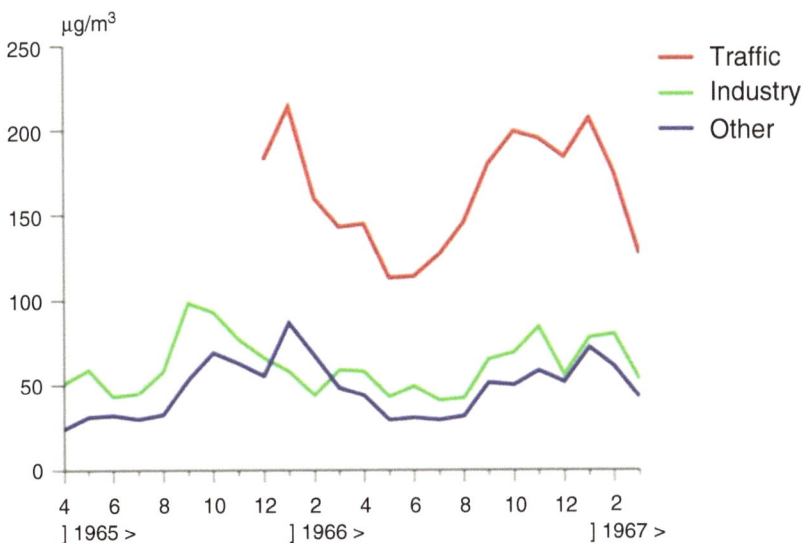

Graph 15.5 Monthly average concentrations of black smoke in Utrecht, 1965–1967
Source: E. Buijsman, *Stof in Nederland, een Reconstructie van Historische Metingen in Nederland* (Houten 2010), 19

Literature

Beenakkers, E.M.T. (1991). *Aandacht van de overheid voor bodembescherming: Sinds wanneer?* Den Haag: WODC.

Bervoets E.M.L. and F.C.A. Veraart (2003). 'Bezinning, ordenning en afstemming, 1940–1970'. In J.W. Schot, H.W. Lintsen, A. Rip and A.A. Albert de la Bruhèze (eds.), *Techniek in Nederland in de twintigste eeuw*. Zutphen: Walburg Pers.

Beukema, H. (2006). *Lichtmetaal op zware grond: 40 jaar Aldel*. Delfzijl: Maritext.

Bieleman, J. (2008). *Boeren in Nederland: Geschiedenis van de landbouw 1500–2000*. Amsterdam: Uitgeverij Boom.

Boer, J.H. de (1958). 'Primaire en secundaire energiebronnen'. In *Bevolkingsgroei en energieverbruik*. Symposium der Universiteit van Amsterdam, gehouden in de zomer van 1957. Assen: Van Gorcum.

Breij, B. (1991). *De mijnen gingen open, de mijnen gingen dicht*. Alphen aan den Rijn: ICOB.

Buijsman, E. (2003). *Er zij een meetnet…: Een geïllustreerde geschiedenis van het luchtmeetnet van het RIV(M)*. Bilthoven: RIVM.

Buijsman, E. (2010a). *De moordende mist: De ramp in Maasvallei bij Luik in 1930*. Houten: Uitgeverij Tinsentiep.

Buijsman, E. (2010b). *Meten waar de mensen zijn: De ontwikkeling van de stedelijke luchtkwaliteit in Nederland*. Houten: Uitgeverij Tinsentiep.

Buijsman, E. (2010c). *Stof in Nederland: Een reconstructie van historische metingen in Nederland*. Houten: Uitgeverij Tinsentiep.

Buijsman, E. (2011). *Een geannoteerd overzicht van publicaties over chemische samenstelling van lucht en neerslag in Nederland*. Houten: Uitgeverij Tinsentiep.

Buijsman, E. (2015). 'Van mosterdgas naar zwaveldioxide: Over de oorsprong van de eerste zwaveldioxidemonitor'. *Studium*, 3(8), 159–162.

Correlje, A. (1998). *Hollands welvaren: De geschiedenis van een Nederlandse bodemschat.* Hilversum: Stichting Teleac/NOT.

Dieren, W. van (1979). *Een grondige zaak: 50 jaar vuilafvoermaatschappij VAM, 1929–1979.* Amsterdam: Vuilafvoer Maatschappij.

Dijkstra, T.J. (2006). *Het bezwaar: De beleving van leefomgevingshinder in de periode 1870–2000 in de Friese havenstad Harlingen.* Groningen: Universiteit van Groningen.

Dinkelman, G.H. (1995). *Verzuring en broeikaseffect: De wisselwerking tussen problemen en oplossingen in het Nederlandse luchtverontreinigingsbeleid (1970–1994).* Amsterdam: University of Amsterdam.

Gales, B. (2000). 'Delfstoffen'. In J.W. Schot, H.W. Lintsen, A. Rip and A.A. Albert de la Bruhèze (eds.), *Techniek in Nederland in de twintigste eeuw – Delfstoffen, energie, chemie.* Zutphen: Walburg Pers.

Gales, B. (2009). 'Gemütlich Am Ofen?' *Zentrum Für Niederlande-Studien – Jahrbuch 2009.*

Hawkins, G. (2013). 'Made to be wasted: PET and topologies of disposability'. In J. Gabrys, G. Hawkins and M. Michael (eds.), *Accumulation: The material politics of plastic.* London: Routledge.

Hölsgens, H.N.M. (2016). *Energy transition in the Netherlands: Sustainable challenges in a historical and comparative perspective.* Groningen: Rijksuniversiteit Groningen.

Hölsgens, H.N.M. 'Resource vulnerability and energy transitions in the Netherlands since the mid-nineteenth century'. *Energy Policy, Forthcoming.*

Homburg, F.V., A. Selm and P.F.G. Vincken (2000). 'Industrialisatie en industrie-complexen: De chemische industrie tussen overheid, technologie en markt'. In J.W. Schot, H.W. Lintsen, A. Rip and A.A. Albert de la Bruhèze (eds.), *Techniek in Nederland in de twintigste eeuw – deel II.* Zutphen: Walburg Pers.

Houven van Oordt, H.C. van der and G. Vissering (1901). *Economische beteekenis van afsluiting en drooglegging der Zuiderzee.* Leiden: E.J. Brill.

Janssen, G.B. (1987). *Baksteenfabrikage in Nederland, 1850–1920.* Zutphen: Walburg Pers.

Kaiser, W. and J. Schot (2014). *Writing the rules for Europe: Experts, cartels and international organisations.* London: Palgrave Macmillan.

Kasteele, R. van de (1949). *Het kunststoffengebied: Chemie, grondstoffen en toepassingen.* Amsterdam: Ahrend.

Kielich, W. (1988). *Ondergronds rijk: 25 jaar gasunie en aardgas.* Groningen: Nederlandse Gasunie.

Knotter, A. (eds.) (2015). *Limburg kolenland: Over de geschiedenis van de Limburgse kolenmijnbouw.* Zwolle: Uitgeverij Wbooks.

Kuijser, J. de (1910). 'Mededeelingen betreffende de vereeniging tot bevordering van rookvrij stoken'. *Water, Bodem Lucht* 1:1–10.

Lintsen, H.W., M. Hollestelle and R. Hölsgens (2017), *The Plastics Revolution. How the Netherlands became a global player in plastics.* Eindhoven: Foundation of the History of Technology.

Luiten van Zanden, J. and R.T. Griffith (1989). *Economische geschiedenis van Nederland in de 20e eeuw.* Utrecht: Spectrum.

Meikle, J.L. (1994). 'Materia Nova: Plastics and design in the U.S., 1925-1935'. In S. Mossman and T. Morris (eds.), *The development of plastics.* Cambridge: CRC Press.

Most, F. van der, et al. (2000). 'Nieuwe synthetische producten: plastics en wasmiddelen na de Tweede Wereldoorlog'. In J.W. Schot, H.W. Lintsen, A. Rip and A.A. Albert de la Bruhèze (eds.), *Techniek in Nederland in de twintigste eeuw – deel II.* Zutphen: Walburg Pers.

Nemery, B., P.H.M. Hoet and A. Nemmar (2001). The Muese Valley fog of 1930: An air pollution disaster. *The Lancet* 357(3), 704–708.

Overbeeke, P. van (2001). *Kachels, geisers en fornuizen: keuzeprocessen en energieverbruik in Nederlandse huishoudens, 1920–1975.* Eindhoven: Technische Universiteit Eindhoven.

Raedts, C.E.P.M. (1974). *De opkomst, de ontwikkeling en de neergang van de steenkolenmijnbouw in Limburg.* Assen: Koninklijke van Gorcum BV.

Redactie (1910–1911). 'De Oprichting Der Vereeniging´ in *Water, Bodem Lucht*, 1: 19–20.

Redactie (1915). 'Luchtverontreininging door eene loodaschbranderij' in *Water Bodem, Lucht*, 5: 89–91.

Redactie (1931). 'Mist in het Maasdal' in *Water, Bodem, Lucht*. 21: 8–9.

Royen, E.J.G. and H. Buiter (1994). *Grofkeramische industrie*. Zeist: Stichting Projectbureau Industrieel Erfgoed.

Schuyt, K. and E. Taverne (2000). *1950: Welvaart in zwart wit*. Den Haag: Sdu Uitgevers.

Siraa, H.T., A.J. van der Valk and W.L. Wissink (1995). *Met het oog op de omgeving: Het Ministerie van Volkshuisvesting, Ruimtelijke Ordening en Milieubeheer, 1965–1995*. Den Haag: Sdu Uitgevers.

Smil, V. (1994). *Energy in world history*. Oxford: Westview Press.

Staal, G. (1981). 'Het wonder, het wantrouwen en de weerstand'. In M. Boot, A. von Graevenitz, H. Overduin and G. Staal (eds.), *De eerste plastic eeuw: Kunststoffen in het dagelijks leven*. Den Haag: Haags Gemeentemuseum.

Tesch, J.W. (1960). 'Volksgezondheidsaspecten van luchtverontreiniging'. *Water, Bodem, Lucht*, 50: 45–48.

Veen, B.A. van (1994). *Glas- en glasbewerkingsindustrie*. Zeist: Stichting Projectbureau Industrieel Erfgoed.

Verbong, G.P.J. (2000). 'Energie'. In J.W. Schot, H.W. Lintsen, A. Rip and A.A. Albert de la Bruhèze (eds.), *Techniek in Nederland in de twintigste eeuw – Delfstoffen, energie, chemie*. Zutphen: Walburg Pers.

Windt, H. van der (1994). 'De totstandkoming van 'de Natuurbescherming' in Nederland'. *Tijdschrift voor Geschiedenis*, 107(3), 485–507.

Chapter 16
The Turn of the Tide. Well-being and Sustainability Around 1970

Frank Veraart and Harry Lintsen

Contents

Abstract Around 1970, welfare and economic growth became increasingly suspect. This chapter analyses and explains how this came about. It provides an inventory of the driving forces and institutional frameworks that shaped the development of well-being. In the period 1910–1970 the government energetically pursued the building of the welfare state. It was supported in this endeavour by a radically pillarised societal midfield. The economy was also under the tutelage of a *dirigiste* government. The six large Dutch multinationals generally supported the government's ambitions regarding the development of well-being. Characteristic for this period was the development of new patterns of consumption and a linear economy. Thanks to the mutual alignment among government, midfield and private enterprise it seemed possible to make well-being.

The monitor is used to evaluate the state of well-being around 1970 from three perspectives. Viewed from the perspective of 1910, the monitor shows how the original agenda was realised. But when viewed through the lens of the new societal visions of 1970, an entirely different image emerges. The increase in welfare had been achieved at the cost of serious environmental pollution and the loss of nature in both the Netherlands and elsewhere. The mounting criticism of the dark side of

© The Author(s) 2018
H. Lintsen et al., *Well-being, Sustainability and Social Development*,
https://doi.org/10.1007/978-3-319-76696-6_16

well-being introduced a period in which ecology, natural resources, energy and climate change received emphatic attention (see Chaps. 17, 18, 19, 20, and 21). From the perspective of 2010 the situation had indeed become serious. In retrospect it appeared that around 1960 welfare, well-being and sustainability were most in balance.

Keywords Government · Midfield · Private enterprise · Welfare state · Linear economy · Monitor · Trade-off

16.1 Aldrin, Dieldrin, Eldrin en Telodrin: Blessing or a Shady Business?

In the Netherlands populations of various bird species (have) gone downhill at a tremendous rate. The decline of the sandwich terns on the island Griend from 40,000 pairs around 1955 to at present maximum 700 pairs is very likely partly due to serious pollution of the North Sea and the Wadden Sea with herbicides and pesticides... This pollution is mainly due to wastewater from chemical firms along the New Waterway...The feeding grounds of the sandwich terns are located a few hundred kilometers away from the factories that emit the poisonous wastewater. Nonetheless seaweed and plankton have been poisoned, and the fish that ate these organisms, and the birds that ate the fish. The 1962 Law on Herbicides and Pesticides that came into force in 1964 gives the government new possibilities to intervene...[1]

The Shell facility at Pernis, in the Rotterdam Harbour, was one of the world's foremost producers of pesticides. Its pesticides Aldrin, Dieldrin, Eldrin and Telodrin were held responsible for the massive bird mortality. In 1967 the toxicological division of the company invited the national press to a press-conference and a visit to the British Shell laboratory at Sittingbourne. In the conservative newspaper, *De Telegraaf*, Shell experts responded to the claims of what they called 'emotionally moved' alarmed individuals.

Today 12,000 people in the world will die of hunger...and over the entire year an estimated 1.3 million. That is an awful reality...The world is on the threshold of the biggest famine in her history...In order to feed the 6 billion inhabitants of this globe, three times more food has to be provided than is now produced...the yield of every square meter of arable land (must) be maximized to the fullest extent possible. And according to the insecticide experts this is only possible if the production of these pesticides – in the eyes of the public often regarded as shady substances – is significantly increased...Unfortunately several recent catastrophes – massive local bird mortality in England and Drenthe – have made the conservation-minded public wary and distrustful, which according to insecticide producers

---

[1] B. Bruins, 'We veranderen de wereld.. met chemische vergiften' in *Het vrije volk: democratisch-socialistisch dagblad* (5-8-1967), 4.

is lamentable because it is precisely the responsible substance Dieldrin, as well as Aldrin and DDT, that have over the course of years contributed so frightfully much to combating world hunger… And as far as bird mortality goes, however lamentable that may be, it pales into insignificance compared to the number of birds that are daily crushed by motorists on European highways.[2]

The debate between the critics of pesticides, who, following Rachel Carson, feared for the poisoning of humans and nature and the proponents from industry and agriculture was framed as the 'pesticide paradox.' It was presented as a dilemma of local environmental protection versus global food production, the offering that had to be brought to mitigate the famines in the world. In the background, of course, there were the interests of Shell, the significance of the agrarian sector, and jobs in the Netherlands.

Shell's charm-offensive was intended to temper the ever-growing criticism of environmental pollution. By the late 1960s Dutch industries were ever more insistently being held responsible for air and water pollution and the degradation of the environment. The 'pesticide paradox' was a typical outcome of vested ambitions and new choices in well-being by governments, the societal mid-field, firms and researchers. In this chapter we will first of all chart the different ambitions in their institutional contexts. After that we take a look at the monitor for 1970. We look at which ambitions were realised, which new problems were articulated and who articulated the problems. Wat does this analysis portend for well-being and sustainability after 1970? New demands, concerns and norms informed new challenges. To conclude, we examine well-being in 1970 in the light of present-day norms. How rosy or how threatening was the situation?

16.2 Synergetic Dynamics: Government, Citizens, Researchers and Entrepreneurs

16.2.1 Government – Makeable Well-being

The First World War was in numerous respects the starting point for new institutional relationships in the Netherlands. Under pressure of the wartime situation, the political 'pacification' was achieved. Crucial disputes like universal suffrage were resolved and with the first general elections the relationships among the political parties were determined for many years. The confessional parties would long dominate both parliament and the government. The socialists and the liberals were constantly frustrated by their power. In the big cities the socialist parties also took part in local government.[3]

---

[2] A. Huguenot van der Linden, 'Landbouw legt het af tegen groei van wereldbevolking als gewas niet wordt beschermd' in *De Telegraaf* (29-07-1967). H.G.S. van Raalte was since 1964 employed at Shell's Toxological Division. S. Howarth and J. Jonker, *Powering the Hydrocarbon Revolution, 1939–1973* (New York 2007), 401.

[3] P. de Rooy, *Ons stipje op de waereldkaart: De politieke cultuur van modern Nederland* (Amsterdam 2014), 151–73.

The government, as it had for years in the past, pursued a welfare state. Economic growth was the basis of national well-being. In the nineteenth century, the government tried to encourage this by investing in infrastructure. This policy was continued into the twentieth century. The national government, province and municipalities all constructed roads. The provinces invested in provincial electrical networks. In the end the national government managed the telephone network et cetera. Education was also a domain in which various governments kept a finger in the pie and that began to consume a major part of the national budget in the twentieth century. A new phenomenon was the influence of the national government on flows of raw materials and on land-management. At the end of the nineteenth century it intervened in agriculture, founded the State Forest Service in 1899, underwrote the Hoogovens in 1920 and concerned itself with local conflicts around the mining of marl, gravel and clay. The national government developed an elaborate (but shifting) set of instruments to regulate and stabilize flows of raw materials and more generic economic processes.

Also new was the fact that the state, besides wanting to be a welfare state, now also had the ambition to become a 'caring-state.' That had already become visible prior to the First World War in its concern for the so-called 'social question.' It commiserated with the fate of the poor, the workers and the vulnerable ill with the Housing Law (1901), the Accident Law (1901) and the Invalidity Law (1913). These laws were followed by, among others, the Unemployment Act (1917), the Illness Law (1930), mandatory socialized health insurance (1941) and a series of laws passed after the Second World War: the Emergency Law Old Age Assistance (1946), followed 10 years later by the General Old Age Law (AOW), the General Widows and Orphans Law (1959) and the Unemployment Assistance Law (1964). The finale was the General Welfare Law that came into force in 1965. With this latter law the government absolutely replaced private initiatives in caring for the poor. Poor relief was no longer a question of charity; it now became a right and a duty for municipal governments to provide support.[4] The 'caring state' had now definitively been founded.

Finally, the state had an age-old task in the domain of public works and, most crucially, water management. In the twentieth century this particularly Dutch aspect of well-being once again demanded important investments. The Zuiderzee Works (1918) and the Delta Works (1953) were the most impressive national projects. The provinces, water boards and municipalities also undertook a wide range of projects aimed at reducing the vulnerability of the Dutch delta.

The state's political ambitions were revealed in the formation of new ministries and the allocation of the government budget. In 1910 the ministry of Public Works, the ministry of Agriculture, Industry and Trade and the ministry of Internal Affairs were active in the expansion of well-being. The latter ministry included budgets for education, public health and social affairs. The costs of these ministries accounted for about 40% of the total government budget. This share grew to about 50% in the interbellum and again after 1960. In 1918 a new ministry of Education, Art and

---

[4] K. Schuyt and E. Taverne, *1950: Welvaart in zwart wit* (Den Haag 2000), 288–306.

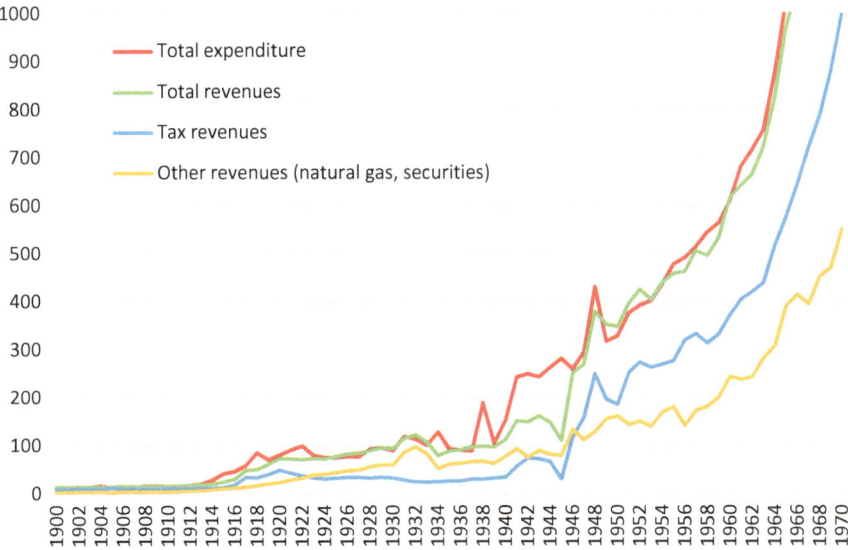

Graph 16.1   Government Finances, expenditures and income in euro per inhabitant, 1900–1970
Source: CBS – Rijksfinanciën vanaf 1900

Sciences was called into being. In that same year the ministry of Trade, Industry and Agriculture founded a department for Labour, which blossomed into the ministry of Social Affairs in 1933. After the Second World War the new ministries of Reconstruction and Public Housing, Transport and Public Works, and Agriculture, Fisheries and Food Supply underscored the new agenda of the Dutch government. In 1951 government dedication to public health crystallized in a new ministry of Social Affairs and Public Health.

Alongside the expansion of the governmental apparatus, budgets increased for policy domains like education, social affairs, infrastructure, public housing and public health.[5] State expenditures increased from a few tens just after the First World War to nearly a hundred euros per inhabitant at the end of the 1930s. After the Second World War government spending again increased steadily (Graph 16.1). These expenditures also began to comprise an ever larger share of the national economy. The share of government spending in the national economy rose from less than 10% at the outset of the twentieth century to more than 40% around 1980.[6]

The steadily growing government intervention was partly the consequence of external crises. Two wars, the Great Depression, and reconstruction cleared the way for new perceptions of the role of government. But these perceptions required support among the population at large as well as among specific societal groups.

[5] Handelingen Tweede Kamer, 'Nota betreffende den toestand van 's lands financiën' 20 september 1920. Via www.rijksbegroting.nl/algemeen/rijksbegroting/archief (1910).

[6] J.L. van Zanden en R.T. Griffiths, *Economische geschiedenis van Nederland in de twintigste eeuw* (Utrecht 1989), 61.

Hence, after the Second World War there was clear social support for a *dirigiste* role for government. The first questionnaires circulated by the Netherlands Institute for Public Opinion (NIPO), founded in 1946, included the question 'What should the government tackle first?' The results showed that according to the population, the government should take the lead in seeing to it that primary needs were met: food, housing and sufficient income.[7] But even more important was support from specific social groups and associations. These articulated social needs, represented their constituencies in politics and possessed the means to exert influence. It then appears that the solid anchoring of politics and the government in the 'societal midfield' was decisive for the success of government policy with respect to well-being. But in precisely that anchorage lay the seeds of new problems around well-being and sustainability.

16.2.2   The Exalted 'Mid-Field'

In the period between 1910 and 1970 the 'societal midfield,' the complex of non-governmental associations based on shared interests and religious convictions, became increasingly densely populated. Confessional groups as well as socialists organised themselves in countless organizations and institutions. Thanks to their supremacy at national and municipal levels they were able to direct significant subsidies to their own organizations; this enabled them to engage in all kinds of social services: aid to the poor, the construction of public housing, public health and education. Though it was true that the Welfare Law of 1965 had compelled confessional groups to surrender their dominance in the fight against poverty to municipalities, other aspects of well-being surfaced in its stead. There were short and direct connections to the political parties. Pressure on the government and the public administration was severe. The pillarised mid-field co-authored the legal framework for well-being or lent its support to such efforts. It was also able to influence what was to count as well-being.

It should not be supposed, however, that the mid-field was composed only of the 'pillars.' Quite the contrary. The economic domain, for example, was also populated by a variety of branch organisations.[8] During the first half of the nineteenth century the Netherlands had seen the rise of a number of agricultural organizations. These would become more numerous (in part via the fragmenting effects of the pillars), would dissolve into a network of governmental organisations, knowledge institutes and enterprises and in this way eventually come to be known as the 'green front.' Toward the end of the nineteenth century, other economic domains sprouted interest

---

[7]The Netherlands Institute for Public Opinion (NIPO) was founded in 1946, with the aim of collecting information on opinions among the populace as an aid to government. J.C.H Blom, *Crisis, bezetting en herstel: Tien studies over Nederland, 1930–1950* (Rotterdam 1989), 184–90.

[8]B. Bouwens and J. Dankers, *Tussen concurrentie en concentratie. Belangenorganisaties, kartels, fusies en overnames* (Amsterdam 2010), 54–57.

and lobby groups of building contractors, butchers, bakers, carpenters and other 'petit-bourgeois' groups. These began locally and were often the basis for national, partly pillarised, organisations. During the First World War they became more numerous. In 1907 the ministry of Agriculture, Industry and Trade counted 35 national and 307 local organisations. After the war their number increased to no less than 337 national and 1329 local organisations. They were in the first place interest groups of entrepreneurs seeking to limit competition among themselves, to offer courses to their members, to exchange information about innovation and to lobby the government. But they were also harnessed to help solve the social problems of their time. During the First World War, for example, the national government worked closely with the Bakers' Union and the Station for Milling and Baking that the latter had founded. Due to the flour shortage, they had to formulate norms for so-called government wheat, search for flour substitutes and develop alternative recipes for baking bread.[9] In this way branch organisations became involved in food quality, housing construction, working conditions and so on. Increasingly they formed a portal for welfare legislation and set themselves up as essential links and consultation bureaus. They also began to play an active role in industrialisation policy after the Second World War.[10]

In addition, the mid-field also sported an increasing number of professional societies that also felt compelled to take some stand on well-being. Accountants, sociologists, home economists, electrical engineers and other professionals organized themselves in the first half of the twentieth century around new academic programs of study. Their professions were extremely dynamic and in search of social and academic recognition. At the end of the nineteenth century, engineers had already shown how that could be done. The members of the Association of Civil (later Delft) Engineers participated in the debate on the 'social question' and loudly proclaimed the relevance of their profession in finding solutions. They did the same in the course of the debates during the interbellum on the rationalisation of the economy and the firm.[11] They argued that this would provide advantages for the entrepreneur, the worker and national welfare. Together with accountants and psychologists, engineers were the shock troops of the rationalisation movement. The movement itself was broader and included politicians, entrepreneurs, leading cadres, civil servants, and advisors. They came together in large numbers for the first time at the Economising (Efficiency)-Congress in Amsterdam in 1923 and organised themselves in the Netherlands Institute for Efficiency, the NIVE. Professionals did not exert their influence on politics and policy by way of the pillars, as did the confessionals and socialists, but via these kinds of networks. The networks of congress organisations, knowledge institutes and laboratories were interwoven with ministries, provincial services and municipal organisations.

---

[9] M. Davids, H.W. Lintsen, and A. van Rooij, *Innovatie en kennisinfrastructuur: Vele wegen naar vernieuwing*, vol. 5, Bedrijfsleven in Nederland in de Twintigste Eeuw (Amsterdam 2013), 75–76.

[10] Bouwens and Dankers, *Tussen concurrentie en concentratie*, 158–159.

[11] H.W. Lintsen, *Made in Holland: Een techniekgeschiedenis van Nederland [1800–2000]* (Zutphen 2005), 171–173.

There was also a category of civil associations not associated with a pillar, economic sector or profession, like the tourist organisation ANWB (1883), the Association for the Preservation of Natural Monuments (1905), the Netherlands Association of Housewives (1912) and the Netherlands Consumer Union (1953). These too became more or less involved in defining and ameliorating the quality of life. The Netherlands Association of Housewives (NVVH), for example, first profiled itself as a think-tank for labour-saving solutions in the household.[12] Together with liberal and socialist women's clubs they took the initiative in finding collective solutions for the more tedious aspects of housekeeping. There were, for example, experiments with putting the family laundry out to collective laundries and delegating cooking to collective kitchens. Special housing complexes were designed with collective facilities like a space for washing and drying, a bath-house, central heating, a central waste disposal chute or a central kitchen. These initiatives did not survive the experimental phase. In addition, the NVVH concerned itself with the quality of household work and household products. Their seal of approval for household products: 'Approved by the NVVH' was probably the best known of all. In essence the NVVH was still an old-fashioned status organisation, because it furthered the interests especially of middle-class 'petit bourgeois' women.

These civil associations had varying kinds of influence. The ANWB fought for highways and found strong partners in the Royal Institute of Engineers and the Rijkswaterstaat (the national public works agency). Hence the organisation was allowed a finger in the pie of the National Road Plan of 1928. Contrariwise, nature associations were hardly able to make a dent in the land consolidation projects that were carried out on a large scale. Nature conservation and environmental problems were neither big issues for the government nor in politics more generally. Despite this, the civil associations concerned with this issue did score regular, if small, successes. The occasion was more often than not a local problem that was taken up by a local interest group. In 1925, for example, the Natural History Society in Limburg protested against marl mining that threatened to consume the local hill, the Sint Pietersberg, a stance that led to the amendment of the permit. Similarly, protests against gravel dredging in Limburg led to the Excavation Law of 1965. Disquiet among Rotterdam's population concerning air pollution was articulated by the Association against Air Pollution in and around the New Waterway (1963) and the Committee for Habitability of the Waterway Region (1968). These organisations mobilized political pressure, resulting in the 1970 Law on Air Pollution. Nature conservation and environmental organisations had to look for support to the 'pillars,' that had access to the political parties, or to the professionals with their networks in the state apparatus. They were the part of the 'mid-field' that made the 'pillarised' and 'professional' Netherlands receptive to environmental issues and nature conservation and that paved the way for the big debate on sustainability that welled up at the end of the 1960s.

---

[12] R. Oldenziel, 'Het ontstaan van het moderne huishouden: toevalstreffers en valse starts, 1890–1918', in: J.W. Schot, H.W. Lintsen, A. Rip en A. Albert de la Bruhèze (eds.), *Techniek in Nederland in de twintigste eeuw* (Zutphen 2001), deel IV, 41–41.

16.2.3   Research for Well-being

The 'professional' Netherlands, that is to say, the higher and academically educated part of the Dutch population with their professional associations, was tightly bound to the modern knowledge infrastructure whose foundations were laid in the First World War. We focus here on the laboratory, that became the icon of the production of innovations and well-being during the interbellum. The laboratory became *the* place where research was done on resources and materials and where experiments were performed on new products and processes. To be sure, innovations still originated on the work floor, in workshops, by dint of the efforts of inventive entrepreneurs and entrepreneurial inventors, but the laboratory was the modern route. As far as we know, there were at least 500 laboratories in the Netherlands around 1940, of which about 31% at the universities, 30% in (semi-) government agencies, and 19% in industry. The rest was private or was run by an association or foundation.[13] The 1932 founding of Netherlands Organization for Applied Scientific Research (TNO), the central government laboratories, was an important decision. After a difficult start, this organisation eventually developed into the biggest public research institute in the Netherlands. In 1965, TNO had four branches: the Industrial Organisation TNO, the Health Organisation TNO, the Nutrition Organisation TNO and the National Defence Organisation TNO with all told nearly 40 research institutes.[14] The Industrial Organisation with 24 institutes was the biggest branch. TNO was a cooperative research organisation, an alliance among the government, industry and the 'professional' Netherlands. Ministerial functionaries, members of sector-organisations and researchers at the (technical) universities were well-represented in all the managerial and consultative organs of TNO. They drew up the collective research agendas. The government and industry financed the research.

Some of the laboratories were dedicated to research and development. This was especially the case for the industrial laboratories. These labs produced a stream of innovative prototypes, including electron tubes, margarines, radios, plastics, chemical insecticides and pesticides, industrial automatons and artificial fibres. The (semi-) governmental laboratories also contributed to this stream. For example, the agricultural test stations investigated the use of artificial fertilizer and the tillage of vegetables. TNO operated in numerous fields and among other things developed new plastics, analysed the behaviour of building constructions, searched for better vaccines and investigated conservation techniques. In addition to research and development the laboratories also dedicated themselves to the important task of testing and evaluating. They developed testing methods, designed standards of quality, collected data on the quality of raw materials, semi-finished products and end-products and evaluated production processes and products.

The researchers contributed to paving the way for a modern economy, one in which the electro-technical and chemical industries were dominant and in which all

---

[13] Davids, Lintsen en Van Rooij, *Innovatie en kennisinfrastructuur*, 100–102.

[14] H.W. Lintsen (eds.), *Tachtig jaar TNO* (Delft 2012), 29.

sectors were suffused with the new principles of efficiency and rationality. They were also partly the trailblazers of the consumer society and experimented with new services like the telephone and electricity as well as with an enormous potential in new consumer goods like vacuum cleaners, washing machines, electric irons, radios and plastic toys. Between the two world wars, parts of the middle class garnered experience with these new commodities. After the Second World War the new consumer goods became available to the masses.

The researchers were also deeply involved in aspects of well-being and the construction of the caring state. TNO made especially important contributions. The Industrial Organisation TNO investigated the quantitative aspects of public housing construction (rate of production, costs of production and labour productivity) as well as the qualitative aspects (problems with moisture, frost damage, thermal resistance etc.).[15] The Health Organisation TNO worked on public health, for example the relation between personal health, housing, work and the environment (soil, water and air).[16] The Nutrition Organization TNO analysed, among other topics, the main components of food like carbohydrates, proteins, vitamins, minerals, fats and fibres. The results were used to produce nutritional charts.[17] Research and testing by TNO often led to norms for working conditions, housing quality, fire safety, food quality and food safety.

In the 1960s the researchers became ever more involved in environmental problems. Knowledge of air, soil and water pollution would in large measure come to shape the debate on environmental problems. New measuring instruments increased the accuracy with which it was possible to chart both old types of pollution and newly discovered ones. Politicians and policy-makers mobilized researchers to define problems, establish norms and provide solutions.

16.2.4 The Linear Economy

By the 1960s economic developments had finally ushered in the welfare state. That followed on several violent shocks: two world wars and an economic recession. After the Second World War, the economy stabilised and grew rapidly. Economic historians qualify the period as an economic wonder. It signified the definitive breakthrough of welfare for the vast majority of the population. Some see its origins in the favourable development of the world economy, the politics of planned wages, industrialisation policy and so on. But viewed over the long term we must conclude that it was a case of the crystallisation of an oversaturated solution. In the Netherlands, as in many West-European countries, the welfare state had been in the works for more than a half-century: technically, economically, politically and culturally. It was just

---

[15] Lintsen, *Tachtig jaar TNO*. (2012): 44.

[16] Lintsen, Tachtig jaar TNO. (2012): 133.

[17] Lintsen, Tachtig jaar TNO. (2012.: 144.

a question of waiting for favourable economic circumstances to enable the entire structure to blossom – and these were certainly present in the 1960s.

In this period, the economy and society had a strongly corporatist bent. Netherlanders worked together in many fields in associations, foundations, councils and other public and private organs. Corporate bodies were present in agriculture, public housing and healthcare. Entrepreneurs cooperated in branch organisations. Cartels were important instruments in ordering markets and regulating competition.[18] A corporatist tradition of consultations lay at the foundation of Collective Labour Agreements (CAOs) and the Social-Economic Council (SER). It was also visible in TNO's structure. The corporatist model (the polder or Rijnlands model as it would later be called) was based on the belief that cooperation was better than competition for economic growth and social progress. In this model the government fulfilled a key role. It was *dirigiste* in the creation of the caring state and a safe hydraulic structure, coordinating in the rise of the welfare state and a follower in the emergence of the consumer society.

Another characteristic of the economy was the dominance of six multinationals: Philips, Shell, DSM, Akzo, Unilever and Hoogovens. They played a big part in the growth of the economy and employment. They also shared in the ambitions for well-being in the Netherlands. Philips, for example, invested in housing, schooling, health care, recreation and culture for its employees.

Characteristic of this economy dominated by multinationals was the complexity and international character of the chains of production and consumption. The chains were long (they included multiple production processes), became denser (numerous production processes co-existed within one organisation) or differentiated (chains split up into multiple chains). For example, biochemical research at Unilever led to the interchangeability of vegetable and animal oils and fats. Margarine could be produced from whale blubber, fish oil or palm oil. Chains of raw materials fragmented into different basic products (commodities) that were traded as bulk goods on the international market. This changed the position of food processing firms. Where it once was a purveyor of added value in a chain of foodstuffs, it now became a hub in a network. This enabled them to acquire important positions of power in the diverse international flows of products. These developments transported sustainability issues around the production of food across the national borders. While in the Netherlands 'not a penny too much' was paid for margarine, in the Congo vast swaths of the jungle was cleared for palm oil plantations in order to satisfy the demand for oil 'on the global market.'

The linear nature of the economy was also an important determinant of sustainability. Materials were acquired at one location, separated and processed at other places, and consumed at a third location, after which they degenerated into waste products or emissions. This had already long been the case for the use of turf as fossil fuel in the Netherlands. Coal, oil and natural gas made no difference here. The combustion of fossil fuels was by definition linear. With the increasing use of

---

[18] Bouwens and Dankers, *Tussen concurrentie en concentratie. Belangenorganisaties, kartels, fusies en overnames,* 158–159.

artificial fertiliser and the flushing of faecal matter through sewers at the end of the nineteenth century a similar linear chain developed for food. In the course of the twentieth century other substances like paper, glass and plastics were added to the list. At the start of the chain, the linear economy dotted the landscape with empty gravel pits, marl pits and closed mines, and at the end with huge garbage mounds.

The period was marked by a short effort, born of necessity, to retreat back to the circular economy. During the Second World War disrupted supplies of raw materials encouraged re-use and substitution. Wood replaced energy suppliers like coal and gas. Soap was fabricated from bones. Paper, textiles, glass, metals, rubber, hair and many other materials were collected in order to be recycled.

16.3   The Monitor for 1970: Development of Well-being and Sustainability

How can we judge the culmination of this period – around 1970 – in terms of well-being and sustainability? In this section we will address this question from three perspectives (Table 16.1). The perspective of 1910: to what extent was the societal agenda of the first decades of the twentieth century realized? The perspective of 1970: which elements were at the time praised by politics and policy, what were the contemporary concerns and what was the agenda for the future? Finally, we look at the developments from a present-day perspective. How do we judge the results of the period 1910–1970 by today's norms?

16.3.1   Perspective 1910: The Agenda at the Turn of the Century

The further suppression of extreme poverty (securing the vital necessities: food, clothing and housing) had been supplemented as a social-economic agenda with concerns about the quality of food, housing, health and labour. Around 1970 the aims of 1910 had been largely realised. Extreme poverty had been abolished. Much had been accomplished in regard to food quality, housing quality, public health and working conditions. The population had also profited from increased economic and material welfare. People lived longer and were healthier, had jobs and time for themselves.

These aims were achieved by means of the large-scale employment of domestic and foreign natural capital. Increased exploitation of natural capital had been an explicit goal around 1900. 'Wastelands' had been reclaimed and land had been

Table 16.1 Dashboard well-being and sustainability 'here and now' in 1970 from the perspectives of 1910, 1970 and a present-day perspective

Dashboard well-being 'here and now''						
Theme	Indicator	Unit	1970	Perspective 1910	Perspective 1970	Present day perspective
Population		million inhabitants	13,0			
Material Welfare and Well-being						
Consumption, income	Consumptive Expenditures per capita / constant prices	index (1850=100)	340	+	+	+
	Income inequality, general	Gini coefficient 0-1	0,36	+	+	+
	Gender income inequality	% difference hourly wage M/F	29%	+	O	–
Subjective well-being	Satisfaction with life	score 0-10	7,4	?	+	+
Personal Characteristics						
Health	Life expectancy	years	75	+	+	+
Nutrition	Height	cm	182	+	+	+
Housing	Housing quality	% slums	6	+	–	–
	Public water supply	m³/capita	109	+	+	+
Physical Safety	Murder victims	Number per 100.000 inhabitants.	0,7	–	–	–
Labour	Unemployment	% workforce.	1,6	+	+	+
Education	Level of Education	years	9	+	+	+
Free time	Free time	hours / week.	47,9	+	+	+
Natural Environment						
Biodiversity	MSA	% original biodiversity	66	+	–	–
Air quality	SO_2	kg SO_2 / capita	21	+	–	–
	Greenhouse gas emissions	ton CO_2 /capita	10,1	+	+	–
Water quality	Public water supply	m³/capita	109	+	–	–
Institutional environment						
Trust	Generalised trust	% population with adequate trust	?	–	–	–
Political Institutions	Democracy	Democracy-index 0-100	39	+	+	+

(continued)

Table 16.1 (continued)

Dashboard well-being 'later'							
Theme	Indicator	Unit	1970	Perspective 1910	Perspective 1970	Present day perspective
Natural Capital							
Energy	Energy consumption	TJ /capita	0,16	+	−	−
Non-fossil fuels	Domestic consumption	ton/capita	9,4	+	−	−
Biodiversity	MSA	% original biodiversity	66	+	−	−
Air quality	SO_2 emissions	kg SO_2/capita	21	+	−	−
	Greenhouse gas emissions	ton CO_2/capita	10,1	+	−	−
Water	Public water supply	m³/capita	109	+	+	−
Economic Capital:							
Physical capital	Economic capital stock/capita	index (1850=100)	518	+	+	+
Financial capital	Gross national debt	% gdp	48	+	+	+
Knowledge	Supply of knowledge capital	index (2010=100)	30	+	+	+
Human Capital:							
Health	Life expectancy	years	75	+	+	+
Labour	Unemployment	% workforce	1,6	+	+	+
Educational level	Schooling	years	9	+	+	+
Social Capital:							
Trust	Generalised trust	% population with adequate trust	?	−	−	−
Political institutions	Democracy	democracy index 0–100	38,5	+	+	+

Dashboard well-being 'elsewhere'							
Theme	Indicator	Unit	1970	Perspective 1910	Perspective 1970	Present day perspective
Material Welfare							
Consumption, income	Development aid	% gdp	0,6	+	O	O
Natural capital							
Natural capital	Import of raw materials	ton/capita	8,6	+	O	−

Legend

+	Not problematic or not problematized
−	Generally acknowledged as problematic
O	Under discussion: different opinions about the scale and nature of the problems
?	Unknown

Source: See note 24 of Chap. 2

wrested from the sea. Coal, salt and marl were mined and sand, gravel, oil and gas extracted from the subsoil. Imported metal ores, crude oil, grains, natural oils and fats were incorporated into Dutch production processes. Increasing exploitation was accompanied by increasing energy consumption. The political and policy establishment of 1910 would have looked at these indicators with great satisfaction. That would also have been the case for the indicators of economic and human capital and of democratic quality, while they would have been concerned about the political turbulence around 1970 (social capital) – as they were about political struggles in their own time. Politics and policy in those days had no goals for biodiversity and emissions, so that little can be said from this perspective about the declining biodiversity and increasing emissions. On the other hand the political elite of 1910 would have deeply regretted the changed relationship to the former colonies (development aid) and would have rejoiced at the increased exploitation of foreign natural capital (import raw materials).

16.3.2   Perspective 1970: Environmental Problems in New Babylon

These developments were viewed in an entirely different light from the perspective of 1970. Up to the end of the 1950s the Netherlands, a country faithful to traditions and customs, had been maintained by a system of pillarisation. A few individuals foresaw a great change. Automation and robotisation would create a New Babylon, as Cobra artist Constant Nieuwenhuys proclaimed in 1959. The future would see the rise of 'homo ludens,' the creative, playful human.[19]

Ten years later these intimations of change had become the fertile soil for diverse forms of protest against authority, industrialisation and modernisation. Compared to surrounding countries, the turnaround was gradual and without extreme violence, either on the side of the opposition or on that of the authorities.[20] Confessional, socialist and liberal parties remained at the centre of power, but cloaked with a new 'progressive' élan. And they had to co-exist with new political parties and movements. Around 1970, the Pacificist-Socialist Party (1957), the Farmers' Party (1963), Provo (1965), the Democrats '66 (1966), the Political Party Radicals (1968), the Kabouter (Gnome) Movement (1970) and other political movements attempted to change the existing political and institutional structures on the basis of their own social visions.

The renewed traditional parties and most of the new ones shared one big concern, namely concern about the natural environment and the exploitation of natural capi-

---

[19] J. van der Lans and H. Vuijsje, *Lage landen, hoge sprongen: Nederland in de twintigste eeuw* (Wormer 2003), 185.

[20] J.C. Kennedy, *Nieuw Babylon in aanbouw: Nederland in de jaren zestig* (Amsterdam 1995), 9–22.

tal. The worries focused originally on changes in the landscape (among other things, biodiversity), the use of materials (e.g. the growing mountains of waste), water pollution (especially surface waters) and air pollution (particularly the emission of sulphur dioxide). The exhaustibility of raw materials and energy sources soon surfaced as a major social problem. Other aspects like the safety of nuclear energy and the emission of greenhouse gases were not yet on the agenda.

An important theme at the time was certainly the global opposition between rich and poor. It was essential to support the poor countries. Between 1965 and 1975 the Dutch contribution rose from one million to nearly one billion euros. This made the Netherlands, along with Norway, one of the most generous countries in the world.[21]

16.3.3   The View from 2015–1970 as a Critical Watershed

From a present-day perspective the construction of the caring and welfare state was a great achievement. The Netherlands had become a prosperous country, in which a quality of life had been achieved that pretty much answered to present-day standards. Both the general and material aspects of well-being were at a high level. To be sure, the wage inequality between men and women that prevailed in 1970 would have raised eyebrows in the present day. Average life expectancy had increased and was still climbing. The numerous victims of traffic accidents, the large numbers of smokers and other health risks would no longer be acceptable. On the other hand, the low incidence of crime would be something to long for again. The housing shortage was nearly solved and big investments in this field continued. And the internal arrangements and furnishings of homes increasingly approached present-day standards for heating and sanitation. The spacious homes built in the 1970s are still popular nearly 50 years later. The level of unemployment was lower than what is presently considered acceptable. Access to and quality of education was high.

At the same time a present-day observer would share the big concerns about the natural environment and natural capital. In 1970 biodiversity was in better shape than it is at present. And even then it was anything but hopeful. Salmon and trout had disappeared from rivers and streams due to pollution of surface water. Emissions of poisonous effluents in the Rotterdam Harbour area threatened flora and fauna in the Wadden Sea. Farmers sought to eliminate every kind of weed and insect that threatened their yields. Reclamations and land consolidation created an ever more monotonous landscape. In 1970 natural ecosystems and biodiversity were under extreme pressure.

Air pollution due to sulphur dioxide emssions was significant, but diminishing thanks to the use of filters, de-sulphurisation of fuels and the transition from coal to gas. New insights into air quality would lead to new concerns about public health and air quality. Soot particles, nitrous oxides and fine dust replaced sulphur

---

[21] Kennedy, *Nieuw Babylon in aanbouw*, 73–74.

dioxide as an indicator of air quality. In 1970 there was as yet little concern about climate change; that debate surfaced only in the 1980s after which it began to draw increasing attention. Due to the increased use of energy and materials the emission of 'climate gases' increased. CO_2 emissions at the time were about 21% higher than the current 2020 European goal of 8.6 tons per inhabitant. In the domain of natural capital and the human environment the period around 1970 was indeed a watershed.

16.3.4   The 1960s as the Great Transformation

In the prologue to this book we wrote about the great transformation in well-being: the Netherlands increased the quality of life of its population, but at the cost of depleting Dutch and foreign natural capital. That transformation can be traced over the course of the 1960s. In this part of the book it has been analysed for the three large material flows – the bio and the mineral and fossil sub-soil assets – and from the perspective of the four institutions – government, economy, the societal mid-field and research. And yet the issue still demands a generic explanation. That explanation will have to account for a remarkable phenomenon. The biggest issue in the career of well-being in the Netherlands, namely the problematic quality of life of the population, was solved at the outset of the 1960s with the establishment of the 'caring-state.' But the big problems with natural capital and the natural environment arose *after* that period. As statistics of various material flows make amply clear, the exploitation of natural capital accelerated in the course of the 1960s. This was accompanied by an accelerated growth in emissions that soon reached levels exceeding present-day norms. What is the origin of this 'de-railing?' Did it have to do with the demands that the 'caring state' had to make on natural capital in order to guarantee a certain quality of life for a growing population? Was it a consequence of the welfare state? Could it be attributed to fundamental changes in the economic structure, among other things due to the discovery of the Groningen natural gas field? In the epilogue to this book, when we review the entire period from 1850 to 2010, we will return to these questions. The issue raises another insistent question. At the end of the 1960s contemporaries were aware of the 'de-railing' that was taking place. Around 1970 a critical boundary in sustainability had been passed in various respects. Old and new groups in the societal mid-field signalled problems around the over-exploitation of natural capital and the pollution of the environment and manoeuvred them onto the societal agenda. Despite this, the situation would get worse before it got better. The problems proved nearly intractable. Why was it so difficult to create a sustainable society? That will be the most important question of the next part.

Literature

Blom, J.C.H. (1989). *Crisis, bezetting en herstel: Tien studies over Nederland, 1930–1950.* Rotterdam: Nijgh & Van Ditmar Universitair,

Bouwens, B. and J. Dankers (2010). *Tussen concurrentie en concentratie: Belangenorganisaties, kartels, fusies en overnames. Bedrijfsleven in Nederland in de twintigste eeuw.* Amsterdam: Boom.

Bruins, B. (1967, 5 augustus). 'We veranderen de wereld… met chemische vergiften'. *Het vrije volk: democratisch-socialistisch dagblad*, p. 4.

Davids, M., H.W. Lintsen, and A. van Rooij (2013). *Innovatie en kennisinfrastructuur: Vele wegen naar vernieuwing. Vol. 5. Bedrijfsleven in Nederland in de Twintigste Eeuw.* Amsterdam: Boom.

Howarth, S and J. Jonker (2007). 'Powering the hydrocarbon revolution, 1939–1973'. In J.L. van Zanden, J. Jonker, S. Howarth and K. Sluyterman (eds.) *A history of Royal Dutch Shell.* New York: Oxford University Press.

Huguenot van der Linden, A. (29-07-1967). 'Landbouw legt het af tegen groei van wereldbevolking als gewas niet wordt beschermd' in *De Telegraaf.*

Kennedy, J.C. (1995). *Nieuw Babylon in aanbouw: Nederland in de jaren zestig.* Amsterdam: Boom.

Lans, J. van der and H. Vuijsje (2003). *Lage landen, hoge sprongen: Nederland in de twintigste eeuw.* Wormer: Inmerc.

Lintsen, H.W. (2005). *Made in Holland: Een techniekgeschiedenis van Nederland [1800–2000].* Zutphen: Walburg Pers.

Lintsen, H.W. (eds.) (2013). *Tachtig jaar TNO.* Delft: TNO.

Oldenziel, R. (2001). 'Het ontstaan van het moderne huishouden: toevalstreffers en valse starts, 1890–1918'. In J.W. Schot, H.W. Lintsen, A. Rip and A.A. Albert de la Bruhèze (eds.), *Techniek in Nederland in de twintigste eeuw – Deel IV.* Zutphen: Walburg Pers.

Rooy, P. de (2014). *Ons stipje op de waereldkaart: De politieke cultuur van modern Nederland.* Amsterdam: Wereldbibliotheek.

Schuyt, K. and E. Taverne (2000). *1950: Welvaart in zwart wit.* Den Haag: Sdu Uitgevers.

Zanden, J.L. van and R.T. Griffiths (1989). *Economische geschiedenis van Nederland in de twintigste eeuw.* Utrecht: Het Spectrum.

Part III: The Great Turnabout 1970–2010

Chapter 17
The Point of Departure Around 1970: Overabundance and Discontent

Frank Veraart

Contents

Abstract Around 1970 anti-authoritarian groups rose up against industrialisation and the development of welfare. They were the vanguard of a broader societal sentiment. At the end of the 1980s environmental problems were at the core of social and political debate Despite increasing worries the pattern of consumption barely changed. Contrasting the monitor for 1970 with that of 2010, this chapter sketches the growth of material welfare and the development of quality of life. Smoking, overweight and unemployment became the new societal challenges. The domestic consumption of energy continued to grow. In this period pressure on natural capital, both domestic and foreign, increased dramatically. The Netherlands continued to be dependent on foreign lands for important material flows – in some cases to an extreme extent. 80% of Dutch grain, for example, was still imported. These developments led to shifts in the sustainable development of the Netherlands to foreign countries. The following chapters analyse the societal dynamics in the in the chains of agriculture and foods (Chap. 18), construction and building materials (Chap. 19) and energy and plastics (Chap. 20)

Keywords Protest movements · Protest · Monitor · Well-being · Natural capital · Internationalisation · Sustainability

---

This chapter is written by Frank Veraart with contributions by Fred Lambert.

© The Author(s) 2018
H. Lintsen et al., *Well-being, Sustainability and Social Development*,
https://doi.org/10.1007/978-3-319-76696-6_17

17.1   The Mushroom of the New Society

> How does a new society emerge from the old one? Like a mushroom on a rotting tree trunk.
> An alternative society emerges from the subculture of the existing order. The underground
> society of the rebellious youth surfaces and begins to manage itself independently of the
> still ruling authorities. This revolution is now occurring...

On February 5th 1970 the Amsterdam Gnome Party (*Kabouterpartij*) led by the
former Provo activist Roel van Duijn, announced the establishment of the Orange
Free State (*Oranje-Vrijstaat*). This had its own anthem, its own constitution and, via
the Gnome Party's city council seat, its own 'embassy' to the regular administration. Van Duijn and his comrades no longer wanted to remain in the opposition:

> ...This is the end of the underground, of protest, of demonstrating; from now on we devote
> our energy to building an anti-authoritarian society. We will take what we can use from the
> old society: knowledge, socialist ideals and the best of the liberal traditions. The mushroom
> of the new society feeds itself on the juices of the rotting tree trunk until it is
> decomposed.[1]

Twelve 'peoples departments' were the administrative units that were to put the ide-
als of the new state into practice. For example, the 'people's department of public
works' was charged with cultivating parks and green spaces and with demolishing
highways. The department of 'environmental hygiene' devoted itself to restoring
biological equilibrium and the department 'for need satisfaction' aimed at building
an 'alternative economy that did not have profit as a goal, but the satisfaction of
human needs.' The Orange Free State was anti-authoritarian and pacifistic.[2]

The Gnome movement was rooted in Provo, the anarchist movement that tarted
public administration and authority with provocative, often one-off, demonstrations
in the second half of the 1960s. Provo described itself as a 'first, furious reaction to
the emergence of a technocratic mass-culture.'[3] Provo agitated against the 'petty
folk,' the greedy, hard-working, boring masses devoid of fantasy. The ruling elite
had succeeded in suppressing the rebelliousness of workers with stimulants, drugs
and amusement.[4]

> The proletariat has subjected itself to its political leaders and its TV. It has melted together
> with the old bourgeoisie into a big grey petty mass of addicted consumers.[5]

Demonstrations targeted the monarchy and agitated against modern addictions like
smoking, watching television and the automobile. The movement acquired international notoriety with the detonation of smoke bombs during the marriage of crown-
princess Beatrix. At the end of the sixties the movement lost its momentum. Provo

---

[1] Tasman, *Louter Kabouter, Kroniek van een Beweging 1969–1974* (Amsterdam 1996), 401.

[2] Tasman, *Louter Kabouter*, 403.

[3] Roel van Duijn, *Provo, de Geschiedenis van de Provotarische Beweging, 1965–1967* (Amsterdam 1985), 7.

[4] James C. Kennedy, *Nieuw Babylon in Aanbouw, Nederland in de Jaren Zestig* (Amsterdam 1995), 133.

[5] Roel van Duijn in *Buiten de Perken*, cited in Tasman, *Louter Kabouter*, 18–19.

became just another part of 'rebellious Amsterdam' together with squatters and revolutionary students. This was even exploited as a tourist attraction. As part of a guided tour, and for a fee, tourists could let themselves be harangued by Provos. Authorities gave Provo a podium and took some of its criticisms to heart. This took the sting out of their actions. Provo became a victim of this 'politics of assimilation' and disbanded itself in 1967.[6]

Parts of Provo survived in the Gnome movement. The founder, Roel van Duijn, cherished the ideal of restoring harmony between culture and nature and of searching for the 'ideal marriage between city and countryside.' Humans should see nature as their ally and not as an opponent that had to be conquered. With the founding of the Orange Free State, the Gnome movement wanted to give shape to societal changes and to work out prototypes of alternative ways of life. The first National Squatters day – May 5th 1970 – was a manifestation to demand a solution of the housing shortage. Gnome stores became the urban outlets for organic foods. The Gnome movement acquired a national following in more than sixty Gnome groups. Gnomes participated in the 1970 elections for municipal councils in fifteen places. Besides Amsterdam, Gnomes won seats in the city councils of Leeuwarden, The Hague, Alkmaar, Leiden, Arnhem, and Amersfoort.

But the Gnome movement was not destined for a long life. Internal conflicts took their toll and its influence declined rapidly after 1971. The movement was a typical sign of the times. Experiments and political demonstrations brought the problematic issue of well-being into the public limelight. Many other initiatives followed in the 1970s, each with its own perspective on dealing with nature, the environment and energy. In 1969, Otto Munters, a student at Leiden University, founded the ecological commune, 'De Hobbitstee' in the village of Wasperveen in Drenthe. Students at Wageningen Agricultural University founded the 'Farmers' Group' (*Boerengroep*) in 1971, in which they worked together with small farmers in the search for a more equitable distribution of incomes for farmers within and without the European Community. They also mobilized protests against land reapportionments.[7] In Riethoven in the province of Brabant, the journalist Sietz Leeflang and his wife Anke founded the environmental-experimental farm 'The Small Earth' (*De Kleine Aarde*) in 1972. This project developed experiments with biodynamic agriculture and alternative energy production.[8]

The environment became one of the most important societal issues. In the voter survey held prior to the national elections of 1971 and 1972, this issue headed the list of problematic issues (Table 17.1). Environmentalism was also embraced by

---

[6] Kennedy, *Nieuw Babylon in Aanbouw, Nederland in de Jaren Zestig*, 136.

[7] D. Strijker and I.J. Terluin, 'Rural Protest Groups in the Netherlands' in *Rural Protest Groups and Populist Political Parties*, ed. D. Strijker, G. Voerman, and I.J. Terluin (Wageningen 2015).

[8] Leeflang was a former employee at Philips and worked as science editor for the national newspaper *Algemeen Handelsblad*. He was a prominent critic of nuclear energy and wrote critical articles on the petrochemical industry. Dick Hollander, 'Tegen Beter Weten In'. *de Geschiedenis van de Biologische Landbouw En Voeding in Nederland* (Hurwenen 2012), 99. Egbert Tellegen and Jaap Willems, eds., *Milieu-Aktie in Nederland* (Amsterdam 1978), 155–59.

Table 17.1 Societal problem areas according to the voters' survey

	Unemployment/ jobs	Housing	Environment	Politics	Prices/ Living expense	Economy and financial problems	Defense, war, peace	Crime and violence	Public health	Minorities/ Refugees	Social security/ Welfare services
Most important problems in our country (first answer) as percentage (total = 100%)											
1967	29	21	1	11	8	6	1	4	–	–	–
1971	3	25	19	2	4	–	1	2	2	2	3
1972	10	14	17	3	12	–	1	1	0	2	3
1977	54	5	3	1	6	–	0	3	3	1	3
1981	50	6	2	2	2	4	3	3	0	2	2
1982	60	1	1	4	0	5	6	3	0	2	2
1986	41	0	2	3	0	0	11	6	0	3	5
1989	17	1	43	3	0	0	1	5	3	3	3
1994	24	1	5	3	0	4	1	5	2	– (26)	10
1998	6	1	5	2	0	0	–	16	15	6 (11)	5
2002	1	4	10	2	–	0	–	2	18	0 (1)	8
2003	4	4	12	1	–	4	3	1	16	0 (1)	7
2006	4	1	2	4	0	4	1	6	11	7 (13)	4
Multiple answers possible (percentage of respondents):											
1986	61	n.a.	8	8	n.a.	19	n.a.	16	13	9	16
1989	32	n.a.	58	7	n.a.	7	n.a.	14	12	7	7
1994	41	n.a.	16	12	n.a.	15	n.a.	23	11	51	30
2002	4	4	12	10	n.a.	7	–	22	57	40	4

2006	9	4	8	15	21	12	2	22	23	36	15
2010	10	7	5	19	11	38	1	22	20	22	12
2012	20	9	4	13	22	47	–	13	19	14	14

Note: numbers in brackets in 1994-2006 first choices refer explicitly to category 'refugees' in other years number refer to category 'minorities'

Sources: 1967: Aarts et al. 1992, 68 mentioned in H.T. Siraa, A.J. van der Valk, and W.L. Wissink, *Met het Oog op de Omgeving*, Den Haag 1995, 243.1971–2006: Bojan Todoseijevic, Kees Aarts, and Harry van der Kaap, "Dutch Parliamentary Election Studies," (The Hague 2010), 123. Sources multiple answers: 1986–1994: CBS: Statistisch Zakboek 1995, 401; 2002: G.A. Irwin, J.J.M. Holsteyn, and J.M. de Ridder, 'Dutch Parlementairy Election Study 2002–2003', (Amsterdam 2005), 60-67. 2006–2012: CBS – Nationale problemen volgens stemgerechtigden

political parties. Local initiatives and social concern acquired extra momentum with the publication of *Limits to Growth*, the report of the Club of Rome, that appeared in 1972 and of which half the copies printed worldwide were sold in the Netherlands.[9]

Around 1970 the individual pleasures of modernisation came into conflict with collective needs. In the early 1970s, the Social and Cultural Planning Office noted the conflicts in the domain of the living environment

> Recent developments, to put it in a nutshell, boil down to a scarcity of space (congestion) and increasing erosion of the quality of space as a living environment. Virtually everyone is affected by these developments: it is almost impossible to escape the smells, noise, traffic congestion, 'horizon pollution' and overfull vacation destinations.[10]

According to the researchers, food security for the Netherlands came with a price: an enormous impact on the landscape as well as environmental degradation. The quality of nature and biodiversity declined precipitously due to developments in agriculture.[11] At the same time, increased recreation stimulated a broader appreciation of natural values and of the experience of landscapes. Land was no longer seen as solely a means of production in the food chain, but acquired new functionalities like nature and recreation. After massive bird mortality at the end of the 1960s, the pro-nature associations succeeded in bringing the negative impacts of agrarian developments into the limelight. Debates on pesticides produced the first rents in the fabric of the public-private agricultural bastion.

The environmental aspects of agricultural production also led to questions regarding food safety. Were we not busy poisoning ourselves? Consumer associations took the environmental aspects as the point of departure for a critical analysis of the food supply chain. Did we know where our food came from, what was added to it, and whether all this was healthy for us? – these were the questions. The same kinds of questions were also raised in and around the industrial zones of the Limburg mining region, the Rijnmond area and around the North Sea Canal. Did the smells, the soot pollution and the smog not compromise the health of nearby residents?

Concern about the environment and nature would wax and wane in subsequent decades. It declined in the course of the 1970s and 1980s, only to return with great force at the end of the 1980s and once again at the start of the present century. The results of the environmental movement were also variable. In some domains like sustainable energy and combating air pollution the Netherlands was quite advanced compared to other European countries. After 2000, the Netherlands fell behind in

---

[9] Jacqueline Cramer, 'Milieu', *Elementaire Deeltjes 16* (Amsterdam 2014), 14.

[10] "Sociaal Cultureel Rapport 1974" (Den Haag 1975), 83.

[11] There is little data on the development of biodiversity in the long term. Significant time series data is available for only a few animal and plant species. There is also data on the prevalence of flora and fauna in specific areas. The latter clearly show that the geographical range of many species is getting smaller. In 1989 a report entitled *The State of Nature 2* concluded that: 'On the average, rare species are getting rarer and common species are becoming even more common or are stable. On a regional level this is causing a decline in diversity. Overall at a national level there is an ongoing decline in biodiversity.' cited in R.J. Bink et al., 'Toestand van de Natuur 2' (Wageningen 1994), p. 11; Jan Luiten van Zanden and S.W. Verstegen, *Groene Geschiedenis van Nederland* (Utrecht 1993), 81.

the implementation of alternative energy sources. Whence this variable performance? What were initiatives taken and by whom? What were the drivers and what were the obstacles? In this section we examine the difficult transformations in agriculture and nutrition, and the use of materials and energy between 1970 and 2015. Using the well-being monitor we first provide an overview of changes between 1970 and 2015.

17.2   Well-being 'Here and Now': Transition to a Postmodern Society, 1970 Versus 2010

Halfway through the 1970s a previously unheard of period of powerful economic growth came to an end. Labour-intensive industries, like the textile, tobacco and shoe manufactures suffered from increased foreign competition. Wages had increased substantially, so that the Netherlands was no longer a low-wage country. The growth potential of the most important key technologies, chemicals and electronics, was exhausted. This return to 'normal' coincided with international economic stagnation, reinforced by increasing oil prices and oil crises. The latter impacted in particular on sizeable energy-intensive industries, including chemical firms.[12] Originally the various crises had no direct effect. Exports and investments declined and fell behind international developments, but an increase in consumption nonetheless fostered economic growth. Unemployment remained low in the first few years. It was only at the end of the 1970s and well into the 1980s that the full impact was felt in the Netherlands.[13]

Between 1970 and 2010 the Dutch population increased from 13 to more than 16.5 million inhabitants (Table 17.2). The consumptive expenditures of the Netherlanders increased during this period. After the mid-1980s, in addition to household appliances and a car, Dutch households possessed ever more electronic gadgets like video and cd players and computers. The economic recession at the end of the seventies slowed these developments. The 1990s witnessed a new upsurge of economic prosperity that lasted until the financial and economic crisis of 2008. Increased consumptive expenditures were reflected in continually increasing material welfare in the household (Graph 17.1).

The economic decline of the 1980s had a major impact in the area of income inequality. The decline that commenced in the late 1970s was followed by an increase after the mid-1980s. The reason was the lowering of the minimum wage and the extension of welfare benefits during the recession of the 1980s, followed by the restructur-

---

[12] Jan Pieter Smits, 'Technologie, Productiviteit en Welzijn.' in *Techniek in Nederland in de Twintigste Eeuw, Techniek en Modernisering, Balans van de Twintigste Eeuw*, J.W. Schot et al. eds., vol. 7, 7 vols., *Techniek in Nederland in de Twintigste Eeuw* (Zutphen 2003).

[13] Jan Luiten van Zanden and R.T. Griffith, *Economische Geschiedenis van Nederland in de 20e Eeuw*, (Utrecht 1989), 255. Table 10.1.

Table 17.2 Dashboard well-being 'here and now,' 1970 – 2015

Theme	Indicator	Unit	±1970	±2010 (2000)	Evaluation of the change from a present-day perspective
Population		million inhabitants	13,0	16,5	
Material welfare and well-being					
Consumption, income	Consumptive expenditures per capita/constant prices	index (1850=100)	340	581	⇧
	Income inequality, general	Gini coefficient 0-1	0,36	(0,32)	⇧
	Gender income inequality	% difference hourly wage M/F	29%	19%	⇧
Subjective well-being	Satisfaction with life	Score 0-10	7,4	7,8	⇧
Personal characteristics					
Health	Life expectancy	years	74	81	⇧
Nutrition	Height (military conscripts)	cm	178	(183)	⇧
Housing	Housing quality	% slums	6	<1	⇧
	Public water supply	m³/capita	109	120	⇧
Physical safety	Murder victims	number per 100.000 inhabitants.	0,7	(1,1)	⬇
Labour	Unemployment	% workforce.	1,6	5,0	⬇
Education	Level of education	years	9	(11)	⇧
Free time	Free time	hours / week	47,9	44,7	⬇
Natural environment					
Biodiversity	MSA	% original biodiversity	66	(63)	⬌
Air quality	SO_2	kg SO_2/capita	21	4	⇧
	Greenhouse gas emissions	ton CO_2 capita	10,1	10,6	⬌
Water quality	Public water supply	m³/capita	109	120	⇧
Institutional environment					
Trust	Generalized trust	% population with adequate trust	?	67	?
Political institutions	Democracy	Democracy-index 0-100	39	(39)	⬌

Legend

⇧	Positive development
⬇	Negative development
⬌	Not positive/not negative
?	Unknown or irrelevant

Note: The numbers in brackets are from J.L. van Zanden et al. (ed.), *How was life? Global well-being since 1820* (OECD Publishing 2014) and relate to the year 2000. Numbers for these indicators – measured according to the same methodology – are not available for 2010
Source: See note 23 of Chap. 2

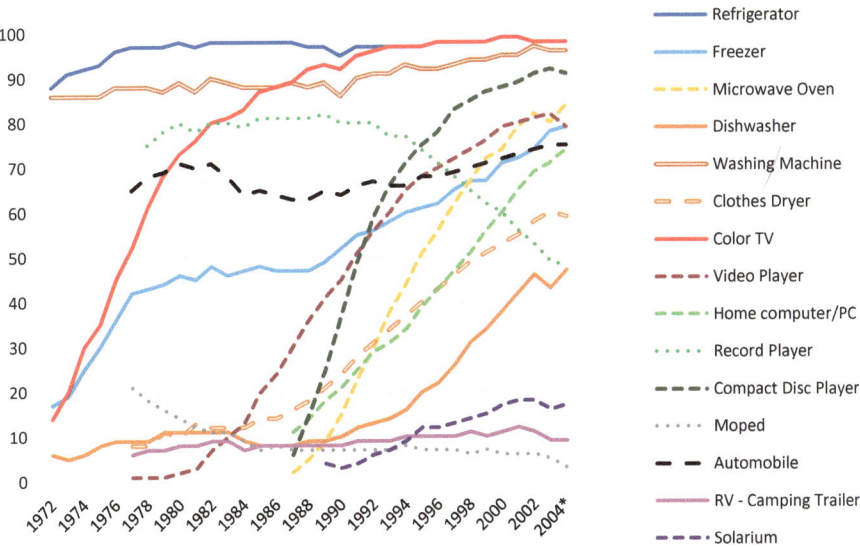

Graph 17.1 Ownership of consumer goods per household (in percent) 1970–2004
Source: CBS – Historie inkomen, vermogen en consumptie/Duurzame goederen; bezit naar huishoudenskenmerken

ing of social security and tax reforms in the 1990s.[14] The end result was that income inequality in 2010 stood at nearly the same level as it had in 1970. The discrepancy in wages between men and women showed a more favourable development. The difference between men and women declined, but the hourly wage difference of 18% in 2010 was higher than the average for the European Union as a whole (16% in 2010).[15]

In the personal sphere the differences between 1970 and 2010 show contradictory tendencies. In 2010, the average Netherlander lived longer, healthier and under ever better living conditions in comparison with 1970. Average life-expectancy and height increased. In 2010 life expectancy in the Netherlands reached 81 years. Men (78.5 years) were above the European average; women, at 82.6, just below. Increasing tobacco and alcohol consumption among women retarded increases in life-expectancy.[16] In general, alcohol use and overweight increased since the 1980s while smoking as an unhealthy habit declined (Graph 17.2).

This period was marked by continuing improvements in housing and utilities. Water, natural gas and electricity networks now penetrated into every corner of the

[14] Wiemer Salverda et al., 'Nederlandse Ongelijkheid Sinds 1980: Loonvorming, Overheidsbeleid En Veranderde Samenstelling van Huishoudens' in *GINI Growing Inequalities´ Impacts* (Amsterdam 2013); Wiemer Salverda et al., 'Growing Inequalities and Their Impacts In The Netherlands' in *Changing Inequalities and Societal Impacts in Rich Countries: Thirty Countries´Experience*, ed. Brian Nolan et al., vol. 20 (Oxford 2014), 459–87.

[15] Szilvia Borbély, *The Netherlands: Gender Pay Gap (GPG)* (Budapest 2016).

[16] Rutger Hoekstra and Jan Pieter Smits, 'Monitor Duurzaam Nederland 2009.' *Monitor Duurzaam Nederland* (Den Haag/Heerlen 2009), 53–55.

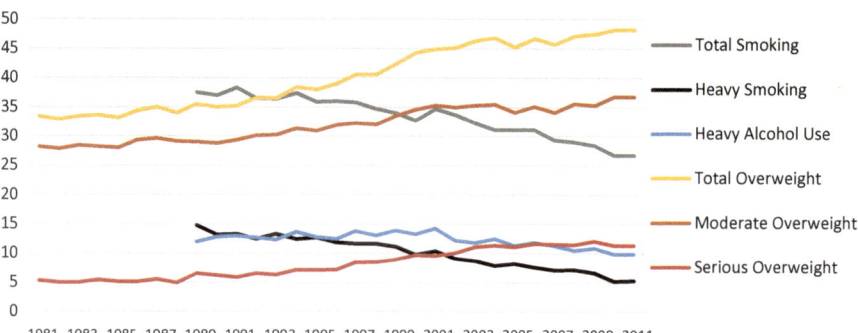

Graph 17.2 Development of unhealthy lifestyle as a percentage of the population 1981–2011
Heavy smoking: >20 cigarettes per day
Heavy alcohol use: 1 × per week, >6 glasses alcohol in one day
Overweight: Body Mass Index (BMI) > 25 kg/m²;
Moderate overweight: BMI: 25–30 kg/m²
Serious overweight: BMI: >30 kg/m²
Source: CBS: Health, lifestyle, health care demand and supply, mortality; from 1900

country. In the 1970s and 1980s, the last slums fell prey to urban renewal projects. Investments in education manifested themselves in among other things extension of compulsory schooling and a growing number of students in vocational and academic educational programs. Netherlanders continued to be longer and better educated.

The economic decline of the 1980s and the recession after 2008 had an equivocal effect on unemployment. After 1980, unemployment quickly exceeded the 'friction' level of 4%. In the first half of the 1980s more than 10% of the workforce was out of work – a percentage that declined only very gradually to about 6% in the early 1990s. A period of economic growth, especially in the information and communications sector and in services, was responsible for a period of low unemployment around the turn of the century. After that unemployment rates again increased gradually (Graph 17.3). The decrease in the amount of free time per week is striking; work in particular took up an ever larger part of the day.[17] Developments between 1970 and 2010 exhibited a deterioration in the areas of employment and free time.

Other indicators of the personal life-world also show negative developments. The chance of becoming a victim of murder, for example, increased.[18] Other kinds of criminality like theft also increased. Theft accounted for about three quarters of the total number of registered crimes.[19] The total number of registered crimes grew from 2,65,000 in 1970 to 1,218,000 in 2006 (Graph 17.4).

---

[17] Mariêlle Cloin et al., *Met Het Oog op de Tijd, Een Blik op de Tijdsbesteding van Nederlanders* (Den Haag 2013), 34–35.

[18] Paul Nieuwbeerta and Ingeborg Deerenberg, 'Trends in Moord en Doodslag, 1911–2002.' *Bevolkingstrends*, Centraal Bureau voor de Statistiek Bevolkingstrends, 1e kwartaal (2005): 56–63.

[19] Marcel Metze, *De Staat van Nederland, Op Weg naar 2000* (Nijmegen 1996), 139–54.

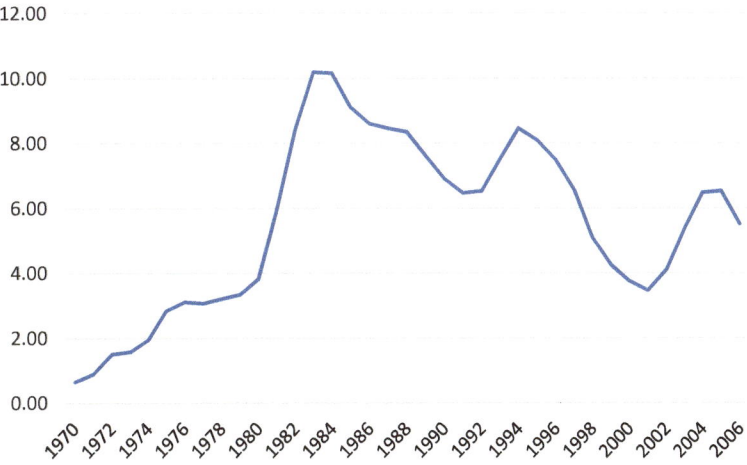

Graph 17.3 Levels of unemployment 1970–2006
Source: CBS – Beroepsbevolking; historie 1970–2006

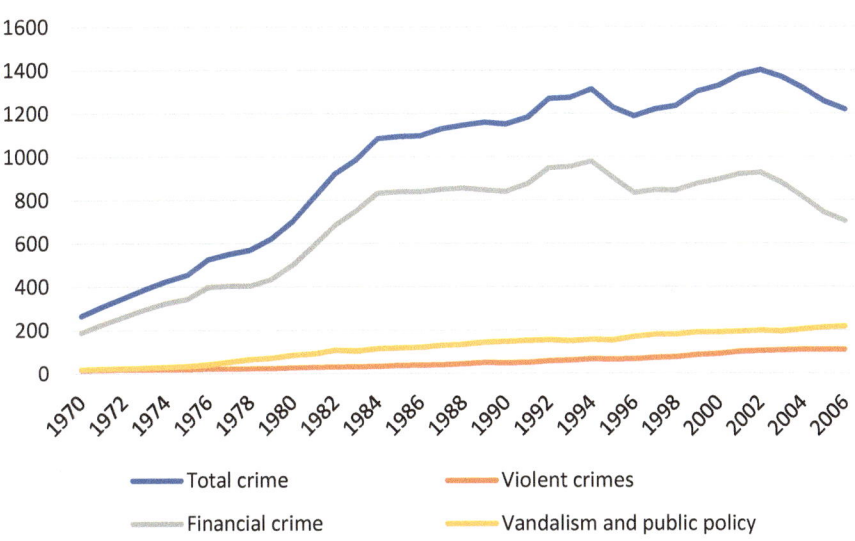

Graph 17.4 Crimes 1970–2006
Source: CBS – Legal Protection and Safety; History

With respect to the natural environment there was an improvement in the domestic situation, but at the same time an entrenchment of the international problems. Air and water quality recovered in the course of the 1980s and thereafter.[20] The decline

---

[20] CBS, PBL, Wageningen UR (2016). 'Vermesting in grote rivieren, 1970–2014' (indicator 0249, version 10, 13 April 2016). www.compendiumvoordeleefomgeving.nl. CBS, Planbureau voor de Leefomgeving en Wageningen UR (2012). Milieucondities in water en natuurgebieden, 1990 – 2010 (indicator 1522, version 04, 20 September 2012). www.compendiumvoordeleefomgeving.nl.

in biodiversity was arrested and after the 1990s there was even evidence of the beginnings of a turnaround, but in general the natural environment in the Netherlands continued to be under serious duress.[21] Per capita emissions of greenhouse gases continued to increase in this period. Measures to combat climate change did little more than stabilise emissions at 1980s levels. The consumption of fossil fuel energy and the per capita contribution to the global climate problem remained as high as ever. Dutch per capita emissions were higher than was the case in the surrounding countries of Belgium, Germany, France, Great Britain, Norway and Denmark.[22]

Comparing indicators of well-being 'here and now' for 1970 with those of 2010 reveals a variegated picture. The problem of extreme poverty, the most important issue of well-being and sustainability in the past, became totally irrelevant to the issue of quality of life in this period. One of the last remnants, the housing shortage and in particular the elimination of slum-dwellings, was solved in this period. Instead of extreme poverty, other issues now came to the fore: like criminality, work, nature, environment and climate change. Improvements were situated mostly in the individual life-world; in the more inclusive societal and global context, indicators exhibited a deterioration or at best a stabilisation.

17.3 Well-being 'Later': Material Growth in a Cleverer and Cleaner Country, 1970 Versus 2010

The economic structure of the Netherlands changed radically after the mid-1970s. A wave of bankruptcies marked the industrial transition. Industries like shipping, textiles and many branches of manufacturing began to founder in heavy weather. The six multinationals in the economy, including Philips and AkzoNobel, relinquished some of their pre-eminence in the national economy. New key technologies like biotechnology, medical technology and information and communications technology became driving forces. Employment opportunities were increasingly found in knowledge-intensive sectors. New investments were called for in knowledge development, education, innovations and infrastructures.

Economic capital developed favourably thanks to the strong growth in investments, though at the cost of an increasing national debt. The growing stockpile of knowledge had a positive effect on economic capital. Investments in research and development acquired their own labels. At the end of the seventies investments were being channelled into 'the information society,' a Pollyanna that by the nineties had

---

[21] CBS, PBL, Wageningen UR (2015). 'Trend fauna – alle gemeten soorten – Living Planet Index Nederland, 1990–2014´ (indicator 1569, version 02 , 29 oktober 2015). www.compendium-voordeleefomgeving.nl. PBL (2014). 'Natuur´ Website Balans van de Leefomgeving 2014. www.pbl.nl/balans2014. Planbureau voor de Leefomgeving, Den Haag.

[22] Data EU: Emission Database for Global Atmospheric Research (EDGAR) – CO2 time series 1990–2011 per capita for world countries (website: http://edgar.jrc.ec.europa.eu/overview.php?v=CO2ts_pc1990-2011) (consulted: 24-11-2016).

morphed into the 'knowledge economy.' Research expenditures by private firms and the government crowded close to the European average. In 2011 the Netherlands found itself in the European middle-bracket, spending a bit more than 2% of its gross domestic product on research.[23]

These investments also had an impact on human capital. Expenditures for higher education increased in the 1990s in step with the European average. Between 1990 and 2010 the number of highly educated Netherlanders increased from 19% to 28% of the population.[24] Thanks to investments and new constraints on compulsory schooling, the number of students following secondary and tertiary (higher vocational and university) education increased – despite the demographic decrease. In 2010 the Netherlands boasted a better educated and (insofar as this can be deduced from life-expectancy) a healthier workforce. Insecurity about employment, however, was greater. Unemployment was higher than in 1970. More flexible work situations via employment bureaus, temporary contracts and part-time work increasingly became the norm.

Indicators for social capital exhibited no shifts. But behind the figures remarkable changes were afoot. While the social climate around 1970 was influenced by 'left-wing, anti-authoritarian' youth groups and environmental groups dedicated to preserving nature and the environment, by 2010 the political climate was dominated by so-called 'right populism' a movement that was deeply sceptical about environmental problems and climate change and that championed new issues like Dutch identity, the migrant problem and the issue of integration. We shall come back to the consequences below.

Natural capital exhibited a bifurcation. On the one hand improvements could be seen. After the 1970s, concern for the environment and the natural surroundings increased. Dutch air and water quality improved and the decrease in biodiversity was turned around. On the other hand, the incessant growth of consumption and production meant undiminished demands on energy, land-use and raw materials. Although in the 1980s there was increasing social and political concern about interregional, international and global environmental problems like acid rain and climate change, energy consumption in this period was still based largely on the combustion of fossil fuels. In 2010 only 4% of the energy consumption was derived from renewable sources.[25] An energy transition was still very much in the initial phases. With that, this period is characterised by an unmitigated negative development as regards the emission of greenhouse gases. The improvement in natural capital thus took place above all on the domestic front (Table 17.3).

---

[23] CBS – Statistics Research & Development, 1970–1994 (via Statline), data R&D personnel en R&D investments. Walter Manshanden et al., 'De Staat van Nederland Innovatieland, R&D: Impuls Voor Economische Groei,' Special Issue, *Strategy & Change* (Den Haag 2013), 40–49.

[24] Data CBS Statline – Sociale Monitor; 1990–2011.

[25] André Meurink and Reinoud Segers, *Hernieuwbare Energie in Nederland 2014* (Den Haag/ Heerlen 2015), 18.

Table 17.3 Dashboard well-being 'later', 1970–2010

Theme	Indicator	Unit	±1970	±2010 (2000)	Evaluation of the change from CBS monitor 2018
Natural capital					
Energy	Energy consumption	TJ /capita	0.16	0.17	⬌
Non-fossil fuels	Domestic consumption	ton /capita	9.4	9.8	⬌
Biodiversity	MSA	% original biodiversity	66	63	⬌
Water	Public water supply	m³/capita	21	4	⬆
Air quality	SO_2 emissions	kg SO_2/capita	10.1	10.6	⬌
	Greenhouse gas emissions	ton CO_2/capita	109	120	⬆
Economic capital:					
Physical capital	Economic capital stock/capita	index (1850=100)	518	1046	⬆
Financial capital	Gross national debt	% gdp	48	59	⬇
Knowledge	Stock knowledge capital	index (2010=100)	30	100	⬆
Human capital:					
Health	Life expectancy	years	75	81	⬆
Labour	Unemployment	% workforce	1.6	5.0	⬆
Educational level	Schooling	years	9.0	(11)	⬆
Social capital:					
Trust	Generalized trust	% population with adequate trust	?	67	⬆
Political institutions	Democracy	democracy index 0-100	39	(39)	⬌

Legend

⬆	Positive development
⬇	Negative development
⬌	Not positive/not negative
?	Unknown or irrelevant

Note: The numbers in brackets are from J.L. van Zanden et al. (ed.). *How was life? Global well-being since 1820* (OECD Publishing 2014) and relate to the year 2000. Numbers for these indicators – measured according to the same methodology – are not available for 2010
Source: See note 23 of Chap. 2

17.4   Well-being 'Elsewhere': An International Trading Power, 1970 Versus 2010

Dutch support for the poorest countries continued to develop in a positive fashion. There was a substantive shift in development aid, in which the autonomy of the developing countries, but also Dutch industrial interests, acquired more leeway. Together with the Scandinavian countries, the Netherlands disbursed the most development aid per unit gross domestic product.[26]

Between 1970 and 2010 economic and material growth translated into an undiminished increase in the import of raw materials. The percentage of raw materials imported from abroad increased in this period from 49% to 53%. 12,500 kilograms of raw materials were imported per capita. This flow of goods imposed an enormous burden on foreign sources of supply as well as on domestic and foreign transport facilities (Table 17.4).

Table 17.4 Dashboard well-being 'elsewhere,' 1970–2010

Theme	Indicator	Unit	±1970 (2000)	±2010	Evaluation of the change from CBS monitor 2018
Material welfare					
Consumption, income	Development aid	% gdp	0.6	0.8	⬆
Natural capital					
Natural capital	Import of raw materials	ton/capita	8.6	12.9	⬇

Legend

⬆	Positive development
⬇	Negative development
⬄	Not positive/not negative
?	Unknown or irrelevant

Source: See note 23 of Chap. 2

---

[26] A. Nekkers and P.A.M. Malcontent, eds., *De Geschiedenis van Vijftig Jaar Nederlandse Ontwikkelingssamenwerking, 1949–1999* (Den Haag, 1999), 339–43.

17.5 Natural Capital and Material Flows: 1970–2010

A number of remarkable changes took place in the development of material flows between 1970 and 2010. Just as in previous periods, the flow of goods and products increased, in this case by no less than 60%. The growth was above all due to growth in biological (60%) and fossil (90%) raw materials. The most striking feature, however, was the enormous growth in the export of raw materials (230%) over against a minimal growth in import (20%) (see Table 17.5). A closer examination of the material flows can shed some light on these changes.

In absolute numbers this period experienced a substantial growth of raw materials. Per capita the growth of minerals and bio materials was modest. Solely the use of fossil resources increased with 50% (Table 17.5). This data concerns resources processed in the Dutch economy (i.e. domestic production and imports added together). In contrast data of domestic material consumption (i.e. domestic production and imports added together minus export) result in other conclusions. According to European reports the consumption of raw materials declined between 1970 and

Table 17.5 Raw materials in the Netherlands, 1970–2010 in kton

	1970	2010	Ratio 1910:2010
Bio raw materials:			
Gross available (kton)	42,400 kton	67,020 kton	1 : 1.6
Bio/capita (ton/cap.)	3.3 ton/cap.	4.0 ton/cap.	1 : 1.2
% import	22%	31%	1 : 1.4
% export	9%	23%	1 : 2.6
Mineral subsoil resources:			
Gross available(kton)	80,120 kton	95,570 kton	1 : 1.2
Mineral/capita (ton/cap.)	6.1 ton/cap.	5.8 ton/cap.	1 : 1.0
% import	38%	58%	1 : 1.5
% export	8%	17%	1 : 2.1
Fossil subsoil resources:			
Gross available (kton)	104,890 kton	199,630 kton	1 : 1.9
Fossil/capita (ton/cap.)	8.0 ton/cap.	12.0 ton/cap.	1 : 1.5
% import	69%	69%	1 : 1.0
% export	22%	47%	1 : 2.1
Total raw materials:			
Gross available(kton)	227,400 kton	362,220 kton	1 : 1.6
Raw materials/capita (ton/cap.)	17.5 ton/cap	21.9 ton/cap.	1 : 1.3
% import	49%	59%	1 : 1.2
% export	15%	35%	1 : 2.3

Remark: Gross available = domestic production + imports
Source: F. Lambert, *Massastromen in Nederland. In de jaren 1850, 1913, 1970, 2010* (researchrapport Technische Universiteit Eindhoven, oktober 2016)

2001 from 14.7 to 13.7 tons per capita.[27] National data show a decrease from 18 to 15.5 tons per inhabitant between 1996 and 2006.[28] These developments suggest a so-called de-materialisation of the Dutch economy. The reduction in the use of materials was above all due to a decreased consumption of mineral raw materials, the use of which declined between 2000 and 2004 from about 10 to 8 tons per inhabitant. In 2011 the Netherlands belonged to the group of European countries with a low per capita domestic consumption of materials.[29] In other words: the resource use per capita serving the economy (i.e. also serving export) tends toward stabilisation. In domestic material consumption per capita there seems a gradual dematerialisation. The situation is however more complex, the analysis of dematerialization ideally should also include complex final products (i.e. cars, machines, electronics). Possibly this dismisses dematerialisation trends, more research on this issue is necessary to come to final conclusions.[30]

Another conclusion is that the Netherlands remained dependent on foreign supplies for important material flows, in some instances to a considerable degree (see Table 17.6). Almost 80% of grain consumed was still imported from foreign sources. Foreign dependency increased for mineral raw materials. Due to its geological structure, the Netherlands had long been dependent on foreign sources for ores and stone. In addition, foreign shipments of gravel – a commodity that was plentiful in the Netherlands – also increased (for the explanation see Chap. 19). Due to the closure of the coal mines, electric power plants were stoked with foreign coals. Petroleum was largely imported from overseas. It was only in natural gas that the Netherlands was a net exporter. The Netherlands was able to supply merely 47% of its material needs domestically (2008).[31]

The effects of the industrial transition and the dematerialisation were visible in the material flows in two ways. A first indication was the relative decline in the processing of domestic raw materials and subsoil resources (see Table 17.7). Firms made progressively less use of domestic natural capital. This national uncoupling pertained particularly to processors of bio and fossil raw materials. These industries were fed by an increasing import of unprocessed raw materials from abroad. In the second place, total imports more than doubled in an absolute sense and exports almost tripled (see Table 17.8). Processed products formed the lion's share. The figures sketch the rough contours of a changing industry that oriented itself less and

---

[27] Helga Weisz, *Development of Material Use in EU-15: 1970–2001, Material Composition, Cross-Country Comparison and Material Flow Indicators* (Vienna 2006), 30.

[28] Sjoerd Schenau et al., *Milieurekening 2008* (Den Haag/Heerlen 2009), 44.

[29] Rita Bhagethoe-Datadin and Roel Delahaye, 'Materiaalstromen En Grondstofafhankelijkheid van de Nederlandse Economie,' in *De Nederlandse Economie 2013*, ed. Hans Langenberg et al., De Nederlandse Economie (Den Haag/Heerlen 2013).

[30] One of the problems is the issue double counting. The Netherlands exports raw materials and processed (semi-finished) products (i.e. parts for the car industries). These partially return with the import of final products (i.e. cars).

[31] Sjoerd Schenau et al., *Milieurekening 2008* (Den Haag/Heerlen 2009), 45–46.

Table 17.6 The ten most prominent raw materials (in kilotons) and the portion thereof imported (in percent), 1970–2010

		1970	2010	1970	2010
		Raw material (kton)	Raw material (kton)	Import (%)	Net import (%)
Bio raw materials	Total, of which	42.200	67.020	22	27
	Milk	8.330	11.290	1	4
	Grain	6.180	9.180	78	81
	Potatoes	5.040		0	
	Raw maize fodder		11.630		3
Mineral subsoil resources	Total, of which	80.120	95.570	38	41
	Clay	8.470		1	
	Sand	22.980	28.360	5	27
	Gravel	27.170	19.900	47	82
	Stone	10.390	17.460	68	81
	Ores		15.030		100
Fossil subsoil resources	Total, of which	104.890	199.630	69	69
	Coal	9.160	22.190	1	100
	Turf				
	Mineral oil	68.920	98.140	97	97
	Natural gas	25.650	77.930	0	23
	Total	**227.400**	**362.220**	**49**	**53**
		197.040	**311.110**	**44**	**64**

Remark: Gross available = domestic production + imports

Source: F. Lambert, *Massastromen in Nederland. In de jaren 1850, 1913, 1970, 2010* (researchrapport Technische Universiteit Eindhoven, oktober 2016)

Table 17.7 Proportion of domestic raw materials and subsoil resources processed in Dutch industry (in percent) 1970 and 2010

	1970	2010
	Industrial processing of domestic raw materials and subsoil resources	Industrial processing of domestic raw materials and subsoil resources
Bio-raw materials	59%	54%
Mineral raw materials	60%	54%
Fossil raw materials	62%	29%

Remark 1: Industry, excluding mineral and fossil subsoil asset extraction

Remark 2: In mineral resources application at building sites, ca. 50% of gravel and 20% of sand usages is assumed for concrete production

Source: F. Lambert, *Massastromen in Nederland. In de jaren 1850, 1913, 1970, 2010* (researchrapport Technische Universiteit Eindhoven, oktober 2016)

Table 17.8 Imports and exports of processed products in kilotons and in percentage of total imports and exports, 1970 and 2010

	1970	2010
Total imports	155.920 kton	359.150 kton
Imports processed products	44.640 kton	144.880 kton
Imports processed products (% of total)	29%	40%
Total exports	95.500 kton	279.500 kton
Exports processed products	62.020 kton	153.480 kton
Exports processed products (% of total)	65%	55%

Source: F. Lambert, *Massastromen in Nederland. In de jaren 1850, 1913, 1970, 2010* (researchrapport Technische Universiteit Eindhoven, oktober 2016)

less to domestic demand and increasingly became a link in international flows of trade and commodities.

The changes in the ten biggest flows of raw and processed products confirm this picture and reveal several core economic activities (see Tables 17.6 and 17.9). Among biological products, the material flows of raw maize fodder and mixed fodders are prominent. Both are largely for domestic consumption. These feed the strongly increased intensive livestock farming sector and its production of animal products like milk and meat. But the most striking change is the growth in the fossil raw materials and petrochemicals sector. In 2010 mineral oil and natural gas were by far the biggest material flows. The refining of mineral oil and gas also produces an enormous flow of organic chemical products like artificial rubber, plastics and other petroleum products. By contrast, the size of the flows of mineral subsoil resources decreased or grew only slightly. In 2010, in contrast to 1970, the Netherlands was no longer involved in big construction projects like the Delta Works or massive housing projects.

After 1970 the Netherlands commenced a new phase in its history and had to find an answer to new problems of well-being and sustainability. The country was confronted with the question how to combine the development of personal well-being with the collective interest in the natural environment and natural capital. It turned out to be a real struggle. We work this out in the following chapters. How could agriculture and food production achieve an equilibrium with natural values and sustainable nutritional patterns? How much room was there for the extraction of mineral resources? How should we deal with the globally shrinking sources of materials and energy? A complicating factor was the increasingly international scope of the issues, like acidification and climate change. These too demanded new answers in the subfields of biological, mineral and fossil material flows.

Table 17.9 The ten most prominent processed products (in kilotons) and the proportions thereof imported and exported (in percent), 1970 and 2010

		1970	2010	1970	2010	1970	2010
		Processed products (kton)	Processed products (kton)	Imports (%)	Imports (%)	Exports (%)	Exports (%)
Food processing industry	Various mixed fodders	13.020	13.978	25	30	6	28
Minerals processing Industry	Coarse ceramics (incl. bricks)	7.960	6.850	4	9	7	8
	Concrete products	11.600	7.600	5	17	4	7
	Concrete and asphalt	26.000	16.000	0	0	0	0
	Cement	6.930	7.210	33	40	2	6
Chemical industry	Petroleum products	75.330	113.570	16	56	56	67
	Inorganic chemicals	7.250		60		20	
	Organic chemicals		15.130		65		85
	Rubber and plastics		9.450		28		77
Metals and machinery industry	Raw steel	5.330	7.030	5	3	3	1
	Rolling mill products	8.020	11.460	41	59	35	57
	Iron slag	5.320		66		1	

Source: F. Lambert, *Massastromen in Nederland. In de jaren 1850, 1913, 1970, 2010* (researchrapport Technische Universiteit Eindhoven, oktober 2016)

Literature

Bhagethoe-Datadin, Rita, and Roel Delahaye. (2013). 'Materiaalstromen En Grondstofafhankelijkheid van de Nederlandse Economie.' In *De Nederlandse Economie 2013*, edited by Hans Langenberg, Rita Bhagethoe-Datadin, Frank Notten, and Marieke Rensman. De Nederlandse Economie. Den Haag/Heerlen: Centraal Bureau voor de Statistiek.

Bink, R.J., D. Bal, V.N. van den Berk, and L.J. Draaijer. (1994). 'Toestand van de Natuur 2.' Wageningen: Informatie- en KennisCentrum Natuur, Bos, Landschap en Fauna, Ministerie van Landbouw, Natuurbeheer en Visserij.

Borbély, Szilvia. (2016). *The Netherlands: Gender Pay Gap (GPG)*. Budapest: WITA GPG.

Cloin, Mariêlle, Andries van den Broek, Remko van den Dool, Jos de Haan, Joep de Hart, Pepijn van Houwelingen, Annet Tiessen-Raaphorst, Nathalie Sonck, and Jan Spit. (2013). *Met Het Oog Op de Tijd, Een Blik Op de Tijdsbesteding van Nederlanders*. Den Haag: Sociaal Cultureel Planbureau.

Cramer, Jacqueline. (1985) 'Milieu'. *Elementaire Deeltjes 16*. Amsterdam: AUP, 2014.

Duijn, Roel van. *Provo, de Geschiedenis van de Provotarische Beweging, 1965–1967*. Amsterdam: Meulenhoff.

Hoekstra, Rutger, and Jan Pieter Smits. (2009). 'Monitor Duurzaam Nederland 2009.' in *Monitor Duurzaam Nederland*. Den Haag/Heerlen: Centraal Bureau voor de Statistiek.

Hollander, Dick. (2012). 'Tegen Beter Weten In.' in *de Geschiedenis van de Biologische Landbouw En Voeding in Nederland*. Hurwenen: 4 Heuvels.

Irwin, G.A., J.J.M. Holsteyn, and J.M. de Ridder. (2005). 'Dutch Parlementairy Election Study 2002–2003.' NIWI-Steinmetz Archive. Amsterdam: Foundation for Electoral Research in the Netherlands, SKON.

Kennedy, James C. (1995). *Nieuw Babylon in Aanbouw, Nederland in de Jaren Zestig*. Amsterdam: Boom, 1995.

Manshanden, Walter, Marcel de Heide, Olaf Koops, and Tom van der Horst. (2013). "De Staat van Nederland Innovatieland, R&D: Impuls Voor Economische Groei." Special Issue. Stratagy & Change. Den Haag: The Hague Centre for Stratagic Studies en TNO.

Metze, Marcel. (1996). *De Staat van Nederland, Op Weg Naar 2000*. Nijmegen: SUN.

Meurink, André, and Reinoud Segers. (2015). *Hernieuwbare Energie in Nederland 2014*. Den Haag/Heerlen: Centraal Bureau voor de Statistiek.

Nekkers, J.A., and P.A.M. Malcontent, eds. (1999). *De Geschiedenis van Vijftig Jaar Nederlandse Ontwikkelingssamenwerking, 1949–1999*. Den Haag: SDU Uitgeverij.

Nieuwbeerta, Paul, and Ingeborg Deerenberg. (2005). 'Trends in Moord En Doodslag, 1911–2002.' *Bevolkingstrends*, Centraal Bureau voor de Statistiek Bevolkingstrends, 1e kwartaal, 56–63.

Salverda, Wiemer, Loes de Graaf-Zijl, Christina Haas, Bram Lancee, and Natascha Notten. (2013). 'Nederlandse Ongelijkheid Sinds 1980: Loonvorming, Overheidsbeleid En Veranderde Samenstelling van Huishoudens.' in *GINI Growing Inequalities' Impacts*. Amsterdam: Gini-research.

Salverda, Wiemer, Christina Haas, Marloes de Graaf-Zijl, Bram Lancee, and Natascha Notten. (2014). 'Growing Inequalities an Their Impacts In The Netherlands.' In *Changing Inequalities and Societal Impacts in Rich Countries: Thirty Countries' Experience*, edited by Brian Nolan, Wiemer Salverda, Daniele Checchi, Ive Marx, Abigail McKnight, István György Tóth, and Herman van de Werfhorst, 20:459–87. Oxford: Oxford University Press.

Schenau, Sjoerd, Roel Delahaye, Bram Edens, Isabel van Geloof, Cor Graveland, Maarten van Rossum, and Kees Jan Wolswinkel. (2009). *Milieurekening 2008*. Den Haag/Heerlen: Centraal Bureau voor de Statistiek.

Siraa, H.T., A.J. van der Valk, and W.L. Wissink. (1995). *Met Het Oog Op de Omgeving, Het Ministerie van Volkshuisvesting, Ruimtelijke Ordening En Milieubeheer, 1965–1995*. Den Haag: SDU Uitgeverij.

Smits, Jan Pieter. (2003). 'Technologie, Productiviteit En Welzijn.' In *Techniek in Nederland in de Twintigste Eeuw, Techniek En Modernisering, Balans van de Twintigste Eeuw*, edited by J.W. Schot, A. Rip, H.W. Lintsen, and A.A. Albert de la Bruheze, Vol. 7. Techniek in Nederland in de Twintigste Eeuw. Zutphen: Walburg Pers.

Sociaal Cultureel Rapport 1974. Den Haag: Sociaal Cultureel Planbureau, 1975.

Strijker, D., and I.J. Terluin. (2015). 'Rural Protest Groups in the Netherlands.' In *Rural Protest Groups and Populist Political Parties*, edited by D. Strijker, G. Voerman, and I.J. Terluin. Wageningen: Wageningen Academic Publishers.

Tasman, Coen. (1996). *Louter Kabouter, Kroniek van Een Beweging 1969–1974*. Amsterdam: Babylon-De Geus.

Tellegen, Egbert, and Jaap Willems, eds. (1978). *Milieu-Aktie in Nederland*. Amsterdam: De Trommel/Vereniging MilieuDefensie.

Todoseijevic, Bojan, Kees Aarts, and Harry van der Kaap. (2010). "Dutch Parliamentary Election Studies, Data Source Book, 1971–2006." DANS Data Guide 7. The Hague: DANS.

Weisz, Helga. (2006). *Development of Material Use in EU-15: 1970–2001, Material Composition, Cross-Country Comparison and Material Flow Indicators.* Vienna: Eurostat/IFF-Social Ecology.

Zanden, Jan Luiten van, and R.T. Griffith. (1989). 'Economische Geschiedenis van Nederland in de 20e Eeuw'. *Aula 190.* Utrecht: Het Spectrum.

Zanden, Jan Luiten van, and S.W. Verstegen. (1993). *Groene Geschiedenis van Nederland.* Utrecht: Het Spectrum.

Chapter 18
Agriculture and Foods: Overproduction and Overconsumption

Frank Veraart

Contents

Abstract In the period 1970–2010, environment, landscape and healthy nutrition were core issues in the supply chain of agriculture and foods. Concern for the environment put pressure on agriculture. Since the 1950s, agriculture had oriented itself to ever higher levels of production. This had seduced farmers into extreme specialisations with consequences for the environment, both domestically and elsewhere. In order to reveal these dynamics, this chapter follows developments in cattle husbandry. In the early 1980s, European measures to restrain overproduction and increasing concern about acidification and over-fertilisation destabilised the established agricultural world.

In the wake of changing ecological insights, new issues emerged in regard to the landscape. Nature management and agricultural interests had to be harmonised. Though the Netherlands laid the basis for the European Natura 2000 directive, it encountered great difficulties in implementing them domestically. Food consumption also presented new challenges; for example the problem of overweight. The chapter analyses how government, private firms and consumers responded to this issue. Consumers appear to have great difficulty grasping the complex issues of sustainable and healthy nutritional patterns.

Keywords Agriculture · Over-fertilisation · Acidification · Overproduction · Landscape · Nutritional patterns · Overweight

© The Author(s) 2018
H. Lintsen et al., *Well-being, Sustainability and Social Development*,
https://doi.org/10.1007/978-3-319-76696-6_18

18.1   The Bolt Out of the Blue

On Friday, November 2 1984, the Minister of Agriculture Gerrit Braks announced:

> that, in view of the manure surplus situation in the Netherlands, and in anticipation of the passage of the new Manure Law and the Law on Soil Protection, it is necessary to forbid the creation and expansion of hog and poultry farms in the Netherlands, or in certain designated regions in the Netherlands.[1]

The interim law was to go into immediate effect and it hit the agricultural organizations like a steamroller. Many farmers rushed to acquire permits to expand their farms and their livestock. In various town halls in the province of Brabant, home to numerous hog farms, civil-servants burned the midnight oil in order to be able to handle all the applications.[2]

The government deemed quick action essential because of the introduction of a European milk quota, aimed at limiting the production of milk. In view of this constraint on dairy farming, many farmers were considering switching to hog and poultry farming. This threatened a huge increase in the volume of animal manure. The Netherlands faced a quandary: 'drowning in the milk or otherwise choking in a manure surplus' according to a parliamentarian who championed the adopted course.

Agricultural organisations denounced the sudden decision. In the words of the secretary of the Society for Agriculture in the province of Gelderland: 'This is certainly a fatal stab in the back for a great number of agrarian enterprises.' According to him it deprived countless young farmers of the opportunity to develop a future: 'In this way the countryside is doomed to death.'[3] The chairman of the Poultry Section of the Agricultural Board saw hog farming as the chief culprit. 'We have to prevent poultry farming from having to foot the bill.'[4] The North-Brabant Christian Farmers' Union (NCB) characterised the measures as 'a raid.' But not all the farmers were in mourning. As a pig breeder from Boekel in Brabant put it: 'There are more than enough pigs. The measure may have a positive effect on the price of pork.'[5]

More legal restrictions in subsequent years provoked many agricultural enterprises to switch to other types of animal husbandry. In 1989, due to manure regulations, the number of pigs had increased by 28% and that of chickens by 16% relative to 1984. The number of sheep had also increased significantly. After the introduction

---

[1] Koninklijke Boodschap, 'Verbod tot vestiging en uitbreiding van varkens- en pluimveehouderijbedrijven in Nederland dan wel bepaalde delen daarvan (Interimwet beperking varkens- en pluimveehouderijen)' Kamerstuk Tweede Kamer 1984–1985 kamerstuknummer 18695 ondernummer 1, 's Gravenhage, 1 november 1984.

[2] Frits Bloemendaal, *Het Mestmoeras* (Den Haag: SDU Uitgeverij, 1995), 7.

[3] P. Jongeling et al. 'Opzetten nieuwe varkens- en pluimveebedrijven verboden' in *Nederlands Dagblad: Gereformeerd Gezinsblad*, 03-11-1984 p. 1.

[4] Anoniem. 'Landbouwschap volslagen verrast', *Leidsch Dagblad*, 3 november 1984, p. 9.

[5] Max Pauwmen, 'Boeren kwaad over beperking varkens- en pluimveeteelt', in *NRC Handelsblad*, 03-11-1984, p. 11.

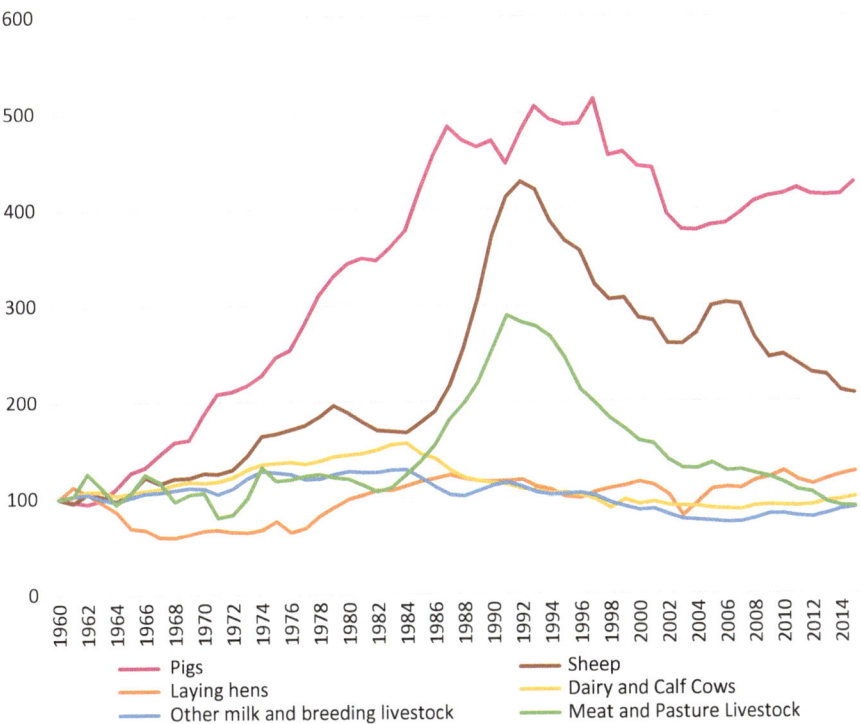

Graph 18.1 Development of most common livestock (excluding broilers) 1960–2015 (1960 = 100)

Source: CBS – Landbouw; from 1851

of the 'super-levy' in 1984 the number of dairy cows decreased and the number of cows destined for beef production increased (see Graph 18.1).

The measures taken by the Minister of Agriculture without prior consultation with the farming world betrayed the changing relationship between the government and the agrarian sector after the mid-1980s. It turned out to be one of the first nails in the coffin of the 'Green Front' with the Agricultural Board (*Landbouwschap*) as a powerful actor. Specialisation within agriculture had also served to fragment the sector into conflicting interest groups. In this respect the recriminations among different livestock sectors were revealing. The strong corporatist character of the agrarian sector as a whole with strong bonds among farmers, administrators and civil-servants began to fall apart, as the further analysis of the manure problem below will show. In addition, agriculture in this period was also confronted by new issues like nature, landscape and biodiversity to which we shall also devote attention. Tensions were not limited to agriculture. In this period the entire food chain was under pressure. The availability of great quantities of cheap food raised new questions for public health, like obesity and heart and circulatory diseases. What handholds were found to combat the new welfare diseases?

18.2   The Shifting Perspective on Agriculture

18.2.1   Specialisation and Over-Fertilisation

The modernisation of agriculture after the Second World War was a great success. Mechanisation, rationalisation and the structural policies of the government led to an increase in yields per hectare and per farm. It proved possible to maintain farmers' incomes at satisfactory levels. Agricultural organisations and the government were happy bedfellows and aimed at increased production and the stimulation of export. The problem of surpluses was accepted as a given. Throughout the 1970s Dutch agriculture was oriented entirely to competition with the other member states of the European Community.[6]

Further specialisation was the guiding principle. Livestock farmers abandoned the mixed farm, that provided for the needs of its own livestock. In 1961 the government rescinded the 'poultry regulations' that had been introduced in 1953 to limit the number of animals to the amount of land available. This opened the door for the so-called non-land-coupled husbandry of especially chickens and pigs. This livestock was now maintained on the basis of purchased (mostly imported) feed. These changes led to an enormous increase in livestock for meat production. The number of broilers increased from 4,5 million in 1960 to almost 50 million in 2015. In the same period the pig population increased from 3 to almost 13 million. (see Table 18.1)

Specialisation also transformed other agricultural sectors. The sale of vegetables and greenhouse export crops like tomatoes and peppers increased, as did that of flower bulbs, while the national fruit production decreased.[7] In the dairy sector, refrigerator tanks, box stalls and milking and feeding robots further industrialised farming. The

Table 18.1   Development of livestock 1960–2015 (× 1000 animals)

	1960	1970	1980	1990	2000	2010	2015
Goats	.	13	.	61	179	353	470
Laying hens	37,886	25,315	37,455	44,320	44,036	47,904	47,684
Broilers	4525	30,060	38,609	41,172	50,937	44,748	49,107
Dairy and calf cows	1628	1896	2356	1878	1504	1479	1622
Bulls	25	37	44	43	37	22	26
Other milk and breeding cattle	1495	1650	1908	1686	1299	1225	1324
Meat and pasture livestock	281	298	336	718	446	322	252
Veal calfs	78	434	582	602	783	928	909
Sheep	456	575	858	1702	1305	1130	946
Pigs	2955	5533	10,138	13,915	13,118	12,255	12,603

Source: CBS - Landbouw; from 1851

---

[6] Jan Bieleman, *Boeren in Nederland, Geschiedenis van de Landbouw, 1500–2000* (Amsterdam: Boom, 2008), 476.

[7] Bieleman, 543–66.

introduction of plastic foils made ensilage an attractive option. With this technique the annual yield of raw fodder could be doubled compared to traditional haymaking. New technologies and materials accommodated an increasingly intensive exploitation of farming land.[8]

The strategy of specialisation and export was of a piece with the common agricultural policy developed within the EEC in the 1960s. A complex funding system shielded a stable internal market from global price fluctuations, among other things by guaranteeing minimum prices for agricultural products. This system worked well in times of scarcity, but by the end of the 1970s had lost its polish. New policy instruments were needed to put an end to the unprecedented growth of milk lakes and butter mountains. In addition, global negotiations under the umbrella of the General Agreement on Tariffs and Trade (GATT) put European policy under pressure.[9] By the end of the 1970s a European quota on milk production was in place. Farmers had to pay a levy on the production of excess milk. The levies set bad blood between domestic policy makers and agricultural organisations. And new environmental constraints only increased the pressure on the agricultural world

The shifting perspective was clearly visible in relation to the problem of over-fertilisation. At the end of the 1960s the Institute for Soil Fertility had already expressed concern about over-fertilisation of the soil. In 1972, the Foundation for Nature and Environment compared the bio-industry with the mythological Augean Stables and lambasted the effects of unregulated spreading of animal manure, especially on the sand grounds. The agricultural world responded irritably and pronounced it a shame that a prosperous sector should be hindered in its development.

The Ministry of Agriculture sought succour in technical solutions. It played down the problems, among other things by pointing the finger at phosphates associated with the use of laundry detergents. The message was that agriculture was not the sole polluter. Nonetheless, the ministry established manure banks. Matching the supply of manure with the demand was only partly successful. In the event, it appeared that by expanding the cultivation of fodder corn the government and farmers could kill two birds with one stone. Not only was this crop easy to cultivate and harvest and excellent as fodder, it also grew exceptionally well on the abundantly fertilised sand grounds. Between 1970 and 2010 the acreage of fodder corn grew steadily from 6000 to 231,000 hectares.

New knowledge about the relationship between acid substances in the air and soil totally undermined the permissive policy of the Ministry of Agriculture. In 1982 researchers at the Wageningen Agricultural University published new findings about the role played by ammonia in acidification.[10] These findings added a new dimension

---

[8] Bieleman, 531–33.

[9] Bieleman, 511; A.H. Crijns, 'De Grote Ommekeer in de Agrarische Sector', in *Geschiedenis van Noord-Brabant, Dynamiek En Expansie, 1945–1996*, ed. H.F.J.M. van den Eerenbeemt, vol. 3, 3 vols., Geschiedenis van Noord-Brabant (Boom, 1997).

[10] N. van Breemen et al., 'Soil Acidification from Atmospheric Ammonium Sulphate in Forest Canopy Throughfall,' *Nature 299*, no. 5883 (October 7, 1982): 548–50.

to the manure problem.[11] The newly appointed Minister of the Environment, Pieter Winsemius, saw acidification

'as one of the biggest problems we have to deal with. … New data … show that acidification in the Netherlands is caused for 20% by NO_x, for 45% by SO_2 and for 35% by NH_3. … We know where the handles … are located, namely road traffic, electrical power plants and most probably intensive livestock husbandry.'[12]

Winsemius hesitated due to a brief controversy at Wagening about the data being presented, but possibly also because he feared reactions from the agricultural sector. In the end, together with prime-minister Ruud Lubbers, Winsemius was able to convince the Minister of Agriculture of the seriousness of the situation. The latter, in the person of Gerrit Braks, took the unexpected and in agricultural circles incomprehensible measure to limit the number of hogs and chickens – the affair with which we began this chapter. The incomprehensibility was among other things rooted in the apparently one-sided manner in which the measure had been conceived. Moreover there had been no prior consultation with the agricultural organisations with the result that the ministry now appeared to be opposed to the powerful agricultural lobby. For the first time, agricultural organisations felt their influence waning in The Hague.

The incident also revealed the vulnerability of agrarian cooperation. There was no consensus in agricultural circles about possible solutions. Specialisation had fomented different interests in different agricultural sectors. The lack of unity finally broke the power of the Agricultural Board and made it possible to introduce new measures that privileged ecological interests above those of production. But, though the power of the Green Front and the Agricultural Board might be broken, a divided sector with many new specific interest groups made the introduction of policy measures more difficult. Due to fraudulent practices by municipalities and farmers, dispensations and lack of cooperation by administrative organs, the number of pigs increased substantially even after restrictive measures came into force.[13]

The Ministry of Agriculture continued to put its faith in technical and bureaucratic solutions that could tackle the manure problem while at the same time permit the growth of animal husbandry. This approach was only partly successful. From 1987 on, funds were pumped into the industrial processing of manure. Promest in Helmond focused on the processing of pig manure. After several difficult years the factory folded in 1990. The industrial processing of chicken manure was more successful. In addition to the production of manure pellets, the late 1990s saw the introduction of manure incinerators. In this process, nitrogen was released into the

---

[11] G.H. Dinkelman, *Verzuring En Broeikaseffect, de Wisselwerking Tussen Problemen En Oplossingen in Het Nederlandse Luchtverontreinigingsbeleid (1970–1994)* (Universiteit van Amsterdam, 1995), 84–90.

[12] P. Winsemius in *Vaste commissie voor milieubeheer*, 1983, p. 34, geciteerd in H. Dinkelman, *Verzuring en Broeikaseffect, de Wisselwerking Tussen Problemen en Oplossingen in het Nederlandse Luchtverontreinigingsbeleid (1970–1994)*, (Universiteit van Amsterdam, 1995), 90.

[13] John Grin, Jan Rotmans, and Johan W. Schot, *Transitions to Sustainable Development, New Directions in the Study of Long Term Transformative Change* (New York: Routledge, 2009), 295–97.

atmosphere and phosphate was reclaimed from the ashes, to be used for non-agricultural purposes.[14]

In 1991, goaded by the European Nitrate Directive, an accounting system for the emission of minerals was set up. In 1998 a mineral denotation system (MINAS) was introduced. In addition, the scale of production was fixed per farm in terms of so-called 'animal-allotments' for pigs (1999) and poultry (2001). In 2002 this was followed by the Manure Disposal Agreements (MAOs). These administrative measures were coupled to financial sanction mechanisms. These measures, however, appeared inadequate to meet the targets of the European Nitrate Directives. In 2003 the Minister of Agriculture, Veerman, characterized the problem succinctly: 'We import feed, we export pigs and keep the mess here. That system has seized up.'[15] In 2006 under European pressure a new Fertiliser Substance Law came into effect. Among other things the law mandated a meticulous accounting of the fertilisation of the soil and of soil use. Thanks to this detailed accounting, Dutch agriculture was granted a European exemption, a so-called derogation, with respect to its manure use.[16]

In addition to measures like these, after 1997 the Ministry of Agriculture also developed plans to reduce the size of especially the hog-farming sector. By purchasing excess livestock, a so-called 'warm restructuring' was initiated. Dramatic epidemics like swine fever (1997, 1998), hoof and mouth disease (2001), and bird flu (2003) accelerated the warm restructuring and eventuated in a reduction of the size of pig and poultry stocks (see Graph 18.1 and Table 18.1). In 2001 the manure problem seemed solved, but this turned out to be only a temporary victory.

With the Fertiliser Substance Law and the derogation arrangement breathing down its neck, the Dutch cattle husbandry sector remained under continuous pressure to meet the targets of the environmental directives. This was a precarious balancing act. In 2014 new European derogation guidelines were negotiated, which emphatically included the dairy sector. The suspension of the milk quota in 2015 felt like 'Liberation Day' for the dairy sector. Many of the dairy farmers had invested in new stalls for more animals. Despite warnings by the ministry and environmental organisations, increased manure production quickly became a problem. The Netherlands even threatened to lose its favourable exemption arrangement and it would be unable to satisfy the demands of the strict general European directives. New ad hoc measures by the ministry and negotiations with the dairy farmers were supposed to bring profitability for the dairy sector and enforcement of environmental directives into alignment again.

The manure question is illustrative of the way in which, from the 1970s on, specialisation and specific interests caused the agricultural conglomerate to fall apart into smaller units, each with its own interests. The Ministry of Agriculture, from the perspective of the farmers, became more of an opponent than the ally it had always been. The Ministry searched for an ecological balance in food produc-

---

[14] Ben Hermans, *De Mestmarathon, Kroniek van Ruim 42 Jaar Nederlands Mestbeleid* (Utrecht: Natuur & Milieu, 2016), 20.

[15] Minister Veerman cited in Hermans, 25.

[16] Hermans, 28.

tion and created strict frameworks within which agriculture was to be practiced. At the same time the ministry continuously kept an eye out for ways to increase scale and profitability.

The disquiet and new directives turned out to provide fertile soil for other forms of agriculture. In crop farming sophisticated technologies were used to implement so-called 'precision agriculture' in which fertilisers and crop protection substances were applied at the right place and in the most optimal amounts.[17] In addition, some farmers followed a new path: organic farming. The number of certified organic farms grew from 835 in 1998 to 1412 in 2014. Organic acreage grew in the same period from 22,268 ha to 49,333 ha. In 2014 this was only a modest 2.7% of the total agricultural acreage.[18] The transition to a more sustainable – not necessarily organic – agriculture is one of the great challenges for the twenty-first century.

18.2.2   Room for Agriculture, Nature and Public Health

In the 1970s a new approach to nature development emerged. Conservation evolved into a new, more ecological approach to nature. The unintended and unexpected spontaneous nature development in the area around the *Oostvaardersplassen* (a marshy remnant within the new polder Flevoland) was in this respect a revelation. Nature conservation now became a matter of restoring plant and animal ecotopes. The 'nature restoration' notion was accorded a wondrously warm reception by nature, environmental and tourist organisations as well as by provincial and national governments and even by agricultural organisations.[19]

In 1975 the Ministries of Agriculture and of Public Housing, Spatial Planning and Environment presented the so-called 'Relation Memorandum' (*relatienota*) in which they described the points of departure for nature and landscape conservation in agricultural landscapes. This policy defined two types of management. In so-called reservation areas agricultural activities were all but banned. In the course of about three decades approximately 100,000 hectares of farming land in these regions were to be purchased by the state and brought under the management of the State Forestry Service and environmental protection organisations. Additionally the ministries designated about 100,000 hectares as so-called 'management zones.' In these zones farmers would be compensated for nature management, like the restoration of hedgerows, planting flowers along drainage canals and protecting

---

[17] Bieleman, *Boeren in Nederland, Geschiedenis van de Landbouw, 1500–2000*, 573.

[18] Compendium voor de Leefomgeving, *Biologische landbouw: aantal bedrijven en areaal, 1998–2014*, (Indicator, 21 mei 2015), www.clo.nl/indicatoren/nl0011-biologische-landbouw; CBS, *Ruim 1,4 duizend biologische landbouwbedrijven* (19-12-2014), www.cbs.nl/nl-nl/nieuws/2014/51/ruim-1-4-duizend-biologische-landbouwbedrijven

[19] Arjen Buijs, Thomas Matthijssen, and Bas Arts, '"The Man, the Administration and the Counter-Discourse": An Analysis of the Sudden Turn in Dutch Nature Conservation Policy,' *Land Use Policy* 38 (2014): 678.

field birds. Constrained in the 1980s by European measures to limit production, farmers were in search of extra income. Despite the fact that little profit was to be had, about half the farmers in the management zones participated.[20]

In 1990 the ecological approach was codified in the first Nature Policy Plan published by the Ministry of Agriculture, Fisheries and Nature Management. Particular attention was called to the changes considered necessary in the domain of agriculture.

> The extension of agricultural production will have to be in agreement with, for example, contributions to protecting the environment, the conservation and development of nature values and the maintenance of the habitability of the countryside.[21]

The Nature Policy Plan entailed a spatial vision of nature development, namely the so-called Ecological Main Structure (EHS). Within this structure, important nature areas were connected with each other via ecological transition zones. One example was the 'wet axis' from the Lauwerszee and the German Ems in the north to the Biesbosch and the Zeeland Delta in the south. The EHS comprised a continuous network of more than 700,000 hectares of nature zones, landed estates, nature development areas and connective corridors. With these plans, the Netherlands, together with Estonia and Flanders, propelled itself into the European vanguard. The EHS together with already existing habitat directives was the direct inspiration for the European Natura 2000 policy, that was codified into new European rules at the turn of the century.[22]

The originally broad support was among other things based on the perception in agricultural circles that the new measures were a further elaboration of the policy embodied in the Relation Memorandum. And the exemplary role that this policy had played in the realisation of the Natura 2000 directives was also presented with a certain pride.

But in 2000 the mood changed abruptly. In that year, basing its case on the new legislation, the conservation association *Das en Boom* (Badger and Tree) commenced legal proceedings against the construction of an industrial park in Heerlen. The construction threatened the habitat of the *Korenwolf*, a type of hamster rare in the Netherlands. The lawsuit stopped the construction of the prestigious industrial park. This example was followed by other nature organisations to put a stop to other undesirable construction and agrarian activities. Between 2001 and 2010 the number of legal actions rooted in the nature conservation laws grew apace.[23]

---

[20] C.M. Volker, *Boeren in Betwist Landschap, Strategische Keuzes van Boeren in een Waardevol Agrarisch Landschap* (Wageningen Universiteit, 1999), 56–59.

[21] *Natuurbeleidsplan*, Pub. L. No. TK 21149, 1989–1990 Handelingen Tweede Kamer 2 (1990), 191.

[22] Rob H.G. Jongman, 'Nature Conservation Planning in Europe: Developing Ecological Networks,' *Landscape and Urban Planning* 32 (1995): 196–183.

[23] Raoul Beunen, Kristof van Assche, and Martijn Duineveld, 'Performing Failure in Conservation Policy: The Implementation of European Union Directives in the Netherlands,' *Land Use Policy* 31 (2013).

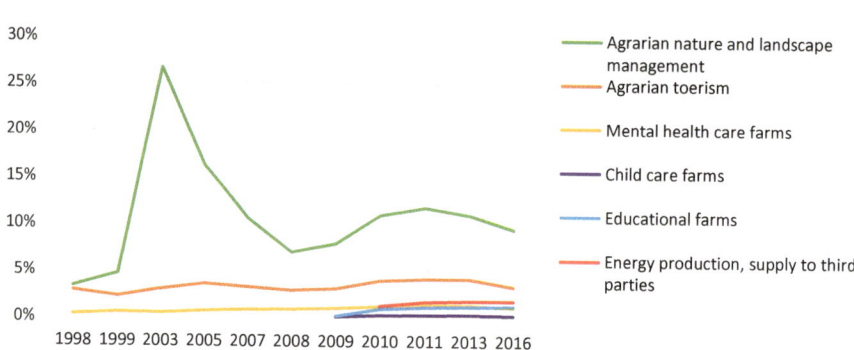

Graph 18.2 Supplemental activities in farming 1998–2016
Source: CBS- Landbouw; bedrijven met verbredingsactiviteiten

The confounded local politicians, real estate developers and agricultural organisations vented their spleen on the Natura 2000 policy. They saw this as an impediment to economic growth. Policy makers mobilised the weapon of 'substantial societal interest' in order to gain the upper hand in lawsuits. Societal unrest and political dissatisfaction undermined the legitimacy of the nature conservation rules. In 2009 prime minister Balkenende wrote a letter to the chairman of the European Commission, Barosso, in which he requested more 'equilibrium among ecological values, economic interests and other uses.'[24] No adjustments were forthcoming, but subsequent Dutch cabinets persisted in seeking weaker regulations. The environmentalist directives were framed as elite and 'left-wing hobbies' that impeded the economic development of the Netherlands.[25]

> 'There is only a very small elite that gets their kicks from the unique nature that you're supposed to be able to find in Natura 2000 zones. You almost need an academic degree to be able to appreciate it,'

as Minister of Agriculture Henk Bleker put it in 2011.[26] Ever fewer farmers saw profit in nature development (see Graph 18.2). On the other hand, nature developers did succeed in finding new partners, for example in the domain of gravel mining (see Chap. 19).

Nature development was only one of the challenges facing the agrarian world. At the end of the twentieth century the countryside acquired new functions. It was no longer the exclusive domain of agriculture. Due to the increasing urbanisation of the countryside, by 2016 some 3 million people lived in rural neighbourhoods. Of these, only 173,000 worked in agriculture. Farmers saw the opportunities in these developments and combined their farming with health care, child care, recreation, education and energy production (see Graph 18.3).

---

[24] 'Balkenende wilde andere natuurwet' in *NRC*, 11 januari 2010.

[25] Buijs, Matthijssen, and Arts, '"The Man, the Administration and the Counter-Discourse": An Analysis of the Sudden Turn in Dutch Nature Conservation Policy.'

[26] Klaas van der Horst, 'Bleker wil van elitenatuur naar boerennatuur', in *Boerderij*, 18 januari 2011.

But the increasing population of the countryside also brought new tensions with it. Circles defining zones of odour nuisance were a spatial illustration of the conflicts between the old and new uses. The Q-fever epidemic of 2007–2011, involving microbes that could be transmitted from humans to animals, exhibited the tensions between economic agrarian interests and the public health of recreational visitors and Dutch residents in the countryside. The spatial challenge facing the large-scale and technologically advanced agrarian enterprises was to integrate intensive animal husbandry within a densely populated country, while also paying due respect to issues of ecology and energy.

18.3   The Wages of Excess – In Search of Tasty Healthy Food

Netherlanders are losing control of their diet. In the 1950s most of the Dutch had adopted a clear nutritional pattern: each day two bread meals and a warm meal, ideally constituted according to the 'disk of five.' Since then, the Dutch eating pattern has changed dramatically. Although food comprises a substantially smaller portion of the household budget, it has by no means become less important. In 1960 more than 30% of the household budget was spent on food. By 1980 this had declined to 16% and by 2011 to a mere 11%. At the same time the attitude to food shifted. For most of the Dutch population, food is not so much a necessity of life as it is a means of expression, an element of a lifestyle: 'You are what you eat.' More luxurious foods became available and new exotic nutritional patterns emerged, visible, among other things, in the increased consumption of meat and dairy products.[27] And also in the increasing patronage of restaurants.

The classic nutritional pattern and the 'disk of five' was consonant with the epoch of the wage-earner model, in which the man of the family earned the money and the woman devoted her labour power to care for the home and family. Women's organisations, businesses and governments aimed at the housewife in their informational and educational campaigns. This model gradually lost its force. In 1973 only 21% of the married women worked for wages; this increased to 44% in 1997.[28] By 2016 women in 60% of the couples had paid employment.[29] Meanwhile the number of single persons increased due to individualisation and an increasingly aged population.

Changes in nutrition were also related to globalisation. The decolonisation of Indonesia and Surinam, labour migration from Southern Europe, Turkey and Morocco and the growing number of foreign vacation destinations introduced new

---

[27] Ronald van der Bie et al., *Smakelijk Weten, Trends in Voeding en Gezondheid* (Den Haag / Heerlen: Centraal Bureau voor de Statistiek, 2012), 40–41.

[28] A.H. van Otterloo, ed., 'Voeding,' in *Landbouw & Voeding*, vol. Voeding, Techniek in Nederland in de Twintigste Eeuw 3 (Zutphen: Walburg Pers, 2000), 282.

[29] CBS Statline: Arbeidsdeelname Paren

food habits. The reliance on potatoes diminished and was replaced in evening meals by rice, pasta and pizzas.[30]

The Dutch diet became much more varied thanks to an ever-more elaborate selection of foodstuffs. The food industry developed a broad range of instant foodstuffs, of consistent quality and homogeneous taste. Conservation techniques and additives regulated taste, color, smell, appearance and shelf life. All of this was aimed at the needs of the consumer and supported by a media offensive controlled by the food sector. The supermarket giant, Albert Heijn, developed the door-to-door magazine *Aller Hande* (1954) available for free in all its stores since 1983. During the 1990s, multinational Unilever started worked with commercial television stations to develop cooking programs.[31] Magazines, television and books became important channels for the diffusion of new eating patterns and knowledge about foods.

18.3.1 Snacks Everywhere

During the 1960s, the food-processing industry made inroads into the ever-expanding household food budget. Riding the wave of the popular fascination with American culture it initiated research into new snacks and new 'eating moments.' Their introduction was not a foregone conclusion. In 1971, Unilever researchers concluded that the consumption of snacks in the Netherlands took place chiefly within the home between the hours of 20:00 and 22:00, during TV viewing and the entertainment of guests. The Dutch hearty breakfast and the early warm meal interfered with earlier snack moments, according to the investigators.[32]

Changing established food patterns turned out not to be a simple task. It required many experiments and much research. Market research became ever more elaborate and aimed at an ever larger variety of subgroups with their own customs and tastes. This made it possible to gradually define new eating moments for a variety of breakfast, lunch, sport, happy hour, desert and casual snacks. From small 'in-betweens' consisting of relatively simple products like potato chips and nuts, snacks developed into more complex products like pizzas, hamburgers and breakfast bars. Gradually the pattern of the day with its three meals transformed into one in which food and drink were consumed throughout the day at different moments (see Graph 18.3).[33]

---

[30] Marlou Schrover et al., 'Lekker,' in *Verandering van Het Alledaagse, 1950–2000*, ed. Isabel Hoving, Hester Dibbits, and Marlou Schrover, Cultuur En Migratie van Nederland (Den Haag: Sdu Uitgevers, 2005), 77–112.

[31] Schrover et al., 'Lekker,' 106–12.

[32] Otterloo, 'Voeding,' 366.

[33] C.T.M van Rossum et al., *The Diet of the Dutch, Results of the First Two Years of the Dutch National Food Consumption Survey, 2012–2016*, RIVM Letter Report 2016–0082 (Bilthoven: National Institute for Public Health and the Environment (RIVM), 2016), 37–39.

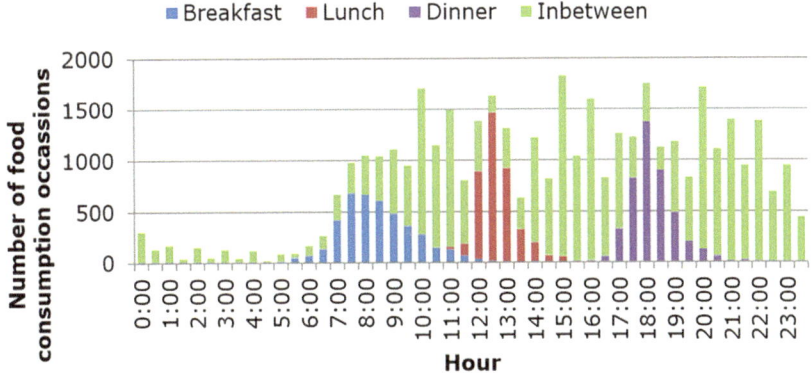

Graph 18.3 Number of food consumption occasions of the Dutch population by hours of the day n = 2237 stratified by age-gender groups, weighted for socio-demographic characteristics, season and day of the week
Source: Van Rossum et al. (2016) RIVM – DNFCS 2012–2014

Outside of the home, eating habits also changed. Restaurants were increasingly frequented. In bistros, French cuisine from the vacation could be re-experienced and in 'Chinese restaurants' the pleasures of the former colony. American culture could be tasted starting in 1969 at Kentucky Fried Chicken (KFC) and from 1971 on at McDonalds.[34]

Despite the increasing number of eating establishments the majority of meals in 2016 (78%) were still eaten at home. But the age group between 9 and 50 also ate more frequently at school or at work (15%–18%), in restaurants or on the road (5%–6%) or at other locations (2%–3%).[35] Food and fast foods were available at more and more places and times.[36] This included not only fast food restaurants in cities, transportation hubs and along highways, but also increasingly coffee corners in garden centres, bookstores and home furnishing stores.

The ever expanding assortment, the increasingly ready availability and increasing welfare had different effects on health. Sufficient and sufficiently healthy food had an unmistakably positive effect on the general health of the Dutch. Longevity and average height continued to increase. The incidence of so-called deficiency-diseases also declined. Malnutrition disappeared, but was rapidly replaced by over-

---

[34] M. van Rotterdam, *De 70's, Alles over de Jaren Zeventig* (Utrecht/Antwerpen: Kosmos, 2005), 85–93; W. Schreurs, *De Jaren Zeventig van Abba Tot Zitkuil* (Amsterdam: Balans, 2016), 57–58. The first McDonalds restaurant in Europe opened its doors in Zaandam in 1971. The concession was set up in cooperation with the Dutch supermarket chain Albert Heijn. In 1975 the cooperation between McDonalds and Albert Heijn was ended.

[35] Rossum et al., *The Diet of the Dutch, Results of the First Two Years of the Dutch National Food Consumption Survey, 2012–2016*, 63.

[36] Anneke H. van Otterloo, 'Healthy, Safe and Sustainablie, Consumers and the Public Debate on Food in Europe and the Netherlands Since 1945,' in *Food Practices in Transition*, by Gert Spaargaren, Peter Oosterveer, and Anne Loeber, Studies in Sustainable Transitions 3 (New York / London: Routledge, 2012), 71.

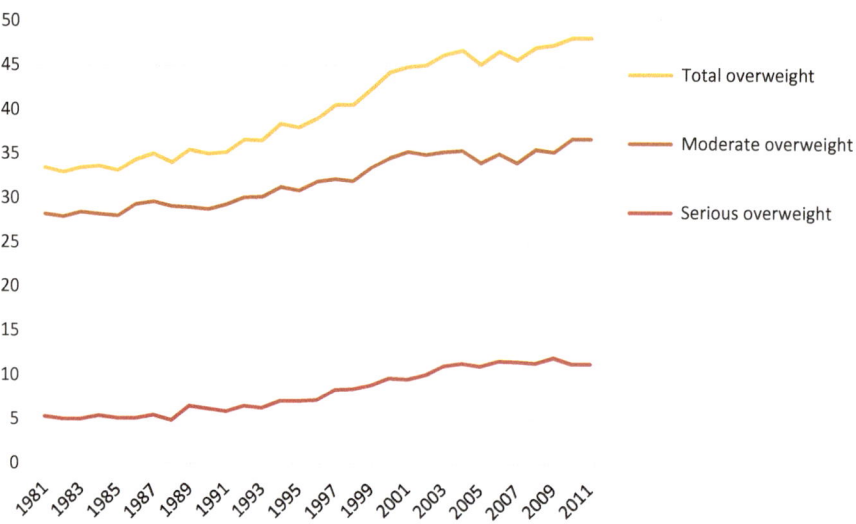

Graph 18.4 Percentage overweight people in the Netherlands, 1980–2011
Source: CBS

consumption (see Graph 18.4). The new food regime of abundant, fat and sweet created new health concerns. Combined with decreasing physical exertion at work this led to increasing overweight, heart and circulatory disease and diabetes. The increasing incidence of different forms of cancer was also attributed to the new eating and lifestyle habits.[37] Governments, business and consumers responded differently to the changes in food consumption and lifestyles.

18.3.2   Government: Informing the Public and Research

In 1983 the Dutch government published a first Memorandum on Food Policy. This was above all a continuation of the existing agenda supplemented with concerns about hygiene. The core was a robust supply of safe foods along with the encouragement of good eating habits.

Until the 1980s the Law on Food Commodities had been the means to guarantee food quality in the Netherlands. In 1982, harmonisation with European food directives was experienced as a step backward. The Netherlands could not turn its back on the more liberal European policy, that was more lenient in regard to the use of additives and the making of health claims by the food processing industry.[38] However, unrest among consumers gave rise to a supplemental European directive

---

[37] C.F. van Kreijl and A.G.A.C. Knaap, *Ons Eten Gemeten, Gezonde Voeding En Veilig Voedsel in Nederland* (Bilthoven: Rijksinstituut voor Volksgezondheid en Milieu (RIVM), 2004), 55–56.
[38] Otterloo, 'Voeding,' 309.

in 1989. With the mandatory publication of ingredients on the packaging it became possible for consumers to make conscious choices.[39]

In more recent decades, the big challenge has been combatting the new welfare diseases. In the 1980s 'reducing fat consumption' was the main issue. Policy aimed above all at preventing heart and circulatory disease and cancer. By the end of the 1990s the government had become more concerned about food habits and lifestyles. The Nutrition Centre and public campaigns targeted both the general public and specific subgroups. Young people were apprised of healthy eating habits through school-based programs. From the 1990s on the government sought to enlist the business community using the instrument of covenants. Research monitored the effects of this food policy.[40]

In 2004 the National Institute for Public Health and Environment measured the health effects of food and lifestyle. The effects were expressed in terms of so-called Disability Adjusted Life Years (DALY), an indicator that cumulated the number of lost and dependent years over a lifetime. From the calculations it followed that a dearth of physical exercise, bad nutritional habits and overweight were the most significant factors in the health and longevity losses. Compared to these factors, the losses due to food allergies, infections or chemical substances were trivial.[41] In the first decades of the twenty-first century government policy therefore shifted from regulating foodstuffs to influencing lifestyles, with food consumption as a specific target. Varied and healthy foods and sufficient attention to physical exertion were the main ingredients.

18.3.3 Private Enterprise: New Functional Foodstuffs

Overconsumption was indirectly the result of the success of the agrarian sector and the food processing industry. This laid the basis for the enormous assortment and the low prices. This did not mean that nutritional problems left the business community cold. For them the problems in a certain sense meant a new challenge to produce tasty foods, but now, for example, containing less fat, sugar and calories.

The 1980s saw a proliferation of research into so-called 'functional foods,' foodstuffs developed in order to promote health. A firm like Unilever already had a lot of experience in this area. As early as the 1960s the company had already developed the cholesterol-lowering margarine Becel, aimed in the first place at heart patients.[42]

---

[39] Otterloo, *Healthy, Safe and Sustainable, Consumers and the Public Debate on Food in Europe and the Netherlands Since 1945*, 71.

[40] Kreijl and Knaap, *Ons Eten Gemeten, Gezonde Voeding En Veilig Voedsel in Nederland*, 104–9.

[41] Kreijl and Knaap, 283–90. The results of the calculations for health loss were as follows for these aspects: unfavourable diet (245.000 DALY), overweight and obesity (215.000 DALY), insufficient exercise (150.000 DALY), food allergies (1000 DALY), food infections (1000–4000 DALY) and chemical substances (500–1000 DALY).

[42] Mila Davids, 'Technology as the New Frontier: Unilever and the Rise of Becel Margarine,' *Journal of Modern European History* 14, no. 1 (2016): 101–18.

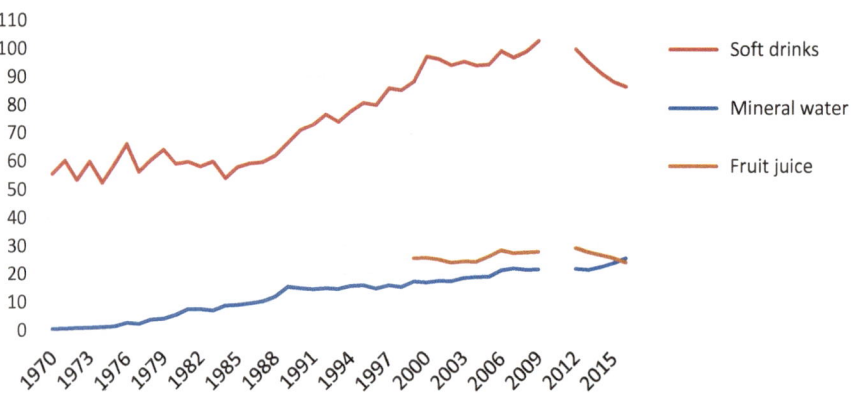

Graph 18.5 Annual consumption of soft drinks, mineral water and fruit juice (in l/cap.), 1970–2015
Source: CBS and after 2011 FWS (CBS Statline – Voedings- en genotmiddelen; consumptie per Nederlander, 1899–2009; Nederlandse Vereniging Frisdranken, Waters, Sappen (FWS) – *Kerngegevens/Basic Statistical information 2016* via www.frisdrank.nl)

In the early 1970s the same firm introduced halvarine, a margarine with half of the usual fat percentage.

The link between food consumption and lifestyle was evident in the consumption of soft drinks. After the introduction of 'family size' bottles in the mid-fifties this market expanded steadily. In 1970 the average Netherlander drank about 55 litres of soft drinks per year. At the end of the 1980s consumption increased markedly and around the turn of the century per capita soft drink consumption was almost a hundred litres per year, nearly a can (30 cl) per day (see Graph 18.5).

With the development of synthetic sweeteners like cyclamate (1951) and saccharine (1958) sugar-free soft drinks could be developed and tentatively marketed. Initially they were aimed above all at the market for diabetics. The Dutch Food Commodities Law was conservative when it came to such artificial additives. But for diabetics exceptions could be made. Only a few manufacturers, at the end of the 1960s, developed sugar-free soft drinks. Advertisements in which these were presented as diet products were challenged in the 1970s by both the Ministry of Public Health as well as by manufacturers of soft drinks containing sugar.

In 1984 European regulations admitted saccharine, cyclamate and the new artificial sweetener aspartane as safe sugar substitutes for sweetening foods. This was the origin of the so-called 'light' products. Gradually sugar-free soft drinks established a position in the soft drink market. The market share grew from 3% in 1985 to 13% in 1992 to 30% in 2015.[43] The soft drink industry also promoted the sale of bottled mineral water by playing on emergent consumer preferences for healthy and low-calorie nutrition. The consumption of mineral water increased from a half litre per

---

[43] Peter Zwaal, *Frisdranken in Nederland, Een Twintigste Eeuwse Productgeschiedenis* (Rotterdam: Stichting BBM, 1992), 314–31.

Table 18.2 Consumption of soft drinks, mineral water and fruit juice (in litres per capita per year)

	Soft drinks		Mineral water	Fruit juice
	Sugared	Light		
1984	53	1	9	
1992	67	10	15	
2012	69	31	22	29
2016	56	26	26	24

Source: CBS and after 2011 FWS

capita per year in 1970 to more than 30 litres per capita per year in 2015, almost as much as light soft drinks (see Table 18.2).

European regulations that came into force at the end of the 1990s expanded the possibilities for the use of additives and vitamin preparations. They also increased leeway in making health benefit claims.[44] Using claims about the reduction of cholesterol, the prevention of heart and circulatory disease and better health through less sugar consumption and the use of natural colour and taste substances, the food industry played to the increasing concerns about health and overweight. In the meantime a lively debate has emerged about the grounding of those claims.

Welfare diseases and a lifestyle of excess have silently been transformed into a new market for the sale of craftily designed foodstuffs.

18.4   Crisis in Nutrition

A portion of the consuming public was extremely critical of the increasing industrialisation and commercialisation of agriculture and foodstuffs. This criticism had already been ventilated in the anarchistic program of the Provos and the Kabouters around 1970 (see Chap. 17). They laid the foundations for numerous so-called macro-biotic stores with organic products. This created an important link between organic farming and consumers.[45]

The sale of organic products was limited until the 1990s, in part because of the high prices consumers had to pay for them. The new regulations for synthetic additives in 1989 and the debates between 1998 and 2002 about allowing genetically modified agricultural products in the EU initiated a new discussion about healthy and sustainable food production. The outbreak of 'mad cow disease' (BSE) in 1987, the compulsory hog slaughter in connection with hoof and mouth disease in 2001 and the subsequent chicken slaughter after outbreaks of avian flu in 2003 added fuel to the fire. Environmental and animal welfare associations joined in the public debate about food safety and the food chain. They provided a perspective on more

---

[44] Otterloo, 'Voeding,' 374.

[45] Dick Hollander, *Tegen Beter Weten In. de Geschiedenis van de Biologische Landbouw En Voeding in Nederland* (Hurwenen: 4 Heuvels, 2012), 105–6.

sustainable food patterns via publicity campaigns aimed at the products available in supermarkets and the behaviour of consumers.

The international ramifications of the food chain make the debate even more complex. Concepts like 'food kilometre,' 'ecological footprint' and the associated 'footprint of our food' are intended to give consumers insight into the problems of the international character of the food chain.[46] Raw materials and food are dragged hither and thither. That costs energy, produces emissions and makes it more difficult to monitor food safety. Food kilometres attempt to express this. Dutch pigs, for example, are transported to the Italian city Parma in order to be processed into Parma ham, a product that can bear this name only if it is produced in that city. The ham subsequently finds its way to the Netherlands. The footprint is supposed to give us an idea of the ecological effects, for example the effects of cattle husbandry on logging, reclamation, draining of swamps and disproportionate water use. Social aspects are often also at issue, like fair trade with the developing world and the working conditions of farm workers. A variant is a calculation by the Dutch Nature and Environment Federation of the size of the claim a consumer makes on the amount of land annually necessary to provide his or her food.[47] For a Dutch person that would be 6.2 hectares, while the rest of the world population makes do with an average of 2.7 and while there is only 1.8 hectare per person available. Their conclusion is that the Dutch use disproportionate amounts of land and that nature is being disproportionately abused. While the concepts and calculations may be contested, they do express concerns about the biological capacity of the earth and the habitability of the planet for future generations.

How is a consumer to get a grip on the complex issues of sustainable and healthy nutritional patterns? Among his resources are the trademarks, certificates and labels.[48] These are supposed to provide him with information about, among other things, environmental effects, animal-friendly production, fair trade and health effects. Despite this, the certificates only partly succeed in winning the confidence of the consumer in sustainable, healthy and responsible food. A comprehensive and unambiguous image of healthy and sustainable food is lacking. The variety is overwhelming. Doubt is often cast on the objectivity of the information provided. The claims made are not always grounded. The organisation providing the information is not always independent. Even information based on extensive scientific research is a cause of confusion. Notorious in this respect are the discussions about E-numbers, a European system for the coding and classification of supplements to food, like dyes, coagulants, anti-oxidants and vitamins.

The consumer faces many dilemmas. In this way eating becomes an act of knowing and conscience, of believing and hoping.[49]

---

[46] L.O. Fresco, *Hamburgers in het paradijs. Voedsel in tijden van schaarste en overvloed* (Amsterdam 2012), 364–369.

[47] http://voetafdruknederland.nl/over-de-voetafdruk/ (geraadpleegd op 20–11-2017).

[48] Fresco, *Hamburgers in het paradijs*, 369–373.

[49] See also the section 'Dilemma's in de menselijke voedselvoorziening' in: Fresco, *Hamburgers in het paradijs*, 362–373.

Literature

Anoniem. (1984, 3 November). 'Landbouwschap volslagen verrast', *Leidsch Dagblad*, p. 9.

Beunen, Raoul, Kristof van Assche, and Martijn Duineveld. (2013). 'Performing Failure in Conservation Policy: The Implementation of European Union Directives in the Netherlands.' *Land Use Policy* 31.

Bieleman, Jan. (2008) *Boeren in Nederland, Geschiedenis van de Landbouw, 1500–2000.* Amsterdam: Boom.

Bloemendaal, Frits. (1995). *Het Mestmoeras*. Den Haag: SDU Uitgeverij.

Buijs, Arjen, Thomas Matthijssen, and Bas Arts. (2014). "The Man, the Administration and the Counter-Discourse': An Analysis of the Suddesn Turn in Dutch Nature Conservation Policy.' *Land Use Policy* 38: 676–84.

Crijns, A.H. (1997). 'De Grote Ommekeer in de Agrarische Sector.' In *Geschiedenis van Noord-Brabant, Dynamiek en Expansie, 1945–1996*, edited by H.F.J.M. van den Eerenbeemt, Vol. 3. *Geschiedenis van Noord-Brabant*. Boom.

Davids, Mila. (2016). 'Technology as the New Frontier: Unilever and the Rise of Becel Margarine.' *Journal of Modern European History* 14, no. 1: 101–18.

Dinkelman, G.H. (1995). 'Verzuring en Broeikaseffect, de Wisselwerking Tussen Problemen en Oplossingen in Het Nederlandse Luchtverontreinigingsbeleid (1970–1994).' Universiteit van Amsterdam.

Fresco, L.O. (2012). *Hamburgers in het paradijs. Voedsel in tijden van schaarste en overvloed.* Amsterdam: Prometheus.

Grin, John, Jan Rotmans, and Johan W. Schot. (2009). *Transitions to Sustainable Development, New Directions in the Study of Long Term Transformative Change*. New York: Routledge.

Hermans, Ben. (2006). *De Mestmarathon, Kroniek van Ruim 42 Jaar Nederlands Mestbeleid.* Utrecht: Natuur & Milieu.

Hollander, Dick. (2012). *Tegen Beter Weten In. de Geschiedenis van de Biologische Landbouw en Voeding in Nederland.* Hurwenen: 4 Heuvels.

Jongman, Rob H.G. (1995). 'Nature Conservation Planning in Europe: Developing Ecological Networks.' *Landscape and Urban Planning* 32: 196–183.

Natuurbeleidsplan, Pub. L. (1990) No. TK 21149, 1989–1990 Handelingen Tweede Kamer 2.

Otterloo, Anneke H. van. (2012). 'Healthy, Safe and Sustainablie, Consumers and the Public Debate on Food in Europe and the Netherlands Since 1945." In *Food Practices in Transition*, by Gert Spaargaren, Peter Oosterveer, and Anne Loeber, 60–85. Studies in Sustainable Transitions 3. New York/London: Routledge.

Rossum, C.T.M van, E.J.M. Buurma-Rethans, F.B.C. Vennemann, M. Beukers, H.A.M. Brants, E.J. de Boer, and M.C. Ocké. (2016). "The Diet of the Dutch, Results of the First Two Years of the Dutch National Food Consumption Survey, 2012-2016." RIVM Letter Report 2016–0082. Bilthoven: National Institute for Public Health and the Environment (RIVM).

Schreurs, Wilbert. (2016). *De Jaren Zeventig van Abba Tot Zitkuil.* Amsterdam: Balans.

Schrover, Marlou, Ineke Mestdag, Anneke H. van Otterloo, and Chaja Zeegers. (2005). 'Lekker.' In *Verandering van het Alledaagse, 1950–2000*, edited by Isabel Hoving, Hester Dibbits, and Marlou Schrover, 77–112. Cultuur en Migratie van Nederland. Den Haag: Sdu Uitgevers.

Breemen, N. van, P. A. Burrough, E. J. Velthorst, H. F. van Dobben, Toke de Wit, T. B. Ridder, and H. F. R. Reijnders. (1982, October 7) 'Soil Acidification from Atmospheric Ammonium Sulphate in Forest Canopy Throughfall.' *Nature* 299, no. 5883: 548–50. https://doi.org/10.1038/299548a0.

Bie, Ronald van der, Brigitte Hermans, Cor Pierik, Lieke Stroucken, and Elma Wobma. (2012) *Smakelijk Weten, Trends in Voeding en Gezondheid.* Den Haag/Heerlen: Centraal Bureau voor de Statistiek.

Kreijl, C.F. van, and A.G.A.C. Knaap. (2004). *Ons Eten Gemeten, Gezonde Voeding en Veilig Voedsel in Nederland.* Bilthoven: Rijksinstituut voor Volksgezondheid en Milieu (RIVM).

Otterloo, A.H. van, ed. 'Voeding.' (2000). In *Landbouw & Voeding*, Voeding: 234–74. *Techniek in Nederland in de Twintigste Eeuw* 3. Zutphen: Walburg Pers.

Rotterdam, Marjolein van. (2005). *De 70's, Alles over de Jaren Zeventig*. Utrecht/Antwerpen: Kosmos.

Volker, C.M. (1999). *Boeren in Betwist Landschap, Strategische Keuzes van Boeren in een Waardevol Agrarisch Landschap*. Wageningen Universiteit.

Zwaal, Peter. *Frisdranken in Nederland, Een Twintigste Eeuwse Productgeschiedenis*. Rotterdam: Stichting BBM, 1992.

Chapter 19
Building Materials and Construction: Sustainability, Dependency and Foreign Suppliers

Frank Veraart

Contents

Abstract This chapter describes the extraction of mineral subsoil resources in a changing context of increasing internationalisation and domestic concern for nature and the environment. The cases are gravel and marl in the province of Limburg and the European inventory of strategic mineral resources. The period around 1970 formed the high point of Dutch building activities. The extraction and production of building materials had an increasing impact on the landscape (see Chap. 14). Bringing laws against excavations and spatial planning to bear, the government increased its control over the extraction activities. The new policy integrated the excavations in spatial planning and landscape goals.

The politicisation of environmental issues led to the harmonisation of the extraction of mineral resources with the local societal requirements for tourism, nature development, and flood control. Intensification of European cooperation positioned domestic extraction within a European economic framework. Higher prices for gravel and other building materials made recycling, among other things, attractive. The mining of gravel shifted in part to surrounding countries where it resulted in local damage to the landscape.

The outsourcing of gravel mining was similar to the overall European offshoring of the mining of mineral subsoil resources. From 2008 the European Commission commenced investigations into strategic mineral resources. The Netherlands followed in its footsteps. Initially, geological, economic and geo-political aspects were the main concerns. In a later phase, environmental issues and working conditions played a role. Developments in the area of well-being and sustainability required not only measures close to home, but also a concern for these issues outside of the Netherlands.

Keywords Mineral resources · Gravel · Offshoring · European Union · Metals

19.1 A Celebration in Margraten

> Justified joy for many in Limburg yesterday after the unexpected decision by minister Smit-Kroes to spare the Margraten plateau. The years-long struggle by the people around and on this beautiful piece of Limburg, municipal officials, and environmentalists, has not been futile...[1]

Despite the potential loss of jobs, the government had rejected a request by the cement company ENCI to expand its operations. As the government explained to the Provincial Estates of Limburg: 'In 1992, in the framework of an emergent Europe, maintaining a national cement industry has declined in value.'[2] The decision came as a surprise:

> ... Yesterday no one had much hope that the cabinet would arrive at this decision. A decision that does however resonate with the bow-wave of interest that has emerged recently at all levels for the improvement of our natural environment. An issue for which even employers and trade unions have fallen into each other's arms. On the Plateau of Margraten the environment has triumphed over industry...

Opinions about digging for mineral subsoil resources like marl, gravel and sand had shifted. Up to 1970, housing construction, the Delta works and other investments under the aegis of reconstruction had been the priorities steering the domestic production of cement and the mining of sand and gravel. After that time, landscape and environment also began to be incorporated into policy-making.

In the first part of this chapter we dwell upon the building programs from the 1970s on. What role did Dutch raw materials play? What factors were responsible for changing views on the domestic production of raw materials? In the second part of the chapter we illustrate these changes using gravel-mining as an example. Changing perspectives provided the impulse for developing a raw materials policy for construction materials.

But worries about mineral raw materials extended beyond marl, gravel, and sand. European assessments and unrest about Chinese policies led in 2010 to questions

---

[1] 'Eigen cementindustrie in open Europa niet nodig', in *Limburgs Dagblad* 19-01-1989, p 21

[2] 'Milieu wint het van economisch belang bij verbod tot afgraven mergelplateau' in *NRC Handelsblad, 19-1-1989*

about the depletion and availability of copper, tin, indium and other metals. Once again the question was raised: what role did (mineral) raw materials play in the Dutch economy? What risks were there for Dutch industrial sectors? What policy was needed for economically vital raw materials? These last questions will be addressed in the concluding part of this chapter.

19.2   The Building Programs

Dutch building efforts experienced an apotheosis around 1970. On the basis of the 1966 Second Memorandum on Spatial Planning, that had predicted 20 million inhabitants by the year 2000, the volume of housing construction stood at an all-time high. In 1972 and 1973, 150,000 new dwellings per year were being realised. They were built above all in the outskirts of cities and in suburban 'growth cores.'[3] A new national road plan – also published in 1966 – stimulated an increase in the construction of national roads, for the most part, limited access highways. 1966 also saw the publication of the Memorandum on Maritime Harbours, that envisioned the improvement of Rotterdam's *Europoort* harbour by means of landfills and its further extension into the sea in the form of the *Maasvlakte*.[4] Coastal defences also acquired a significant boost from the ongoing Delta works.

The enormous pace of construction and the technocratic style of governance that made this possible became a focus of criticism around 1970. Social protest decrying the poor quality of life in cities and emerging worries about the environment turned against the planners and builders of cities, roads and dikes. The protest movement succeeded in convincing the political establishment. Plans were re-evaluated. The symbol of the turnaround became the storm-surge barrier in the Eastern Scheldt (the *Oosterscheldekering*) a marvel of technological prowess that replaced the planned total closure of the estuary, the Eastern Scheldt. This adjustable storm-surge barrier was completed in 1986.[5]

Urban construction aimed at urban renewal and the compact city. In addition to investments in city centres, it included the construction of about 835,000 dwellings between 1995 and 2015 around nexuses of public transport in close proximity to the bigger cities. The plans also included investments in the 'mainports.' Rotterdam's harbour was to be expanded with a second *Maasvlakte* and its throughput capacity

---

[3] H. van der Cammen and L.A. de Klerk, *Ruimtelijke Ordening, van Plannen Komen Plannen, de Ontwikkelingsgang van de Ruimtelijke Ordening in Nederland*, 4e ed. (Utrecht: Het Spectrum, 1999), 147.

[4] Han Meyer, *De Staat van de Delta, Waterwerken, Stadsontwikkeling en Natievorming in Nederland* (Nijmegen: Van Tilt, 2016), 129–32.

[5] A. Bosch and W. van der Ham, *Twee Eeuwen, Rijkswaterstaat 1798–2015*, 2e druk (Asten: Nieuwe Uitgevers, 2015), 261–65.

augmented with the *Betuwe* route, a freight railway line to the German hinterland. Schiphol Airport would get a fifth runway and would be made accessible for high-speed trains.[6]

Perilously high river stages on the Meuse and Rhine in 1993 and 1995 were an incitement to reassess flood risks in the region of the large rivers. In 1995 portions of cities along the Meuse were flooded and 250,000 people were evacuated from the river-region of the province of Gelderland. These crises eventuated in the Delta Plan for the Big Rivers, an accelerated program of levee reinforcements in the river region. At the same time more attention was devoted to the expected effects of climate change, in particular increasing rainfall and sea-level rise. These new insights took shape in the 'Room for the River' program that was implemented between 2005 and 2015. In this program engineers combined the creation of hydraulic infra-structures with environmental policy, landscape development, and spatial planning.

The new political preferences of a conservative national government became visible in 2004 in proposals contained in the so-called Space Memorandum (*Nota Ruimte*). This memorandum brought together different spatial planning domains like transport, nature, agriculture, industry and urban development. However, their coordination was transferred from the national to lower levels of government. The national government was to restrict itself to the delineation of overall structure and core regions in line with European directives.

Sensitivity to climate change and sea-level rise led to increased concern for the coast. A survey held in 2003 revealed that more than 40% of the primary coastal defences did not yet satisfy the Delta norms established in 1961. The 'weak links' along the coast were dealt with after 2006. At various locations the coast was reinforced by sand suppletion in a manner that conserved the ecological and recreational values of the dune landscape. In 2007 the government appointed a new Delta Commission. This analysed the consequences of sea-level rise for the Netherlands and recommended possible countermeasures. In 2009, the Commission's recommendations led to the financing of this sustainability agenda and to a first 'National Water Plan' that would guarantee not only defences against water but also the supply of fresh water. The plan, to be revised every 6 years, ongoingly establishes the norms and the implementation program.[7]

19.3 Grounded Excavation

Sand, gravel, and chalk (including marl) were the most voluminous building materials. Domestic mining of gravel and marl had been concentrated since the mid-twentieth century in the province of Limburg (see also Chap. 14). The mining of

---

[6] Nil Disco and Frank Veraart, 'A Farewell to Big Planning? 1990–2010,' in *Builders and Planners, a History of Land-Use and Infrastructure Planning in the Netherlands*, ed. Jos Arts et al. (Delft: Eburon, 2016), 361–77.

[7] Meyer, *De Staat van de Delta, Waterwerken, Stadsontwikkeling En Natievorming in Nederland*, 150–82.

these kinds of shallow subsoil resources had been regulated since the 1960s in the Excavation Law. New excavation permits thus became embedded in regional structure plans, that also regulated activities like recreation and nature conservation. Digging for shallow subsoil resources thus became increasingly constrained.

In 1974 the Federation of Shallow Subsoil Resource Extraction Industries (FODI) sounded the alarm. It warned that if a conservative excavation policy were maintained it would not be possible to guarantee materials supplies for the building industry, including the public works sector.[8] In response, the Ministry of Transport and Public Works launched an investigation into the demand for raw materials. After 1976 a National Commission for the Coordination of Excavation Policy (LCCO), composed of representatives of the Ministry and the provincial estates, was charged with harmonizing the demand for and the supply of shallow subsoil resources. The LCCO developed 'target plans' for the different provinces in the form of 10-year plans. The first of these 'target plans' appeared in 1978.

The establishment of the LCCO was seen as the first step toward a national excavation policy. The environmental movement also began to take part in the discussion. The Foundation for Nature and Environment presented its own ideas for excavation policy. It pointed to alternatives such as the use of demolition waste, dredging mud, and slag from furnaces. They also pointed to alternative mining sites at sea, an increase in imports and they argued that because of its spatial implications excavation policy should be part and parcel of spatial planning policy.[9] The 1966 Second Memorandum on Spatial Planning, after all, sketched the rough contours of land use, with specific attention to space for recreation and nature (see also Chap. 15).

Against this background, in 1976 the cement industry ENCI developed plans for extending the extraction of marl beyond 1991. Its existing concession on the Sint Pietersberg appeared to have reached its economic limits. This is why the firm now set its sights on extraction on the Margraten Plateau. This option, once rejected in the 1940s, now seemed the economically most opportune. The opposition, however, was fierce. No less than 8000 persons filed objections, in addition to the mayors of the affected villages and the chief superintendent of the Margraten American War Cemetery, who added an unexpected diplomatic loading to the controversy. Succumbing to popular pressure, the Limburg provincial estates postponed a decision on the procedures. ENCI as well as FODI protested and took the matter to court. In 1978 the Council of State, the nation's highest administrative court of appeal, ruled in their favour.

The subsequent renewal of the planning procedures was once again followed by mass protests and by a legal war of attrition involving objections and delaying tactics. Opponents also produced alternative plans and reports. Investigations into other extraction methods and excavation sites were again undertaken. Expectations of shrinking domestic cement production and new production processes involving

---

[8] G.J.A. Sigmond et al., *1974–1984, Tien Jaar: Industrie en Ontgrondingen* (De Steeg, 1984).

[9] G.W. Grondelle, *Ontgrondingen, over de Noodzaak en Mogelijkheden voor een Ander Beleid t.a.v. de Winning van Mergel, Grind, Zand, Klei, Veen, etc.* ('s Gravenland: Stichting Natuur en Milieu, 1978).

fly ash decreased the need for marl. This made further digging at the existing pit a feasible short-term alternative. In 1985 the Council of State ruled that in the context of these developments, ENCI could not, for the time being, start digging on the Margraten Plateau. The Province granted a permit to mine the Sint Pietersberg until 2010. This seemed to guarantee an adequate level of cement production for the coming 20 years. This compromise satisfied both opponents and proponents of marl mining on the Margraten Plateau. The peace, however, was of short duration.[10]

A new government memorandum – titled Grounded Excavation (*Gegrond Ontgronden*) – explored the long-term state of the national raw materials supply. The draft version that appeared in 1987 once again named the Margraten Plateau as a potential mining site. This led to a resurgence of the mass protests. In 1989 mayor Herman Kaiser of Margraten succeeded in enticing the Minister of Transport and Public Works, Neelie Smit-Kroes, to visit the targeted area. The visit was accompanied by a massive peaceful protest and led to the remarkable decision with which this chapter began. Two weeks after the visit the mayor received a phone call from the minister who said: 'you may keep your plateau.' The importance of a domestic cement industry was struck from the text of the definitive version of the memorandum 'Grounded Excavation.' The minister framed cement production in the context of European developments. This would eventually lead to the termination of marl mining.

The publication of the draft version of 'Grounded Excavation' also served as a starting point for negotiations with the various provinces regarding the 'targets' for the supply of shallow subsoil-resources. The Limburg provincial estates succeeded in reaching an agreement about limiting the mining of gravel. Another 80 million tons of gravel would be extracted in Limburg, enough to supply the estimated need for the 1990s and possibly somewhat longer. The new policy assumed that after the turn of the century, demand could be satisfied with alternative materials, mining at sea and imports.[11] Re-use of building materials was also an option.

Re-use of building materials was stimulated by various environmental measures taken in consequence of, among other things, the Waste Materials Law (Afvalstoffenwet) of 1978. In the course of the parliamentary debates on this law, the Christian Democratic parliamentarian A.G.W.J. Lansink introduced the so-called Waste Hierarchy, a desirable sequence in dealing with waste materials: prevention of waste, re-use of 'waste' flows in production, transformation into energy and as final option controlled dumping of non-processable waste.[12] This approach,

---

[10] Thomas Rode, *ENCI's Struggle for the Margraten Plateau, a Clash of Economical and Environmental Interests* (Maastricht University, Faculty of Cultural Sciences, Bachelor's Thesis, December 6, 2009), www.oudsintpieter.com/Presentaties.htm; Thomas Rode, 'Een Wereld van Cement? De Strijd Om Het Plateau van Margraten, 1976–1989,' *Studies over de Sociaal-Economische Geschiedenis van Limburg* 55 (2010): 26–48.

[11] Paul Ike, *De Planning van Ontgrondingen* (Rijksuniversiteit Groningen, Geo Pers, 2000), 43–44.; B. De Jong, 'Ontgronden in Nederland een steeds groter probleem', in *De Ingenieur,* nr 11 (november 1989), p. 7–11.

[12] Tweede Kamer der Staten Generaal, Zitting 1979–1980, Rijksbegroting voor het jaar 1980, 15,800 Hoofdstuk XVII, Departement van Volksgezondheid en Milieuhygiëne, nr. 21, 'Motie van het lid Lansink c.s.', voorgesteld 1 november 1979.

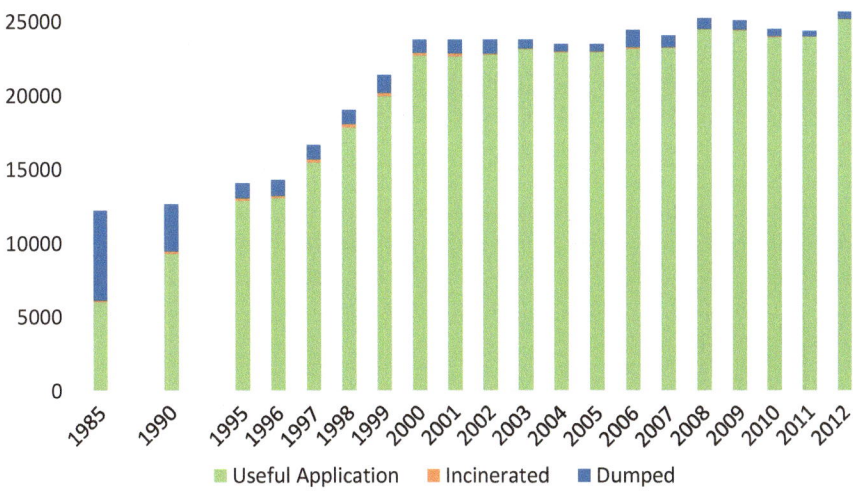

Graph 19.1 Available construction and demolition waste 1985–2012 (million kg)
Source: *Compendium voor de leefomgeving*

that became known in the Netherlands by the name 'the Ladder of Lansink' became more nuanced in subsequent years. The first National Environmental Policy Plan (NMP) published in 1989 committed itself to reducing waste flows by employing so-called integral chain management. Industries were challenged to think through the environmental consequences of supplies, production, consumption and waste processing. This was supported by supplementary measures.

In the building sector, the emphasis was on the re-use of construction and demolition waste. Research at TNO and universities inventoried possibilities for re-use of waste in concrete, asphalt and foundation materials. In addition, new designs were promoted that needed fewer materials.[13] The private sector played a big role in these activities. It was therefore hardly surprising that initiatives started with the re-use of demolition waste and not with prevention of use by re-using buildings. Prevention was the economically less attractive option.

Ministries, the building industries, and knowledge institutes worked together on preparing the so-called Building Materials Decision (Bouwstoffenbesluit) that was enacted in 1995. This stipulated the possibilities, norms, and rules for re-use.[14] In that year, too, a prohibition against the dumping of re-usable building materials came into force. This stimulated an enormous increase in the re-use of construction and demolition wastes. In 1985 half of the wastes were re-used; after the turn of the century this increased to 97% (see Graph 19.1). In surveys of supplies of construction and demolition waste granulates it was proudly announced that, after Hong

[13] Jacqueline Cramer, *Milieu*, Elementaire Deeltjes 16 (Amsterdam: Amsterdam University Press, 2014), 35–48.

[14] R.T. Eikelboom, E. Ruwiel, and J.J.J.M. Gouwmans, 'The Building Material Decree: An Example of a Dutch Regulation Based on the Potential Impact of Materials on the Environment,' *Waste Management* 21 (2001): 295–302.

Kong, the Netherlands was the frontrunner in the use of these recycled raw materials.[15] Despite these achievements, critical voices were also heard. Re-use appeared above all to be so-called 'down-cycling.' Construction and demolition wastes served as fill material and as granulate for road foundations. In the use of concrete the primary raw materials still prevailed.[16] For the time being this meant a continued significant need for gravel and masonry sand.

Regulations in the area of waste management and industrial developments in the area of integral supply chain management provided the basis for a policy program launched in 2013, 'From Waste to Raw Material.' These were the first steps on the road to a transition, embraced by the government and environmental organisations, toward a circular economy.[17]

19.4 Passing the Buck or Market Forces

In 2002 a commission under the chairmanship of the former state secretary for Public Housing, Spatial Planning and Environmental Policy (VROM) D.K.J. Tommel, formulated policy recommendations for the Structure Scheme for Shallow Subsoil Resources. The commission, entirely in line with the political climate of the time, recommended more leeway for market forces and less central steering by the state. The freshly elected Balkenende cabinet enthusiastically embraced the recommendations. The planning and coordination of shallow subsoil resources would no longer be a task for the government. Aspects of excavation policy, just like other spatial planning issues, were subsumed in the Space Memorandum of 2004. This memorandum expressed the new direction of policy. Government regulation was to be transformed into over-all planning. 'Decentralize where possible, centralize where necessary' became the new slogan according to which the state shifted responsibilities for spatial planning to the provinces and municipalities.

For excavation policy, this meant the gradual abandonment of coordination via targets and the liberation of market forces. The new policy underscored the importance of local extraction in order to prevent the displacement of environmental problems, needless transport, and excessive energy consumption. Policy aimed at so-called 'multifunctional' extraction, that united extractive imperatives with local needs.

> It is expected of the excavating industry that it will orient itself to the development of qualitatively good and societally responsible projects in close cooperation with the affected parties. (…) This means that in the course of extraction use must be made of the chances offered by excavations for the realisation of other desirable social functions like nature

---

[15] U. Hofstra et al., *Scenariostudie BSA-Granulaten, Aanbod en Afzet van 2005 Tot 2025* (Sittard: INTRON / Expertisecentrum Bouwstoffen Rijkswaterstaat, 2006), 15.

[16] Evert Mulder, *Kringbouw, naar een Duurzame Grondstoffenvoorziening in de Bouw* (Apeldoorn: TNO Industrie en Techniek, 2008)., p. 36–37.

[17] Cramer, *Milieu*, 91.

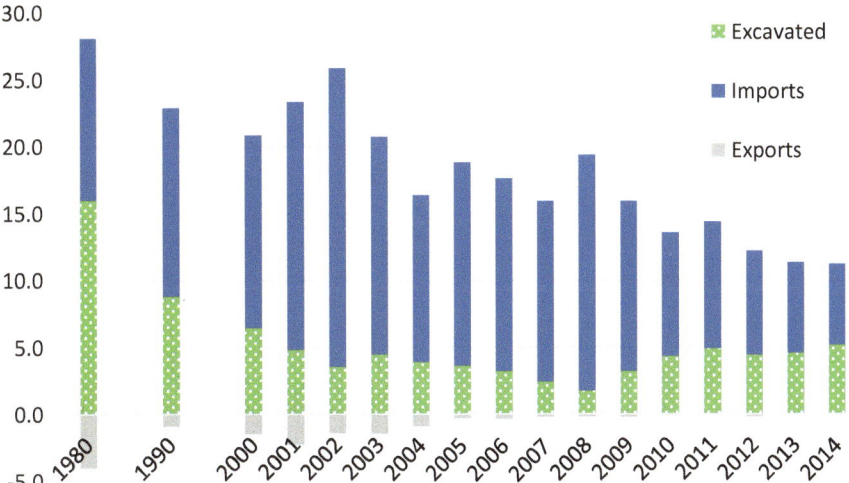

Graph 19.2 Excavation, import, and export of gravel 1980–2011 (million tons)
Source: Rijkswaterstaat, Waterdienst en Compendium voor de leefomgeving (2000–2014)

development, recreation, waterfront residences, water management, the dredging of shipping channels. In this way projects can be realised that improve spatial quality and that
provide the Netherlands with raw materials.[18]

This policy shift put an end to the promised reduction in gravel extraction. In
Limburg the news was followed with eagle-eyes. In 2009, Limburg's newspapers
headlined: 'Dredgers return' and 'Open door for gravel mining.'[19] Nature conservation organisations took a more positive view of the news. They were of course
already struggling with the right-populist policies of the first Rutte cabinet. Budgets
for nature development projects had been cut by 70%, environmental regulations
relaxed and elements of the European Natura 2000 network were postponed or terminated.[20] Multi-functional gravel mining offered unexpected chances for nature
development. The former opponents, dredging companies and nature conservationists, became new partners in the creation of wet nature development projects.[21]

As a rule, the additional demands placed on gravel and sand extraction led to
price increases, to a decline in domestic production, and increasing imports of sand
and gravel, particularly from Germany (see Graph 19.2 and Fig. 19.1). Germany

---

[18] Ministries of VROM, LNV VenW en EZ, *Nota Ruimte, ruimte voor ontwikkeling*, deel 4: text
after parliamentary consent, Den Haag 2006, p. 165.

[19] Cited in Bert Hammes and Jan Hensels 'Baggermolen keert terug' in *Dagblad de Limburger,*
October 5th 2009; Bert Hammes en Jan Hensels, 'Deur open voor ontgrindingen' in *Dagblad de
Limburger* October 5th 2009; Bert Hammes en Jan Hensels, 'Grind roept emoties op' in *Dagblad
de Limburger* October 9th 2009.

[20] Arjen Buijs, Thomas Matthijssen, and Bas Arts, "The Man, the Administration and the Counter-
Discourse:' An Analysis of the Sudden Turn in Dutch Natures Conservation Policy,' *Land Use
Policy* 38 (2014): 676–84.

[21] Sander Heijne, 'Struinnatuur in ruil voor grind', in *De Volkskrant*, August 10, 2011, p. 18.

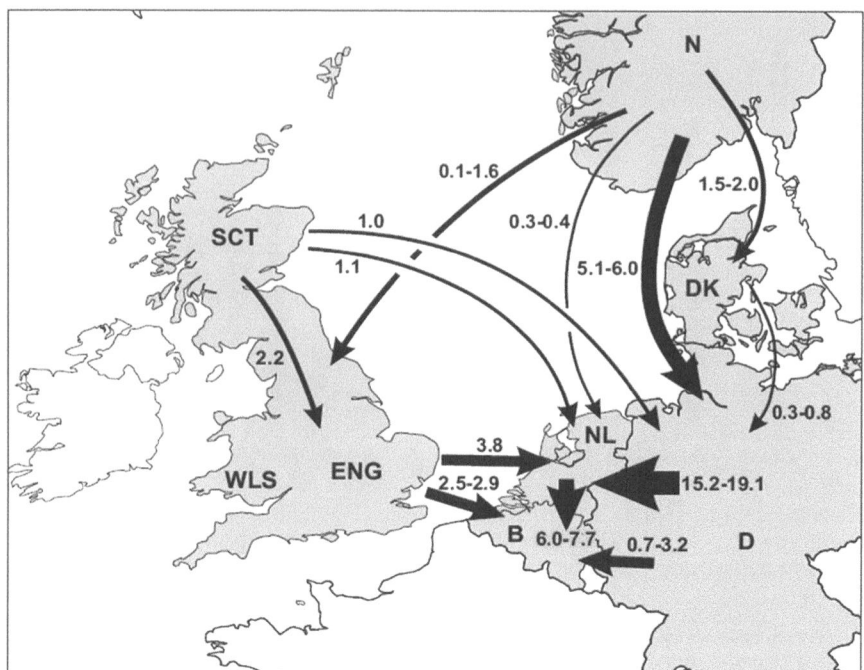

Fig. 19.1 Net flows of trade in sand, gravel and crushed stone among the countries around the North Sea

Source: M.J. van der Meulen, T.P.F. Koopmans, and H.S. Pietersen, "Construction Raw Materials Policy and Supply Practices in Northwestern Europe," *Aardrijkskundige Mededelingen*, Industrial Minerals – Resources, Characteristics and Applications., 13 (2003): 19–30

was not pleased. The Netherlands was simply transferring its sustainability problems to a neighbouring country. But the Tommel Commission redefined the problem. German irritation about Dutch policy shifts was seen as an indirect encouragement for Dutch policy:

> The commission views resistance in Germany as a positive development for Dutch policy. Whereas in the Netherlands there seems to be a consensus that we want to work toward the development of qualitatively good and socially responsible excavation projects, in Germany the debate about goals is still in full swing. The commission expects that in the long run, due to social pressure, more demands will be made on projects, which will ultimately lead to a decline in the acquisition of cheap sand and gravel from Germany. (…) A possible price increase will, in point of fact, provide new possibilities for multi-functional projects and the utilisation of secondary raw materials.[22]

---

[22] D.K.J. Tommel, G. Blom, and M.A. van Weel, eds., *Slotadvies van de Commissie Taakstellingen en Flankerende Beleid Voor Beton- en Metselzandvoorziening* (Ministerie van Verkeer en Waterstaat, Interprovinciaal Overleg (IPO), December 17, 2009), 18.

The phasing out of a directive role for the national government was concluded in 2012 with a parliamentary debate. In regard to the diplomatic pressure from Germany, Minister M.H. Schulz van Haegen (Infrastructure and Environment) wrote:

> Via diplomatic channels, the cabinet has received signals from the government of Nordrhein-Westfalen that it is not happy with this. The Netherlands does not unnecessarily want to burden its neighbours with the spatial demands of raw materials acquisition (…) I share the expectations of the Tommel Commission that new projects have the potential to replace a substantial share of these imports. But this is dependent on market and price developments and on the demands that are imposed in Germany on the execution of sand mining projects and on the processing after extraction.[23]

In the eyes of the Dutch government, increasing German exports to the Netherlands were the result of a free market. The lack of German sustainability measures made its shallow subsoil-resources attractive for the Dutch market. The remarks could be read as a covert bit of advice to German politicians. The Dutch government was not averse to more stringent demands and higher prices. It reasoned that these provided more possibilities for recycling and for more expensive multi-functional excavation.

The so-called 'structure schemes' for the shallow sub-soil resources with their estimates of demand and agreements about levels of supply were unique for the Netherlands. In contrast to surrounding countries, these kinds of strategic explorations were unusual in the Netherlands.[24] After the turn of the century, the Dutch government traded in its national perspective on mineral raw materials for a broader European perspective. Research undertaken at the behest of the European Commission revealed that the Netherlands was the largest net importer of mineral shallow sub-soil resources (see also Fig. 19.1).[25] In 1910, the top layer of Dutch roads consisted of asphalt made with Belgian and Scottish crushed rocks. Concrete was to a significant degree composed of gravel from Germany. By contrast, the foundations of buildings and constructions consisted primarily of domestic construction and demolition waste.

---

[23] M.H. Schultz van Haegen (minister of infrastructure and environment), in letter to parliament 'Slotrapportage afbouw rijksregierol bij ontgrondingen', March 5th 2012, p. 7. Kamerstukken 2012.

[24] Environmental historian Henk van Zon mentions among other things strategic surveys of coal in Germany and Great Britain from the eighteenth to the twentieth centuries. See Henk van Zon, *Geschiedenis en Duurzame Ontwikkeling, Duurzame Ontwikkeling in Historisch Perspectief, Enkele Verkenningen*, Vakreview (Nijmegen: Universitair Centrum Milieuwetenschappen, 2002), 36–53.; Another unique exception is the dispute around the zinc mine in Kelmis that in the course of the determination of the boundaries of the Kingdom of the Netherlands in 1820 led to a neutral territory administered jointly by The Netherlands (and after 1830 Belgium) and Prussia.

[25] M.J. van der Meulen, T.P.F. Koopmans, and H.S. Pietersen, 'Construction Raw Materials Policy and Supply Practices in Northwestern Europe,' *Aardrijkskundige Mededelingen*, Industrial Minerals - Resources, Characteristics and Applications., 13 (2003): 19–30.

After the turn of the century, local judgements about environment, economy and spatial planning determined the development of the national mining of raw materials. Despite the existence of large quantities of sand, gravel, and marl, the domestic mining of mineral raw materials languished. The Netherlands trusted to international trade for its supplies of raw materials. But the question was what the limits and risks were of this outsourcing of raw materials supplies.

19.5 'Toward a Strategy for Raw Materials'

This question began to nag in 2008, but at an international European level. The European Union (EU) had become increasingly dependent on countries outside the Union for its raw materials:

> On the one hand, the EU has many raw material deposits. However, their exploration and extraction are facing increased competition for different land uses and a highly regulated environment, as well as technological limitations in access to mineral deposits. On the other hand, the EU is highly dependent on imports of strategically important raw materials which are increasingly affected by market distortions. In the case of high-tech metals, this dependence can even be considered critical in view of their economic value and high supply risks.[26]

European leaders saw access to raw materials as strategic and necessary prerequisites for remaining competitive and for economic growth and jobs. The coming of specialised high-tech industry and services from the 1980s on had altered the material basis of the economy. On the one hand this mitigated the growth of material flows, while on the other hand ever higher demands were placed on exotic material properties and on purity. Scarce materials were the basis of a number of specific specialised applications. Rare molybdenum (Mo) for example, was used in the production of steel alloys for the chemical, the offshore and the automobile industries. Antimony, also rare, was applied in fire arresting compounds, in catalysts for plastics and in batteries.[27]

Materials scarcity proceeds from various causes. A number of these materials are geologically rare. In addition, mining plays an important role. Supply and demand are co-determinants of the scale of mining. Economically viable mining depends on the required purity and other effort required in the course of mining and purification. Economic profitability is varies with minerals that are by-products of other mineral

---

[26] "*The Raw Materials Initiative - Meeting our Critical Needs for Growth and Jobs in Europe,*" Communications from the Commission to the European Parliament and Council COM (2008) 699 (Brussels: Commission of the European Communities, April 11, 2008), 2.

[27] Th. Henckens, *Managing Raw Materials Scarcity, Safeguarding the Availability of Geologically Scarce Mineral Resources for Future Generations* (Universiteit Utrecht, 2016)., 74–79 and 110–112.

extraction. Due to all these factors, the debate concerning the degree of scarcity and depletion of materials is often left undecided.[28]

At the outset of the 1970s the Club of Rome had already put the issue of the depletion of raw materials squarely on the global agenda. After that, at least for mineral raw materials, the debate receded into the background. At the beginning of this century the issue received renewed attention in political circles. In November 2008 the European Commission launched the Raw Materials Initiative. A year later a European working group showed how immensely vulnerable Europe was in regard to imports of raw materials. These imports were a result of changes in international market relations due to the concentration of extraction in specific regions. This geographic concentration was not simply a geological phenomenon. Social pressure had shifted mineral extraction in Europe to other continents.

The analyses acquired increased importance because of the imposition of a quota on the extraction of rare earth metals by the Chinese government in 2009. An expected shortage of materials that threatened the production of mobile phones, flat-screen TVs, hybrid automobiles, solar panels and other developments in communications technology and sustainability, rapidly became world news.[29]

In 2010 the European Ad Hoc Working Group on Defining Critical Raw Materials presented the first results. It identified a group of 41 economically important minerals, of which 14 were potentially risky. The 'rare earths' and the platinum group metals (PGM) were risky due to quasi-monopolies by respectively China and Russia; the transition metals Tungsten and Niobium – much used in the fabrication of stainless steel – due to the enormous economic impacts.[30]

In response to the European investigation, the Dutch government requested Statistics Netherlands and TNO (the National Applied Science Research Institute) to do a study of materials crucial for the Dutch economy. The first results were reassuring. Direct use of the 41 'European' minerals was relevant for only a small fraction of the Dutch economy and the economy had no great degree of direct dependence on the 14 most crucial materials.[31]

But subsequent analyses of the production chains revealed a much greater dependency. Critical materials were used as raw materials in only 7% of production processes. These raw materials were much more often worked into metals (52%) or as

---

[28] A pessimistic current among scientists assumes depletion; more optimistic scientists stress that higher prices will lead to the discovery of new mining locations and that technological developments will lead to the adoption of alternative materials. *Ibid.* 16–18.

[29] Piet Depuydt, 'Chinezen beheersen de zeldzame aardmetalen: Tekort aan onmisbare grondstoffen zet verhouding in de wereld op scherp', in *NRC*, sectie economie, 16–1-2010, p. 13.

[30] *Critical Raw Materials for the EU, Report of the Ad-Hoc Working Group on Defining Critical Raw Materials* (Brussels: European Commission, Enterprise and Industry, 2010).

[31] *Critical Materials in the Dutch Economy, Preliminary Results* (The Hague / Heerlen: CBS Center for Policy Related Statistics, 2010).

Table 19.1 Share of the critical raw materials Indium, Neodymium en Copper in the Dutch economy in 2011

		Indium	Neodymium	Copper
Import of goods	Million EUR	20,914	553	33,809
Re-exports	Million EUR	15,686	343	22,314
Exports	Million EUR	26,902	555	36,008
Domestic imports	Million EUR	5229	210	11,495
Domestic export	Million EUR	11,217	212	13,694
Domestic use:				
Consumption	%	15	19	12
Investments	%	39	8	22
Intermediate use by industry	%	46	73	66
Total domestic imports	%	100	100	100
Added value	Million EUR	938	268	2889
As % gdp	%	0.2	0.1	0.6
Jobs	× 1000	22.3	7.3	70.4
Jobs % of Dutch Total	%	0.3	0.1	0.9
Import value Total NL	Million EUR	375,393		
Export value Total NL	Million EUR	398,757		
Total gdp	Million EUR	513,525		
Jobs (LISA) NL	× 1000	7976		

Source: Korteweg (2011) p. 92

intermediate products (37%).[32] A so-called 'quick scan' analysed the material flows of Indium, Neodymium and Copper for the Dutch economy.[33] Indium was applied in solar panels and LEDs. Neodymium was imported into the Netherlands chiefly as an ingredient of permanent magnets, used in windmills because of their low maintenance demands. The use of Neodymium had an added value of 268 million euros and was related to 7300 jobs. The impact of Indium at 938 million euros and 22,700 jobs, was even greater (see Table 19.1). The report showed the impact these specific materials had in the Dutch economy. But the report did not commence with the economic interpretation, but with geopolitical concerns.

> The West no longer determines international trade and the political and economic balance of power has shifted in favour of emerging economies like China, India and Brazil.

The report pointed out the increasing popularity of state capitalism, with China as an example. The new 'multi-polar' relationships in the world were characterised by instability and ad hoc relations. The increased complexity might possibly lead to

---

[32] Derk Bol and Ton Bastein, *Critical Materials and the Netherlands - a View from the Industrial-Technological Sector* (Delft: M2i Materials Innovations Institute / TNO, 2012), 5.

[33] *Rem Korteweg, Op Weg Naar een Grondstoffenstrategie, Quick Scan ten behoeve van de Grondstoffennotitie* (The Hague: The Hague Centre for Strategic Studies / TNO / CE Delft, 2011).

friction among countries. The scarcity of raw materials could become one of the most important bones of contention.

The report speculated on chances for the Netherlands in the areas of recycling, pre- and post-consumption. Much might be gained by applying supply chain approaches and 'cradle-to-cradle' principles to the flows of critical materials. And according to the report critical materials could also play a role in development aid. 'The raw materials sector, if well-organised, is a catalyst for economic growth in African countries.'[34] How and if these recommendations were entangled, was not made clear. The 'quick scan' seemed above all to be a pragmatic summing up of opportunities.

The reports were the most important source of inspiration for the raw materials memorandum that was presented to parliament in 2011 by no less than four government ministries. In the multi-polar world 'security regarding the supply of raw materials had also for the Netherlands to a certain extent become an economic and safety concern.' According to the politicians, it would be obligatory to interfere with the market if that proved necessary. But the scarcity of raw materials also offered possibilities, like urban mining. Waste should be seen as a raw material and it would be necessary to invest in so-called raw-materials roundabouts. The memorandum also noted that sustainable extraction and processing of raw materials also offered opportunities for development aid.[35]

New research by TNO into materials in the Dutch economy proposed another addition to the three familiar vulnerabilities. In addition to the geological-economic risks (reserves and production),[36] the geo-political risks (stability of nations), and price fluctuations, TNO also pointed to the reputational damage for firms in the use of materials. A number of critical materials were appropriated with little respect for the environment or norms for well-being. This could possibly lead to reputational damage for companies that applied these materials. Suppliers to the transport equipment industry ran the greatest risk, according to TNO. This was due to their use of gold, tantalum, and tin, the mining of which in African countries was controversial.[37] With this last point the report made visible the impact of economic activities on well-being and the environment elsewhere in the world. Well-being was not only a concern for governments, but also a strategic factor for companies.

---

[34] Rem Korteweg, *Op Weg Naar een Grondstoffenstrategie, Quick Scan ten behoeve van de Grondstoffennotitie* (The Hague: The Hague Centre for Strategic Studies / TNO / CE Delft, 2011).

[35] '*Grondstoffennotitie*' aangeboden aan de Tweede Kamer op 15 juli 2011 door de ministers van Buitenlandse Zaken, Economische Zaken, Landbouw & Innovatie en staatsecretarissen van Infrastructuur & Milieu en Buitenlandse Zaken.

[36] A recent dissertation pointed to the development of 'absolute' scarcity and to the depletion in the short term of a number of metals like antimony (20 year supply), gold (40 years), zinc (80 years) and molybdenum (80 years). Henckens, *Managing Raw Materials Scarcity*.

[37] Ton Bastein and Elmer Rietveld, *Material in the Dutch Economy, a Vulnerability Analysis* (Delft: TNO, 2015), 42.

19.6   New Issues in Well-being

New issues in well-being, both 'here and now', 'later' and 'elsewhere' emerged in relation to the acquisition both of abundant mineral raw materials like gravel and sand as well as scarce materials like molybdenum and indium. Questions concerning trade-offs among economy, environment, and spatial planning; on the displacement of problems from the Netherlands to abroad; on the dependence of Europe on the rest of the world. But also questions about the responsibility of private enterprise and governments. Where did these begin and end? And in addition, also questions about scale. What scale was best suited to taking action in a world of global flows of trade? And finally questions about initiative and engagement: where did initiative lie in a manifold 'multi-polar' world?

Finding a balance proved to be the big challenge and was at times no sinecure. For example in 2013, the Foundation 'The Green Accountancy Chamber' – a critical follower of Dutch environmental policy – issued a report on Neodymium. This material was used in the permanent magnets of windmills. In the course of extracting this rare earth metal in Baotou in China, radioactive and poisonous by-products were deposited in a lake measuring 120 km^2.

> This is the deadly and sinister side of the windmill manufacturers that we don't like to be reminded of.[38]

The report became popular above all under critics of wind energy. The VVD (the right-wing People's Party for Freedom and Democracy) asked questions about Baotou in parliament. Minister of Economic Affairs, Kamp, (also VVD) shared the concerns about people and the environment. He also offered China assistance via the Special Emissary in combatting the harmful side-effects, but no further steps were taken.[39] In the provincial estates of North Holland, North Brabant and Gelderland the PVV (ultra-right populist Party for Freedom) got this issue on the agenda. In Gelderland in 2014 a proposal was submitted to ban the use of Neodymium in windmills. After legal evaluation, the proposal was withdrawn by the provincial government in 2015.[40]

---

[38] *Windenergie in Nederland, de Dodelijke Keerzijde van Windenergie* (Apeldoorn: De Groene Rekenkamer, 2013).

[39] Tweede Kamer der Staten Generaal, Aanhangsel van de Handelingen, vergaderjaar 2013–2014, '*Vragen gesteld door de leden der Kamer, met daarop door de regering gegeven antwoorden*' no 1031

[40] Besluitenlijst GS Gelderland 10 februari 2015, nr 3 (2014–002183) '*Beleid over gebruik Neodymium in windmolens*' (http://applicaties.gelderland.nl/asp2008/besluitenlijst/c-lijst.asp?AgendaID=789).

Literature

Anonymous, (1989a, 19 January). 'Eigen cementindustrie in open Europa niet nodig', in *Limburgs Dagblad*, p 21.

Anonymous, (1989b, 19 January). 'Milieu wint het van economisch belang bij verbod tot afgraven mergelplateau' in *NRC Handelsblad.*

Bastein, Ton, and Elmer Rietveld. (2015). *Material in the Dutch Economy, a Vunarebilty Analysis.* Delft: TNO.

Bol, Derk, and Ton Bastein. (2012). *Critical Materials and the Netherlands – a View from the Industrial-Technpological Sector.* Delft: M2i Materials Innovations Institute/TNO.

Bosch, A., and W. van der Ham. (2015). *Twee Eeuwen, Rijkswaterstaat 1798–2015.* 2e druk. Asten: Nieuwe Uitgevers.

Buijs, Arjen, Thomas Matthijssen, and Bas Arts. (2014). "'The Man, the Administration and the Counter-Discourse': An Analysis of the Sudden Turn in Dutch Nature Conservation Polict." in *Land Use Policy* 38: 676–84.

Cammen, H. van der, and L.A. de Klerk. (1999). *Ruimtelijke Ordening, van Plannen Komen Plannen, de Ontwikkelingsgang van de Ruimtelijke Ordening in Nederland.* 4e ed. Utrecht: Het Spectrum.

Cramer, Jacqueline. (2014). *Milieu.* Elementaire Deeltjes 16. Amsterdam: Amsterdam University Press.

CBS, (2010a). *Critical Materials in the Dutch Economy, Preliminary Results.* The Hague/Heerlan: CBS Center for Policy Related Statistics.

CBS, (2010b). *Critical Raw Materials for the EU, Report of the Ad-Hoc Working Group on Defining Critical Raw Materials.* Brussels: European Commission, Enterprise and Industry, 2010.

Disco, Nil, and Frank Veraart. (2016). 'A Farewell to Big Planning? 1990–2010.' In *Builders and Planners, a History of Land-Use and Infrastructure Planning in the Netherlands*, edited by Jos Arts, Ruud Filarski, Hans Jeekel, and Bert Toussaint, 351–437. Delft: Eburon.

Eikelboom, R.T., E. Ruwiel, and J.J.J.M. Gouwmans. (2001). 'The Building Material Decree: An Example of a Dutch Regulation Based on the Potental Impact of Materials on the Environment.' in *Waste Management* 21: 295–302.

"ENCI's Struggle for the Margaten Plateau, a Clash of Economical and Environmentel Interests." Universiteit Maastricht (Faculteit Cultuurwetenschappen) Bachelorscriptie, December 6, 2009. www.oudsintpieter.com/Presentaties.htm.

Grondelle, G.W. (1978). *Ontgrondingen, over de Noodzaak En Mogelijkheden Voor Een Ander Beleid T.a.v. de Winning van Mergel, Grind, Zand, Klei, Veen, Enz.*, 's Gravenland: Stichting Natuur en Milieu.

Henckens, Theo. (2016). *Managing Raw Materials Scarsity, Safeguarding the Availability of Geological Scares Mineral Resources for Future Generations.* Universiteit Utrecht.

Hofstra, U., B. van Bree, R. de Wildt, and J. Neele. (2006). *Scenariostudie BSA-Granulaten, Aanbod En Afzet van 2005 Tot 2025.* Sittard: INTRON / Expertisecentrum Bouwstoffen Rijkswaterstaat.

Ike, Paul. (2000). *De Planning van Ontgrondingen.* Rijksuniversiteit Groningen, Geo Pers.

Korteweg, Rem. (2011). *Op Weg Naar Een Grondstoffenstrategie, Quick Scan Ten Behoeve van de Gronstoffennotitie.* The Hague: The Hague Centre for Stratagic Studies/TNO/CE Delft.

Meulen, M.J. van der, T.P.F. Koopmans, and H.S. Pietersen. (2003). 'Construction Raw Materials Policy and Supply Practices in Northwestern Europe'. in *Aardrijkskundige Mededelingen,* Industrial Minerals – Resources, Characteristics and Applications., 13: 19–30.

Meyer, Han. (2016). *De Staat van de Delta, Waterwerken, Stadsontwikkeling En Natievorming in Nederland.* Nijmegen: Van Tilt.

Mulder, Evert. (2008). *Kringbouw, Naar Een Duurzame Grondstoffenvoorziening in de Bouw.* Apeldoorn: TNO Industrie en techniek.

Rode, Thomas. (2010). 'Een Wereld van Cement? De Strijd Om Het Plateau van Margraten, 1976–1989.' in *Studies over de Sociaal-Economische Geschiedenis van Limburg* 55: 26–48.

Sigmond, G.J.A., A.A. Veerbeek, F. Fokke, D.J.van Herwaarden, L.S.de Jonge, and J.A.J. Kemps. (1984). *1974–1984, Tien Jaar: Industrie En Ontgrondingen*. FODI, De Steeg.

"The Raw Materials Initiative – Meeting Our Critical Needs for Growth and Jobs in Europe." Communications from the Commission to the European Parliament and Council COM (2008) 699. Brussels: Commission of the European Communities, April 11, 2008.

Tommel, D.K.J., G. Blom, and M.A. van Weel, eds. (2009, 17 December). *Slotadvies van de Commissie Taakstellingen En Flankerende Beleid Voor Beton- En Metselzandvoorziening*. Ministerie van Verkeer en Waterstaat, Interprovinciaal Overleg (IPO).

"Windenergie in Nederland, de Dodelijke Keerzijde van Windenergie." Apeldoorn: De Groene Rekenkamer, 2013.

Zon, Henk van. (2002). *Geschiedenis En Duurzame Ontwikkeling, Duurzame Ontwikkeling in Historisch Perspectief, Enkele Verkenningen*. Vakreview. Nijmegen: Universitair Centrum Milieuwetenschappen.

Chapter 20
Energy and Plastics: The Slow Transition

Harry Lintsen

Contents

Abstract This chapter, exploring fossil subsoil resources, focuses on two domains: energy and plastics. The energy section analyses the difficult transition to renewable energy sources. The focus here is on electricity because promising renewable energy sources like biomass, windmills and solar panels contribute above all to the supply of electricity. There is, moreover, a close relationship among oil, natural gas and electricity.

Dutch electricity supply was long trapped in tensions among the policy of the provincial electricity suppliers, the energy policy of the national government (in particular the Ministry of Economic Affairs) and the environmental movement, with as main issues decentralised electricity generation, the inclusion of nuclear power, the role of domestic natural gas and energy-saving. Privatisation and liberalisation are setting the electricity sector completely on its head. There is now more room for other forms of electricity generation, in particular decentralised generation and heat-power coupling. Opportunities for renewable energy sources have increased,

---

This chapter is written by Harry Lintsen with contributions by Rick Hölsgens and Ben Gales.

among other things thanks to international agreements ('Paris') in connection with climate change.

The plastics sector too has undergone dramatic changes in this period. The *production* of bulk plastics and artificial fibres still takes place in the Netherlands, but hardly at all by *Dutch* firms. The plastics *processing* industry, that consists above all of small and medium-sized firms (up to 50 employees) has developed into the Netherlands' most innovative sector. The attitude toward plastics has become ambivalent. They have shaped a life of comfort, ease, luxury, sport, and games. At the same time they are a source of litter, waste, 'plastic soup' and micro plastics.

Keywords Energy · Oil · Gas · Coal · Electricity · Nuclear energy · Environment · Decentralised generation · Heat-power coupling · Wind · Solar cells · Plastics · Plastic soup · Microplastics

20.1   The Trial

On June 24th, 2016 the court of The Hague issued a remarkable ruling: 'The [Dutch] State has to ensure that in 2020 the emission [of greenhouse gases] in the Netherlands is at least 25% lower than in 1990.'[1] The ruling was world news. Foreign newspapers, radio stations and websites like the Guardian, El Pais, the BBC and ABC Australian Radio covered the story. The BBC noted that

> the judgement was unprecedented in Europe, and unexpected. It pushes the Dutch government to honour its commitment to cut emissions.[2]

A foundation called Urgenda, directed by Marjan Minnesma, and almost 900 private co-plaintiffs wanted to use a court decision to force the state to do more about the emission of greenhouse gases. But the question was whether a judge was competent to pass such a judgment. In a democracy, was it not parliament that was responsible for dealing with issues like this?

An essential characteristic of the rule of law – so argued the court – was that an independent judge could (and sometimes had to) judge the actions of political organs like the government on the point of legal protection. That, it considered, was the case here. The state was legally obligated to take measures against climate change. After all, in all probability climate change has serious and life-threatening consequences for people and the environment:

> … the Netherlands will be confronted with higher average temperatures, changing precipitation patterns and rising sea levels … [with] heat waves and extremes of precipitation …

---

[1] Rechtbank Den Haag, zaaknummer C/09/456689/HA ZA 13-1396, uitspraak 24-06-2015, *ECLI:NL:RBDHA:2015:7145.*

[2] 'Netherlands ordered to cut greenhouse gas emissions' in *BBC News*, 24 June 2015.

dangerous situations on the lower reaches of the rivers ... increasing salinity in the coastal zones and less available water for agriculture ...[3]

In addition there are the global consequences of melting ice, desertification, a decline of biodiversity, threats to food production, and other big problems. In both national and international law, the precautionary principle then prevails: measures have to be taken, despite the fact that complete scientific certainty is lacking. The possible consequences are simply too serious. This has led to international treaties to combat climate change, signed, among others, by the Netherlands.

The court further pointed to the agreements made within the European Union. These have encumbered the state with the obligation to reduce greenhouse gas emissions in 2020 by 25–40% of 1990 levels. Present government policy will lead to at most a 20% reduction. According to the court this is unacceptable.

Meanwhile, the State has filed an appeal. Urgenda did not expect otherwise. Its strategy is ultimately to plead its case before the European Court. If Urgenda wins there, it will have consequences not only for the Netherlands, but also for the other member states. That could mean a breakthrough in the difficult energy transition. All the member states can then be compelled to take far-reaching measures.[4]

It is easy to see why Urgenda speaks of a difficult energy transition. At the moment, climate change is the most important argument for drastic cutbacks in the use of fossil fuels (coal, oil and gas). But controversy about the problems of fossil raw materials has been around for much longer. In the 1960s, air pollution, particularly due to coal combustion, was the main issue. The Club of Rome report in 1972 put the issue of depletion of fossil fuels squarely on the agenda. The oil crisis of 1973 fueled fears of foreign dependency, particularly on the Middle East. These issues continued to reappear in various guises and with shifting urgency in subsequent debates. The climate issue became dominant during the 1990s.

Suggestions for alternatives to fossil raw materials also go back a long way. Research into windmills, solar panels, biomass and other renewable energy sources goes back to the 1970s. But the harvest of a half-century of innovation, debate and policy seems rather meagre: around 4–5% of the total energy demand is supplied by 'sustainable,' 'green,' or 'renewable' energy. With respect to the implementation of renewable energy sources, the Netherlands is suspended somewhere near the bottom of the list of European Union member states.

Fossil energy sources have played an extremely important role in history. They contributed greatly to the fight against poverty. But since 1970 other issues are at stake and the Netherlands is searching for alternative ways to produce its energy. History reveals that radical changes often require a long gestation period. After that, things can nonetheless move quickly. Is the Netherlands now on the threshold of such a pivotal moment? It may also be possible that the process will grind to a halt and that the transition stagnates. This chapter will summarise the history of fossil and renewable sources of energy in the Netherlands over the past decades and make

---

[3] Rechtbank Den Haag, zaaknummer C/09/456689/HA ZA 13-1396, uitspraak 24-06-2015, *ECLI:NL:RBDHA:2015:7145.*

[4] Cox, H. (2011), *Revolutie met recht.*(pp 288) Maastricht: Stichting Planet Prosperity Foundation.

an effort to characterise the present phase in the transition. We set off with an overview of the supply of and demand for fossil energy sources.

20.2 The Energy Balance and the Energy Mix

A large stock of fossil raw materials is available in the Netherlands. In terms of the unit of energy 'joule,' the amount is about 10,000 PJ (petajoule or 10^5 joule, 2011). Two raw materials dominate: oil and natural gas. Oil is mainly imported, though 83% of it is exported again, immediately or after refining. Seventy-five percent of the natural gas is domestically produced and more than half is exported. The actual domestic consumption of fossil fuels is therefore significantly lower. Of the available 10,000 PJ the Netherlands itself consumes roughly 2500 PJ. Not all fossil raw materials are destined for energy production. Fifteen percent is used to produce plastics and other chemical products. In discussions on sustainability these substances are a distinct topic. In this chapter, we shall deal with plastics in a separate section.

Energy production thus relies on a remaining 2100–2200 PJ. Part of the energy flow (about 30%) undergoes an important intermediate transformation before it reaches the end-user, namely the conversion of almost all the coal and part of the natural gas into electricity. Ultimately, the flow of energy (including the electricity) is delivered to four important categories of end-users. The industry is the biggest consumer of energy, followed by transport, services (and agriculture) and households (Graph 20.1). Within the industry, chemicals, metals and foods are the most important energy consumers. In transport, road traffic claims the biggest share.[5]

Since the 1970s, energy supply in the Netherlands has been composed of a mix of chiefly oil and natural gas. While coal has a smaller share, it has still not entirely disappeared (Graph 20.2). Over this period, the transformation of fossil fuels into electricity steadily increased. The flow of energy from oil is chiefly consumed in transportation, while natural gas, coal and electricity find their way mainly to households, industry and services.

Other raw materials and energy sources are only a small part of this story. Renewable energy sources (including biomass, water power and geothermal energy) appear as extremely thin lines in the graphs. They are clearly overshadowed by their big competitors. On the other hand, their share is growing steadily – if slowly. Let us look more closely at the dominant energy sources. We focus on electricity, because promising, renewable sources of energy like biomass, windmills and solar panels contribute above all to the electricity supply. Moreover, there is a close relationship between oil, natural gas and electricity.

---

[5] An interactive overview of the energy balance and final consumption of the Netherlands 1975–2015 can be found at the website of the International Energy Agency (IEA):

Energy balance 1975–2015 at http://www.iea.org/Sankey/#?c=Netherlands&s=Balance

Energy final consumption 1975–2015 at http://www.iea.org/Sankey/#?c=Netherlands&s=Final%20consumption

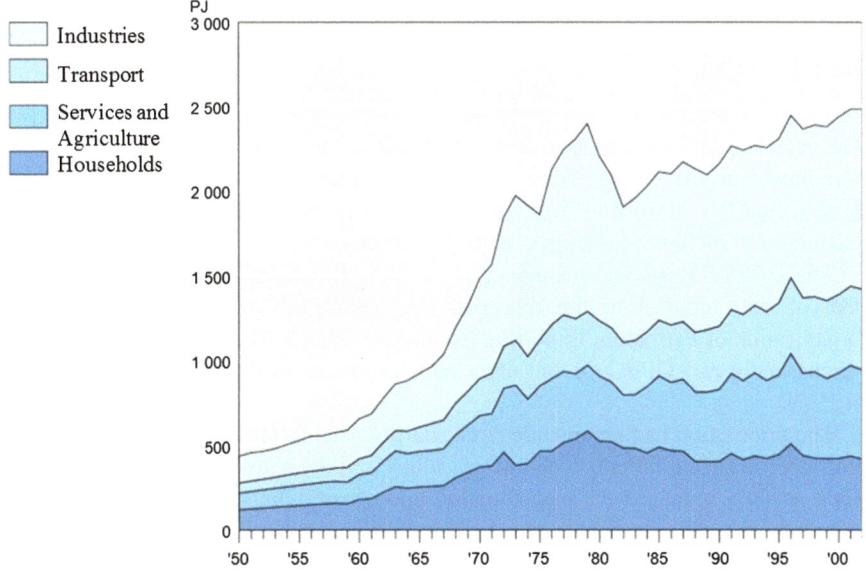

Graph 20.1 Energy consumers, 1950–2002
Source: *Statistisch Jaarboek 2004, 295*

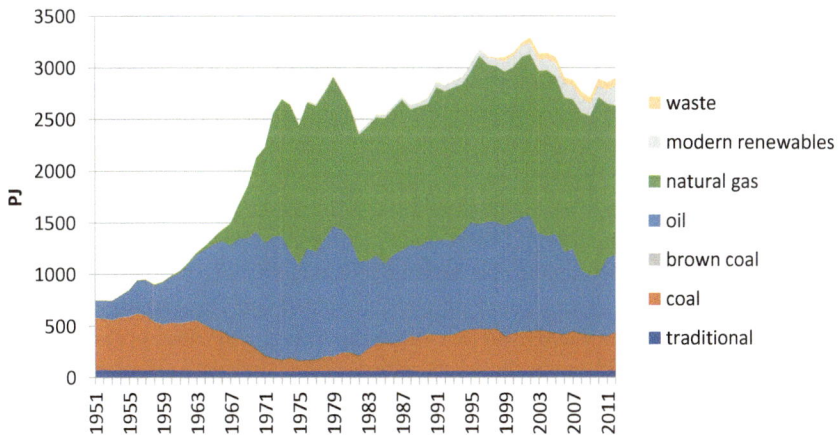

Graph 20.2 Energy consumption by sources of energy, 1951–2012
Source: B. Gales and H. Hölsgens, *Energy transitions in the Netherlands: Sustainability challenges in a historical and comparative perspective* (Groningen, 2016) 11 and appendix 1

20.3   Fossil

20.3.1   Oil

Around 1960 oil overtook coal in the Netherlands and 10 years later it became a dominant source of energy. Worldwide, five big American oil companies (Chevron, Exxon, Gulf Oil, Mobil and Texaco) and two European (Shell and BP) controlled oil production.[6] Shell was the largest in the Netherlands.

The 1973 oil crisis was a game changer. The oil companies surrendered a good deal of their control to the oil producing countries, partly organised in the Organisation of Petroleum Exporting Countries (OPEC). They were still privileged purchasers of raw OPEC oil, but that status too succumbed to the second oil crisis of 1979.

The price of oil had meanwhile risen sharply. Vacillating oil prices would thereafter exert a strong influence on energy policy and economic growth. In the 1990s Shell adroitly managed the new situation and became the world's most profitable company. But the company's position would weaken again around 2000 due to declining oil prices and the lack of takeovers. Growth in production stagnated. It appeared increasingly difficult and more expensive to develop new oil and gas fields. At the outset of the century, shale gas and shale oil occasioned a revolution in the energy supply.[7] They turned the world of oil, dominated by the OPEC cartel, on its head once again. Oil prices, that had meanwhile increased again, were once again under pressure and declined from 2014 on.

The Netherlands – even with such a powerful player as Shell – has no influence on the supply and the price of oil. And that holds to a great extent for the demand as well. Most of the oil is destined for motorised traffic. At the end of the 1950s and the beginning of the 1960s the demand for mobility increased explosively.[8] Initially it concerned mostly the increasing use of mopeds and only later the automobile. In the course of the 1970s, the government attempted to curb the explosion and saw an ally in the economic recession of the 1980s. But the number of passenger cars continued to increase exponentially, as did the number of kilometres driven per year and the average commuting time per week. Controversies about fuel for automobiles had no effect. Problems with the emission of carbon monoxide and other polluting gases were solved in the 1970s with the introduction of the catalyser. Threats to public health from lead in gasoline were eliminated by the development of lead-free gasoline in the 1980s.[9] Since the 1990s, fine dust from, among other sources, diesel

---

[6] See for the following: K. Sluyterman, 'Concurreren in turbulente markten 1973–2007' in Geschiedenis van Koninklijke Shell, deel 3, (Amsterdam, 2007: Boom) pp. 5–9; 93; 239.

[7] Shale gas and shale oil are mined from shale rock formations, consisting primarily of clay minerals.

[8] G. Mom, R. Filarski, Van transport naar mobiliteit. De mobiliteitsexplosie [1895–2005], (Zutphen, 2008: Walburg Pers), pp. 265, 374.

[9] H. Lintsen, T. Van Helvoort,. and R. Van Veen. De kracht van de katalysator. De magie van het onderzoek, (Eindhoven, 2014: Stichting Historie der Techniek) pp. 17–29.

engines, has become an issue. Such debates have little influence on the oil market. But the production of natural gas in the Netherlands is another matter.

20.3.2   *Natural Gas*

As little influence as the Dutch government had on the supply of oil, so strongly did it determine the supply and demand of natural gas. Intervention in sources of energy by the government was not new. It had decisively intervened in coal mining at the beginning of the twentieth century. Now it was the turn of the natural gas located in the subsoil of the province of Groningen. The concession for extracting natural gas in Groningen had been granted in the early 1960s to the Netherlands Oil Company (NAM) in which Shell and Esso (later Exxon) cooperated.[10] But the government retained a firm hold of developments in the field. The costs of production as well as the income were managed by the *Maatschap Groningen* in which the state partici- pated for 40% via the State Mines (DSM). Initially 75% of the profits were claimed by the state. In the 1970s this would increase to 95%. The *Maatschap Groningen* sold the gas to the *Gasunie,* in which the government had a share of 50%, both directly and via DSM. As we saw, the *Gasunie* built an elaborate natural gas net- work. Households switched from coal, fuel oil and municipal gas to natural gas for heating and cooking. Low prices stimulated firms to use natural gas and to invest in energy-intensive production processes, for example the aluminium smelter at Delfzijl. Energy-intensive branches like the greenhouse garden-farming industry began to use gas instead of fuel oil. Electricity companies switched from coal to natural gas. In 1970, the Netherlands supplied 92% of the internationally traded gas in Europe.

The oil crisis was a watershed for natural gas as well. The Netherlands restricted production of natural gas and acquired new competitors on the European market in the form of the Soviet Union, Norway and Algeria. The government lowered the rate of extraction in the Groningen field and encouraged the exploitation of smaller gas fields. Government policy aimed on the one hand at a leading role for natural gas in the Dutch energy supply, and on the other hand at maximization of the proceeds from gas sales in the short and long term. This made the government dependent on the oil price, from which the gas price was derived.

The European discussion over increased leeway for market forces in the European gas market made it necessary to restructure the natural gas supply system. The con- sequence was that the *Gasunie* was split up in 2005. In its new embodiment, the *Gasunie* has become the owner of the gas network with the state as sole stockholder. Selling gas – the *Gasunie's* other activity – has now devolved to Gas Terra, in which Shell and Exxon each participate for 25% and the state for 50%. In this way an

---

[10] See chapter 15 and also: Sluyterman, *Concurreren in turbulente markten*, pp. 225–229; Correljé, A. C. van der Linde en Th. Westerwoudt, *Natural Gas in the Netherlands. From Cooperation to Competition?* (Amsterdam, 2003: Oranje-Nassau Groep).

independent gas transport network has come into being which every supplier can use to bring its gas to market. Gas Terra is only one of the suppliers and the gas comes not only from the Groningen field, but also from smaller fields and from foreign sources. The transformation meant further limits to governmental influence on the energy supply.

Of the original quantity of Groningen gas – amounting to about 2600 billion m³ – less than 800 billion m³ of retrievable gas remains in the subsoil (2013).[11] Production has vacillated over recent decades and for several years hovered around 50 billion m³.[12] Since then, production has declined significantly. In 2017 the government decided to reduce gas extraction to about 22 billion m³ (or 770 PJ). That was in part a response to the growing protest in Groningen against the increasing number of earthquakes in the gas fields and dissatisfaction about the way the earthquake damage was being compensated.

In several instances over the past decades, gas extraction in the Netherlands has become controversial. A long struggle ensued over plans to extend gas mining into the Waddenzee. Only after gas mining was combined with nature development, did the government see fit to condone the development in 2004.[13] At present, gas mining is the scene of an ongoing struggle between residential and environmental interests on the one hand and economic interests and those of energy policy on the other.

20.4   Electricity, Natural Gas and Coal[14]

The introduction of renewable energy sources seemed to have the greatest chance of success in the field of electricity supply, but it turned out to require a major effort to acquire a toehold in this sector.

Until the early 1970s the electricity sector had a relatively stable organisation. Provincial (and a few municipal) companies produced electricity. They worked together in the Cooperating Electricity Production Companies (SEP). Together they had complete control over the facilities and wished for as little governmental interference as possible. With the coming of natural gas, the sector became involved in the industrial and energy policy of the government. The transition in electricity production from coal to gas went reasonably smoothly. By the mid-1970s, 80% of the electricity would be generated with natural gas. Conflicts emerged above all around

---

[11]*Aardgas in Nederkand.* http://aardgas-in-nederland.nl/de-toekomst-van-aardgas/aardgasreserves-en-verbruik/ consulted 19 April 2017.

[12]NAM, *Bron van onze energie.* http://www.nam.nl/feiten-en-cijfers/gaswinning.html consulted 19 April 2017.

[13]J.J. De Jong, E.O. Weeda, E.O., Th. Westerwoudt, A.F. Correljé, *Dertig Jaar Nederlands Energiebeleid. Van bonzen, polders en markten naar Brussel zonder koolstof,* (Den Haag, 2005: Clingendael International Energy Programme), pp. 153–155.

[14]This section based on: G. Verbong and F. Geels. 'The ongoing energy transition: lessons from a socio-technical, multi-level analysis of the Dutch electricity system (1960–2004)'. *Energy Policy* 35 (2007) pp. 1025–1037.

the introduction of nuclear energy, a darling of the Ministry of Economic Affairs. The Ministry was convinced that the future belonged to nuclear energy. It was essential to build up a new industrial sector in the Netherlands around this new technology. In this context, the construction of a nuclear power plant in the Netherlands itself would be very welcome. But the SEP charted its own course. One of the associated companies bought such a power plant in Germany to be built in Borssele, in the province of Zeeland.

In this period, the electricity sector experienced next to no trouble from the emerging environmental movement. Resistance to nuclear power was still modest. Air pollution was a hornets' nest, but this was mitigated by technical measures and the declining use of coal. The sector was totally disinterested in small-scale experiments with sun and wind energy, let alone in the utopian dream of 'small is beautiful.'

Once again, the oil crisis of 1973 initiated fundamental shifts. A year later the first *Energy Memorandum* appeared. This circumscribed the autonomy of the electricity producing companies. The Ministry of Economic Affairs acquired a decisive voice in the construction of nuclear power plants and the use of fuels in conventional power plants. The wide-ranging memorandum was inspired by two basic principles: reliability of the supply and the cheapest possible energy for big industry. To be sure it also took note of the environmental problems of electricity generation and addressed the depletion of raw materials, but these were clearly subsidiary issues. Only marginal attention was paid to renewable energy sources.

The electricity supply companies were also faced with a new phenomenon: the deployment of gas turbines and consequently the growth of decentralised energy production. Big industry already generated its own local electricity in order to partly provide for its own needs. With the advent of natural gas it began to install gas turbines to handle peak loads and thanks to a new type of gas turbines, gas became dominant in decentralised energy production.[15] This also generated waste heat (that the firms recycled into their production processes) and hence so-called 'heat-power coupling' was born.

The second oil crisis of 1979 led to the second *Energy Memorandum*. Saving energy became a top priority in addition to reliability and cheap energy. In this respect, heat-power coupling offered excellent opportunities. Surplus heat could be employed in the neighbourhood of the industrial plant and surplus electricity (on the basis of power) could be returned to the electricity grid. The government decided to provide extra natural gas for decentralised electricity production. But the electricity companies were far from happy with the 'uncontrolled' delivery of electricity by big industry. They were ill-prepared and paid only a low rate. The government had to intervene in order to enable local producers to connect up with the electricity grid on realistic terms.

In this period the government also began to champion nuclear power. It was, however, faced with a broad protest movement. Nuclear power would go on to

---

[15] It should be noted that the electricity companies too installed gas turbines in order to accommodate peak loads. Gas turbines are easier to start up than conventional power plants.

dominate the debate on energy supply for years to come, in the process overshadow-
ing other crucial issues like the scarcity of raw materials and the introduction of
renewable energy sources. Resistance to nuclear power also contributed to a certain
hesitancy within the environmental movement to address the climate issue. Fighting
climate change would only strengthen the hand of the proponents of nuclear power,
who framed this technology as the ultimate solution to global warming.[16]

The further diffusion of nuclear power was thwarted due to massive protests and
the accident with the Chernobyl nuclear power plant in 1986. Expensive oil was,
however, quickly replaced as fuel for power plants by cheaper coal. That in turn
demanded extra measures to deal with the environmental effects.

The electricity-producing sector was hard pressed. During the recession of the
1980s, privatisation and efficiency were the key concepts; de-regulation and liberalisation those of the 1990s. The European Union supported the latter policy and
was able to enforce it after the Treaty of Maastricht in 1992. The electricity sector
tried to retain a grip on developments by advancing proposals for restructuring. But
to no avail. The government was already working on a new Electricity Law. This
came into force in 1989 and after modifications was superseded by the Electricity
Law of 1998.

The Electricity Laws led to a reshuffling and splitting up of the grids of the old
electricity companies. The high tension grid has become state property and is managed by a new organisation, Tennet. The grids for medium and intermediate tension
have become the property of public grid managers in the hands of municipalities
and provinces. They deliver electricity to users who can in turn remit electricity
from proprietary windmills and solar panels to the grid. At the moment, there are a
limited number of regional grid managers in the Netherlands, like Endiner, Liander,
and Stedin, all of which also deliver gas.

In addition, the laws have created a new set of actors. The old provincial and
municipal companies have been privatised and transformed into commercial electricity companies. These buy electricity as well as produce it themselves. The biggest companies are Nuon, Essent, Eneco and EPZ. They are the owners of the power
plants. In addition, the commercial electricity companies initiate joint ventures,
make deals with firms that generate electricity locally and own windmill parks.
End-users are free to choose from among these suppliers. Market mechanisms have
come to replace the internal planning mechanisms of the original electricity sector.

The new structure provided opportunities for decentralised electricity generation. The Ministry of Economic Affairs especially supported heat-power coupling
schemes by means of subsidies and tax breaks. These schemes either focused on
power (and hence electricity) with heat as a by-product, or focused on heat for the
heating of office buildings, hospitals and agricultural greenhouses with electricity as
a by-product. However, the rapid increase in the number of such schemes occasioned a crisis in the planning and monitoring of electricity production. This was
exacerbated by the development of electricity production on the basis of wind and

---

[16]Duyvendak, W. *Het groene optimisme. Het drama van 25 jaar klimaatpolitiek*, (Amsterdam, 2011: Uitgeverij Bert Bakker), pp. 65.

solar power. Coordinating supply and demand has become much more complicated with the advent of decentralised electricity production. 'Smart grids' are currently among the 'trending topics' in the sector.

20.5   Environment, Depletion and Climate[17]

The driving force behind the changes in the electricity sector was initially the Ministry of Economic Affairs, seeking to get a grip on the national electricity supply in connection with its interests in natural gas and nuclear power. Privatisation and liberalisation would only later begin to play a role and the European Union would also become a new actor in the background. The Ministry pursued low energy costs for industry, reliability of the supply, and decentralized generation. Environmental problems, depletion of fossil fuels and climate change were secondary considerations. The introduction of renewable sources of energy was a marginal aspect of energy policy. The societal debate on sustainability barely made a dent, at least not in the energy policy of the Ministry of Economic Affairs.

The debate came in waves.[18] At the beginning of the 1970s, protest groups, environmental organisations and political parties were up in arms above all around the issues of air pollution, the depletion of raw materials and nuclear power. The debate on nuclear power polarised the country and dragged on interminably. Air pollution was a quite different matter. The focus here was on sulphur dioxide (SO_2) and its effects on public health. The ministry of Public Health, Spatial Planning and Environmental Management (VROM) took the lead on this issue. The reduction of SO_2 emissions in those years was a success. That was partly the outcome of legislation and the implementation of technologies like gas scrubbers to remove sulphur, but above all a beneficent side-effect of the transition from coal to natural gas. Concern about the depletion of fossil fuels moved the government to pay ongoing attention to energy conservation. The Ministry of Economic Affairs adopted energy conservation as one of the cornerstones of its energy policy. Subsidies and so-called 'multi-annual agreements' were mobilised to enrol private industry. Regulation, permits and energy taxes comprised the stick in the event these carrots failed to produce adequate results.[19] The recurring high energy prices formed a favourable backdrop. This story too turned out a success.

The debate ebbed away but resurfaced with renewed force in the course of the 1980s. Nuclear power and acidification and their connection with fossil raw

---

[17] See for this section the elegant overview of the history of sustainable energy in the Netherlands: G. Verbong. G, et al. *Een kwestie van lange adem. De geschiedenis van duurzame energie in Nederland.* (Boxtel, 2001: Aeneas uitgeverij).

[18] See also: J. Cramer, *Milieu.* (Amsterdam, 2014: Amsterdam University Press).

[19] J.J. De Jong, E.O. Weeda, E.O., Th. Westerwoudt, A.F. Correljé, A.F. *Dertig Jaar Nederlands Energiebeleid. Van bonzen, polders en markten naar Brussel zonder koolstof,* (Den Haag, 2005: Clingendael International Energy Programme), pp. 106.

materials were the key issues. The accident at Chernobyl put an end to the nuclear dream of Economic Affairs and the nuclear power lobby. The debate on acidification shifted attention from public health to the environment and from sulphur dioxide to ammonia emissions (SO_2 and NH_3). In this case too, it proved possible to take measures to keep emissions within legally defined norms.

From the 1990s on fine dust and greenhouse gas emissions were the most important issues. The problem of fine dust (airborne particles smaller than 10 micrometers) re-oriented the discussion once again to public health. With greenhouse gas emissions the debate about fossil raw materials acquired a new aspect: climate change. The biggest concern was CO_2 emissions but these proved far more resistant to legal and technical measures than the other emissions. Moreover, the concern was global, leading to international agreements about reduction that the Netherlands could hardly ignore. In the short term – up to the year 2020 – it seems likely that the government can satisfy the norms with 'classic' policy measures (energy conservation, energy covenants and so on). For the long term, however, more ambitious measures will have to be taken.

At certain moments it could seem that there was broad support for taking measures in energy production. At other times support has proven to be quite limited and riddled with weak spots. One of the weak spots is that the debate is constantly plagued by uncertainties. In each period there has been fundamental debate about the reserves of fossil raw materials and about the effects of air pollution on public health and the environment.

The situation with respect to climate change and CO_2 emissions is no different. Some question the analyses that predict global warming, consider the severity of the problem grossly exaggerated and find political support for their views.[20] Others argue that the influence of CO_2 emissions on global temperatures is still imperceptible and that humans do not have the power to fiddle with the 'climate dials.'[21] The court of law that ordered the Dutch state to reduce CO_2 emissions acknowledged that there is scientific uncertainty about '...the question when and to what degree, which specific effects will become manifest, and also about the effectiveness and possible negative consequences of certain precautionary measures.'[22] The court, however, considered the fact that there is consensus among climate scientists and within the international policy field about the serious consequences of CO_2 emissions. The precautionary principle thus justified the ruling that appropriate measures be taken.

---

[20] In this spirit the PVV (Party for Freedom) – the biggest but one party in the Second Chamber in 2018 – submitted a motion in which the government '... in view of the fact that the climate treaty is bad for the Dutch economy ... is requested to dump the climate treaty in the wastebasket' Motion by parliamentarian Madlener (19 May 2016) *Kamerstuk 31,793 nr. 150, Tweede Kamer der Staten-Generaal.* Vergaderjaar 2015–2016.

[21] A prominent representative of this line of thought in the Netherlands is S. Kroonenberg, emeritus professor at the Technical University Delft. See among other publications his book: S. Kroonenberg, S. *De menselijke maat: de aarde over tienduizend jaar.* (Amsterdam, 2006, (revised edition in 2008): Uitgeverij Atlas).

[22] Rechtbank Den Haag, zaaknummer C/09/456689/HA ZA 13-1396, uitspraak 24-06-2015, *ECLI:NL:RBDHA:2015:7145.*

20.6   Renewable Energy Sources[23]

CO_2 emissions can be reduced by, among other things, saving energy or closing coal-fired power plants. Renewable energy sources play a special role. They provide an alternative for fossil raw materials, but their practical implementation has long left much to be desired. This is hardly surprising because these are new technologies that have to buck up against the prevailing mature energy technologies. Their future depends on investments in research and development, practical experiments, anchorage in laws and regulations and the building of networks. All these activities take much time and face numerous obstacles. Two examples – wind energy and solar panels – make this clear.

20.6.1   Wind

The wind energy story goes back to the 1970s. For a long time the technology was regarded with great sympathy. The ministry of Economic Affairs invested tens of millions of euros in research and development. A first National Research Program on Wind Energy was initiated (1976). This was followed up by a second National Development Program on Wind Energy (1981), an Integral Program on Wind Energy (1986), a Wind Plan (1989) and the Implementation of Wind Energy in the Netherlands (1992). The Energy Research Centre Netherlands (ECN) became the core of a national research network. The institute opted for large wind turbines. The small wind turbines championed by the environmental movement were pretty much ignored by the ECN and in the end lost the race against their bigger brothers.

But things went anything but smoothly with the big wind turbines. Researchers and designers struggled with technical shortcomings like broken blades and with what turned out to be a dead-end strategy. They invested in two-bladed turbines, while three-bladed turbines eventually became the dominant design. The most important obstacles, however, were the relations with the electricity sector on which the wind-energy developers were dependent for their practical trials. The creation of a wind-energy park at the beginning of the eighties gave rise to a host of conflicts with the electricity companies. The latter wanted complete control of the wind-park and managed to disengage the project from the national research program. It suffices to note that one of the turbine manufacturers abandoned the project and the other nearly went bankrupt.

One of the issues was the coupling of wind energy to the electrical power network. According to ECN, the electricity sector grossly underestimated the wind turbine capacity that could be coupled to its network without endangering the stability of the electricity supply. Moreover, the sector was willing to pay only modest amounts for supplying electricity to the network. That created problems for local

---

[23] This section is based on: Verbong and Geels, 'The ongoing energy transition', pp. 1033–1035.

wind turbine initiatives. And the costs of coupling a turbine to the network also played a role. The actual installation of windmills therefore lagged far behind the goals that had been set. And due to these problems a wind turbine industry also failed to develop. Manufacturers had too little experience with wind turbines and got too little feedback from actual practice.[24] The obstacles disappeared after the privatisation and liberalisation of the sector and the founding of public network managers and private electricity companies. However, dependence on subsidies did not disappear. In terms of price, wind energy was no match for fossil energy sources.

Furthermore, after 2000, wind energy began to lose its societal base. According to critics, the actual implementation of wind energy was too much of a top-down affair managed by researchers and policy-makers or by individual farmers seeking to earn extra income. That created a backlash. Local residents increasingly agitated against the noise, shadow flicker and the ugly aspect of the high wind turbines, while nature conservationists protested against turbines as 'bird choppers' and 'landscape polluters.' Politicians viewed windmills as a hardly cost-effective means to reduce CO_2 emissions. In a study the introduction of wind energy at the start of the new century was characterised as:

> ... possibly the most painful policy domain... Despite the best of intentions and ambitious policy aims, this domain is characterised by a continual process of 'pushing and shoving' leading in the end to laborious outcomes. And then once again to ambitious new resolutions.[25]

Recently, the tide has turned. In 2013 the government, employers, the unions and nature and environmental organisations signed the Energy Agreement for Sustainable Growth.[26] By 2023, 16% of all energy would have to be generated in sustainable ways. One of the strategies to achieve this aim was the installation of large wind turbine parks at sea.

In 2023 minimally 4450 MW of generating capacity will have to be installed at sea, of which around 35–40% will produce electricity due to wind fluctuations. (By comparison: the capacity of many gas-powered electricity plants runs between 300 and 700 MW).[27] And the government has even bigger ambitions for the period to follow: up to 2030 an additional 1000 MW per year.[28] And the number of land-based windmills will also have to be increased substantially. The strategy requires signifi-

---

[24] Davids, M., Lintsen, H., Van Rooij, A., *Innovatie en kennisinfrastructuur. Vele wegen naar vernieuwing.* (Amsterdam, 2013: Boom),. pp. 179–183.

[25] J.J. De Jong, E.O. Weeda, Th. Westerwoudt, A.F. Correljé, *Dertig Jaar Nederlands Energiebeleid. Van bonzen, polders en markten naar Brussel zonder koolstof,* (Den Haag, 2005: Clingendael International Energy Programme), pp.226.

[26] *Energieakkoord voor duurzame groei* (2013) Den Haag: SER.

[27] The biggest gas-fired power plant has an output of 1275 MW and the biggest coal-fired plant of 1560 MW. See: https://nl.wikipedia.org/wiki/Lijst_van_elektriciteitscentrales_in_Nederland consulted 22 mei 2017

[28] M. Niekoop 'Tweede Kamer opent deuren voor meer offshore wind in 2023', *Linkin,* 1 maart 2017. https://nl.linkedin.com/pulse/tweede-kamer-opent-deuren-voor-meer-offshore-wind-2023-mike-niekoop

cant subsidies, but thanks to, among other things, economies of scale and the learning curve these turn out to be lower than estimated.[29] While wind energy is still not uncontroversial, it seems that a tipping point has been reached: wind energy will be implemented at a large scale.[30]

20.6.2   Sun

Every new source of energy follows a different trajectory. With solar cells, for example, the ambitions were completely the inverse of those for wind. Initially the government and policy makers saw little potential in solar cells, while nowadays they are the promise of the future in electricity supply. The solar cell (photovoltaic cell or PV cell) was considered ill-suited to the Netherlands. The country had too little sunshine due to its geographical location and frequent cloud cover. Nonetheless, academic researchers used their freedom to investigate solar cells. They also formed a lobby group that aimed to convince politicians, policy makers, industrialists and the environmental movement of the future of PV on the basis of demonstration projects.

The about-face in perception occurred at the beginning of the 1990s. PV acquired substantial support and became a serious option. Shell, for example, integrated solar cells into its future scenarios and the multinational predicted that PV would become an important energy source by the middle of the twenty-first century. Other parties like the ministry of Economic Affairs, Greenpeace and Nuon encouraged the adoption of solar cells by households, among others. The budget for PV increased substantially in these years.

However, PV was definitely more expensive, certainly in comparison with fossil energy sources, but also in comparison with other renewable energy sources. Despite the improvement of the solar cell, the gap remained. PV scored poorly in an evaluation by the government at the end of the 1990s. It barely contributed to CO_2 reduction and the costs were high. Implementation of PV stagnated.

---

[29] J. van den Berg, 'Buitengaats pionieren met miljarden', *De Volkskrant* 8 mei 2017, 6. Zie ook: C. Grol and B. van Dijk 'Shell gaat tweede grote Borssele-windpark aanleggen' in het *Financieel Dagblad*, 12 december 2016, https://fd.nl/ondernemen/1179351/shell-gaat-tweede-grote-borssele-windpark-aanleggen. For five wind parks at sea (two off Borssele and three off the Holland shore) a maximum of € 18 billion in subsidies had been estimated. The estimate in 2016 was € 6 billion. The decrease is also the result of low interest rates, cheap steel prices and the availability of cheap offshore material due to the malaise in the oil and gas sector.

[30] See for example the debate between minister H. Kamp of Economic Affairs and Dercksen (PVV) in the Second Chamber on Dec. 19, 2016. Handelingen TK 2016–2017, 2, 19 December 2016. Or the reactions in response to an article by C. Grol and B. van Dijk 'Shell gaat tweede grote Borssele-windpark aanleggen' in het *Financieel Dagblad*, 12 december 2016, https://fd.nl/ondernemen/1179351/shell-gaat-tweede-grote-borssele-windpark-aanleggen

Recently PV is once again on the rise thanks to technological developments, subsidies,[31] and so-called *salderen* (charging private consumers on the basis of delivered energy *minus* energy returned to the network by local PV or wind sources). The contribution to the electricity supply is growing, but the technology has not yet reached the stage of large-scale implementation that wind power has achieved.

The two examples exhibit uneven and unpredictable trajectories.[32] Future promises were often too ambitious. Technical problems were too easily made light of. Implementation took longer than expected. Factors largely beyond the control of the parties concerned also played a role: the oil price, an economic recession, concern for the environment, liberalisation, etc.

The development of renewable sources was often stimulated by technology. Researchers focused on technical designs and technical challenges and paid less attention to the political process and social acceptance. They failed to adequately anticipate resistance by the environmental movement and local groups.

The building of well-functioning networks among research institutes, industry, societal organisation and the government proved to be essential.[33] Those networks were necessary to protect the new and vulnerable technologies against their harsh environments. Within these networks, the proponents of wind energy and solar cells had to organise subsidies, create favourable preconditions for innovations, organise practical experiments in sheltered environments, exchange knowledge and above all create optimistic images of advantages and disadvantages, costs, and profits. If a critical party became obstructive, as was the case with wind energy and the former electricity sector, the process could stagnate for years. The lack of stability of the networks and the changing role of the actors was remarkable. Policy makers had a tendency to change their strategies when results were disappointing and learning processes became more laborious than hoped for. The government often provided meagre guarantees for the longer term. Investors were uncertain and careful. Societal organisations tended to jump ship if the technology stimulated too much social protest.

In the short term investments in fossil energy sources will remain dominant. Low energy costs, market forces, reliability and diversification still have a high priority in the energy supply; climate, public health and environment take a back seat. To be sure there are clear EU norms for reducing CO_2 emissions and those of other greenhouse gases, but the solution will partly be sought within existing technologies and institutional structures. CO_2 storage is an option. Possibilities for energy saving, heat-power coupling and decentralised electricity generation are not yet exhausted. There are new chances for fossil fuels, in particular for gas. 'If coal was the fuel of

---

[31] In particular the subsidy measure 'Stimulering Duurzame Energie (SDE+)', Rijksdienst voor Ondernemend Nederland, 2017: http://www.rvo.nl/subsidies-regelingen/stimulering-duurzame-energieproductie-sde

[32] Verbong and Geels, 'The ongoing energy transition', pp. 1035.

[33] Also see: B. Verhees, R. Raven, F. Veraart, A. Smith and F. Kern, 'The development of solar PV in the Netherlands: a case of survival in unfriendly contexts', *Renewable and Sustainable Energy Reviews* 19 (2013), 275–289.

the nineteenth century and oil the fuel of the twentieth, then natural gas is predestined to become the fuel of the twenty-first century' as Jeroen van der Veer, president of Shell, predicted in 2002.[34]

Still, it remains to be seen whether this prediction is correct. Wind energy has acquired momentum and will experience significant growth in the coming decade. Work will continue on the expansion of other renewable energy sources. Societal support is clearly not lacking. The Second Chamber aims to have sustainable energy account for 30% of the energy supply by 2030.[35] This will only be the beginning, because the norms for CO_2 reduction for 2050 are stringent.

We shall return to this point in the conclusion. First we investigate another function of 'fossil.' Fossil raw materials are not only sources of energy, but also raw materials for chemical products. We focus on plastics, because they are an important example of this use of fossil resources.

20.7   Fossil and Plastics: Prelude to a Second Revolution[36]

After the Second World War, plastics became a revolutionary force. They rapidly penetrated into the capillaries of society with plastic products in the living room, the kitchen, the playroom, the bathroom and the bedroom, with fibres in textiles, films in packaging, laminates in construction, coatings in the paint industry and with plastic parts in machines and plastic applications in medicine, agriculture, transportation and the office. The Netherlands emerged as one of the international leaders in the production, processing and use of plastics.

The oil crisis of 1973 initially led to panicked reactions in the plastics sector. Companies began to hoard, fearing a scarcity of raw materials for plastics. Demand increased enormously but subsequently took a nosedive with 1975 as a low point. Economic stagnation, rising prices for plastics and more efficient use were the causes of the hectic movements. 'It will be clear…' noted the trade journal *Plastica* summarizing the mood, '…that the entire complex of events during and after the oil crisis has thoroughly unsettled the plastics industry of the fatherland.'[37] In retrospect, the oil crisis caused little more than a ripple in the ongoing increase of plastics production and the use of plastics. It happened once again during the credit crisis of 2008. Up to the present day, production and use of plastics have continued to increase in the Netherlands, although the spectacular growth of the 1960s has never since been achieved.

---

[34] Citation in: Sluyterman, *Concurreren in turbulente markten*, pp. 235.

[35] Passed motion submitted by parliamentarians Jan Vos (PvdA) en Van Veldhoven (D'66), nr. 511 (30196). Handelingen TK 2016–2017, 49, 7 February 2017.

[36] See for this section: Lintsen H., Hollestelle M., Hölsgens R. (2017) *The plastics revolution. How the Netherlands became a global player in plastics.*

[37] W. Bongers. 'De Nederlandse kunststofindustrie in 1975', *Plastica* 29 (1976), 239.

20.7.1   The Heterogeneous Sector

Still, much has changed in the plastics sector. The production of bulk plastics and
artificial fibres still takes place in the Netherlands, but not or rarely by *Dutch* companies. The 'classic' big plastics producers, Shell, DSM and AkzoNobel, have withdrawn from the fray. Shell still supplies raw materials for plastics. DSM has oriented
itself to technical polymers and AkzoNobel to coatings. In the 1980s and 1990s the
strategy was no longer oriented to diversification and the creation of broad con-
glomerates, but to core activities and the attainment of leadership positions in specialised markets. Profit margins on bulk plastics were too small and production
suffered from extreme ups and downs. Currently, foreign companies have stepped
into the breach as active bulk producers. Companies like General Electric, Dow
Chemical and DuPont pioneered this strategy in the 1960s. Other big firms like the
Saoudi SABIC, the Japanese Shin-Etsu and the international LyondellBasell are
newcomers to the Netherlands.

The plastics *processing* industry consists above all of small and medium-sized
firms (up to 50 employees) and has developed into the most innovative sector in the
Netherlands. It is extremely heterogeneous. Many firms in the plastics processing
sector are specialised in one way or another, for example in products like sliding
roof systems, in materials (among others PVC), in technologies (for example injec-
tion moulding), or in market sectors like automobiles and construction. There is
variety in age (many firms are not older than 40 years), in ownership forms (many
are a BV, a private limited liability company, among which a number of family
companies, some are an NV – a public company – or are part of a conglomerate) and
variety in origins (founded as a plastics processing company or the continuation of
a metalworking firm, a tool and die firm etc.).

The heterogeneity of the plastics sector has its positive and negative aspects. The
advantage is that the sector can operate flexibly in national and international markets. The disadvantage is that the sector is hard to organise and that policy makers
have difficulty getting a handle on it. Among other things, this plays a role in the
many public controversies around plastics. In the field of energy problems the lay of
the land is reasonably evident and it is possible to call large actors like the Ministry
of Economic Affairs and the big electricity companies to account for their policies.
In the plastics sector this is much more difficult.

20.7.2   Controversies

In the 1950s and 1960s, plastics had an ambivalent image. The material had an aura
of progress and modernity, but was also associated with low quality, litter and waste.
These negative associations continued to rear their ugly heads in subsequent years.
Other issues also surfaced with the emergence of the environmental movement.
PVC became the bogeyman of the 1980s. According to critics, PVC production

released carcinogenic substances. As packaging material, this plastic threatened food safety. When burned in waste incinerators, dangerous quantities of poisonous dioxin escaped. In the same period a debate emerged on the dangers of additives in plastics to health and the environment. Later on this debate broadened to include sustainability, which also included energy production and the exhaustibility of fossil raw materials.

Recently plastics waste is once again high on the agenda due to problems with litter, 'plastic soup' and micro-plastics. Unimaginable quantities of plastics end up globally in rivers, along the coasts, and at sea. Birds and other animals perceive floating bits of plastic as food, consume it, are weakened and die. Plastics waste is concentrated in so-called *gyres*, large circular movements fed by multiple ocean currents. These result in enormous garbage dumps, many times larger than the Netherlands.

A portion of the marine plastics degrades into miniscule particles, providing a substrate for the growth of organic material and attracting poisonous substances like dioxin. They cover the ocean floor and enter the food chain via fish. Micro-plastics also end up in the water and sediments of sewage treatment plants because they are for example constituents of toothpaste, shampoos and cosmetics. The modern human body contains minimal but measurable quantities of plastics. As the *Washington Post* announced as long ago as 1972, 'Every human is a little bit 'plastic.'[38] What are the implications for health and behaviour? We do not yet know. 'Under the 10 micro-meters they penetrate through your cell membranes, enter into your bloodstream and travel through your body…' according to Heather Leslie, a researcher in the field of micro-plastics at the Free University of Amsterdam.

> …we still have to investigate how many of these particles you have to take in before you begin to experience severe discomfort, but laboratory tests already show that very fine plastic particles can damage cells and tissue and that they can lead to all kinds of infections.[39]

The problems have an international dimension. Plastics are produced, exported, imported, used and disposed of worldwide. Within the Netherlands and the European Union this attracts some attention. Internationally there are hardly any organisations that articulate the problem. This is why a solution for one of the most pressing issues – marine pollution and the 'plastic soup' – still remains far distant. *The Ocean Cleanup*, founded by the Delft student Boyan Slat, is one of the few organisations that is preparing for one of the biggest clean-up operations in history.[40]

---

[38] Quoted in: S. Freinkel, *Plastic. A Toxic Love Story* (Boston, 2011: Houghton Mifflin Harcourt), 89.

[39] D. Cohen, 'Geplastificeerde maatschappij. Er wordt te veel van de burger verwacht', *De Volkskrant* 24 september 2016, bijlage Vonk, 2.

[40] In 2013 Boyan Slat started a project to capture floating plastic trash in the oceans with the aid of inflatable barrier arms. The project generates much enthusiasm and receives financial support from all over the world. Slat works together with students, engineers, oceanographers and industry experts. Feasibility studies are underway. The first pilot project was launched in 2016. See: www.theoceancleanup.com/ (Geraadpleegd op 8 maart 2017.)

The problems with the 'plastic soup' and the micro-plastics can be traced back to the linear nature of the plastics supply chain. A large proportion of the plastics ends up in garbage dumps, leaks into the environment or is burned. In Europe that percentage is 74%, of which garbage dumbs and the environment accounts for 38% and burning for 36% (2012).[41] In the Netherlands the percentage is 67% but a much smaller proportion ends up in garbage dumps and in the environment – 7% – and 60% is burned. The advantage of burning is in any case that the plastics don't end up in the environment and that they produce useful energy. The disadvantage is that finite resources are depleted.

The problems could be partly solved if degradable bio-plastics were to replace the current ones, so that the waste products could be taken up in the environment. Wageningen University and Research are investigating this, but an important contribution from 'green' plastics should not be expected anytime soon. Degradable bio-plastics cannot compete with the majority of plastics and are used only in niches.

The solution for the time being would have to be sought in the closing of the supply chain, that is to say in the nearly complete recycling of plastics. In the Netherlands some 33% is recycled. Increasing this percentage will be a big challenge in view of the heterogeneous nature of the plastics sector and the absence of influential organisations prepared to assume responsibility for the problem. It could be the prelude to a second plastics revolution.

20.8   Fossil as a Janus-Head

Fossil raw material is the Janus-head of well-being. It has two faces like the head of the Roman god Janus. As an energy source it was an important factor in modern economic growth, the elimination of poverty and the achievement of welfare. As raw material for plastics it shaped a life of comfort, ease, luxury, sport and games. At the same time it was a source of uncertainties and controversies. In the field of energy supply, dependence on foreign suppliers was an issue as early as the end of the nineteenth century. In recent decades this has been augmented by air pollution, depletion of raw materials and climate change. In the new domain of plastics litter became an issue from the 1950s on, now augmented by concerns over 'plastic soup' and micro-plastics.

Will the dark side of fossil lead to a transition? This is highly unlikely in the short term for issues like 'plastic soup' and micro-plastics. Little urgency is felt, nationally or internationally, to undertake action on these points. Nonetheless there is a possibility for a slow but radical change in the domain of plastics if the tendency to recycling can be sustained.

The energy situation is another story. Climate change is currently the crowbar for the energy transition. Will this transition continue at a slow pace in the coming

---

[41] *Plastics – The facts 2013. An analysis of European latest plastics production, demand and waste data* (z.pl 2013). The text is published by Plastics Europe, Association of Plastics Manufacturers.

decades with 'fossil' continuing to remain dominant? Or will there be a radical breakthrough in which renewable energy sources rapidly become the major suppliers? Renewable energy sources now have a long gestation period behind them of some four decades. That is not unusual for a transition. It was no different with the rise of coal and steam and it was also the case with the transition to oil.

Much happens in the course of such a gestation period. In the fields of science and technology research is undertaken and elaborate experiments carried out. A base of support is organised among societal organisations. The government is wooed with expectations and solicited for favourable conditions. Niches are identified in the economy, so that the new options can come to fruition in a somewhat protected environment. The creation of social networks is essential to achieve synergy among all these efforts and to enable the creation of new institutions.

The question is whether wind energy, solar cells and other renewable sources of energy have reached the stage where they can compete with oil, gas and coal under equal, but newly formed, circumstances. That stage seems to have been reached for wind energy, although the technology is still dependent on subsidies. For solar cells this is not yet the case, but there is still hope.[42] The new technologies are not only in competition with the classic energy sources. There is also a struggle among the alternative options, for example with biomass, heat-power coupling and geothermal energy. It is still unclear what kind of mix will surface in the end.

The conditions for a sustainable energy transition seem favourable at the moment, especially because important actors are lending their support. Business leaders like the chairman of the Employers' Association, the director of the National Railways and even the director of the *Gasunie* have spoken out in favour of the transition.[43] The ABP, the Netherlands' biggest pension fund, recently announced that it wants to commit itself seriously to sustainability. Chairman of the board, C. Wortmann-Kool, is pushing for a climate law and in this she is not alone: 'We have to provide our clients with long-term security and be able to offer gains, so we are not going to sit back waiting on fickle subsidy policies. The cabinet has to provide clarity for a period of ten to twenty years.'[44] The Netherlands Bank calls on the government to

---

[42] See for example the analysis by the Rabo Bank: 'Zonne-energie (fotovoltaïsche zonnepanelen', *Rabobank Cijfers & Trends* 40(2016/2017), 23 May 2017. https://www.rabobankcijfersentrends. nl/index.cfm?action=branche&branche=Zonne-energie_fotovoltaische_zonnepanelen

[43] See, for example: 'Top Nederlands bedrijfsleven op excursie naar Noordpool', *Financieel Dagblad* 2 mei 2017, https://fd.nl/ondernemen/1200342/top-nederlands-bedrijfsleven-op-excursie-naar-noordpool. For example also: 'Uniek paar: Marjan Minnesma en Hans de Boer', *Vroege Vogels Radio* 12 februari 2017, https://vroegevogels.vara.nl/nieuws/uniek-paar-marjan-minnesma-en-hans-de-boer

[44] See: C. Wortmann-Kool, 'Nieuw beleggingsbeleid van het ABP is breuk met verleden', https://www.apg.nl/pdfs/abp-pensioendoc_2016.pdf. The chairman of the board of the General Citizens' Pension Fund (ABP) indicates that it is difficult for the ABP to invest in sustainable projects because it often concerns projects that are too small for the ABP and because too many projects still return too little profit. See: N. Trappenburg and J. Groot, 'Pensioenfonds zoekt grote groene projecten', *Financieel Dagblad* 8 mei 2017, https://fd.nl/economie-politiek/1200685/abp-duurzaam-beleggen-in-nederland-is-lastig. Typical of the situation is that in 2016 the ABP invested €2 billion extra in fossil energy: 'ABP belegde € 2 mrd extra in fossiele energie', *Financieel Dagblad* 15 mei 2017, https://fd.nl/beurs/1201824/abp-belegde-2-mrd-extra-in-fossiele-energie

'commit itself in a timely manner to a credible and feasible path toward a CO_2-neutral economy'[45] For this moment in time, it is hard to image a more powerful statement from one of the most influential organisations in the Netherlands.

Literature

Anonymous, (2015, June 24). 'Netherlands ordered to cut greenhouse gas emissions' in *BBC News*

Anonymous, (2017, May 23), 'Zonne-energie (fotovoltaïsche zonnepanelen)' in *Rabobank Cijfers & Trends* 40 (2016/2017). Retrieved from: https://www.rabobankcijfersentrends.nl/index.cfm?action=branche&branche=Zonne-energie_fotovoltaische_zonnepanelen

Anonymous, (2017, May 2), 'Top Nederlands bedrijfsleven op excursie naar Noordpool' in *Financieel Dagblad*. Retrieved from: https://fd.nl/ondernemen/1200342/top-nederlands-bedrijfsleven-op-excursie-naar-noordpool.

Anonymous, (2017, Februari 12) 'Uniek paar: Marjan Minnesma en Hans de Boer', *Vroege Vogels Radio*. Retrieved from: https://vroegevogels.vara.nl/nieuws/uniek-paar-marjan-minnesma-en-hans-de-boer.

Anonymous, (2017, May 15), 'ABP belegde € 2 mrd extra in fossiele energie', *Financieel Dagblad*. Retrieved from: https://fd.nl/beurs/1201824/abp-belegde-2-mrd-extra-in-fossiele-energie.

Cox, H. (2011), *Revolutie met recht,* pp 288. Maastricht:Stichting Planet Prosperity Foundation

Sluyterman, K. (2007). 'Concurreren in turbulente markten 1973-2007'. *Geschiedenis van Koninklijke Shell, deel 3*, pp. 5-9; 93; 239. Amsterdam: Boom

Mom, G. and R. Filarski (2008). *Van transport naar mobiliteit. De mobiliteitsexplosie [1895–2005]*, pp. 265, 374. Zutphen: Walburg Pers

Lintsen, H., Van Helvoort, T. en Van Veen, R. (2014). *De kracht van de katalysator. De magie van het onderzoek*, pp. 17–29. Eindhoven: Stichting Historie der Techniek.

Lintsen, H., Hollestelle M., Hölsgens R. (2017a) The plastics revolution. How the Netherlands became a global player in plastics. Eindhoven: Stichting Historie der Techniek.

Correljé, A. C. van der Linde en Th. Westerwoudt (2003), *Natural Gas in the Netherlands. From Cooperation to Competition?*. Amsterdam: Oranje-Nassau Groep.

De Jong, J.J., Weeda, E.O., Westerwoudt, Th., Correljé, A.F. (2005a) *Dertig Jaar Nederlands Energiebeleid. Van bonzen, polders en markten naar Brussel zonder koolstof,* pp.153-155. Den Haag: Clingendael International Energy Programme

Verbong, G. and Geels. F (2007), 'The ongoing energy transition: lessons from a socio-technical, multi-level analysis of the Dutch electricity system (1960-2004)' in *Energy Policy* 35, pp. 1025–1037.

Internationally, important investors are pulling out of 'fossil.' That will doubtless set other (smaller) investors to thinking. Examples: the Rockefeller family. http://www.cbsnews.com/news/rockefeller-family-is-exiting-the-oil-business/; https://www.theguardian.com/environment/2014/sep/22/rockefeller-heirs-divest-fossil-fuels-climate-change; http://www.telegraph.co.uk/finance/newsbysector/energy/oilandgas/11114591/Rockefeller-family-sells-out-of-fossil-fuels-and-into-clean-energy.html

Influential investment advisor Bloomberg has long been crystal clear: A single example: https://www.bloomberg.com/company/new-energy-outlook/en; https://www.bloomberg.com/professional/blog/sustainable-investing-strategy-reality/

[45] G. Schotten, S. van Ewijk, M. Regelink, D. Dicou en J. Kakes, *Tijd voor Transitie. Een verkenning van de overgang naar een klimaatneutrale economie* (Amsterdam, 2016: de Nederlandsche Bank).

Duyvendak, W. (2011), *Het groene optimisme. Het drama van 25 jaar klimaatpolitiek*, pp. 65. Amsterdam: Uitgeverij Bert Bakker.

Verbong. G, et al. (2001). *Een kwestie van lange adem. De geschiedenis van duurzame energie in Nederland.* Boxtel: Aeneas uitgeverij.

Cramer, J. (2014) *Milieu.* Amsterdam: Amsterdam University Press.

De Jong, J.J., Weeda, E.O., Westerwoudt, Th., Correljé, A.F. (2005b) *Dertig Jaar Nederlands Energiebeleid. Van bonzen, polders en markten naar Brussel zonder koolstof,* pp.106. Den Haag: Clingendael International Energy Programme

Kroonenberg, S. (2006, revised edition in 2008) *De menselijke maat: de aarde over tienduizend jaar.* Amsterdam: Uitgeverij Atlas.

Davids, M., Lintsen, H., Van Rooij, A. (2013) *Innovatie en kennisinfrastructuur. Vele wegen naar vernieuwing.* pp. 179-183. Amsterdam: Boom.

De Jong, J.J., Weeda, E.O., Westerwoudt, Th., Correljé, A.F. (2005c) *Dertig Jaar Nederlands Energiebeleid. Van bonzen, polders en markten naar Brussel zonder koolstof,* pp.226. Den Haag: Clingendael International Energy Programme

Niekoop, M (2017, March 1) 'Tweede Kamer opent deuren voor meer off-shore wind in 2023' in *Linkin,* retrieved from https://nl.linkedin.com/pulse/tweede-kamer-opent-deuren-voor-meer-offshore-wind-2023-mike-niekoop.

Berg, van den, J. (2017, May 8), 'Buitengaats pionieren met miljarden' in *De Volkskrant*

Grol, C., van Dijk, B. (2016, December 12), 'Shell gaat tweede grote Borssele-windpark aan-leggen' in *het Financieel Dagblad,* retrieved from https://fd.nl/ondernemen/1179351/shell-gaat-tweede-grote-borssele-windpark-aanleggen.

Verhees, B., Raven, R., Veraart, F., Smith, A., Kern, F. (2013). 'The development of solar PV in the Netherlands: a case of survival in unfriendly contexts' in *Renewable and Sustainable Energy Reviews* 19, 275–289

Lintsen H., Hollestelle M., Hölsgens R. (2017b) *The plastics revolution. How the Netherlands became a global player in plastics*, Eindhoven: Stichting Historie der Techniek

Bongers, W. (1976) 'De Nederlandse kunststofindustrie in 1975' in *Plastica* 29, p. 239

Freinkel, S. (2011), *Plastic. A Toxic Love Story*, Boston: Houghton Mifflin Harcourt, p. 89.

Cohen, D. (2016, September 24), 'Geplastificeerde maatschappij. Er wordt te veel van de burger verwacht' in *De Volkskrant*, bijlage Vonk, 2

Wortmann-Kool, C. 'Nieuw beleggingsbeleid van het ABP is breuk met verleden', https://www.apg.nl/pdfs/abp-pensioendoc_2016.pdf.

Trappenburg, N. and Groot, J. (2017, May 8), 'Pensioenfonds zoekt grote groene projec-ten' in *Financieel Dagblad*. Retrieved from: https://fd.nl/economie-politiek/1200685/abp-duurzaam-beleggen-in-nederland-is-lastig.

G. Schotten, S. van Ewijk, M. Regelink, D. Dicou en J. Kakes (2016), *Tijd voor Transitie. Een verkenning van de overgang naar een klimaatneutrale economie*, Amsterdam: de Nederlandsche Bank.

Chapter 21
The Tensions Between Well-being and Sustainability. Well-being and Sustainability Around 2010

Harry Lintsen and Frank Veraart

Contents

Abstract This chapter describes, first, the development of well-being between 1970 and 2010 from the perspective of the efforts of the societal midfield, the national government, and the business community. In the second place, the situation around 2010 is evaluated from the perspective of 1970 and a present-day perspective. From the perspective of 1970 material welfare and well-being have developed in a positive sense between 1970 and 2010. Problematic from this perspective is the increase in criminality and unemployment. In the Netherlands the crisis of nature and environment seems to be past its deepest point.

The present-day perspective is described using the position of the Netherlands in European rankings for different themes and indicators. It confirms what has been said above in previous analyses: the Netherlands has a high quality of life. But this can be contested for the chances for future generations, especially in relation to human and natural capital. In addition, the Netherlands is more than ever intertwined with international supply chains for her resources and energy. In this way the Netherlands has displaced a part of its sustainability problems to foreign lands. On the other hand, international supply chains – as in the meat sector – have led to more landscape damage and environmental problems within the Netherlands.

Keywords Midfield · National government · Business community · Environment · Sustainability · Monitor · Tradeoff

© The Author(s) 2018
H. Lintsen et al., *Well-being, Sustainability and Social Development*,
https://doi.org/10.1007/978-3-319-76696-6_21

21.1   The Netherlands as a Temporary Global Leader in Climate Policy

> What we are presently experiencing is not the destruction of the earth in one blow, but in a silent drama. Our world suffers from deforestation, desertification, pollution and the poisoning of air, soil and water, extinction of animal and plant species, erosion of the ozone layer that serves to protect us from dangerous radiation, and an increase in temperature with threatening consequences, such as sea-level rise. Slowly the earth is dying and the unimaginable – the end of life itself – is nonetheless becoming imaginable.[1]

With these sombre words Queen Beatrix opened her 1988 Christmas Speech. By the end of the 1980s it looked like environmental problems were accumulating and increasing in scope. Well-being with respect for nature, the environment and climate were the challenges for the period between 1970 and 2015.

In 1988 the National Institute for Public Health and Environment (RIVM) had published the Netherlands' first future study on the environment, *Concerns for Tomorrow* (*Zorgen voor Morgen*). Recommendations by the RIVM and by the 1987 World Commission on Environmental Development, chaired by Gro Harlem Brundlandt, inspired a first National Environmental Plan. This was broadly supported by the government. Despite this, Prime Minister Ruud Lubbers' second cabinet fell due to several of the Plan's proposals: the elimination of the default tax restitution for costs of commuting and an increase in excise tax on diesel.

In the 1989 elections, almost all the party platforms prominently flagged plans for sustainable development. The environment had captured the popular imagination. No less than 42% of the electorate considered this the most important issue. Brundlandt's approach in particular, in which sustainable development and economic growth were considered compatible, found favour among political parties. Wim Kok, the leader of the Labour Party, promised to solve environmental problems 'within one generation.' Targeted levies (ecotax) and regulations would encourage sustainable behaviours. The Christian Democrats (CDA) and the Liberals (VVD) focussed especially on the emerging concern for the climate problem. Candidate (and incumbent prime-minister) Lubbers (CDA) announced an 8% reduction in carbon dioxide (CO_2) in the coming cabinet term. The VVD published an 'action plan for the protection of the atmosphere.'[2] But policy for the climate problem was relatively new. The way this issue was dealt with reveals the then current political and societal order and its ruling ambitions.

On November 6 and 7, 1989, environment ministers from 68 countries assembled at the Dutch seaside resort of Noordwijk to discuss the contours of a global climate treaty. The plans for this meeting had arisen in 1988 in the course of the Toronto Conference on the Changing Atmosphere. At that conference it was also

---

[1] Christmas Address 1988, Koning Beatrix cited in Carla van Baalen et al., eds., *Koning Beatrix aan het Woord, 25 Jaar Troonredes, Officiële Redevoeringen en Kersttoespraken* (Den Haag: Sdu Uitgeverij, 2005), 445.

[2] W. Duyvendak, *Het Groene Optimisme, Het Drama van 25 Jaar Klimaatpolitiek* (Amsterdam: Bert Bakker, 2011), 39–52.

decided to promote additional research in this area and for this purpose to found the International Panel on Climate Change (IPCC).

The meeting in Noordwijk took place during the final two days of Ed Nijpels' term as minister of the environment. He proudly announced that 'The Netherlands wants to be the leader in the area of the climate treaty *in spe*, and with respect to reduction of CO_2 pollution.'[3] This was the first time that climate change and measures to reduce CO_2 emissions had been spoken of at the ministerial level.

The issue was also remarkable because of the ambivalent attitude of the environmental movement. The Foundation for Nature and the Environment (*Stichting Natuur en Milieu*) had been invited to the Noordwijk conference only at the very last minute. It submitted a proposal to reduce CO_2 emissions by 20% by the year 2000. At the same time it argued that nuclear energy 'could not be an alternative in the struggle against the greenhouse effect.'[4] Nuclear energy had been one of the big environmental issues of the 1970s and 1980s. The Chernobyl disaster in 1986 had closed the door on this option, or so it seemed. But with growing concern for the climate problem, the environmental movements were concerned lest nuclear energy threatened to become an option again.

In the wake of the conference newspapers headlined: 'Disappointment after climate conference. Great powers block solid agreements.' The Dutch hosts proved unable to forge a climate treaty at the Noordwijk meeting. The United States and Japan, in particular, first wanted to await results of the IPCC investigations. The meeting did settle on an agenda for international climate agreements, such as the different roles of developed and developing countries, the importance of forests and targets for levels of emission reduction with 1990 as a reference year.[5]

The meeting occurred at a peculiar moment in time for the Netherlands. Due to the transfer of political power, the election period of 1989 marked the temporary high-water mark of Dutch societal and political concern for environment and sustainability in the period 1970-2015. Throughout the 1980s, the country had seemed to be buried under an ever-growing avalanche of environmental problems. After concern about air and water pollution in the early 1970s, a number of 'pollution scandals' called attention to the pollution of soil and groundwater. In 1982 there were 4253 reports of pollution nationwide, of which 1246 were classified as highly urgent.[6]

In the same period there were growing concerns about acidification. These were originally focussed on the emission of sulphur dioxide and the trans-border problem

---

[3] 'Nijpels hoopt op afspraken tijdens milieuconferentie', NRC 4-11-1989, 7.

[4] 'Milieuconferentie begint in sfeer van onenigheid, Nederland vraagt om hulp voor Derde Wereld' in Nederlands dagblad 07-11-1989.

[5] Duyvendak, *Het Groene Optimisme*, 53–54.

[6] Investigations into the scope of the problems gained momentum after the pollution scandal in Lekkerkerk (1980). A first overview was prepared at the behest of Minister L. Ginjaar for the interim-regulations on soil-decontamination. See Klaas Bouwer, Jacques Klaver, and Marianne de Soet, *Nederland Stortplaats, een Milieukundig en Geografische Visie op het Afvalprobleem* (Nijmegen: Ekologische uitgeverij, 1983), 64–65.

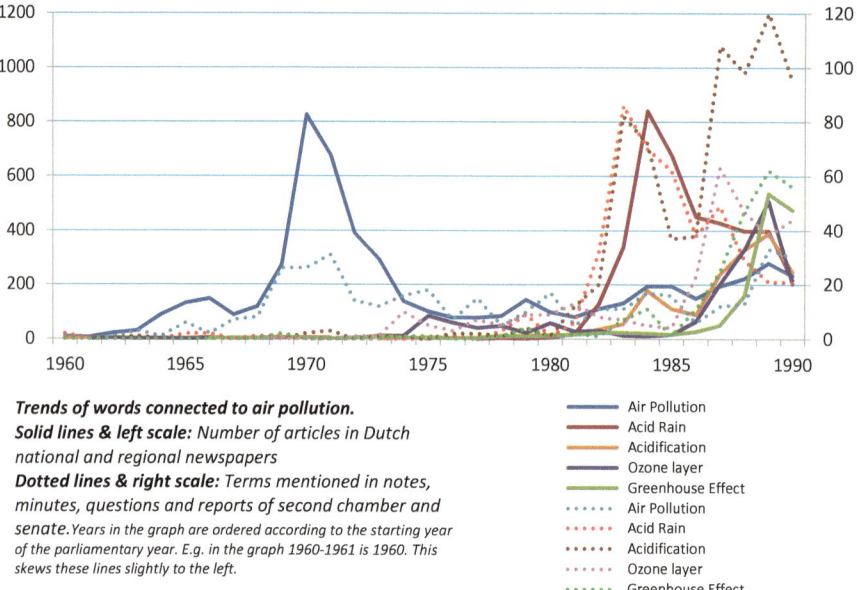

Trends of words connected to air pollution.
Solid lines & left scale: *Number of articles in Dutch*
national and regional newspapers
Dotted lines & right scale: *Terms mentioned in notes,*
minutes, questions and reports of second chamber and
senate. Years in the graph are ordered according to the starting year
of the parliamentary year. E.g. in the graph 1960-1961 is 1960. This
skews these lines slightly to the left.

—— Air Pollution
—— Acid Rain
—— Acidification
—— Ozone layer
—— Greenhouse Effect
· · · · · Air Pollution
· · · · · Acid Rain
· · · · · Acidification
· · · · · Ozone layer
· · · · · Greenhouse Effect

Graph 21.1 Topics related to air pollution in Dutch newspapers and in the Transactions of the Second Chamber of Parliament 1960–1990
Note 1: after 1990 the newspaper data set is incomplete
Note 2: Regarding the newspaper data it may be the case that newspaper articles appearing in both regional and national newspapers are counted twice. In the parliamentary and senate notes, minutes and reports there is a distinction between loose remarks and full reports. The content of these articles, notes, etc. was not checked. Articles might, for example, review foreign news or suffer from other anomalies. This is especially problematic with small numbers
Source: Royal Library – online Newspaper database Delpher-Kranten (www.delpher.nl) and online database governmental notes of parliament and senate (www.statengeneraaldigitaal.nl)

of acid rain. With improved knowledge and more accurate measurements, acidification due to agricultural ammonia emissions also became part of the picture.[7] The latter led to increased interest in eutrophication of the soil by nitrogen and phosphorous deriving from agriculture and households. By the end of the 1980s the international problems of the erosion of the ozone layer and climate change had become prominent issues in the press and in politics (see Graph 21.1).

With its National Environmental Policy Plan and the debate on climate problems the Netherlands regarded itself as world leader in environmental policy. Was this justified? And did it retain this position? What happened to the attention for climate and environment after 1990?

---

[7] Ed Buijsman, 'Gisteren, Vandaag, Morgen, Een Terugblik op het Probleem van Zure Regen,' *Studium* 4 (2008): 251–68; G.H. Dinkelman, *Verzuring en Broeikaseffect, de Wisselwerking tussen Problemen en Oplossingen in het Nederlandse Luchtverontreinigingsbeleid (1970-1994)* (Universiteit van Amsterdam, 1995).

21.2   Turbulence and Crisis

21.2.1   Tense Years: 1970 – 1990

In the 1970s and 1980s the Netherlands had more worries than only those concerning environment and sustainability. The economy was in bad shape. The economic wonder of the 1950s and 1960s with its almost unbroken growth had begun to lose momentum. Increasing wage levels and prices in the course of the 1960s had already compromised Dutch competitiveness.[8] The expansion of world trade ground to a halt in response to extreme fluctuations in currency markets, stock exchanges and markets for raw materials. A wave of global inflation flooded the economy. The first oil crisis of 1973 brought further wage and price increases.[9] The business community suffered. Turnover stagnated. Profits declined. The proportion of those employed in industry declined drastically and unemployment rose. After the second oil crisis of 1978 the economic situation declined even more dramatically, to recover only at the end of the 1980s.

The economic situation was hardly favourable to measures aimed at improving the environment or sustainability. Despite this, the government maintained that an environmental policy including investments by the business community was essential in order to change existing 'dirty' production processes and to repair past environmental damage. However this created even more problems for the business community. It complained loudly about the costs of the environmental measures.

Environmental policy was not the only cause of increasing costs for the business community. During the 1970s the welfare state also expanded.[10] The General Law on Labour Incapacity (*Algemene* Arbeidsongeschiktheidswet) insured not only employed workers but also non-employed workers against the consequences of unfitness for work. The minimum wage and the budgets for social security were increased. The number of Netherlanders dependent on social services increased, in part due to growing unemployment. The minimum wage and welfare benefits were coupled to wage levels in the market sector. This all contributed to increased costs for business: higher wages and increased taxes and welfare contributions.

The government was itself also confronted with the consequences of its policies. In the course of the 4 years after 1973, government expenses increased from 45% of the Gross Domestic Product to 51%. After 1974 the state budget showed a chronic deficit. The gradual decline in the magnitude of the national debt came to a halt in 1977.

Fundamental tensions among the welfare state, the economy and sustainability became apparent in the 1970s. By expanding welfare services, the government

---

[8] J.L van Zanden and R.T. Griffiths, *Economische geschiedenis van Nederland in de 20 ᵉ eeuw* (Aula, Utrecht 1989), 48–49.

[9] Van Zanden and Griffiths, *Economische geschiedenis van Nederland*, 255–258.

[10] See for the following: J. Peet en E. Nijhof, *Een voortdurend experiment. Overheidsbeleid en het Nederlandse bedrijfsleven* (Uitgeverij Boom, Amsterdam 2015), 199–205.

hoped to shore up the purchasing power of the unemployed, the ill and the poor and in this way to stimulate the economy. If that appeared still to work in the 1970s, it proved impossible to maintain in the long run. Government largesse was financed by proceeds from the sale of natural gas, by government loans and by an increase in the budget deficit. But workers stuck by their hard-won rights. Massive protests followed on a proposal to terminate automatic price compensation for wage earners in the market sector. Environmental groups demanded rigorous environmental measures. Employers pointed to the structural economic problems caused by government policy. In many firms profits were inadequate to enable production to continue or to invest in expansion. The business community argued that society was much more concerned with distributing welfare than with creating it.

21.2.2   The Dynamic Midfield: Breakdown and Reconstruction

Societal contradictions increased in the 1970s, ultimately leading to an about-face. The economic miracle after the Second World War had exacted offers from the Dutch population. Parsimony ruled the day. There had been little opportunity for consumption. The reconstruction of the country, the modernisation of the economy and the Delta Works had exacted enormous investments. The resulting tensions had been absorbed by the pillarised social structure. The necessary consensus had been created within the pillars, with their organisations in the fields of work, enterprise, education, sport and social life. This is where the shock of modernisation had been absorbed.[11]

But by the second half of the 1960s de-pillarisation had set in. The influence of the church and religion on social life declined. The membership of the confessional parties declined and their political dominance began to wane. This also meant the end of the soundless achievement of a societal consensus. In the course of the 1970s the number of strikes began to increase after a long period of quiescence. Workers' associations increasingly went their own way. Labour contract negotiations and bargaining in the Social-Economic Council no longer 'automatically' led to a consensus.

A new societal midfield emerged but did not replace the pillars. To have done so it would have had to have been less specifically oriented to the issue of the environment. Groups like the national *Association for the Preservation of the Wadden Sea* (1965), the *Foundation for Nature and the Environment* (1972) and the *Central Action-Committee Rijnmond* (1970) were oriented above all to exercising political and administrative influence. Groups like *Action Last Hope* (1970) and the *Association for Environmental Defence* (1970) used demonstrations and publicity to

---

[11] K. Schuyt and E. Taverne, *1950. Welvaart in zwart-wit* (Den Haag, 2000: Sdu Uitgevers), 40–41.

advance the environmentalist cause. Associations like *De Kleine Aarde* (1971) and ecologically inspired working communities used concrete examples to try to win the broader public for an environmentally friendlier and energy-conserving lifestyle.[12]

These organisations became highly visible in the collective protest against nuclear energy. In the 1980s they were able to mobilise thousands of people for protest demonstrations at the experimental reactor at Dodewaard and the uranium enrichment facility Urenco in Almelo.[13] Another broadly supported protest aimed to prevent the closure of the East Scheldt and the expansion of marl mining. In the course of these engagements the environmental movement gradually acquired a new role. They increasingly became sources of information and expertise in the area of environmental technology. This enabled them to make substantive contributions to proposals and solutions.

In the 1980s idealism gave way to pragmatism. In cooperation with governments and the business community, the overthrow of capitalism was abandoned as an aim, to be replaced by a transition to ecological modernisation. This was inspired in particular by the sustainability concept as worked out in the Brundtland report. At the same time the environment became a more central concern for businesses and governments, who found useful partners in the environmental organisations with their new knowledge and expertise.[14]

In the 1980s, strategies of cooperation were reintroduced into the economy. In 1982 representatives of employees and employers, under the benevolent eye of the government, signed the famous 'Wassenaar Agreement.' A long period of polarisation thereby came to an end as a new era of compromise began to take shape. Workers were now prepared to relax wage demands in return for shorter working hours. The expectation was that this would both improve the competitive position of the business community as well as increase employment opportunities. In the same spirit, the relationship between social policy and the economy was redefined. Guarantees of social security were not rejected out of hand, but limited to what was necessary and to what was manageable from an economic and financial point of view.[15]

---

[12] J. Cramer, *Milieu* (Amsterdam University Press, Amsterdam 2014), 21–22.

[13] In 1980, 15,000 people demonstrated at the Dodewaard nuclear power plant. In 1987, 25,000 people protested against the underground storage of nuclear wastes in Gasselte and 40,000 at the uranium enrichment facility in Almelo. Duyvendak, *Het Groene Optimisme,* 62.

[14] Cramer, *Milieu,* 53–54.

[15] Peet en Nijhof, *Een voortdurend experiment,* 223. At the national level there was the tri-polar consultation among employers, employees and the government. In the 1990s this would become known as the 'polder' model and come to be seen as the basis of the *Dutch miracle*: a well-oiled economy, growing employment, decreasing budget deficits and a less-costly welfare system. B. Bouwens and J. Dankers, *Tussen concurrentie en concentratie. Belangenorganisaties, kartels, fusies en overnames* (Uitgeverij Boom, Amsterdam 2012), 205.

21.2.3   The National Government: In Search of a New Role

In regard to environment and energy, the government had long pursued a directive policy. The Ministry of Economic Affairs imposed its will on energy policy by advocating a central role for energy conservation and nuclear power. A new ministry, the Ministry of Public Housing and Environmental Hygiene, founded in 1971, developed legal instruments to compel the business community to take environmental measures.[16] These measures were primarily aimed at managing local environmental pollution, initially of air and water and by the early 1980s of the soil as well. Problems were defined chiefly as public health risks. The policy was inspired by research conducted by the National Institute for Public Water Supply, the Institute for Waste Research, and the National Institute for Public Housing that had been investigating air quality since the 1960s. Norms, permits and subsidies were mobilised to force a reduction of emissions and energy consumption.

At local levels this strategy was certainly successful. End of pipe technologies decreased emissions.[17] And as far as SO_2 emissions were concerned: the transition to natural gas also had an effect. Nonetheless, there was increasing unease about the environment and energy. Policy makers began to focus on biodiversity and damage to the landscape. Problems became more and more global in nature. Acid rain, climate change and destruction of the ozone layer were not limited to the Netherlands. The national government once again had to reconsider its strategy in regard to environmental and energy issues.

The environmental issues were transferred to the newly-formed ministry of Public Housing, Spatial Planning and Environmental Management. At the end of the 1980s, a report, *Concerns for Tomorrow* (*Zorgen voor Morgen*) and the National Environmental Policy Plan opted for drastic reductions in the emissions of many substances and the prevention of future environmental damage. Toughening up technical specifications for firms, devices and vehicles was no longer enough. It was necessary to pursue an integral strategy for energy supply, agriculture and traffic and transport.[18] Spatial planning was also subjected to an integral approach. The

---

[16] H.T. Siraa, A.J. van der Valk and W.L. Wissink, *Met het Oog op de Omgeving. Het Ministerie van Volkshuisvesting, Ruimtelijke Ordening en Milieubeheer, 1965–1995* (Den Haag: SDU Uitgeverij, 1995), 244–67.

[17] Cramer, *Milieu*, 18–19.

[18] Cramer, *Milieu*, 29-31. The shift in environmental policy can also be interpreted in another way, namely from *end of pipe* policy to policy based on the precautionary principle to a *no-regret* policy. The precautionary principle was the point of departure for the policy entailed in the report *Concerns for Tomorrow* (*Zorgen voor Morgen*, 1988). A much-used definition states that new technologies may not be implemented without precautions if there is a chance of serious or irreversible damage even when there is no scientific certainty about risks. See, among others: A. Reichow, *Effective regulation under conditions of scientific uncertainty: How collaborative networks contribute to occupational health and safety regulation for nanomaterials* (Enschede, UT Twente, 2015) (dissertation), 37–38.). The actual application of this approach was often highly problematic and there was political concern about the possible adverse effects on competitiveness of new environmental regulations. In the 1990s the precautionary principle made way for the *no regret* policy of ecological

first National Nature Policy Plan of 1989 featured the so-called Ecological Main Framework, in which nature zones were connected to one another by means of connective corridors.[19] The national plans translated into spatial programs like the VINEX (1990–2010) and the Space Memorandum (2004).

Progressive European unification also inspired important changes. This brought with it a uniform set of environmental regulations in the various member states. Environmental regulations now increasingly came from 'Brussels' instead of 'The Hague.' It should be noted that these were partly based on Dutch experience. The Dutch Ecological Main Framework inspired the European ecological network of the Natura 2000 program.[20] European norms regarding air quality (fine dust norms) bore the mark of the Dutch National Institute for Environmental Management (RIVM) throughout the 1980s and 1990s.[21]

In another respect, too, there was a change of course. The directive policies were gradually abandoned. Government policy now more often than not aimed at cooperation with the business community and the societal organisations. Self-regulation within business sectors was emphasised. In the 1980s the national government was busy re-inventing its societal role. The change of course was also possible thanks to the changing attitude of the business community toward environmental and energy issues.

21.2.4   The Business Community: In Search of Its Own Responsibility

Originally the business community reacted defensively and strategically to the criticisms it had to endure as the 'chief suspect' in the environmental crisis.[22] It delegated representatives to government commissions charged with developing environmental regulations. It arranged representation in the directorates of water

---

modernisation, such as measures to limit the effects of climate warming. (Jeroen P. van der Sluijs, Rinie van Est, and Monique Riphagen, *Room for Climate Debate, Perspectives on the Interaction between Climate Politics, Science and Media* (The Hague: Rathenau Instituut, 2010), 18–20.)

[19] Hans van der Cammen and Len de Klerk, *Ruimtelijke Ordening, van Grachtengordel tot Vinex-Wijk* (Utrecht: Het Spectrum, 2003), 252–53.

[20] See also chapter 18. The Netherlands, together with Flanders and Germany, were pioneers in the development of ecological networks. In Rob H.G. Jongman, 'Nature Conservation Planning in Europe: Developing Ecological Networks' *Landscape and Urban Planning* 32 (1995): 196–183. Per Högselius, Arne Kaijser, and Erik van der Vleuten, *Europe's Infrastructure Transition, Economy, War, Nature*, vol. 4, Making Europe: Technology and Transformations, 1850–2000 (Basingstoke / New York: Palgrave Macmillan, 2016), 261–65.

[21] See also chapter 20; T. Arnoldussen, *The Social Construction of the Dutch Air Quality Clash. How Road Expansions Bit the Dust Against Particulate Matter* (Eleven International Publishing, The Hague 2016), 125–64.

[22] See for the following among others: H. Lintsen and J. Korsten, *De veerkracht van de Brabantse economie. De Kamers van Koophandel en de kracht van netwerken [1840–2015]* (Stichting Zuidelijk Historisch Contact & Uitgeverij Verloren, Hilversum 2017), 134–139. The Chambers of Commerce represent the business community in many environmental issues.

boards that developed policy on the pollution of surface water. It made an inventory of the problems entrepreneurs ran into due to environmental regulations, the need to process wastes and the associated costs. It for example protested against '... the disproportional increases that are the consequence of the new levies associated with the implementation of the Law on Surface Water Pollution...'.[23] But in the course of the 1970s this attitude began to change.

> In the previous decade one could roughly say that in societal consciousness concern for the environment was most highly valued. This was possibly not unjustified as a counterweight to the increased appreciation for the aspect of welfare in the period immediately after the Second World War. It seems to me that the time is now ripe for a balanced approach in weighing these priorities...[24]

In this way the entrepreneur and freshly minted chairman of the Eindhoven Chamber of Commerce, H. Schellens, expressed the feelings of the business community in 1976. At the end of the 1980s, representatives of the business community took an additional step. A sound environmental policy was considered to be highly important for 'the preservation of an attractive residential, working and living environment' and hence also for a favourable climate for attracting new firms.[25] The new Law on Environmental Management of 1993 thus enjoyed a positive reception. 'The business community is aware of its own environmental responsibility and does not seek to avoid it...,' but it did request from the government 'the necessary room for manoeuvre to operate in this manner. On an international level, too, it was necessary to seek a fit between norms and requirements.'[26] Contacts with the governments were intensified.

In addition to consultations, the business community also undertook its own initiatives in the area of environment and energy. '...Not only because the government and society have given it an important priority, but precisely also from the viewpoint of efficiency, quality control, cost control, use of raw materials, working conditions, shop floor risks and market image.'[27] Firms set up their own internal environmental care departments or employed external environmental services.[28]

---

[23] *Jaarverslag Kamer van Koophandel Eindhoven* 1975, 28 en 65.

[24] *Jaarverslag Kamer van Koophandel Eindhoven* 1975, 68.

[25] *Beleidsplan Provinciaal Samenwerkingsorgaan Kamer van Koophandel Noord-Brabant Beleidsplan 1989,* 11 in Archief Provinciaal Bureau Kamers van Koophandel in Noord-Brabant, BHIC 1157, map 44.

[26] *Jaarverslag Kamer van Koophandel Noordoost Brabant 1993,* 27–28. The chamber informed the business community in its area both collectively and individually about the new law and its consequences. In Oss, for example, it cooperated with the municipality. *Jaarverslag PSK over de periode 1 juli 1992 tot 1 juli 1992,* in Archief Provinciaal Bureau Kamers van Koophandel in Noord-Brabant, BHIC 1157, map 43.

[27] *Jaarverslag Kamer van Koophandel Noordoost-Brabant* 1991.

[28] Lintsen en Korsten, *De veerkracht van de Brabantse economie,* 138–139.

21.2.5   Toward a New Order and New Approaches After 1990

A more pro-active business community and fewer directive interventions by the government fit perfectly in the new era in which liberalisation of the economy had become the watchword.[29] The economic crisis at the end of the 1970s and the beginning of the 1980s cleared the way for a reassessment of private initiative in a free market. The American president Ronald Reagan and the British prime minister Margaret Thatcher gave the international example as they advocated privatisation and de-regulation. The government should no longer play an active role in the steering of the economy. State enterprises were better off in private hands. Private firms could in many cases perform government tasks more efficiently.

The Dutch government toed the international line. It sold its shares in companies like KLM and the Hoogoven steel mills and withdrew from public utilities like the postal and telephone services. Suppliers of energy and waste processing facilities were privatised. The government also undertook severe cutbacks and chiselled away at social security. The Competition Law of 1998 sought to create more room for market dynamics. It prohibited agreements among companies to suppress competition, a measure that meant the end of cartel-formation in the Dutch economy.[30] The *dirigiste* tradition in agriculture also made way for government at a great distance from this sector. Agriculture increasingly became just another economic sector. But not all forms of societal coordination were abandoned. One mode did fit with the liberal model, namely the covenant. This instrument enjoyed immense popularity in the area of energy and environment.

The government used covenants to create a constituency for environmental goals in the short and long term.[31] A covenant emerged out of consultations among the government, the business community and societal groups. Out of this came voluntary, but by no means non-committal, agreements – often at the level of a specific branch. These were congruent with general policy frameworks established by the political system. Branch-organisations shouldered part of the responsibility for policy and oversaw the implementation. The government supported progress by providing permits, subsidies and other 'supplementary policy.'[32] Environmental organisations partnered with the business community and contributed concrete and realistic solutions.

The shift to self-regulation also stimulated the business community to undertake other initiatives like 'sustainable entrepreneurship' (later also called 'socially responsible entrepreneurship'). Firms pursuing this course formulated a vision of sustainability, coupled this to a corporate strategy, established a monitoring and

---

[29] See for the following: Bouwens and Dankers, *Tussen concurrentie en* concentratie, 203–207.

[30] The economic-political policy also resulted in the dismantling of long-established coordinating organisations such as the Agricultural Board (*Landbouwschap*, 2001), the Chambers of Commerce as independent regional organisations (2014) and the industry and product boards (2015).

[31] Bouwens and Dankers, *Tussen concurrentie en concentratie*, 213–215; Peet and Nijhof, *Een voortdurend experiment,* 236.

[32] Bouwens en Dankers, *Tussen concurrentie en concentratie*, 213; Cramer, *Milieu*, 40–41.

reporting system and infused the vision into their corporate culture. The financial sector followed with sustainable investing. Sustainability funds ranked firms on the basis of their achievements on economic, social and environmental dimensions.[33]

There was also another shift. Policies had long aimed at suppressing emissions of noxious substances. From the 1990s on, prevention became the main strategy: designing production and other processes in such a way that the emission of environmentally polluting substances was kept to a minimum. With 'sustainable entrepreneurship' the focus was on the production process of the individual firm. So-called 'integral chain management' shifted attention from discrete production processes to production-consumption chains and aimed at redesigning the entire chain from an environmentally conscious perspective.[34]

The transition to a preventive and integral approach also had to do with the globalisation of environmental and energy issues. The emergence of new industrial economies in countries like China, Brazil and India extended environmental problems beyond the previously industrialised countries. More than ever, environmental issues were transported to other places ('elsewhere') and as such became a global concern. Two issues dominated the agendas of international organisations and national governments: climate and raw materials.

These issues became prominent in the Netherlands as well.[35] For example, in 2007 the government set a goal for 2020: a 30% reduction in CO_2 emissions, 20% energy saving and 20% sustainable energy, all relative to 1990. An encompassing program called 'Clean and Thrifty' was set up. Nine ministries were involved. Municipalities and provinces signed the climate agreement and made deals with the national government regarding their contributions to the goals. Branch organisations agreed to an overarching covenant and ten sectoral covenants. Funds were made available for research and innovation.

The program was inspired by the transition approach. This went a step further than the integral strategy. A society-wide, long-term vision on sustainability formed the point of departure for short-term actions. All the societal actors had to be involved. There had to be sufficient space for experiments and learning. It was accepted that the course of a transition process could not be fixed beforehand. Actions at the micro-level had to influence structural changes at the meso-level, which also entailed processes at the macro-level. New images of the future revealed the fundamental character of the transition. Changes in the domains of energy and raw materials would form the basis of a *third industrial revolution*. The social fabric of 2050 would be interwoven with a new economic system that would be based on the re-use of products and resources and the conservation of natural resources.

---

[33] Cramer, *Milieu*, 63–64.

[34] Cramer, *Milieu*, 46.

[35] For the following see: Cramer, *Milieu*, 81–82; 91; 109–111.

Material flows would then become closed cycles and energy sources become renewable. The Netherlands would have to pursue the road to a *circular economy*. We will deal extensively with this last image in the epilogue to this book.

What were the results of the 'Clean and Thrifty' program?[36] None of the aims for 2020 in regard to CO_2 reduction, energy saving and sustainable energy will be achieved. The feasibility of a reviewed and revised energy agreement' signed in 2013 is also in doubt (chapter 20). One of the reasons was that the system of emissions trading ended up a failure. The system was developed at the European level. It gave businesses rights to CO_2 emissions and the possibility to trade these rights. It would have encouraged the reduction of CO_2 emissions had the price remained at € 30–€ 40 per ton of CO_2, but the price dropped to less than €5 in 2014.

Another important reason was the change in the political climate. After a change of government in 2010, the 'Clean and Thrifty' program was put on the back burner. Terrorism, migrants and economic crises drove sustainability issues to the background. Concern for the environment and sustainability once again became a political choice. According to many in the field, the government did not show itself at its most trustworthy. Subsidy policy changed continually. Governments repeatedly failed to achieve the goals they set.[37]

But relationships in the field had also changed. Citizens lost interest in the environment as new societal issues emerged. The professionalisation of environmentalism also played a role. The environment had become the property of experts in industries, the government and environmental organisations. After the turn of the century a new popular movement emerged that was independent of the established institutions and that developed initiatives like local energy cooperatives and small businesses for sustainable products.

It is within this context that we now evaluate the quality of life in the Netherlands for the last sample date of this book. We do that using the monitor for 2010 and from two perspectives: from the perspective of 1970 and from a present-day perspective.

---

[36] Cramer, *Milieu*, 82–84.

[37] In October 2017, the Planning Bureau for the Environment, for example, expressed doubts whether the Rutte-III cabinet would be able to meet its own climate goals. The Bureau states: 'The emission goal of 49% proposed by the Rutte-III cabinet fits with the ambitions of the Paris accord to limit temperature rise to well under 2 degrees. The measures analysed here achieve roughly half of the necessary reduction in emissions. A transition policy is in place but can acquire more momentum. In order to achieve the aimed-for reduction of 49% in 2030, it will therefore be necessary to work out additional measures in a new climate and energy agreement.' *Analyse regeerakkoord Rutte-III: effecten op klimaat en energie* (Uitgever: Planbureau voor de Leefomgeving, Den Haag 2017), publicatienummer: 3009, 33.

21.3   The Monitor for 2010: Development of Well-being and Sustainability

21.3.1   Perspective 1970: Progress

How should well-being and sustainability around 2010 be evaluated from the perspective of 1970? Which themes were then on the societal agenda and what is their present state? We arrive at the following brief summary.

An evaluation from the perspective of 1970 shows a mixed verdict. Material welfare has increased and the inequality between men and women has declined. Netherlanders are older, taller and heavier. Dwellings and facilities are of higher quality. All this with more free time. In these respects the image is positive.

But unemployment levels have, grosso modo, increased. During the economic recession at the end of the 1970s these increased for the first time since the 1930s. After that, unemployment followed the ups and down of the economy. The recessions impacted on social inequality. Due to the lowering of the minimum wage in the 1980s, the chronic pressure on welfare budgets and restructuring of the social security system income inequality grew between 1985 and 1990 and again around the year 2000.[38]

From the perspective of 1970 the participation of the populace in politics is also disappointing. The democratisation of society and the levelling of power differentials were important issues around 1970. In this respect the Netherlands has shown no improvement. Quite the contrary. The Dutch demonstrate less quickly, attend consultation meetings less frequently and have less recourse to political parties. They are, moreover, less frequently members of political parties, trade unions, conservation societies and consumer organisations (in terms of percentage of population).[39] These trends are not a sign of greater satisfaction, but rather of lower levels of participation and commitment. There is mistrust of politics and dissatisfaction with the existing institutions.[40] The linkages between politics and social networks were lost with the end of pillarisation. New linkages have yet to emerge in their stead. For example, more Netherlanders are active politically via the internet and more frequently boycott products because they cause environmental damage or involve child labour. In addition there is evidence that small-scale local forms of citizens' initiatives are emerging. The populace certainly appears to be sensitive to

---

[38] W. Salverda et al., *Nederlandse ongelijkheid sinds 198. Loonvorming, overheidsbeleid en veranderde samenstelling van huishoudens* (Uitgever: Amsterdam Institute for Advanced Labour Studies/ University of Amsterdam, Amsterdam 2013).

[39] R. Bijl et al. (ed), *De sociale staat van Nederland 2015* (Uitgever: Sociaal en Cultureel Planbureau, Den Haag 2015), 221–222.

[40] See among others the following publication, with which we opened this book: P. Schnabel, 'Feiten en gevoel', in: C. van Campen et al. (eds.),*Sturen op geluk. Geluksbevordering door nationale overheden, gemeenten en publieke instellingen* (Uitgever: Sociaal en Cultureel Planbureau, Den Haag 2012), 17–23.

societal issues but possibly have more faith in their own initiatives than in those of politics and the government.

Concern for the environment and nature seems to have bottomed out and is again on the rise. Local environmental problems have been rigorously dealt with. Nature restoration projects have commenced. The numbers of seals, beavers, badgers and birds of prey – symbolic for the restoration of natural habitats – have increased. But measured against the quantity of all animal species the comeback is still quite modest. In any case, the decline seems to have been arrested.

International and global environmental problems were hardly at issue around 1970. That was not the case for the problems with raw materials that became a major concern in the light of the Report of the Club of Rome. This concern persists, but is now overshadowed by the climate problem.

21.3.2 Perspective 2010: New Issues

How should we evaluate the present situation from a present-day perspective? We could look with the gaze of politicians, policy makers, opinion leaders or business leaders. An evaluation could also proceed from the priorities of the Dutch population, as they can be deduced from surveys. Here we choose the approach used by Statistics Netherlands (CBS) in which the situation in the Netherlands is compared with the other countries of the European Union. This is done on the basis of the indicators of the 'well-being monitor' (Table 21.1). These provide an image of important societal issues. We assume that the political and policy establishment also considers them important. We also assume that the Netherlands is doing well if it occupies a place near the top of the EU rankings. If the Netherlands finds itself near the bottom, we regard the situation as problematic.[41]

According to the monitor, the quality of life 'here and now' is, by European standards, very high. The Netherlands is in good shape with respect to material welfare, health, life-expectancy, housing, free time and various other indicators for the personal situation. However, after making an inventory of all relevant indicators, Statistics Netherlands and the three Dutch planning bureaus offer the following caveat: '… we create the quality of life in the here and now in a manner that makes it difficult for future generations to create welfare.'[42] The present generation partly shifts the solution of the sustainability problems it has created to the future. The bureaus also note that the present welfare of the Netherlands contributes to sustainability problems in foreign countries (see below).

---

[41] An approach of this nature can be found in *Monitor Duurzaam Nederland 2011* (Centraal Bureau voor de Statistiek, Den Haag/Heerlen 2011) en *Monitor Duurzaam Nederland 2014. Indicatorenrapport* (Centraal Bureau voor de Statistiek, Den Haag/Heerlen 2014).

[42] *Monitor Duurzaam Nederland 2011*, 51.

Table 21.1 Dashboards well-being and sustainability in 2010 from the perspectives of 1970 and a present-day perspective (EU position)

Theme	Indicator	Unit	2010 (2000)	Perspective 1970	NL in EU 2010
Population	Million inhabitants		16.5		
Material welfare and well-being					
Consumption, income	Consumptive expenditures per capita, constant prices	index (1850=100)	581	+	+
	Income inequality, general	Gini coefficient 0–1	0.32	+	+
	Gender income inequality	% difference hourly wage M/F	19%	+	−
Subjective well-being	Satisfaction with life	Score 0–10	7.8	+	+
Personal characteristics					
Health	Life expectancy	year	women 83.0 men 79.3	+	women o men +
Nutrition	Height (military conscripts)	cm	(183)	+	?
Housing	Housing quality	% slums	<1	+	o
	Public water supply	m³/capita	120		
Physical safety	Murder victims	number per 100.000 inhabitants.	(1.1)	−	+
Labour	Unemployment	% workforce.	5,0	−	+
Education	Level of education	years	(11)	+	−
Free time	Free time	hours per week.	44.7	+	?
Natural environment					
Biodiversity	MSA	% original biodiversity	(63)	−	?
Air quality	SO₂	kg SO₂/ capita	4	+	+
	Greenhouse gas emissions	ton CO₂/capita	10.6	+	−
Water quality	Public water supply	m³/capita	120		
Institutional environment					
Trust	Generalised trust	% population with adequate trust	67	−	+
Political institutions	Democracy	Democracy-index 0–100	(39)	o	+

Note: The numbers in brackets are from J.L. van Zanden et al. (ed.), *How was life? Global well-being since 1820* (OECD Publishing 2014) and relate to the year 2000. Numbers for these indicators – measured according to the same methodology – are not available for 2010.

(continued)

Table 21.1 (continued)

Dashboard well-being 'later'			2010	Perspective	NL in EU
			(2000)	1970	2010
Natural capital	**Indicator**	**Unit**			
Energy	Energy consumption	TJ /capita	0.17	–	–
Non-fossil fuels	Gross domestic consumption	ton/capita	9.8	–	?
Biodiversity	MSA	% original biodiversity	(63)	–	?
Air quality	SO$_2$ emissions	kg SO$_2$/ capita	4	+	+
	Greenhouse gas emissions	ton CO$_2$/capita	10.6	+	–
Water	Public water supply	m^3/capita	120		
Economic capital:					
Physical capital	Economic capital stock/capita	index (1850=100)	1046	+	?
Financial capital	Gross national debt	% gdp	59	+	o
Knowledge	Stock knowledge capital	index (2010=100)	100	+	o
Human capital:					
Health	Life expectancy	years	Women 83.0 Men 79.3	+	Women o Men +
Labour	Unemployment	% workforce	5,0	–	+
Educational level	Schooling	years	(11)	+	–
Social capital:					
Trust	Generalised trust	% population with adequate trust	67	–	+
Political institutions	Democracy	democracy index 0–100	(39)	o	+

Note: The numbers in brackets are from J.L. van Zanden et al. (ed.), *How was life? Global well-being since 1820* (OECD Publishing 2014) and relate to the year 2000. Numbers for these indicators – measured according to the same methodology – are not available for 2010.

Dashboard well-being 'elsewhere'					
Material Welfare	Indicator	Unit	2010	Perspective 1970	NL in EU 2010
Consumption, income	Development aid	% gdp	0,8	o	o
Natural capital					
Natural capital	Import of raw materials	ton/capita	12,5	o	–

(continued)

Table 21.1 (continued)

+	Not problematic/was not problematised from the perspective of 1970. Present-day perspective: The Netherlands is in good shape because it belongs among the nine highest-scoring EU member states.
–	Problematic from the perspective of 1970. Present-day perspective: the situation is problematic because the Netherlands belongs among the nine lowest-scoring EU member states.
O	Under discussion: different opinions about the scale and nature of the problems. Present-day perspective: No evaluation in terms of problematic or not problematic. The Netherlands scores in the middle group of the 27 EU member states.
?	Unknown

Remark 1: For present days perspective the table indicates the Netherlands position within the EU-27 (see legend)

Remark 2: Public water supply is left blank; this indicator became irrelevant as an indicator for water quality

Source: See note 23 of Chap. 2; The position of The Netherlands in EU derived from *Monitor Duurzaam Nederland 2014. Indicatorenrapport* (Den Haag 2014), 158–169. Numbers from years between 2010 and 2013

　

The problems pertain above all to the capitals, in other words to the resources for 'later.' This does not concern economic or social capital – these are in reasonable and even good shape – but rather natural and human capital.

In previous chapters human capital has barely been mentioned. An important issue is the level of education of the Dutch populace. To be sure, the level of education measured as years of schooling has increased, but the number of school dropouts is substantial and the quality of primary and secondary education is declining, judging by the scores of Dutch pupils on international comparative tests.[43] A well-educated populace is important for labour productivity and the material welfare of a country. In addition, education also to a great extent determines the quality of an individual's personal life.[44] The well-educated get the better jobs, have a higher income, are healthier and live longer. The less-educated are vulnerable in the labour market, change jobs more often, have to work more years and have a materially less secure existence. Social inequality is not only a question of income but also manifests itself in these kinds of aspects. Nowadays school diplomas are the main determinant of this dimension of inequality.

Of the four capitals, natural capital is the most problematic.[45] In the 1990s, with its research and policy experience, the Netherlands set the tone for European policy that was rolled out from 2000 on, but the Netherlands itself, paradoxically enough, seemed barely able to meet the norms it had implicitly set for itself. The Netherlands

---

[43] *Monitor Duurzaam Nederland 2011*, 58.

[44] Also see: *Kwaliteit van leven in Nederland* (Centraal Bureau voor de Statistiek, Den Haag 2015), 56–58.

[45] *Monitor Duurzaam Nederland 2011*, 51.

is among the EU countries with the highest per capita energy consumption, the lowest investment in renewable energy sources and the highest per capita emissions of greenhouse gases. The nutrient surplus in the soil is among the biggest in Europe. Despite many improvements, Dutch surface water nowhere meets the European quality norm. And the Netherlands still has little forest and nature compared with other European countries. Biodiversity in the countryside and in nature preserves is chronically under pressure.

There are also problems with the degree to which the Netherlands places a burden on nature and environment 'elsewhere.'[46] It belongs among the EU countries with the biggest per capita import of bio-raw materials and mineral and fossil subsoil resources. The raw materials that are imported into the Netherlands from developing countries are a special problem. To be sure, this trade creates incomes in developing countries, but an analysis by the World Bank shows that these local incomes are generally spent in consumption and frequently to the benefit of a small elite. There is also the problem of depletion, loss of biodiversity and often also of a one-sided economic structure.[47] The Netherlands has shifted some of its sustainability problems elsewhere.

On the other hand, Dutch exports and hence the Dutch products consumed in other countries contribute to problems in the Netherlands. For example the massive export of meat is associated with sustainability problems in cattle husbandry and the bio-industry.

This short evaluation confirms what has emerged in this and preceding chapter: The most import sustainability issues for the coming decades are energy, raw materials and biodiversity. In the Epilogue we summarise the historical process that has led to these problems. We also investigate, from a historical perspective, the concept of 'circular economy,' a notion that is regarded as the most important option for a way out of our present-day predicament.

Literature

Anonymous, (1989a, november 4). 'Nijpels hoopt op afspraken tijdens milieuconferentie' in *NRC*

Anonymous, (1989b, november 7). 'Milieuconferentie begint in sfeer van onenigheid, Nederland vraagt om hulp voor Derde Wereld' in *Nederlands dagblad*

Arnoldussen, T. (2016). *The Social Construction of the Dutch Air Quality Clash. How Road Expansions Bit the Dust Against Particulate Matter.* 125–64. The Hague: Eleven International Publishing.

---

[46] *Monitor Duurzaam Nederland 2011*, 60–63.

[47] *Monitor Duurzaam Nederland 2011*, 62–63.

Bijl, R. et al. (2015). *De sociale staat van Nederland 2015*. 221–222. Den Haag: Sociaal en Cultureel Planbureau.

Bouwens, B. and Dankers, J. (2012). *Tussen concurrentie en concentratie. Belangenorganisaties, kartels, fusies en overnames*. 205. Amsterdam: Uitgeverij Boom

Bouwer, K., Klaver, J., and de Soet, M. (1983). *Nederland Stortplaats, een Milieukundig en Geografische Visie op het Afvalprobleem*. Nijmegen: Ekologische uitgeverij.

Buijsman, E. (2008). 'Gisteren, Vandaag, Morgen, Een Terugblik op het Probleem van Zure Regen,' *Studium* 4: 251–68

Centraal Bureau voor de Statistiek (2011). *Monitor Duurzaam Nederland 2011*. Den Haag/ Heerlen: Centraal Bureau voor de Statistiek

Centraal Bureau voor de Statistiek (2014). *Monitor duurzaam Nederland 2014: Indicatorenrapport*. Den Haag/Heerlen: Centraal Bureau voor de Statistiek.

Centraal Bureau voor de Statistiek (2015). *Kwaliteit van leven in Nederland*. 56–58. Den Haag: Centraal Bureau voor de Statistiek

Cramer, J. (2014) *Milieu*. Amsterdam: Amsterdam University Press.

Dinkelman, G.H. *Verzuring en Broeikaseffect, de Wisselwerking tussen Problemen en Oplossingen in het Nederlandse Luchtverontreinigingsbeleid (1970-1994)*. Amsterdam: Universiteit van Amsterdam.

Duyvendak, W. (2011) *Het Groene Optimisme, Het Drama van 25 Jaar Klimaatpolitiek*. Amsterdam: Bert Bakker.

Högselius, P., Kaijser, A. and van der Vleuten, E. (2016). *Europe's Infrastructure Transition, Economy, War, Nature, vol. 4, Making Europe: Technology and Transformations*, 1850–2000. Basingstoke / New York: Palgrave Macmillan.

Jongman, Rob H.G. (1995). 'Nature Conservation Planning in Europe: Developing Ecological Networks' in *Landscape and Urban Planning* 32: 196–183.

Kamer van Koophandel (1989). *Beleidsplan Provinciaal Samenwerkingsorgaan Kamer van Koophandel Noord-Brabant Beleidsplan 1989*, 11 in Archief Provinciaal Bureau Kamers van Koophandel in Noord-Brabant, BHIC 1157, map 44.

Kamer van Koophandel (1991). *Jaarverslag Kamer van Koophandel Noordoost-Brabant* 1991.

Lintsen, H. and Korsten, J. (2017). *De veerkracht van de Brabantse economie. De Kamers van Koophandel en de kracht van netwerken [1840-2015]*. 134–139. Hilversum: Stichting Zuidelijk Historisch Contact & Uitgeverij Verloren

Peet, J. and Nijhof, E. (2015). *Een voortdurend experiment. Overheidsbeleid en het Nederlandse bedrijfsleven* (Uitgeverij Boom, Amsterdam 2015), 199–205.

Planning Bureau for the Environment (2017). *Analyse regeerakkoord Rutte-III: effecten op klimaat en energie* 3009, 33. Den Haag: Planbureau voor de Leefomgeving.

Reichow, A. (2015). *Effective regulation under conditions of scientific uncertainty: How collaborative networks contribute to occupational health and safety regulation for nanomaterials*. 37–38. Enschede: UT Twente

Salverda, W. et al. (2013). *Nederlandse ongelijkheid sinds 198. Loonvorming, overheidsbeleid en veranderde samenstelling van huishoudens*. Amsterdam: Amsterdam Institute for Advanced Labour Studies/ University of Amsterdam

Schuyt, K. and Taverne, E. (2000). *1950. Welvaart in zwart-wit*. 40–41. Den Haad: Sdu Uitgeverij

Siraa, H.T., van der Valk, A.J. and Wissink, W.L. (1995). *Met het Oog op de Omgeving. Het Ministerie van Volkshuisvesting, Ruimtelijke Ordening en Milieubeheer, 1965–1995*. 244–67. Den Haag: SDU Uitgeverij.

van Baalen, Carla et al. (2005). *Koning Beatrix aan het Woord, 25 Jaar Troonredes, Officiële Redevoeringen en Kersttoespraken*. Den Haag: Sdu Uitgeverij.

van Campen, C. et al. (2012) *Sturen op geluk. Geluksbevordering door nationale overheden, gemeenten en publieke instellingen*. 17–23. Den Haag: Sociaal en Cultureel Planbureau.

van der Cammen, H. and de Klerk, L. (2003). *Ruimtelijke Ordening, van Grachtengordel tot Vinex-Wijk.* 252–53. Utrecht: Het Spectrum

van der Sluijs, J. P., van Est, R. and Riphagen, M. (2010). *Room for Climate Debate, Perspectives on the Interaction between Climate Politics, Science and Media.* 18–20. The Hague: Rathenau Instituut

van Zanden J.L., Griffiths, R.T. (1989). *Economische geschiedenis van Nederland in de 20ᵉ eeuw.* 48–49. Utrecht: Aula.

Epilogue: Well-being and Sustainability 1850–2050

Chapter 22
The Long-Term Development: In Search of a Balance

Harry Lintsen and Jan-Pieter Smits

Contents

Abstract The chapter summarises the development of well-being and sustainability in the Netherlands between 1850 and 2010. It commences by establishing that any summary has a normative dimension. Issues relating to quality of life must consistently be analysed from a historical (contemporary) and a present-day perspective.

The summary shows the great transformation of a society with extreme poverty and a circular economy into a welfare society with a linear economy. Present-day sustainability issues, including climate change, resource depletion, raw materials dependency, and worrisome biodiversity have their roots in this transformation. Well-being and sustainability were in balance during only a brief period. Around 1960 the old historical challenge of extreme poverty had been solved and the quality of life as seen through contemporary eyes was reasonably in order, while the claims on nature and the environment were still modest. The balance would be shattered in the course of the 1960s.

Gradually, sustainability has become the new historical challenge. At the same time society is confronted with the so-called welfare paradox: despite the high level of welfare, there is much unrest among the populace.

© The Author(s) 2018
H. Lintsen et al., *Well-being, Sustainability and Social Development*,
https://doi.org/10.1007/978-3-319-76696-6_22

Keywords Extreme poverty · Circular economy · Linear economy · Sustainability
· Welfare paradox

22.1 The Historical Challenge

'Global extreme poverty for the first time under 10%' according to the newspaper
Nieuwe Rotterdamse Courant on October 5, 2015.[1] For the first time in human history the scope of extreme poverty was expected to fall below 10% of the global
population, a feat to be achieved by the end of 2015. While the reliability of the
figures leaves room for doubt (data are lacking, for example, on a number of poor
countries) the decline of extreme poverty is clearly evident. Only 25 years earlier –
at the outset of the 1990s – the percentage had been nearly 40%! And the ambitions
are hardly modest. 'This is the best story in the world today,' according to the president of the World Bank, 'these projections show that we are the first generation in
human history that can end extreme poverty.'[2]

In the Netherlands, extreme poverty – on the basis of a poverty line of $ 1.90 per
day (in 2015) as defined by the World Bank and the United Nations – fell below 10%
at the end of the nineteenth century. That was for the first time in its history. But history shows that the struggle against extreme poverty is not a linear march to victory.
At the beginning of the nineteenth century 40%–50% of the Dutch population also
lived below the poverty line. That percentage declined to some 21% around 1850, only
to increase again thereafter to about 40%. After 1870 a declining trend set in. Wars and
economic recessions, among other factors, influenced the levels of poverty and would
continue to do so throughout the twentieth century, even though in that century the
percentage of the extremely poor in the Netherlands would never exceed 10%.

Another issue is also at play in the struggle against poverty. The World Bank and
the United Nations aim to put an end to extreme poverty by 2030. The experience of
the Dutch with the abolition of extreme poverty shows that that is by no means the
end of poverty altogether. Extreme poverty is defined on the basis of a purchasing
power (of $ 1.90 per day) that is sufficient to provide a person with just the bare
necessities: adequate means to feed and clothe himself and to protect himself, more
or less, from the elements. That says little about the quality of dwellings, health

---

[1] S. Klumpelaars and M. Somers. 'Extreme armoede wereldwijd voor het eerst onder 10 procent',
NRC   5   oktober   2015,   retrieved   from   https://www.nrc.nl/nieuws/2015/10/05/
wereldbank-extreme-armoede-wereldwijd-voor-het-eerst-onder-10-procent-a1412636
[2] World Bank Group. *World Bank Forecasts Global Poverty to Fall Below 10% for First Time;
Major Hurdles Remain in Goal to End Poverty by 2030*, retrieved from http://www.worldbank.org/
en/news/press-release/2015/10/04/world-bank-forecasts-global-poverty-to-fall-below-10-for-
first-time-major-hurdles-remain-in-goal-to-end-poverty-by-2030. Consulted 20 January 2017.

care, education and labour. In the Netherlands in the last quarter of the nineteenth century, the conception of poverty broadened to include problems like the many slums, poor hygiene, miserable working conditions, and dubious food quality. Around 1960 these problems had been solved to a great extent.

That does not mean that poverty has disappeared in the Netherlands. Every generation defines poverty anew. Every year, the National Institute for Budget Information establishes a minimum budget that is just sufficient to meet basic costs (food, clothing, housing, internet etc.) and supplementary costs for social participation (vacation, sport, entertaining etc.).[3] In 2014 this minimal budget for a family with two children was € 1830 per month or about $ 16.40 per day. 1.2 million Netherlanders (7.6% of the population) lived below this poverty line.

Banishing extreme poverty is seen as an historical challenge. In 2015, 193 government leaders proclaimed that extreme poverty should disappear by 2030.[4] It was the first point of a broader United Nations agenda for sustainable development. In the Netherlands, the abolition of extreme poverty within the country had already been high on the social agenda at the beginning of the nineteenth century. History shows that the struggle had to be pursued on many fronts and that it demanded perseverance and patience. It included a social struggle, that is, a struggle around the distribution of material welfare. It was also a struggle between the Netherlands and the colonies about the way in which the colonies would be exploited. History also shows that banishing poverty involved trade-offs: the abolition of extreme poverty has sown the seeds of new problems. Those problems are first and foremost associated with the environment, the finiteness of raw materials, the quality of the landscape and greenhouse gas emissions. Or, in terms of this study: They are above all associated with the exploitation of natural capital in the Netherlands and elsewhere in the world.

To get a grip on this, the present study is built around three analytical points of entry (Chap. 1). First of all, use has been made of a new instrument developed by Statistics Netherlands (CBS), the Personal Wellbeing Index (aka the Monitor Well-Being). The instrument also provides an antidote to the misapprehension that societal progress is first and foremost a question of economic growth. It does this by showcasing three dimensions of well-being: quality of life 'here and now,' 'later' and 'elsewhere.' 'Here and now' includes issues like income, social inequality, health, education, environment and democracy. In this study the question of poverty is positioned in relation to these issues. The dimension 'later' reveals what a society confers on later generations in terms of natural, economic, human and social capital. The dimension 'elsewhere' reveals the effect of domestic activities on societal development in other countries.

---

[3] S. Hoff, J.M. Wildeboer Schut, B. Goderis and C. Vrooman (2016), *Armoede in kaart*. Den Haag: Sociaal en cultureel planbureau.

[4] United Nations, *Sustainable Development Goals. 17 goals to transform our world* (1. No Poverty). http://www.un.org/sustainabledevelopment/sustainable-development-goals/

In the second place, the institutional context plays an important role in our analyses. The developments revealed by the monitor are not abstract processes and the struggle against poverty is not an abstract struggle. They are the work of social groups that in the course of things abolish old institutions and create new ones. In this study the institutional context is broken down into state, market, the societal midfield (civil society), and technology (including science and innovation). These four contexts largely shape the activities of historical actors and are among other things the foundation of poverty (for example as supports for social inequality). Institutional changes are essential, but are often difficult to realise.

In the third place, this study has granted pride of place to natural capital. It sees natural capital as the foundation of a given quality of life. The way a society deals with natural capital (soil, air, water, subsoil resources) in large part shapes social structure in the 'here and now' and 'elsewhere.' It also has a great impact on the quality of life 'later.' Trade-offs, as the study reveals, are essential in this context. Modern agricultural methods, for example, encourage greater productivity and are important for food production and food security, but at the same time influence biodiversity and have various environmental impacts. In this study, the role of natural capital is portrayed by dividing the capital into three types of raw materials and concomitant material flows: bio-raw materials (agriculture and foods), mineral subsoil resources (building and infrastructure) and fossil subsoil resources (energy and plastics).

We will summarise well-being and sustainability between 1850 and the present-day from the perspective of these three analytic approaches. We subsequently orient ourselves to the future from the point of view of natural capital. With respect to well-being this is one of the most important new historical challenges. Maintaining a certain level of well-being in the Netherlands requires a substantial investment in natural resources, here and elsewhere. Eliminating poverty elsewhere makes extraordinary demands on global natural capital, certainly if humanity wants to pursue this project for generations to come. Moreover, successive generations will continue to define well-being and sustainability in new ways and thus continue to make significant demands on natural capital.

22.2   Well-being and the Normative Point of Departure

An historical evaluation of poverty and well-being is determined by the norms and preferences cherished by us in the present or by contemporaries in earlier times. That can be illustrated with a number of graphs that portray the quality of life between 1850 and 2010 from different perspectives.

With Graph 22.1a we started in the prologue. It is based on ten indicators from the monitor.[5] These are related to the four themes of quality of life: prosperity (including income), personal characteristics (such as health), natural environment (including air quality) and institutional environment (such as democracy). The graph provides a historical evaluation of the quality of life, with the well-being indicators between 1850 and 1960 being given the maximum weight in the calculations, while they are participating in average weight after 1960. On the other hand, the indicators of the natural environment until 1960 have been given a minimal weight, while after 1960 they are included maximal in the calculation. This picture roughly corresponds with the picture that contemporaries until 1960 put maximum value on the growth of prosperity and economy and considerably less on the envi-

---

[5] The Graphs 22.1a–c, compare the development of GDP per capita with an index for broad prosperity. This index consists of a weighted average of a number of sub-series concerning the quality of life (see table below). The various series move in different bandwidths. In order to perform the weighing in a technically clean way, all series are first normalized (they have all been given the same bandwidth) before applying the weighting factors.

Theme	Indicator	UNIT	22.1a		22.1b	22.1c
			1850–1960	1960–2010		
Welfare and well-being						
Consumption, income	Consumer expenditures per capita, constant prices	Annual expenses per capita. Index: 1850 = 100	2	1	2	0
	Income inequality, general	Gini coefficient: 0–1	2	1	2	0
Personal characteristics						
Health	Life expectancy	Years	2	2	0	0
Education	Level of education	Years	1	1	0	0
Labor	Unemployment	% workforce	1	1	0	0
Natural environment						
Biodiversity	MSA	% of original biodiversity	0	2	0	2
Air quality	SO$_2$	Kg SO$_2$/capita	0	2	0	2
	Greenhouse gas emissions	Ton CO$_2$/capita	0	2	0	2
Institutional environment						
Physical safety	Murder victims	Number per 100.000 inhabitants	1	1	0	0
Political institutions	Democracy	Democracy-index 0–100	1	1	0	0

Graph 22.1 The
development of well-being
as viewed from different
normative frameworks,
1850–2010 (compressed
and composite index
1850 = 100). (**a**)
Historically differentiated
preferences. (**b**)
Preferences: welfare and
consumption. (**c**)
Preferences: nature and
environment
**Source: See note 5 of this
chapter**

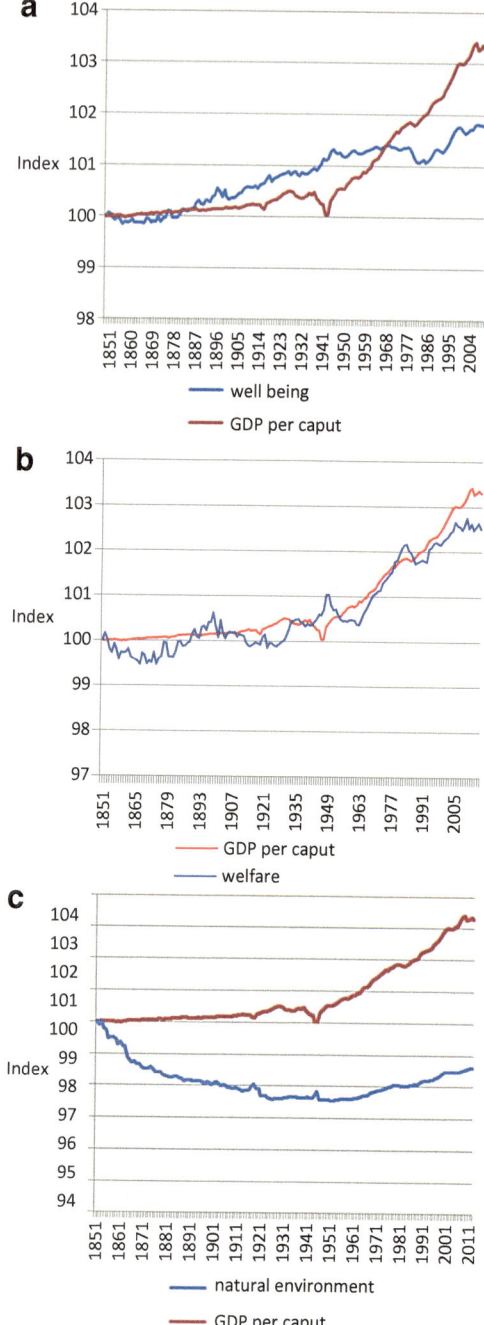

ronment and nature, while the environment and nature after 1960 repeatedly appeared as priorities on the societal agenda..

What does such a normative framing tell us? After an initial decline in quality of life, a reversal sets in after 1870 with an increase that is stronger than the growth of the GDP. After the 1950s, the rate of increase in quality of life slows down, declining absolutely in the 1970s and 1980s and increasing again at the end of the twentieth century, though the increase lags far behind economic growth.

How important normative frameworks are for evaluating the development of well-being is shown by the following two graphs. Graph 22.1b assumes a societal agenda that is only concerned with economic growth and health. It is the only graph in which throughout the entire period there is a direct link between economic growth and quality of life (which is of course hardly surprising). In Graph 22.1c are nature and the environment to be valued most highly after 1850, then well-being would decline over a long period and only rebound again during and after the 1960s, but it would never exceed the level of 1850.

In the following overview we shall have to indicate clearly the norms and preferences underlying the evaluation of the development of well-being and sustainability.

We started the study with a periodisation in which 1910 and 1970 formed the sample years, namely 1850–1910, 1910–1970 and 1970–2010. Historical research leads to a modified periodisation that will here be further developed, namely 1850–1900, 1900–1960 and 1960–2015, in which 1900 and 1960 are the watershed years.

Around 1900 poverty gets redefined in the Netherlands and the new definition is anchored in legislation, after which the struggle against poverty enters a new phase. Around 1960 the use of natural capital appeared to be increasing exponentially, at which point the problem of environment and nature entered a new phase.

22.3 Around 1850: Extreme Poverty in Context

The well-being monitor for 1850 shows that the Netherlands was rich, but not prosperous. Around 1850 the country was among the richest in the world, but the wealth was unequally distributed. Hundreds of thousands of Netherlanders lived in extremely poor circumstances, regularly suffered hunger, were reduced to beggary or the dole in order to stay alive, lived in slums or on the streets clothed in rags. Some contemporaries might possibly speak of a prosperous country. Elsewhere in the world it was often more miserable. It is estimated that the extremely poor constituted about 85% of the population living south of the Sahara, about 77% of the Asian population, 69% of South Americans and 35% of West Europeans.[6] In the Netherlands, the percentage came to 21%. Nonetheless, few Netherlanders were acquainted with the situation elsewhere. Around the mid-nineteenth century they

---

[6] J.L. Van Zanden, J. Baten, P. Földvari, and B. van Leeuwen, *The changing Shape of Global Inequality 1820–2000. Exploring a new dataset* (Utrecht 2011), working paper no.1, table 12.

saw themselves confronted in their own country with cholera, failed harvests, high food prices and flooding rivers.

In the eyes of the contemporary bourgeoisie extreme poverty was one of the most important social problems of their time. In present-day terms it was one of the key issues of well-being. In retrospect this historically stubborn problem could have been solved at the time by means of a more equitable distribution of wealth. If income inequality had been at the same level as now, then the incidence of extreme poverty would have fallen far under 10%. The necessary institutions (for example trade unions) that might have consolidated a struggle for more equality, simply did not exist. Moreover, the system of poor relief ensured adequate survival chances for the poor *and* most of the time for enough social tranquillity.

Another option to eliminate poverty was economic growth. It would have required an annual economic growth of 3.6% between 1820 and 1850. Given the existing natural capital of the Netherlands in combination with the technology of the day, that turned out to be impossible. Under King William I, the Netherlands achieved no more than 2.3% annual growth and even that was a considerable accomplishment. The King had done everything he could to modernise the economy and technology. He succeeded only in part. Old corporatist structures and institutions like protected local markets, urban autonomy and guild-like organisations persisted. Innovations in agriculture, industry and transportation were limited. The use of steam power as a key technology remained stuck in an introductory phase. King William I did succeed in realising economic growth with new colonial policy. He mobilised the Dutch East-Indies as a source of profit by using the cheap mass labour of Javanese farmers to generate income for the Dutch state and economy.

The basis of Dutch wealth was its natural capital: its situation in a delta at a crossroads of shipping and trade routes; the availability of turf as cheap source of energy and the exploitation of agricultural lands for the high-value production of dairy produce (butter and cheese) and industrial crops (among others, madder). And there was more than enough potential to intensify the exploitation of domestic natural capital. In the 'high' Netherlands, agriculture was only marginally integrated into national and international markets. The delta's infrastructure left much room for improvement. Dutch natural capital also contributed to biodiversity. The country had a rich variety of agricultural systems in addition to its wealth of original ecosystems like dunes and mudflats. But natural capital also had a downside. The Netherlands was a vulnerable country. The nation was forced to wage a never-ending battle against water. It was also confronted with an immense environmental problem, namely the urban detritus of human and animal faeces and food remnants. Organic waste was a major cause of the high mortality rates and the low life-expectancy. The combination with excessive and saline water exacerbated the problem in the low parts of the Netherlands.

Three developments provided a new perspective for well-being and sustainability around 1850. The liberalisation of international trade created opportunities for economic growth. The political revolution and the new constitution of 1848 put an end to the autocratic reign of William I and laid the basis for new constitutional relationships. The societal midfield, i.e. civil society, became much more dynamic thanks to the rise of a young generation of professional physicians, engineers and architects.

22.4   1850–1900: Manifold Dynamics

Economic growth was the motive force behind the assault on extreme poverty in this period. Initially, however, extreme poverty increased because agriculture and a small farming and urban elite profited from the liberalisation of international trade. After 1870 modernisation of the economy and growing welfare became linked. They were strongly influenced by the introduction of coal and steam technology.

From 1870 on, economic growth in the Netherlands definitively became dependent on fossil subsoil resources.[7] Up to the middle of the twentieth century this meant coal and after the Second World War oil and natural gas. Economic growth around 1850 was still largely dependent on classical sources of energy like muscle power, turf, wind and water. Human and animal muscle power were the largest single source of energy, contributing about 38% of the total (against 16% for coal). By 1900 muscle power had declined to 20% (against 63% for coal). In the twentieth century physical labour as a source of energy would become almost irrelevant: in 1950 it dipped under 1% of the total energy consumption for the first time.[8]

The modernisation of the economy was also associated with a revolution in the food supply chain, the mass flow of bio-raw materials. Agriculture in the 'high' Netherlands became part of national and international markets and began to specialise in cattle husbandry, among other things. Heathlands were reclaimed. Farmers started to use artificial fertiliser. Food production became mechanised with the coming of bread and flour factories, dairies and slaughterhouses. Production and consumption in the food chain lost their circular character and acquired the characteristics of a linear economy. In the mixed farming sector, animal husbandry with its manure production was no longer subservient to crop farming; instead, crop farming became subservient to animal husbandry. Urban organic waste lost its function as fertiliser and was replaced by artificial fertiliser. Massive amounts of artificial fertiliser and cattle fodder were imported, while butter, cheese, cattle, meat and eggs left the country. Agricultural production was uncoupled from the local and regional production of raw materials.

The supply chain in construction, in other words the massive flow of mineral raw materials, kept much of its circular character and its national orientation. Sand, gravel and clay were mined and processed in the Netherlands. They were embodied in buildings, dikes, roads etc. and after use broken down into materials that were re-used. Production and use in this supply chain led to a substantial improvement in the Dutch transportation infrastructure (among others, railways and canals, including the North Sea Canal) and in this way contributed to economic growth. They also led to a considerable amelioration of the water management situation and thereby to

---

[7] Here it must be noted that turf – also a fossil fuel – played a crucial role during the lengthy period of economic growth of the Netherlands during its 'Golden Century.'

[8] Calculations based on appendix 1 uit: B. Gales and H. Hölsgens, 'Energy consumption in the Netherlands (1800–2012), in: H. Hölsgens, *Energy transitions in the Netherlands: Sustainability challenges in a historical and comparative perspective* (Groningen 2016), 207–217.

the safety of the Netherlands in its struggle against water. An enormous effort went into the improvement of the rivers.

The modernisation of the economy was deeply rooted in a dynamic societal midfield. Periodicals propagated a belief in progress. Organisations like the Society for Industry zealously advocated modern technology. A new generation of professionals created a network of trade schools, laboratories, knowledge institutes and professional associations. A new generation of politicians dedicated themselves to creating a modern society, but from different ideological perspectives: a progressive liberal, a confessional or a socialist perspective. In that connection they worked on a unique social structure, namely 'pillarisation': networks of social organisations for education, labour and entrepreneurship on the basis of a single mentality (liberal, confessional, socialist) that were anchored in political culture.

22.5   Around 1900: Poverty Defined Anew

The dynamic societal midfield defined the issue of poverty more broadly than it had been in the past. It was no longer only about the survival of the poor, but also about their quality of life. Cellar-dwellings, sod-huts, slums or – in general – dwellings of one or two rooms without hygienic facilities were no longer acceptable. A diet consisting chiefly of potatoes and grain products was considered inadequate for a healthy body. Having work was no guarantee for a humane existence. Long working hours, low wages, female and child labour were unworthy of a modern society. Labour ought to contribute to personal development. Safe machines ought to prevent dangerous working situations. Public health required public hygienic facilities. Public housing had to start complying with minimal standards of quality like adequate light, air and space. Domestic appliances were to lighten heavy housework. Foods had to be varied and safe.

It was not only about improving the material condition of the people, but also about inculcating bourgeois values. The bourgeois elite insisted that in addition to the construction of better public housing a rigorous inspection of family life was also necessary. Next to the introduction of a bath and shower also exercises in bodily hygiene; next to providing qualified work also training in labour discipline and doing one's duty; next to improvement in foodstuffs also education for a healthy diet.

Around 1900, the broader definition of poverty acquired a footing in social legislation directed at, among other things, labour and housing. This initiated a new phase in the development of well-being, one that would lead to the emergence of the caring state in the Netherlands.

New values around nature and environment also emerged in the periphery of the dynamic midfield. These still had little political influence. It would take more than a half-century before they had social consequences. In this period the exploitation of natural capital, both from domestic sources and abroad, would increase significantly.

22.6 1900–1960: The Historical Challenge Realised

Though it is true that around 1900 a start had been made with social legislation, many questions had to be worked out more precisely or addressed by additional legislation in the subsequent decades. Netherlanders invested a lot of energy in debates about the minimum floor area of workers' housing, the eight-hour working day, a proper diet, a minimum wage, mandatory health insurance, a mandatory collective old-age pension and many other social issues. In the course of this period and these debates they defined the minimal demands on quality of life in the Netherlands. That took place in the framework of new social relations.

The dynamics of the societal midfield had eventuated in a corporatist social structure that would long remain dominant in the twentieth century. Trade unions, employers' associations, professional societies and the 'pillars' had pride of place in this new constellation. Dutch citizens were connected in manifold ways, partly within separate worlds, but always with connections to politics and the government. Within this constellation, social inequality decreased. Income differences declined and societal incomes were redistributed by means of government subsidies and collective measures. Declining inequality together with economic growth were the primary trends responsible for the near-elimination of extreme poverty and of poverty as seen from the modified perspective of 1900.

Solving the question of poverty was no straightforward process. Two world wars and the economic crisis of the 1930s were a considerable setback. It is significant that around the wars, life expectancy decreased by no less than 10 years. That was also the case during the First World War, even though the Dutch themselves were not at war. The social disruption in Europe claimed victims far beyond the battlefield. Popular welfare had a fragile basis in this period. The national government was burdened with the task of managing these vulnerabilities by means of interventions in the economy and social life.

These interventions concerned, among other things, the exploitation of natural capital, seen as the most important prerequisite of economic development and wellbeing. Food security had to be built in to agricultural policy. State mines served to secure the supply of coal. The founding of the Hoogovens iron and steel plant was intended to make the Netherlands independent of steel imports. Governmental management of the mining of sand and gravel was necessary to solve the post-Second World War housing shortage and complete the Delta Works. In short, the Netherlands had to become self-sufficient with respect to its crucial raw materials.

Intervention by the state was not the only change in the exploitation of natural capital and the concomitant supply chains of production, consumption and use. Compared to the nineteenth century the supply chain had become significantly more complex. That was due in part to the rise of new technologies that revolutionised existing supply chains or that facilitated new supply chains altogether, particularly in the domains of electrical engineering and chemistry. More often than in the past, for example, agricultural raw materials were mechanically processed (pressed, sifted, cut and rasped) and chemically treated (with a variety of methods of extraction). The substances thus isolated could be recomposed into new products by means of mixing, kneading, pressing and spraying (as in the fabrication of margarine). There were also all kinds of new techniques by means of which foodstuffs could be roasted (for example in the case of coffee), baked (bread, among others), fermented (as with cheese), pasteurized (for example, milk) etc. More and other kinds of processing became common in the case of mineral sub-soil resources (for example the making of concrete and concrete products) and fossil sub-soil resources (such as the production of plastics from coal and petroleum).

This processing could take place in different plants in sequence (lengthening of the chain) or within a single firm at the same location (intensification of the chain) or it could be split up into different chains (differentiation of the chain). Some companies like Unilever ceased to be simply a link in a chain, but developed into a node in a network of international and domestic flows of matter. These flows came together in the firm and left it as a variety of products destined for domestic and foreign markets. New kinds of organisations like inspectorates and research institutes became part of the supply chains and their networks.

A characteristic of the dynamics in these chains is also the increase in the mass flows through them. This was inherent to an economy of mass production and mass markets, dominated by large firms and multinationals like Philips, Shell and DSM. They had to ensure that the complex flows of raw materials, semi-products, goods and products were skilfully managed. Cooperative organisations did the same in agriculture as did family firms in the medium and small business sector. Cooperation in the form of cartels was part of what was considered the essential regulation of the economy. As might be expected in a corporatist state structure, the government supported this strategy.

Mass production presupposed mass consumption. This link was already in the making prior to the Second World War. The middle class experimented with a wide range of new products that were on offer: the gas stove, the refrigerator, the automobile, the radio, to name just a few. It gained experience with a broader range of food products, new services (like telephony) and white goods (for example the washing machine). Its daily life changed as a result of new ideas about comfort, hygiene, beauty, adventure, personal development and the quality of life. This middle class was the vanguard of the new consumer society.

A fragment of the middle class was also the vanguard for the protest generation of the nineteen-sixties, that would address the consequences of mass production and mass consumption for natural capital across a broad front. Up to then, the increasing exploitation of natural capital had incited incidental resistance, mostly locally oriented and focused on detail problems. The smouldering unease about land reclamations in the 1930s was resolved with the acquisition of parcels of heathland by the Society for the Preservation of Natural Monuments. The delving of marl for cement production incited limited opposition in and around Maastricht. Gravel mining caused opposition in communities in South Limburg. Chemical herbicides and pesticides were criticised by agronomists immediately after the Second World War. Air pollution due to coal combustion was a nuisance that was addressed by the Netherlands Association against Water, Soil and Air Pollution. Serious opposition to polluted air emerged in Rotterdam in 1960 when that led to smog – but only because it was considered a public health hazard.

More or less universal disquiet surfaced around only one issue. Intensive hunting of whales for the production of soap and margarine at the start of the twentieth century resulted in a rapid decline of the whale population and encouraged concerted international political action by the 1930s. But the agreements were hardly enforceable in the international constellation of the time. They were not in the first place inspired by concern about the extinction of species, but above all by concern about economic losses due to a shortage of whales.

A perennial issue in the Netherland was water pollution.[9] This was so in the nineteenth century around the issue of increasing salinity and water polluted by urban wastes. In the twentieth century it acquired a new dimension with discharges of industrial waste, the use of agricultural chemicals and the introduction of household detergents. The Netherlands Association against Water, Soil and Air Pollution asked for legal restrictions as early as 1919. A research institute, the Government Institute for the Purification of Waste-water (RIZA), was founded the very next year. In subsequent decades the Rijkswaterstaat appropriated the issue. The motivation was not preserving the environment but safeguarding agriculture and the food and water supply. But policy to combat water pollution failed to crystallise. Incidents like the massive fish starvation on the Hollandse Ijssel in 1959 due to the dumping of toxic wastes had little effect. Industrialisation triumphed over the struggle against pollution.

The opposition possessed no common denominator like 'environment,' 'energy,' or 'sustainability' that could channel political action or coordinate campaigns. It was never a mass movement, but only mobilised a small group of concerned citizens and committed professionals like chemists and agronomists.

22.7 Around 1960: Well-being, Sustainability and Economic Growth in Balance

After the Second World War, the Netherlands immediately initiated a fast-paced process of reconstruction. The associated normative framework was for the most part in place. The basic principles for quality of life were undisputed. Reconstruction proceeded along the lines of existing networks among government, business and trade unions. Social relations remained pillarised, but the parties were willing to cooperate and agreed on the ideal of a caring and welfare state.

By 1960, that ideal had been all but realised and with that the classic historical challenge of well-being had been substantially met. Extreme poverty had been all but banished and the basic facilities to guarantee quality of life for the poor, workers and the aged were in place. A few issues remained that continued to make the issue

---

[9] A. Bosch and W. van der Ham, *Twee Eeuwen Rijkswaterstaat. 1798–2015* (revised edition, Asten 2015). 182–187.

of well-being problematic. One of these was the housing shortage that had risen to new heights due to war-damage and under the onslaught of a burgeoning post-war population. The government, public housing associations and building companies worked hard on this problem.

Another aspect of well-being that made great strides was the securing of the vulnerable Delta. This specifically Dutch issue had culminated in a worthy response to the threat of the sea in the form of the Zuiderzee Closure Dike and the Delta Works. That did not mean that the struggle against water was over. As early as 1956, Rijkswaterstaat concluded that the river dikes were too low and began to formulate new safety norms for the rivers. But for the moment the Delta Works had priority.

How was natural capital faring? The creation of the caring state had only been possible thanks to an intensified exploitation of natural capital. The increase in exploitation of natural capital in the first half of the twentieth century was roughly the same as in the second half of the nineteenth century. For example, the use of fossil raw materials between 1850 and 1910 increased by approximately 3.2% per annum and between 1910 and 1960 by 2.7%. Around 1960, however, there was a trend break: Between 1960 and 1975, the annual growth rate suddenly rose to 7.0% (Table 22.1), after which the growth started to fall again. A trend break around 1960 can also be observed for the other two mass flows (Graph 22.2).[10] The sixties and early seventies show a strong growth of the bio-raw per annum materials and mineral raw materials with a strong decline after 1975. It is remarkable that the growth figures of the Dutch population lag behind those of the raw materials. The pressure on natural capital is not only related to population growth (Table 22.1).

Similar shifts can be seen in emissions and the use of substances like artificial fertiliser. CO_2 and SO_2 emissions increase gradually up through the 1950s, after which they exhibit a steep increase (Graphs 22.3 and 22.4). The same is the case for the use of artificial fertiliser (Graph 22.5). Remarkably, present-day norms for CO_2

---

[10] Graph 22.2 consists of various long-term datasets in the field of fuels, agricultural and mineral products. The trend lines are calibrated with the detailed datasets from the study of the reference years 1850, 1913, 1970 and 2010 (see: F. Lambert, *Massastromen in Nederland. In de jaren 1850, 1913, 1970, 2010* (researchrapport Technische Universiteit Eindhoven, oktober 2016). The trend for fossil fuels is made up of data from the final use of coal, lignite, peat, petroleum and gas (taken from Rick Hölsgens, *Energy Transitions in the Netherlands*, Dissertation University of Groningen, 2016). For agricultural products an average was taken for a number of long-term production statistics (derived from Statistics Netherlands Historical series: Agriculture, from 1851). These are the production statistics of starch (potatoes), milk and sugar beet (from 1895). The trend of these agricultural products was indicative of the normalized trends including cereal and meat production. The trend line mineral products is formed by the consumption of cement. This trend was indicative of the normalized consumption of fill sand, industrial sand, masonry bricks, sand-lime bricks and gravel. The datasets of mineral materials derived from CBS: Minerals extraction, CBS: Production statistics for Sand-lime-brick industry 1961; CBS History minerals and industry; Cement production: A. Heerding, *Cement in the Netherlands* (IJmuiden: Cement factory IJmuiden, 1971); N. Smit Kroes, *Basis Ontgronden*, policy memorandum on surface mineral resources for the long term, House of Representatives of the States General (Parliamentary document number 21.100, under numbers 1 and 2), session year 1988–1989 and Cement & Betoncentrum (http://www.cementenbeton.nl/).

Note: Graph 2.1 and Graph 19.5 are based on the same datasets of Graph 22.2.

Table 22.1 Growth in the use of bio-raw materials, mineral subsoil resources and fossil subsoil resources in the Netherlands, 1850–1910, 1910–1960, 1960–1975 and 1975–2010 (percentages)

	1850–1910	1910–1960	1960–1975	1975–2010
Bio-raw materials	1.6	1.8	3.3	1.0
Mineral subsoil resources	2.9	3.5	4.1	0.5
Fossil subsoil resources	3.2	2.7	7.0	1.3
All raw materials	**2.3**	**2.7**	**5.1**	**1.0**
Dutch population	**1.1**	**1.3**	**1.2**	**0.6**

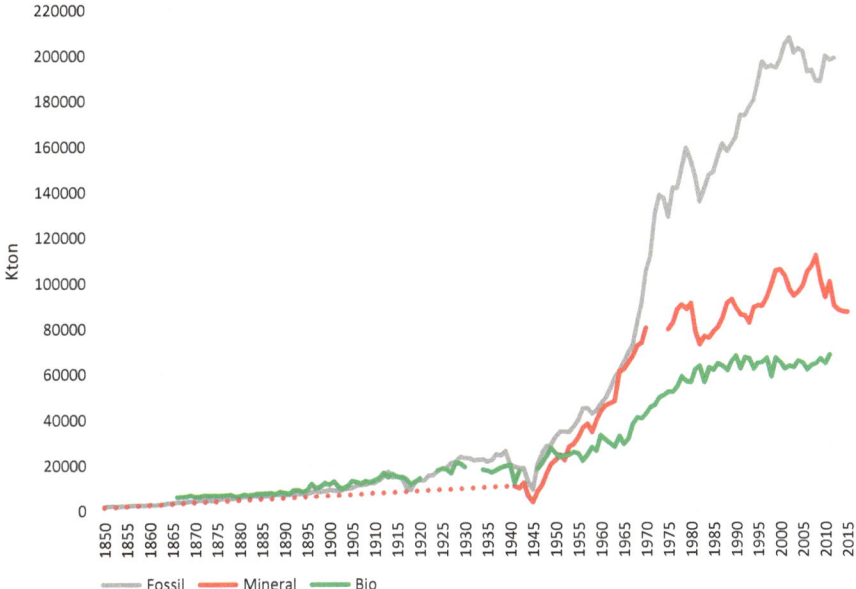

Graph 22.2 Estimation of the trend in bio-raw materials, mineral sub-soil assets and fossil sub-soil assets, 1850–2010 (kton)
Source: See note 10 of this chapter

emissions and the use of artificial fertilisers are not, or if so only marginally, exceeded up to the time of the trend shift. By contrast, SO_2 emissions had already exceeded present-day norms as early as the 1950s.

On the basis of the historical account we can advance the hypothesis that by comparison with the rest of Dutch history, well-being and sustainability were best in balance around 1960, at least from the point of view of the issue of poverty. The historical challenge of poverty had been resolved with the realisation of the caring state, while demands on natural capital were still relatively modest by present-day standards. Not only had material wealth increased substantially, but numerous aspects of individual quality of life (health and level of education) were in excellent shape, while social stability was high. Reasonable economic growth (on the basis of mobilisation of natural capital) and increasing social equality laid the groundwork.

498 H. Lintsen and J.-P. Smits

Graph 22.3 CO$_2$ emissions in kiloton, 1850–2010, set out against the EU sustainability norm for 2030

Source: H. Hölsgens, *Energy transitions in the Netherlands: Sustainability challenges in a historical and comparative perspective* (Groningen 2016), figure A.V.1, 231

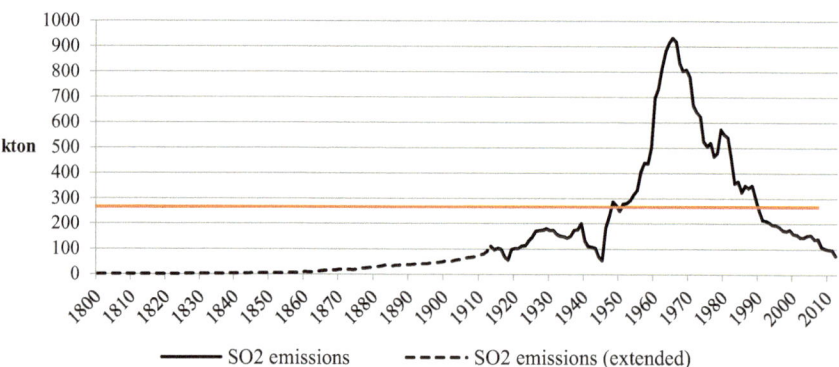

Graph 22.4 SO$_2$ emissions in kiloton, 1800–2010, set out against the EU sustainability norm

Source: H. Hölsgens, *Energy transitions in the Netherlands: Sustainability challenges in a historical and comparative perspective* (Groningen 2016), figure 4.1, 90

Up to around 1960, well-being increased at the same rate as economic growth (or even forged ahead of it, see Graph 22.1a).

This hypothesis demands additional research. Were environmental problems up to then – as this study postulates – above all local and limited in scope? Did the big changes in the landscape and the associated decline in biodiversity take place primarily after the Second World War with land consolidation, urbanisation and the coming of the automobile? Et cetera.

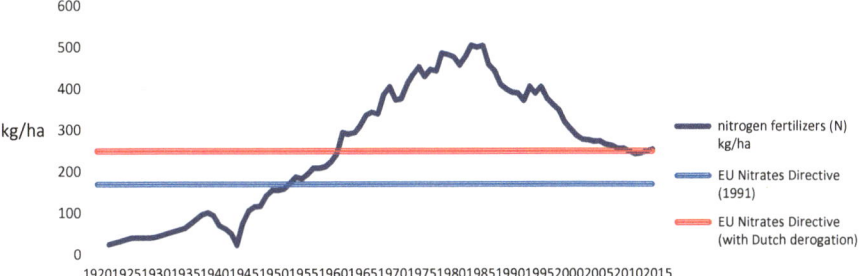

Graph 22.5 Use of fertilizer, 1850–2010, compared with sustainability standards of the European Union

Remark: There are norms for the agricultural use of nitrogen. Since 1990 the general norm has been the EU nitrate guideline that stipulates a maximum of 170 kilograms of nitrogen per hectare. Exceptions to this can be granted. The Netherlands has such an exception (a derogation) that includes the provision that a precise accounting is to be kept for the various types of soil. The higher norm (with the derogation) is 250 kg/ha. The Netherlands itself has worked out norms for different crops and for five different soil types (clay, sand – northern, western and central, sand southern, loess and peat).

Source: CBS (*Agriculture and Fisheries 1899–1999*) and reworked CBS (*Nitrogen in animal manure, artificial fertiliser and gaseous losses, 1990–2015*)

The image of a balance among economic growth, well-being and sustainability needs some correction. 'Balance' suggests an equilibrium for an extended period. But a number of factors in the 1950s were responsible for a rapid disruption of the equilibrium. The Netherlands were confronted with a strong population growth that had to be accommodated. The country placed its bets on industrialisation as a strategy for employment and economic development. Moreover, during the interbellum a middle class had explored a modern lifestyle that also became an attractive perspective for the working class. After the Second World War the Netherlands strove to realize a welfare society. This new dynamic demanded strong economic growth. From the 1960s on, well-being lagged far behind the significantly higher economic growth that set in and that required a considerably larger mobilisation of natural capital (Graph 22.1a). The trend reversal around 1960 thus had its roots in the preceding period.

22.8 1960–2015: Well-being, Sustainability and Economic Growth Out of Balance

In this period, growth in well-being was out of kilter with that of the economy and the rapidly increasing GDP. That had various causes. An important cause was the rapidly changing relationship with the natural environment. On the one hand the Dutch were confronted with phenomena like strong algae growth due to the leaching of fertilisers (eutrophication), the impacts of pesticides including massive bird starvation and incidents of smog-formation due to the emission of SO_2 and smoke. On the other hand, norms regarding nature and the environment changed and these phenomena began to be defined as problems. The idea took root that the natural environment was suffering inordinately under the onslaught of economic growth and that many important subsoil resources were characterised by finite supplies. The alarm set off by the Club of Rome in the 1970s was loud and widely heard. A new normative framework for well-being began to take shape, one which revealed that the norms for the exploitation of the environment and nature were being seriously violated. If humanity did not succeed in rigorously changing its production and consumption patterns, then vital ecological limits would soon be transgressed.

An entirely different cause for laggard well-being was the massive unemployment and wage moderation in the 1980s and during the crisis of 2008. These depressed welfare levels and household consumption. Social inequality in terms of income and wealth increased.

But while unemployment, wage moderation and a deferred increase in consumption had an adverse effect on well-being, the health and educational levels of the Dutch population continued to increase. These aspects of quality of life appeared to be relatively unaffected by the vagaries of economic cycles.

The dramatic decline of well-being in the domains of safety and democracy is remarkable. Having reached a peak in the mid-1950s, indicators for this category subsequently exhibited a steady decline. To an important degree, this development can be attributed to the process of de-pillarisation that set in during the 1960s. Within the pillarised structure the elites of the different religious and ideological pillars were in close touch with their constituencies though numerous societal organisations. This made the societal agendas propagated by the elites sufficiently recognisable by the grassroots constituencies of the pillars. At the same time the system of pillarisation was characterised by a large measure of social control. The literature speaks of a 'decently pillarised society.' From the 1960s on, religion no longer provided a fixed normative framework for Netherlanders.

This process intensified toward the end of the twentieth century, when political party-formation on the basis of traditional ideological orientations began to become less important. This did not mean that citizens were bereft of social visions, only that these were experienced less and less in the context of a group and on the basis of established religious or political viewpoints.

De-pillarisation also contributed to a process of individualisation. From the viewpoint of well-being, this had many positive effects. More than ever, citizens could make choices in complete freedom. But the process also had its dark sides. The self-evident connection of political elites with their constituencies that had previously been the rule became weaker. In the last quarter of the twentieth century the contours of a new societal midfield had become visible, but the connection with political parties remained weak. This was one of the reasons that citizens could no longer see their reflections in 'Hague politics.' Trust in political institutions declined visibly, especially after the turn of the century (and particularly after the assassination of the populist politician, Pim Fortuyn, in 2002). The erosion of traditional societal bonds characteristic of the 'decent pillarised society' and the concomitant

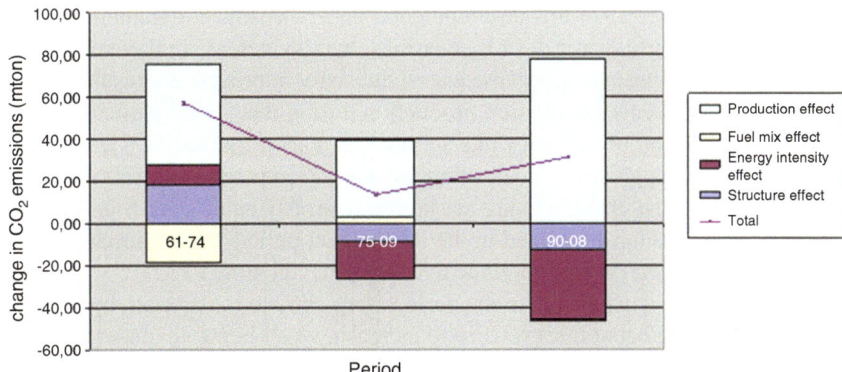

Fig. 22.1 Results of an Index Decomposition. Analysis of the growth of CO_2 emissions, 1960–1974, 1975–1989 and 1989–2008. (change in CO_2 emissions in megatons)
Source: R.P. van der Helm, R. Hoekstra en J.P. Smits, *Economic growth, structural change and carbon dioxide emission: The case of The Netherlands 1960–2008*. Article presented at the 18th International Input-Output Conference (2010)

decline in social cohesion and control, also brought in its wake increasing crime rates, more suicides and a strong increase in alcohol consumption (that by international standards had always been relatively low in the Netherlands).

Looking back, another important issue manifested itself in this period, namely the issue of climate change. Initially the issue led a covert existence among contemporaries. Nowadays, however, climate change due to greenhouse gas emissions is seen as one of the most pernicious problems in the domain of nature and environment. The figure below shows how the emission of one of the most important greenhouse gases, CO_2 can be explained (Fig. 22.1).

As one would expect, the increase in CO_2 emissions can be explained to a large extent by the growth of production (the production effect). Production is of course to a significant degree associated with the burning of fossil fuels and hence with CO_2 emissions. In the period 1961–1974 – when CO_2 emissions increased significantly – this growth was also caused by two other factors. First, by energy-intensity, that is the amount of energy necessary to produce a unit of GDP. Up to 1974, a higher energy-intensity appears to have been a driver of the increase in CO_2 emissions. In addition, the structure effect also played a big role. This should be taken to mean that the growth in emissions cannot be explained only with reference to economic growth in the various sectors of the economy. Emissions are exacerbated by a shift of the economic centre of gravity in the direction of more energy-intensive activities (industries using relatively large amounts of energy in the production process). Due to the active industrialisation policy that was pursued in the first post-war decades, the share of energy-intensive branches like petrochemicals and iron- and steelmaking in the GDP increased strongly up to the early 1970s. The discovery of the Groningen natural gas field amplified the production and structure effects f on CO_2 emissions.

In the period after 1974 this developmental pattern changed. Economic growth remained the strongest driver of CO_2 emissions, but the increase in these emissions was mitigated to some extent because energy intensity improved as a result of technological developments. In addition production during this period shifted increasingly in the direction of activities like services, in which traditionally less energy was consumed. Though these two developments have tempered a further increase in CO_2 emissions, the latter still remains at a high and unsustainable level. Significantly, industrial emissions have declined in the most recent period while households and especially the transport sector continue to emit more and more CO_2.

In retrospect we can say that in the period from 1960 to the present, by contrast with earlier periods, economic growth and increasing well-being no longer marched to the same drummer. Though certain specific forms of welfare suffered, well-being in general continued, grosso modo, to increase. But only thanks to the depletion of natural capital and at a cost in terms of pollution that has become increasingly evident. If we look at the quantity of CO_2 that has been emitted since 1960 and how much it will cost to reduce these emissions to levels that have been agreed upon in international climate accords, then we can only conclude that we will be facing costs estimated at about 12% to 25% of the GDP in 2010. To a significant degree, current and coming generations will be paying the price for the energy-intensive industrialisation and the concomitant growth in prosperity since 1960.

22.9 Around 2015: Sustainability as New Historical Challenge

It was only quite recently –some 50 or 60 years ago – that well-being and sustainability were still reasonably in balance: the classical issue of poverty had been solved while claims on natural capital remained reasonably modest. Since then, the monitor for well-being manifests a deep duality: a considerable increase in material welfare, while well-being lags behind and sustainability is seriously compromised. The perennial question since then has been: to what extent does economic growth contribute to quality of life in the 'here and now' and is the claim on natural resources not so exorbitant that the costs will have to be paid for by lower quality of life 'later' and 'elsewhere'? What can we learn from history in this connection?

As early as the 1960s and 1970s there was already broad public concern about environmental problems and the finiteness of raw materials. Groups of citizens mounted forceful protests and environmental groups flexed their muscles. In some areas progress was rapid, for example in the area of water and air pollution. In other areas it took decades before results could be seen, for example, biodiversity, the decline of which has been halted (although many still consider it to remain at a pitifully low level). And in yet other fields like energy, the yield is even more meagre – even after all this time. Sustainability is a stubborn problem.

Early environmental successes were related to the corporatist state order with its 'pillars,' cartels, multi-national corporations and professional organisations that formed the institutional framework of the Dutch striving for well-being until well into the twentieth century. Environmental policy could be rapidly implemented in those cases where the established political parties embraced the environmental issues, the professional organisations possessed the requisite expertise and the proposed measures did not overly disturb existing relations of production and consumption. Legislation and subsidies encouraged the large-scale construction of sewage treatment plants, the installation of smokestack gas scrubbers and the cleaning up of polluted industrial sites. These 'end of pipe' solutions had a very beneficial effect on the environment. In the 1980s the Netherlands belonged to the European vanguard in addressing environmental issues and was a creative force in the conceptualisation of EU environmental guidelines and measures.

But the corporatist structure also impeded progress on the road to sustainability, in particular when it appeared necessary to intervene in existing institutions. The problem of excess manure demanded breaking through the 'green front,' a coalition of parties in the agricultural sector that was united as one man against fundamental changes in agricultural policy. The Cooperating Electricity Producing Companies (the SEP, disbanded in 2000) long blocked experiments with alternative means of generating electricity (using windmills and solar panels). Until the 1990s, the Rijkswaterstaat found it very difficult to integrate ecological values into their water management policies. To a certain extent old institutions had to be demolished in order to give solutions to sustainability issues a chance.

That said, the erosion of the corporatist state structure would not in the end lead to a breakthrough in the problem of sustainability. The national state was weakened by the emergence of the EU. The liberalisation of the economy, the retreat of the state and the abolition of corporatist organisations created great rents in the social fabric. Environmental improvement did proceed, partly thanks to EU guidelines, but in three key areas little or no progress was made.[11]

---

[11]The *Compendium voor de Leefomgeving* (Compendium for the Environment) provides the following time series data along with norms and conclusions:

Air quality
Fine dust in NL decline since 1990. Fine dust in Rotterdam decline since 1970. Currently under the EU norm.
See: http://www.clo.nl/indicatoren/nl0243-fijn-stof-pm10-in-lucht
Nitrous oxides NOx in NL and Rotterdam, decline since 1973. Currently under the EU norm
http://www.clo.nl/indicatoren/nl0231-stikstofdioxide
Acidification and large-scale air pollution emissions. Decline since 1980. Currently under EU norms, except for concentration NH_3 that is somewhat above the norm.
http://www.clo.nl/indicatoren/nl0183-verzuring-en-grootschalige-luchtverontreiniging-emissies
Water / Fertiliser load
Fertiliser load large rivers decline since 1970. Currently under EU norms, except for Meuse.
http://www.clo.nl/indicatoren/nl0249-vermesting-in-grote-rivieren
Lakes and ponds 1980-2014. Concentration nitrogen, phosphor, algae and transparency above the targets.
http://www.clo.nl/indicatoren/nl0503-vermesting-van-meren-en-plassen-

The CBS monitor for 2014 is clear about the present situation: 'Of the four forms of capital, the preservation of natural capital in the Netherlands has the bleakest outlook.'[12] Four capitals (the economic, social, human and natural capital) must safeguard well-being of future generations. The way in which we leave behind the natural capital at this moment, is worrying in that respect. CBS defines three main problems in particular[13]:

- Energy and climate: The Netherlands consumes much fossil fuel (compared to other EU countries) has high per capita emissions of greenhouse gases and has an extremely low share of renewable energy sources.
- Exhaustion of raw materials: Over the past decades the Netherlands has largely exhausted its own supply of natural gas. It belongs to the group of European nations making the biggest claims on raw materials from the least developed countries in the world. Finally, the country contributes to the exhaustion of those raw materials whose total depletion may be expected in the present century, like petroleum, antimony and zinc.[14]
- Nature and environment: In a number of ways the Netherlands is still not compliant with EU environmental norms. Natural habitats and biodiversity are (compared to other EU nations) still under considerable pressure. The country also places a great environmental burden on other countries, in particular developing countries.

Environmental pressure on Nature (acidification / fertiliser load / dessication)
'The environmental conditions in water and nature preserves have improved, but are often still insufficient for sustainable conservation of biodiversity. Policy is aimed at achieving conditions that make the sustainable conservation of plants and animal species possible.'http://www.clo.nl/indicatoren/nl1522-milieudruk-op-natuur
Green growth
From: CBS, *Green growth in the Netherlands* (The Hague, 2015)
Direct environmental pressure by the Dutch economy has declined. All the environmental efficiency indicators pertaining to wastes and emissions get a green score, while the economy is growing. One can therefore speak of an uncoupling of economic growth and environmental effects. The Netherlands is an average (or poor) performer compared to other countries. http://www.download.cbs.nl/pdf/green-growth-in-the-netherlands-2015.pdf

[12] *Monitor duurzaam Nederland 2014. Indicatorenrapport* (Den Haag 2014), 32–33.

[13] See also: CBS, *Meten van SDGs: een eerste beeld voor Nederland* (Den Haag 2016), 27 and CBS, *Internationaliserings-monitor 2015, vierde kwartaal* (Den Haag 201%), 12–17. SDG refers to Sustainable Development Goals.

[14] For zinc and antimony see: Th. Henckens, *Managing raw materials scarcity* (Dissertation Utrecht University, 2016), 27–33.

22.10   Around 2015: The Welfare Paradox

Well-being is a vulnerable prosperity and not just for future generations.When we compare the achievements of the Netherlands in the domain of well-being to those of other countries, it is obvious that the Netherlands scores well on a large number of aspects. As far as the sense of happiness is concerned, the 2016 *World Happiness Report* pegs the Netherlands at no less than seventh place in a field of 156 nations. At the same time, parts of Dutch society are oppressed by great worries and feel sombre about the future. Is this a question of 'emotion,' and can we dismiss this unease as simple 'cantankerousness,' or is there perhaps more at issue?

This paradox is also visible in the CBS monitor for 2014.[15] The general level of well-being has been high and stable over a long span of time and the Netherlander has on average a bit more to spend than 10 years ago; still, satisfaction with life has declined (though still remaining high by international standards). How can we explain this welfare paradox? The historical analysis presented in this book can give us more insight into this issue. Four factors that have played a role in this study immediately spring to mind:

22.10.1   Social Inequality

A society can experience a welfare-increase, but that says nothing about how it is distributed. Halfway through the nineteenth century the Netherlands was the richest-but-one nation in the world, while a large part of its population balanced around the bare minimum of existence. This kind of social inequality can have a great influence on the way citizens experience welfare, as shown by the literature on the economics of happiness. It seems that people's satisfaction with life not only depends on their objective welfare, but above all on whether there are people in their immediate sur-roundings that are much more prosperous. People with a low level of welfare can be more satisfied if people in their environment enjoy a similar quality of life, than they would be if they were more prosperous but those around them even more so.

Even though at the moment in the Netherlands extreme poverty such as prevailed in the nineteenth century has ceased to exist, nonetheless the contours of a new societal duality are becoming visible. A recent report by Statistics Netherlands (CBS) demonstrates that the schism runs along the distinction between the less and the better educated. The less-educated have a lower chance of employment, are

---

[15] *Monitor duurzaam Nederland 2014. Indicatorenrapport* (Den Haag 2014), 29 and figure 2.2.2, p. 31.

often less healthy and report a lower degree of satisfaction with their existence. These satisfaction scores seem to be lower than might be expected on the basis of the actual material welfare. The level of schooling appears to have deep consequences for very many aspects of quality of life.

22.10.2   Insecurity and Vulnerability

Satisfaction with life often depends not so much on objective welfare factors, but rather on the fear of losing a given level of welfare. In the past there have been periods of political uncertainty and disruptions of the global economy that resulted in extensive societal unrest (think for example of the Great Depression of the 1930s). At the present time the Netherlands is suffering greater uncertainty than it has been for a long time. Processes of globalisation and flexibilisation, even though generating an enormous economic momentum, also increase feelings of insecurity among large parts of the population. The Monitor Sustainable Netherlands reports that people are above all concerned about keeping their jobs and incomes. The young and less-schooled are especially vulnerable. Half of the youthful population up to the age of 25 and more than 35% of the less-educated employees had a flexible (i.e. temporary) labour contract in 2014. This is 50% more than in 2003. These groups are facing big obstacles in ensuring pensions and taking out mortgages. The welfare paradox can thus be partly explained by increasing uncertainty and the fear of losing ground in the future.

22.10.3   Changing Preferences

A certain measure of critical awareness in society is quite healthy. This enables problems and blockages to be identified, making it possible to work on solutions. In the period 1850 to the present this was a perennial phenomenon. In 1910, for example, when the grinding poverty of the mid-nineteenth century had been solved, much higher demands began to made on quality of life and social criticism consequently began to address new issues. Every generation determines what *it* sees as quality of life. As soon as old welfare problems have been solved, a new standard for quality of life begins to take shape. Societal dissatisfaction is therefore not as such a negative thing, as long as problems are tackled in an active way and citizens do not turn their backs on society.

22.10.4   *Channelling the Societal Debate*

One of the essential aspects of the institutional quadrants that have been used as an analytical tool in this book concerns the way in which 'civil society' (the societal midfield) and the government are connected. Halfway through the nineteenth century the Netherlands was confronted with big social problems that did not admit of an easy solution. This situation improved after the 1870s as a result of the rise of a broad societal midfield and pillarisation in particular. The political elites of the various societal groups were closely tied to their constituencies via the diverse organisations in civil society. In this way concerns and wishes that circulated in society could be quickly and clearly communicated to the political level. At the same time, social acceptance of the policies was high.

The process of de-pillarisation was the' main factor in increasing the distance between government and citizens. The result is that at present a substantial segment of the electorate feels that it is not being heard. The fact that a significant number of political decisions have been delegated to Brussels has not increased the involvement of citizens in political processes. As in earlier historical junctures, it is now essential to find new forms in which the state and society can reciprocally fortify one another. This is absolutely essential given the fundamental issues, the Grand Challenges, now facing the world in the domain of well-being and sustainability.

Literature

Bosch, A. and van der Ham, W. (2015). *Twee Eeuwen Rijkswaterstaat. 1798–2015* (revised edition). 182–187. Asten.

CBS (2016). *Meten van SDGs: een eerste beeld voor Nederland.* 27. Den Haag: CBS

CBS (2016). *Internationaliserings-monitor 2015, vierde kwartaal.* 12–17. Den Haag: CBS.

Dijkgraaf, E. et al. (2009). *Effectiviteit convenanten energiebeleid.* Rotterdam: Planbureau voor de Leefomgeving (PBL)

Grin, J., van de Graaf, H., and Vergragt, P. (2002). 'Naar een derde generatie milieubeleid: een mogelijk sturingsconcept' in *Beleid en Maatschappij.*

Helm, R.P. van der, R. Hoekstra en J.P. Smits. (2010). *Economic growth, structural change and carbon dioxide emission: The case of The Netherlands 1960–2008.* Article presented at the 18th International Input-Output Conference, Sydney.

Henckens, Th. (2016). *Managing raw materials scarcity.* 27–33. Utrecht: Utrecht University.

Hoff, S., Wildeboer Schut, J.M., Goderis, B. and Vrooman, C. (2016), *Armoede in kaart.* Den Haag: Sociaal en cultureel planbureau.

Klumpelaars, S. and Somers, M. 'Extreme armoede wereldwijd voor het eerst onder 10 procent', *NRC* 5 oktober 2015, retrieved from https://www.nrc.nl/nieuws/2015/10/05/wereldbank-extreme-armoede-wereldwijd-voor-het-eerst-onder-10-procent-a1412636.

Lambert, F. (2016). *Massastromen in Nederland. In de jaren 1850, 1913, 1970, 2010.* Eindhoven: Technische Universiteit Eindhoven, research report, open access.

Monitor duurzaam Nederland (2014). *Indicatorenrapport.* 29 and figure 2.2.2, p. 31, Den Haag: Monitor duurzaam Nederland.

United Nations, *Sustainable Development Goals. 17 goals to transform our world* (1. No Poverty). http://www.un.org/sustainabledevelopment/sustainable-development-goals/

van Zanden, J.L., Baten, J., Földvari, P., and van Leeuwen, B. (2011). *The changing Shape of Global Inequality 1820–2000. Exploring a new dataset.* Table 14, 42. Utrecht:

World Bank Group. *World Bank Forecasts Global Poverty to Fall Below 10% for First Time; Major Hurdles Remain in Goal to End Poverty by 2030*, retrieved from http://www.worldbank.org/en/news/press-release/2015/10/04/world-bank-forecasts-global-poverty-to-fall-below-10-for-first-time-major-hurdles-remain-in-goal-to-end-poverty-by-2030. Consulted 20 January 2017

Chapter 23
Contemporary Problems of Well-being and How We Got Here

John Grin

Contents

Abstract In this chapter we analyse the historical genesis of the contemporary crisis of natural capital as groundwork for exploring possible routes for future development up to 2050. The central question is: How did currently problematic couplings between GDP growth, well-being, resource use, energy use and pollution emerge?

© The Author(s) 2018   509
H. Lintsen et al., *Well-being, Sustainability and Social Development*,
https://doi.org/10.1007/978-3-319-76696-6_23

In this connection, we analyse three value chains (grain-for-meat; gravel-for-construction; plastics-for-domestic-quality of life) asking the following sub-questions: How did practices of production, consumption and distribution shape (1) specific value chains into non-circularity, and (2) yield a vast expansion of material flows, especially after 1960? What problem definitions and value orientations have guided those developments and how have they been institutionally and spatially embedded?

A fourth case is devoted to energy, in particular to electricity-for-households: How did practices of production, consumption and distribution shape the evolution of energy intensity in households? What problem definitions and value orientations guided those developments and how were they institutionally and spatially embedded?

The results of the analyses are used for reflections on future strategies to promote sustainability (see final Chap. 24).

Keywords Crisis · Natural Capital · Value Chains · Grain · Meat · Gravel · Plastics

23.1 Exploring Contemporary Issues

By 1960 the process of state formation and societal modernization that took off around 1850 had attained its goals: more well-being and greater democratic participation. Moreover, the underlying practices respected the available stock of natural capital. That is, until about 1960. After that, the continuous growth in GDP meant a steep increase in resource and energy consumption (Graph 22.2). This did not go unnoticed. A better-educated populace demanded more political and social say, and began to hold government and business responsible for the massive ecological problems that surfaced in the course of the 1960s.[1] By 2010 social unrest posed a serious threat to Dutch democracy while in the background the country's natural capital was being consumed at alarming rates.

In this chapter we analyse the historical genesis of the contemporary crisis of natural capital as groundwork for exploring possible routes for future development up to 2050. We focus on dominant problem definitions and value orientations and the associated institutional embedding – as argued in Chap. 1. These are also salient because they co-shape future developments, the substance of Chap. 24.

For the same two reasons, we also zero in on the long-term "landscape" trends that have shaped practices (Chap. 1) especially their spatiality. Preceding chapters showed that the spatial dimension of practices deeply informs present-day

---

[1] U. Beck, *The re-invention of politics. Rethinking Modernity in the Global Social Order* (Cambridge: Polity Press, 1997).

dilemmas – as emphasised in recent transition studies.[2] In sum, the central question of this chapter will be:

Q: How did currently problematic couplings between GDP growth, well-being, resource / energy use and pollution emerge? What problem definitions and value orientations have guided that development and how have these been institutionally embedded?

23.1.1 Material Flows

While a few key resources like oil and phosphates are dwindling rapidly, increasing demand on resources in general is savaging landscapes and biodiversity at extraction locales. And though achievements like the tenfold reduction of per capita SO_2 emissions, approaching 1850 levels, are hopeful, newer problems like soil acidification due to surplus manure and the pollution of the oceans by micro-plastics are still far from being solved. Such problems are rooted in two developments. The first is the way economic supply chains evolved since 1850 – especially their degrees of differentiation and specialisation. This resulted in an economy of non-circular chains of production and consumption. When, around 1960, this non-circularity intersected with a second phenomenon, a vast expansion of production and consumption, massive problems ensued (cf. Graph 22.2).

As a mercantile country traditionally poor in resources, the Netherlands has long imported crucial resources, while finished products are domestically consumed or exported. Depending on the spatial and institutional specifics of the economic supply chains involved, repercussions therefore quickly extend beyond the 'here and now' and into the 'elsewhere' and/or 'later.'

---

[2] M. Hodson, S. Marvin, Can cities shape socio-technical transitions and how would we know if they were? (*Research Policy*, 39: 2010), 477–485.

P. Späth, H. Rohracher, 'Energy regions: The transformative power of regional discourses on socio-technical futures' (*Research Policy*, 39,4: 2010), 449–458.

P. Späth, Philipp, H. Rohracher, *Beyond Localism: The Spatial Scale and Scaling in Energy Transitions* (Padt 2014), R. Raven, J. Schot, F. Berkhout, 2012: Space and scale in socio-technical transitions (2012), *Environmental Innovation and Societal Transitions*, 4, 63–78.

B. Truffer, L. Coenen, Environmental Innovation and Sustainability Transitions in Regional Studies (2012), *Regional Studies*, 46,1: 1–21.

A. Switzer, L. Bertolini & J. Grin, Understanding transitions in the regional transport and land-use system: Munich 1945–2013 (2015), *Town Planning Review*, vol. 86 no. 6. p. 699–723., A. Switzer, L. Bertolini, J. Grin, Transitions of Mobility Systems in Urban Regions: A Heuristic Framework (2013), *Environmental Policy and Planning*, 15,2: 141–160.

J. Grin, N. Frantzeskaki, V. Castan Broto, L. Coenen, 'Sustainability Transitions and the City' (2015), Linking to transition studies and looking forward chapter 22 (conclusions) in: N. Frantzeskaki, V. Castan-Broto, L. Coenen (ed.s), *Urban Sustainability Transitions*. J. Grin, N. Frantzeskaki, V. Castan Broto, L. Coenen, *Sustainability Transitions and the City* (2015), Linking to *transition studies and looking forward* chapter 22 (conclusions) in: N. Frantzeskaki, V. Castan-Broto, L. Coenen (ed.s), *Urban Sustainability Transitions*.

Hence the following sub-questions will inform our detailed examinations of several supply chains (grain-for-meat; gravel-for-construction; and plastics-for-domestic-quality of life):

> *Q1: How did practices of production, consumption and distribution shape (i) specific value chains into non-circularity, and (ii) yield a vast expansion of material flows, especially since 1960? What problem definitions and value orientations have guided those developments and how have they been institutionally and spatially embedded?*

23.1.2   Fossil Energy Use

As Graph 22.4 indicates, air pollution due to the rapid increase in fossil energy use since the 1950s was largely solved by first generation "end of pipe" environmental policies[3] not to mention the early 1970s transition to natural gas. But whatever the size of remaining fossil stocks, current rates of (increase of) fossil energy consumption cannot be environmentally sustained throughout the present century – certainly not in view of climate change and certainly not if developing countries seek to emulate western industrialisation since 1850.

Depletion of fossil resources and climate change are serious problems for both 'later' (inasmuch as the consequences of depletion and the worst consequences of climate change are yet to come) and 'elsewhere' (inasmuch as the consequences of climate change will hit the global South hardest).[4] That said, climate change and fossil resource depletion are also serious problems in the 'here and now.' Fossil fuel extraction has had undesirable local effects both in the Netherlands and elsewhere: its spatiality has had and will continue to have, important implications for energy security and financial flows.

Increased energy consumption is directly to blame. Graph 22.2 shows how energy consumption skyrocketed in the 1960s. This follows a similar pattern of growth in overall production and consumption and thus speaks to our first question. The *coupling* between material production and energy consumption obviously reflects the very nature of industrial production: using mechanical energy to augment labour productivity. And inasmuch as prior to 2010 that energy was largely fossil energy, the correlation between production and CO_2 emissions also appears evident. Between 1961 and 1974 the rapid, large-scale introduction of natural gas

---

[3] J. Grin, H. van de Graaf. P. Vergragt, *Een derde generatie milieubeleid: Een sociologisch perspectief en een beleidswetenschappelijk programma*, (2003), Beleidswetenschap, jrg. 17, nr. 1, p. 51–71.

[4] Brauch, H. Günter, Ú. Oswald Spring, J. Grin, C. Mesjasz, P. Kameri-Mbote, N. Chadha Behera, B. Chourou, H. Krummenacher, *Facing Global Environmental Change: Environmental, Human, Energy, Food, Health and Water Security Concepts.* (2009), Hexagon Series on Human and Environmental Security and Peace, vol. 4 (Berlin – Heidelberg – New York: Springer-Verlag).

Brauch, H. Günter, Ú. Oswald Spring, J. Grin, C. Mesjasz, P. Kameri-Mbote, N. Chadha Behera, B. Chourou, H. Krummenacher, *Coping with Global Environmental Change, Disasters and Security – Threats, Challenges,*

mitigated CO_2 emissions. At the same time, energy-intensive industries, most notably the utilities and the chemical sector, became more important in the Dutch economy. After 1974, levels of CO_2 emissions were the combined effect of more efficient production (in response to the oil crises) and greater volumes of production. For most sectors in this period, increases in export were the main drivers of increased CO_2 emissions. The big exception was the utilities where household consumption was responsible.

Below, therefore, we will seek answers to the following sub-questions as we analyse cases of energy use in specific chains:

> Q2: How did practices of production, consumption and distribution shape the evolution of energy intensity in particular value chains? What problem definitions and value orientations guided those developments and how were they institutionally and spatially embedded?

23.2 Analytical Approach

We will analyse four value chains: grain-for-meat; gravel-for-construction; plastics-for-domestic-quality of life and electricity-for-households. These cases cover much of the increase in well-being since 1850: first the satisfaction of basic needs and, especially since the 1960s, increasing convenience, comfort and fun. They also cover the spectrum of contemporary problems, partly because they exhibit different patterns of spatiality in material flows. The first three chains exemplify organic, mineral and fossil resources, respectively, selected on the basis of volume and on above-average import shares for their respective categories. These three cases are used to answer Q1 above, explaining how production (and with it, emissions) increased in the three sectors. The last two cases – exemplifying the consumption and production of electricity – show how CO_2 emissions evolved and will be used to answer Q2.

23.3 Grain-for-Meat

23.3.1 Breaking Circularity

Recall the two dynamics underlying contemporary sustainability problems: first, the non-circularity of economic chains and, second, their sheer volume, especially since 1960. Non-circularity in the grain-meat chain[5] may be summarised by noting that by 2010 the Netherlands

---

[5] Denkgroep Duurzame Veehouderij, *Toekomst voor de veehouderij. Agenda voor een herontwerp voor de sector*, (2001) Advice to the Dutch Agricultural Minister.

Van Zeijts, H, M.M. van Eerdt, W.J. Willems, G.A. Rood, A.C. den Boer, D.S. Nijdam, *Op weg naar een duurzame veehouderij. Ontwikkelingen tussen 2000 en 2010.* (Den Haag/Bilthoven, 2010), Planbureau voor de Leefomgeving (PBL).

- imported 80% of its available grain, nearly half of which was used for cattle fodder. Soy (products) alone accounted for 9 Mton (around 30% of EU flow), mainly
(75%) from Brazil and Argentina.[6]
- exported 63% of its available meat, while it imported somewhat less than a third
- crop production was wedded to artificial fertilisers, which largely replaced animal manure. Hence nutrients from surplus manure end up as effluents in soil,
surface waters and oceans.

As such, this chain is exemplary for the Netherlands a node of food supply chains[7]
(contrary to e.g. England which is more an endpoint). The classical Dutch pattern is
exporting processed foodstuffs and importing raw materials like cattle fodder and vegetable oils; this fits seamlessly with what we have called "a chemical-analytical perspective on agricultural and fisheries products as raw materials for foods." (Chap. 13).

Experts associate this pattern with nutrient depletion both globally[8] and in the
source countries, where it also despoils landscapes and ruins soils. In the Netherlands,
the dominance of animal over crop production has had similar consequences. For
one thing, it entails the long-range transportation of animals and meat (Dutch pigs
travel to Italy, and then Parma ham is exported everywhere, including the
Netherlands) which stresses the animals, augments CO_2 emissions and contributes
to the spread of pandemics[9] The sheer size, transnational nature and complexity of
the production chain, the second core element, has only exacerbated all these
problems.

How did practices of production, consumption and distribution give rise to a non-
circular grain-for-meat chain; what problem perceptions and values animated these
practices, and how were they institutionally and spatially structured?

---

[6] E. Hees, *Voedsel, grondstoffen en geopolitiek. Rapportage aan het Platform Landbouw, Innovatie & Samenleving* (Culemborg 2013), 23–28.

[7] See ch. 13 and WRR, *Naar een voedselbeleid*, Den Haag/Amsterdam 2014), rapport nr. 93, 60–64.

[8] D. Cordell, *The Story of Phosphorus:Sustainability implications of global phosphorus scarcity for food security* (2010), Linköping: Linköping University Press.

M. de Ridder, S. de Jong, J. Polchar, S Lingemann, *Risks and Opportunities in the Global Phosphate Rock Market* (The Hague, 2013), The Hague Centre for Strategic Studies (HCSS). Rapport No 17 | 12 | 12.

M. Heckenmüller, N. Daiju G. Klepper, *Global Availability of Phosphorus and Its Implications for Global Food Supply: An Economic Overview* (2014), Kiel Working Paper No. 1897. University of Kiel.

[9] Zeijts et al., *Op weg naar een duurzame veehouderij* (2010).

23.3.2 How the Netherlands Embedded Itself in the Grain-for-Meat Chain

From 1850–2010 around half of the available grain in the Netherlands was used for cattle fodder.[10] But increasingly, that grain has been imported. Since 1970, around 80% of the total has been imported, an increase compared to 1913 (68%) and especially to 1850 (21%). What lies behind this?

Specialisation was the chief culprit. This had long been the rule in the western provinces, but on the sand grounds the transition occurred between 1850 and 1913 with a shift from crop cultivation to cattle husbandry at the then still 'mixed' farms. The process (cf. Chap. 8) was triggered by the international trade liberalisation of the 1840s, the first effect of which was to substitute commercial gain for self-sufficiency as the basic normative orientation of farming on the sand grounds and around the large rivers. But it also toppled grain prices to the point of undermining mixed farming and encouraging the transformation of crop fields into grazing pastures. Given the new orientation to commercial gain and the price effects of trade liberalisation, agriculture shifted toward exporting expensive meat and importing relatively cheap grain. Industrialising England and Germany were especially enamoured of 'luxury' Dutch meat. By 1913, the Netherlands was exporting 31% of its available meat, while importing 28% – a tremendous increase compared to 1850, when these figures were 4% and 2% respectively.

Thus crop farming increasingly focused on serving livestock farming, while the latter increasingly became a value chain in and of itself, de-coupled from crop production. Here lies the cradle of the Dutch position as a node in international food chains.

As discussed in Chap. 8, the international agricultural crisis of the 1880s was a very rough patch for the Dutch primary sector. Still, this only accelerated the developments just sketched. The US was already decisively on the track to modernisation and, at the cost of soil depletion, was producing vast quantities of grain that could be transported cheaply to the East coast by the new railway lines. From there, steamship companies and older trade networks (the Dutch had long been global grain traders) shipped it to Europe. This seriously depressed European grain prices, further stimulating the trend that set in during the 1840s and reinforcing the orientation to commercial gain.

The spatial and institutional embedding of these practices and their normative orientation was consolidated by ever-cheaper international transportation as well as by the shift from the commons to agricultural co-operatives with their collective bulk purchasing and selling power. Between 1850 and 1913, the total amount of available grain increased by a factor of 3.3 while the amount used for cattle farming increased by a factor of 4.5; at the same time, the share of imports more than tripled. All told, these developments yielded a shift from a circular chain centred around

---

[10] F. Lambert, *Massastromen in Nederland in de jaren 1850, 1913, 1970, 2010* (Eindhoven 2016), open access researchreport, 11, 17, 25, 35.

mixed farms working for domestic or local production to a de-localized, non-circular meat production chain.

In the wings, a newly interventionist government was also keen to maintain and expand its position by serving the national interest. After the hygienist movement had revealed the dangers of unreliable food, food quality had also become part of the normative repertoire of the food sector – with a vigilant government looking over its shoulder. Following several scandals and disputes, the government mobilised the *Codex Alimentarius*, developed in the late nineteenth century by the hygienists, to define standards (1902) for meat, among other things to safeguard the export market. By 1919, chronically unreliable food supplies during World War I had made a Food and Drugs Act inevitable, a move that further bolstered the position of the Netherlands as an exporter of high-quality foods, including meat.

23.3.3   How Nutrients Came to Leak from the Grain-for-Meat Chain

The commercialisation of farming between 1850 and 1913 also became possible thanks to the adoption of artificial fertilisers that gradually replaced the mix of animal manure, urban waste and faeces, fish and animal waste and Guano. Chili Salpeter, imported from Latin America, was a key artificial fertiliser. Other fertilisers were imported from especially Belgium and Germany, which surpassed the Netherlands both as users and as importers of artificial fertilisers well into the 1920s. (Chap. 8) By then, though, domestic Dutch production of artificial fertilisers was also rapidly increasing (see Graph 13.2). This shift was triggered by World War I, when nitrates became critical military resources, especially in those countries from which the Netherlands imported its artificial fertilisers. (Threats of) food shortages and the emergence of new political movements compelled the government to intervene. To compensate for the closing down of international trade, it forced merchants, co-operatives and importers to work together in acquisition associations to supply the superphosphate factories established in the previous decade with scarce nitrates.

These wartime changes in the institutional quadrants persisted into the post-war period. In the following years, the new networks created in the market would promote the rapid growth of the fertiliser sector; and government, once involved, continued to promote fertiliser production by supporting (semi) state enterprises like the State Mines (coal) and Hoogovens (iron and steel). These firms employed new fertiliser production methods, developed internationally during the War. The Netherlands thus swiftly developed the portion of the international market allocated to it by the European nitrogen cartel created to regulate destructive competition. By the end of the decade, domestic producers began to dominate the national market (Table 13.4). The wartime crisis of international trade had ultimately yielded strong networks in the national institutional quadrants.

Such rapid growth in production could of course not have occurred without an equally dynamic demand for fertilisers. This was partly due to the establishment, by the early 1920s, of collective purchasing by agricultural co-operatives. Another factor was the drive for self-sufficiency during the First World War, which enabled production to exceed momentary demand, though with due consequences for prices. While by the early 1930s the fertiliser cartel had managed to stabilise prices, fertilisers had still become much cheaper: from 18 guilders per 100 kg in 1922 down to 5 in 1932. Demand was also stimulated by a knowledge infrastructure that transmitted new insights seamlessly to agricultural practice, and equally seamlessly fed farmers' experiences back into further R&D. Goaded by the holy grail of increased productivity, the same knowledge infrastructure also promoted other innovations that multiplied the benefits of artificial fertilisers, like novel plant varieties and cultivation methods. (Chap. 13).

These developments led to higher agricultural productivity and thence to greater well-being: more income for farmers, and more domestically produced and affordable meat for the population at large (from 32 kg/capita/year around 1913 to no less than 63 kg in 1970). The radical differentiation of the supply chain and the massive expansion of production following on the demise of mixed farming grew out of the institutional arrangements just discussed. They were the seedbed for the institutional framework that nurtured post-1945 agricultural modernisation: the so-called Iron Triangle for policy making and the OVO knowledge infrastructure. This entailed a vast increase in knowledge and in the capital intensity of farming, again promoting further specialization: from mixed farming through animal husbandry and poultry farming towards either chickens for meat or chickens for eggs. Modernisation also increased the scale of production, but that process started earlier and deserves separate analysis.

23.3.4 Expansion of Material Flows in the Grain-for-Meat Chain

By 2010, the volume of meat production was 34 times that of 1850 while meat consumption per capita was 2.5 times as high. The stage for this increase was set between 1913 and 1970. By 1913, meat production had nearly quadrupled relative to 1850 though – due to exports and doubling of the population – per capita domestic consumption had increased by only 12.5%. By 1970, however, meat production had again increased by a factor of 4.7. The bulk of this increase was due to post-1945 modernisation policies. This increase was large enough to effect a per capita increase in meat consumption of 75% (1913–1970) despite another doubling of the population and increased exports.

All in all, 1970 marked the end of a long struggle to provide sufficient meat; a struggle that had been a driving force in food production since 1850. Food security had become political dogma around World War I, when socialist movements tried to seize power by exploiting societal unrest about food scarcity. The established parties countered by arranging emergency imports of food from England, leading to a lasting consensus on the need to increase domestic food production – an orientation that would inspire emergent corporatist arrangements for decades to come.

The most dramatic increase in food production occurred between 1945 and 1970 through a self-reinforcing process of modernisation, aimed at increasing GDP by producing affordable food domestically, while freeing agricultural labour for industry. This process became self-reinforcing due to its drivers: a persistent cultural inclination towards Americanisation (especially among young farmers) and the demonstrably increasing returns that encouraged imitation and provided investment capital. The 'logic' of investments in modernisation – increased capital intensity – increased scale – increased scale – increased returns – new investments[11] was equally inescapable. Modernisation was nurtured within an efficient knowledge infrastructure put together by post-war Minister Sikko Mansholt. From the early twentieth century co-operatives had provided capital for investments even to small farmers and the increased scale was facilitated by structural policies (land reclamation and improved water management) introduced by Mansholt shortly after WW II. All these elements were seamlessly connected through the iron triangle, a corporatist policy-making machinery shaped by Mansholt and rooted in the consensus that had emerged earlier and reinforced during the hardships of WW II.

Second, these years also witnessed the emergence of mass consumption, including food consumption. Not only did virtually the entire population now have access to essential foods, like meat, but also to a range of new foodstuffs – foreign food and items like chicken, convenience and luxury food – that had been unavailable till then. This was promoted by the landscape trends of increased traveling, decolonisation, the spread of refrigerators in households and retail stores and increased leisure time; as well as by the supermarket and the car. Institutionally, these factors were stimulated by governmental measures, especially subsidies for imported cooling and freezing equipment, the liberalisation of the retail sector, the liberalisation of wage increases, the issuing of the disk-of-five guidelines for a healthy diet and the introduction of a five-day working week.[12]

By 2010, with an 81 kg/capita/year meat consumption, the Dutch occupied a 44th place among 170 countries. While this is only slightly more than recommended by the World Health Organization, urban lifestyles are still nurturing an upward trend. This yields both health and ecological problems that have stimulated national policies and to which we will return in Chap. 24.

---

[11] J. Grin, *Changing governments, kitchens, supermarkets, firms and farms: the governance of transitions between societal practices and supply systems* (2012), p. 35–56 in G. Spaargaren, P. Oosterveer, & A. Loeber (Eds.), *Food practices in transition: changing food consumption, retail and production in the age of reflexive modernity* (New York), Routledge.

[12] Chapter 13; J. Grin, 'Changing governments, kitchens, supermarkets, firms and farms: the governance of transitions between societal practices and supply systems', in G. Spaargaren, P. Oosterveer, & A. Loeber (Eds.), *Food practices in transition: changing food consumption, retail and production in the age of reflexive modernity* (Routledge 2012), 35–56.

23.4 Grain-for-Construction

23.4.1 Contemporary Problems and Historical Trends

Between 1850 and 2010, flows of mineral assets steadily increased, absolutely by a factor of 32, per capita by a factor of 16. While the trend became visible between 1850 and 1910, it acquired momentum over the next 60 years, levelling off only after 1970. Within the category of mineral assets, gravel is a revealing example: not only is it an important flow in its own right (constituting between 15–35% of total mineral flows since 1910), but also growing by a factor of nearly 400 between 1850 and 2010. Most of this increase took place between 1910 and 1970, with the rise of (gravel-rich) concrete as construction material. As to spatiality: in recent years the share of imported gravel exceeded 80%, twice the average for all mineral assets.

Gravel is a revealing example for understanding contemporary resource problems because its main environmental problem derives from its extraction: damage to the local landscape and ecosystems. The environmental effects of its actual use are less important.[13] This implies that ecological damage is proportional to total gravel recovery. But the precise location where harm is being done ('here' or 'elsewhere') depends on how much is mined domestically vs. how much is imported. Between 1970 and 2010 the combined effect of reduced use and a higher share of import led to a fourfold decline in the amount of gravel extracted in the Netherlands, while the amount extracted 'elsewhere' (mainly the three neighbouring countries) has increased by 25%.[14]

In terms of the two issues noted in research question Q1 (non-circularity of economic chains, and the sheer volume of mass flows), we note the following:

- mass flows of gravel increased by a factor of 24 between 1850 and 1910; while a nearly equal increase occurred over the following 60 years. After 1970, gravel use declined by just over 25%.
- until around 1900, the gravel chain was largely circular: obsolete buildings, dikes and road infrastructure were demolished and reduced to re-usable materials.
- between 1850 and 1910, the share of imports quadrupled, reaching 67% of the total or twice the mean percentage for mineral flows. Between 1910 and 1970, this share declined to 47%, just 1.25 times the average, only to exceed 80% by 2010 in spite of an overall decrease in gravel use.

---

[13] The production of concrete and concrete products, an important use for gravel, contributes 1,1% and 1,7%, respectively, to energy use and CO_2 emissions; t0068e main problem is particulate matter emissions, constituting more than 10% of annual Dutch emissions.

M. Bijleveld, G.C. Bergsma, M. van Lieshout, *Milieu-impact van betongebruik in de Nederlandse bouw Status quo en toetsing van verbeteropties* (Delft, 2013), CE Delft. Publicatienummer: 13.2828.24

[14] F. Lambert, *Massastromen in Nederland,* 25, 35.

To explain these features, we will discuss construction practices (of buildings, hydraulic engineering objects and road infrastructures), which were the main destinations for gravel, emphasising their institutional and spatial contexts.

23.4.2 Volumes of Gravel: From Historically Vast Increases of Mass Flows to Gradual Decline

In 1850, most of the 8800 ktons of material flows (58%) comprised organic matter; subsoil mineral assets constituted a mere 28%. Of those, clay, sand and gravel were the main bulk construction materials and largely extracted domestically; only gravel was imported to any degree (16%, see Table 3.1). By 1910, both absolute gravel flows and the level of gravel imports had increased dramatically. Despite the enormous increase in imports, domestic extraction continued to increase over those six decades by a factor of 24, from 50 to 1200 kton. In the next 60 years, it would increase again by a factor of 22, reaching more than 27,000 kton in 1970. This vastly growing demand for gravel derived both from an increase in construction efforts and from changes in the nature of construction.

23.4.3 Increasing Construction Efforts

Hydraulic engineering and transportation infrastructure, especially canals and railways, were the two sectors responsible for the initial increases in construction volumes. While the former was ordained by the geographic realities of living in a delta, infrastructure construction aimed to modernise an ailing economy. Around 1850, this was a critical issue both for the King and, after 1848, the national government, both of whom championed increasing welfare and thus supported modernisation and infrastructure construction.

Housing construction was another sector that required ever more building materials. In the final decades of the nineteenth century, growing societal concern about poor housing as a factor limiting popular well-being led to an enormous increase in urban housing construction. This was further stimulated by a number of long-term trends: population growth accelerated after 1880 as mortality fell in the wake of improvements in health care and nutrition.[15] The upshot was a doubling of the population from 3.1 million in 1850 up to 5,9 million in 1910. In 1900, the housing shortage stood at 9% for the country as a whole (Graph 14.2), and even higher in the industrialising and rapidly growing cities. There, hygienists as well as a new caste

---

[15] Ekamper, P. R. van der Erf, N. van der Gaag, K. Henkens, E. van Imhoff, F. van Poppel, *Bevolkingsatlas van Nederland – Demografische ontwikkelingen van 1850 tot heden.* (2003), NIDI. Main results in English at https://www.nidi.knaw.nl/en/research/pd/120504

of experts acquired a new audience as they voiced their concerns over the housing conditions of especially the lower class.

These groups promoted housing efforts according to the new hygienist norms. Their experiments helped shape the 1901 Housing Law. This law would become the cornerstone of a typically welfare-state institutional arrangement in housing and it immediately triggered a further acceleration in housing construction. As Graph 14.3 shows, annual housing production increased from around 15,000 in the late nineteenth century up to 20,000 around 1900 and more than 30,000 after 1905. As a consequence, the housing shortage fell to 3% by 1940, only to rise again to nearly 20% after the destruction of the Second World War. Around 1970 it was back to 3% again and fell to 2% by 2010 (Graph 14.2). Public surveys conducted between 1967 and 1972 revealed it to be perceived as the biggest social problem (cf. Chap. 15, note 41). Key drivers of the shortage were two long term trends: population growth (population again doubled between 1910 and 1970) and individualisation (leading to a demand for more rooms per household).

In consequence, housing production increased throughout most of the twentieth century. From next to nothing during World War I it rose to around 45,000 per annum during the interbellum, only to collapse again between 1940 and 1945. Between 1950 and the early 1970s housing production tripled: from 50,000 to 150,000 per annum; in subsequent decades it would hover at around 100,000. These efforts were coordinated by the institutional arrangement that had emerged with the 1901 Housing Act. Government played a key role. It pushed for high production rates and was itself responsible for a significant share of the construction, aided and abetted by housing corporations and private parties (Graph 14.1). It also defined norms for the quality of dwellings (including the number of rooms and quality of construction) as well as for rent rates.

23.4.4 Changing Construction Practice: Concrete and More Spacious Dwellings

Two developments in housing design increased the demand for gravel in housing construction. The first was the use of concrete, especially after 1910. Concrete solved problem definitions prevailing since 1900: the need for cost-effective improvements in lower class living conditions, and the need to increase welfare through economic modernisation. Concrete also had symbolic appeal: it resonated with the cultural inclination toward progress.

The second development was an increase in the size of dwellings and in the number of rooms per dwelling, signs of increasing well-being and modernistic individualisation. The trend was driven by government policy and enabled by increasing national income. Between 1948 and the mid-1970s, the average floorplan increased from about 55 to around 70 m^2 (Table 14.2) Also, while around 1900 only one-third of the homes had more than three rooms, this increased to more than two-thirds by 1930. By 1947 more than 75% of the homes had at least four rooms, and by 1970 more than 90%. In that year, more than a third even had six rooms or more (Table 14.3).

23.4.5   More Concrete, More Gravel

In light of these developments, concrete production grew apace. Concrete production took off in earnest during the 1920s, to climb to more than 11 million tons in 1970, still largely domestic in origin and consuming roughly 20% of the total gravel production.[16] By the mid-1950s, largely due to demand from the housing sector, it had already climbed to 7 Mt, reaching 10 Mt by 1960, 16 Mt in 1965 and more than 20 Mt in 1970, at which point output remained constant until 1980, when decline set in.[17] By 2010, 28% less gravel was being consumed than had been in 1970. While the decline in housing construction after the mid-1980s (from more than 100,000 units per year down to around 80,000) is clearly responsible, gravel consumption had already stabilised a decade earlier, due to the completion of the Delta Works and stagnation in the housing market (Chap. 19).[18]

23.4.6   The End of Circularity and the Increasing Share of Import

The evolution of the gravel-for-construction chain is typical of many Dutch chains (Chaps. 12 and 14): core demand practices (e.g. infrastructure and housing) become a central node in a network of specialised actors embedded in increasingly ramified supply chains. While gravel continues to flow directly and in its pristine form to the workplace, increasing quantities came to flow through separate chains, especially concrete, that also required coordination with other chains (sand and cement). These chains differ in terms of their circularity and the share of imported gravel. Both features were shaped by the tensions between demand, domestic gravel extraction and processing capacity and concerns about the local side- effects of extraction.

Why did the circularity of 1850 disappear? One key factor was the vast increase in demand. While around 1850 raw building materials were usually recovered in the course of demolishing obsolete constructions, this became untenable with the increased pace of construction of new infrastructures and the housing boom of the

---

[16] Rough estimates of gravel use in concrete production by JG, assuming that gravel constitutes roughly 50% of the mass of concrete – based on the traditional '1,2,3 rule' for volumes and the roughly equal densities of cement, sand and gravel.

[17] CBS figures, cited by Ike (1985). While river gravel production had declined from 3 to about 0,65 Mton during the Second World War, it returned to pre-war levels within about a year, and then increased steadily up to 7,5 Mton over the next ten years. B.C.E. Janssen, 'Het veranderend gelaat van Nederland. Grindwinning in Midden-Limburg', *Tijdschrift Nederlands Aardrijkskundig Genootschap*, 1965, deel 82, 376–380.

Similarly, in Limburg, where much of the Dutch gravel extraction was concentrated, extraction increased immediately after WW II, stabilised around 1965 and declined again around 1975. Cf Provincie Limburg. Letter on 'Afronding Grindfonds' d.d. 4/10/2005. Janssen, 'Het veranderend gelaat', 378.

[18] Provincie Limburg. Letter on 'Afronding Grindfonds' d.d. 4/10/2005.

twentieth century. Especially after the Second World War, with population growth, a housing shortage and a desire for more spacious homes driving the demand for construction, the volumes of material needed for new construction vastly exceeded what could be procured from the demolition of redundant facilities. Circularity had ceased to be a meaningful option.

What did remain throughout post-war reconstruction was strong governmental intervention in allocating scarce resources and ensuring a steady supply of subsoil mineral resources, especially gravel and marl as a basis for cement. In consequence, the national government took a firm stance when protests erupted in Limburg against the ENCI's proposed expansion of marl extraction near Maastricht. While the protest was unusual for the time, the outcome was more typical of the era: the ENCI got its concession for more cement production, although within strict spatial limits and subject to other mitigating measures.

Around 1950, a striving for national autarchy (after two wars and a crisis) intersecting with the perceived problem of insufficient housing made it necessary to expand domestic production of gravel (and sand). But extracting sand and gravel from beneath arable land ran afoul of efforts to solve another key problem of the time: insufficient food production. It also incited local protests against despoiling of the landscape. Gravel dredgers and the Province of Limburg eventually agreed on increased levels of extraction, but on condition of refilling the exhausted gravel pits and making them suitable for agriculture again – all paid for by a levy on the gravel. These kinds of compromises were institutionalised in 1965, with a Law on Subsoil Excavations coming into force by 1971. In the interval, nature and recreation became prominent issues. This brought alternative solutions for domestic gravel extraction to the fore, like turning empty gravel pits into economically viable recreational lakes. After 1970, local resistance finally got the upper hand. The national government capitulated by cutting back on domestic gravel extraction, banking on sources elsewhere within the EC. (Chaps. 16 and 19).

These vagaries of domestic gravel production and demand determined levels of import. Domestic gravel consumption increased 20-fold between 1910 and 1970. Nonetheless, the import share decreased from 67% to 47%. And after the reorientation in 1970, in spite of a 28% decline in gravel consumption, imports again increased to 82% by 2010.

We thus see how the volume of domestic gravel extraction, and with it levels of import, have responded to successive dominant problem definitions: increasing welfare, well-being and a striving for autarchy and the management of landscape quality. While it was the breakdown of circularity and increased gravel demand that incited more intensive domestic extraction and the despoiling of local landscapes, the increasing reliance on import implied that more and more of that damage was being exported elsewhere (see Sect. 23.4.1).

23.5   Plastics-for-Daily-Life

23.5.1   *Exploring Contemporary Problems*

Plastics are an interesting example of the shift to processed products that occurred between 1910 and 1970 (Table 12.6). The shift originated with organic materials and mineral subsoil assets, but after 1945 the breakthrough of plastics brought fossil assets into the fold. The volume of plastics subsequently skyrocketed over the past half century: production in the Netherlands increased by more than a factor of 6 between 1970 and 2010; consumption by a factor 4 to 5.[19]

While plastics have many industrial uses, households and consumer packaging are also an important market. Plastics were part and parcel of the new well-being achieved after 1960, when most people felt that their primary needs were satisfied. As discussed in Chaps. 14 and 16, plastics brought comfort, luxury and design products to the homes of all classes, as well as a new emphasis on convenience (packaging in retail and shopping practices!). While a contrapuntal critique persisted in seeing plastics as 'cheap' and 'artificial', the new products doubtless fostered more equality in the domain of 'second order well-being.' However, from a contemporary environmentalist point of view, plastics remain far from unproblematic, especially after their useful life.[20]

Plastics are seen as virtually synonymous with the 'waste society' – and for good reason. About half of the amount of plastic consumed by Dutch households is wasted – the same as for the EU as a whole.[21] Also, the share of plastics in Dutch domestic waste flows increased threefold between 1970 and 2010. This is the outcome of a steady increase until 2005 followed by a policy-driven decline after 2008.[22] Symbolically and materially, plastic waste is exemplary for the inefficient husbanding of natural resources. Or, as the World Economic Forum[23] puts it "Plastic packaging is an iconic linear application with USD 80–120 billion annual material value loss."

---

[19] F. Lambert, *Massastromen in Nederland*, 27, 37.

[20] Energy consumption is not the most crucial plastics problem. While plastics do belong to the most energy-intensive sectors, they account for only a few percent of Dutch CO_2 emissions. Hoekstra en Van der Helm and Lintsen et al. argue that their production is less energy-intensive than that of the products substituted. In addition, plastic materials are relatively light, which makes some of its applications (e.g. in cars) less energy consuming. R. Hoekstra en R. van der Helm, Paper prepared for the 18th International Input-Output Conference (Sydney 2010), June 20-25th.

H. Lintsen, M. Hollestelle and R. Hölsgens (2017), *The Plastics Revolution. How the Netherlands became a global player in plastics* (Eindhoven 2017, open-access).

[21] Plastics Europe, *Plastics – the Facts 2012. An analysis of European plastics production, demand and waste data for 2011. Plastics Europe* (Brussels 2012), 9.

[22] It was 17% (increase by factor of 3,4 by 2010) and 14% (factor 2,8) for 2011–2015, when the share was constant at 14%. Rijkswaterstaat, Samenstelling huishoudelijk restafval. http://afval-monitor.databank.nl/Jive/Jive?cat_open=landelijk%20niveau/Samenstelling%20van%20huishoudelijk%20restafval

[23] World Economic Forum, Ellen MacArthur Foundation and McKinsey & Company, *The New Plastics Economy — Rethinking the future of plastics* (2016), retrieved from http://www.ellenmacarthurfoundation.org/publications

Second, plastic waste that is not properly disposed of and processed may cause environmental harm[24] Although there are increasing concerns about micro-plastics on land, little is known about them. More is known about damage to marine eco-systems, especially about entanglement and ingestion by fish. According to WEF et al., 2016, the oceans contain a mass of plastics amounting to 1/5 of the total mass of fish; they expect the quantities to be equal by 2050. Other known impacts concern changes in habitats and the transport of alien species, and human and animal health effects due to additives in plastics. While these problems also affect the Dutch marine environment,[25] they are unevenly distributed over the globe,[26] most notably affecting the Pacific.

In sum, contemporary problems with plastics are for here and now as well as for later and elsewhere; and they are a consequence of the vast waste flows produced. So it is relevant to understand the origin and increase of these waste flows. As above we explain the enormous growth of production and consumption, certainly fundamental to the increase of waste flows and we discuss the nature and evolution of the non-circularity of the chain.

23.5.2 How the Chain Became So Vast: The Plastics Revolution

The first plastics were developed at the end of the nineteenth century, and even in those early days their universal appeal was foreseen by, especially American, prophets of mass consumption. Until World War II, consumer products would constitute the most important market. By 1920, as the US started to export the new lifestyle, Berlin and Paris were becoming its European showcases[27] – which may explain why Germany and France were early adopters of plastics – But wartime developments were at least as pertinent: novel systems, like plastics use in radar, or the use of plastics as substitutes for silk parachute textile after Japan entered the war (Chap. 15).

Prior to 1970 the Netherlands was certainly not a frontrunner in plastics consumption, joining the fray only after 1950. The same goes for production. In 1950, a mere 17 kton was produced, but that swiftly increased: to 239 kton by 1963,

---

[24] DG Environment of the EU (2011). Science for Environment Policy, In-depth Reports, Plastic Waste: Ecological and Human Health Impacts.

World Economic Forum, *The New Plastics Economy*.

[25] H.A. Leslie, M.D. van der Meulen, F.M. Kleissen, A.D. Vethaak, *Microplastic litter in the Dutch marine environment: Providing facts and analysis for Dutch policymakers concerned with marine microplastic litter* (Delft 2011).

[26] World Economic Forum, The New Plastics Economy.

[27] R. Oldenziel, A. de la Bruhèze, *Theorizing the mediation junction for technology and consumption* (2009) in A. de la Bruhèze and R. Oldenziel (eds) *Manufacturing technology, manufacturing consumers: The making of Dutch consumer society*, (Amsterdam: Aksant).

900 kton by 1970, 1785 kton by 1975 and up to nearly 9445 kton in 2010.[28] Thus by 1963, the country was in the sub-top of producers (after the US, Germany and Japan) both in absolute and per capita terms, and by 1975 to the global top in terms of per capita production (106 kg/capita, compared to 78 and 45 for Germany and the US, respectively.[29]

That Dutch consumers were 'late adopters' is not that strange, given the country's culture of conservative domesticity that also retarded their adoption of mass consumption until well into the mid-twentieth century.[30] But how then was it possible for the Netherlands to catch up so swiftly after the war? For one thing, the 1950s saw a 'silent revolution' in 'culture and mentality'[31] (augmented by a cultural inclination towards 'Americanisation.' That shift, together with higher incomes, a five day working week, car mobility, TV and the emergence of the supermarket, created an *avant garde* in the Dutch middle class ready for the thrills of mass consumption[32](see also Chap. 22). These developments had been actively promoted by the Dutch government and by Marshall Plan projects, for example excursions by housewives and retailers to the US as well as exhibitions on the American way of life.[33]

Supply-side factors also help explain the swift growth of the chain. Frontrunners in industry and in the R&D sector shared the disposition to modernisation, Americanisation and mass consumption. That they were prepared to cater to the needs of the consumer market is evident from the nature of the plastic products: crockery, statues, medals, chains, dolls as well as furniture and clothing that incorporated the lower classes in the world of design and good taste. This coordination of supply and demand clearly helped the chain's swift expansion. And the underlying expectations about the future of mass consumption indeed stimulated firms to invest in further developments. (Chap. 15).

Yet, at a deeper level, earlier developments were responsible for the ability and preparedness of firms to join the plastics revolution. To be sure, Dutch firms had played no meaningful role in developing and producing plastics (excepting Philips' production of Bakelite). Three circumstances in particular enabled Dutch industry to catch up. Spatiality helped: DSM was located on top of the Limburg coal mines, and was very knowledgeable about fossil resources and ways to process them. Also, the globally important port of Rotterdam was a nodal point for the global oil trade,

---

[28] The number for 1950 is an estimate, inferred from Graph 3.1 in Lintsen et al., which shows that consumption and production were roughly equal. Numbers for 1963 and 1975 were taken from tables 3.2 and 3.3 in Lintsen et al.

[29] Numbers from table 3.1 in Lintsen et al., 2016.

[30] R. Oldenziel, M. Hård, *Consumers, Tinkerers, Rebels: The People who Shaped Europe. (Series: Making Europe: technology and transformations, 1850–2000).* (New York, 2012), Palgrave McMillan.

R. Oldenziel et al. *Theorizing the mediation junction for technology and consumption.*

[31] P. Luykx, P. Slot, *Een stille revolutie? Cultuur en mentaliteit in de lange jaren vijftig.* (Hilversum, 1997): Uitgeverij Verloren.

[32] J. Grin, *Changing governments, kitchens, supermarkets, firms and farms.*

[33] R. Oldenziel, K. Zachman, *Cold War Kitchen. Americanization, Technology, and European Users.* (Cambridge, 2008, The MIT Press).

making it an interesting plastics production site for multinational firms like Shell, DSM, Dow Chemical and ICI, that had developed much of the basic knowledge of polymer production. Other firms, like the AKU (predecessor of AkzoNobel), invested in delivering semi-finished products. Second, the Dutch government sponsored the establishment of a Plastics Institute within TNO – the national public-private R&D agency. The Institute not only became a centre of expertise in plastics production and processing, but also played a main role in the development of material, product and process standards. Finally, other firms using plastics at small scales before the war – like the Draka cable factory, where plastics were used for insulation – were readily able to connect to this open network, promoting an efficient, economically viable sector. (Chap. 15).

The upshot was a self-organizing 'platform'-like institutional arrangement of polymer producers, processing firms and a supportive, public-private R&D infrastructure. Its structure mimicked plastic production itself, which is essentially a matter of recombining and processing a limited set of basic polymers in virtually endless ways. The platform's links with transport nodes and the participation of major multinational firms facilitated the import of key materials and of products not (yet) domestically produced, as well as promoting export. The share of imports in plastics flows rapidly declined. In 1950 import and export were roughly equal to each other. By 1963, the import share was still 44%, but by 1970 it had declined to 23%. Over the same period, exports increased dramatically,[34] further stimulating the sector: export shares were 52% (1963), 70% (1970) and 81% (1975) and despite absolute increases in volume, they continue to fluctuate around the 1975 level, with e.g. 77% in 2010 and 83% in 2014. One key factor behind the rapid expansion of both production and export after 1966 was the abundance of cheap domestic natural gas, a big competitive advantage in this energy intensive industry (Chap. 15).

Here, as in our other two cases, a traditional Dutch supply chain structure emerged as a complex web of specialized, differentiated suppliers and clients, well-embedded in trading activities and the associated institutional and spatial networks. By 2014 it included some 125 producers and more than 1300 intermediate processors. The mostly small and medium-sized enterprises comprising the sector were ranked first in innovation power among all SME sectors. The total turnover of the sector was 17.5 billion euro, i.e. about 2% of Dutch GDP.

The dark side of this significant contribution to economic welfare was increasing waste flows. Levels of waste production were more than proportional to domestic consumption. Less is destined for construction, and more for product categories with a shorter life cycle: applications in cars and electronics devices (two expressions of 'the new well-being' that evolved after 1960) and, most notably, an increasing use of plastics packaging, that inevitable handmaiden of modernisation.[35] As a consequence, in 2014 no less than 40% of the plastics waste flow derived from short-lived applications.

---

[34] Percentages here and in the paragraphs below are taken from Lintsen et al. (2016, Graph 3.1 and tables 3.2, 3.3; for 1950, 1963 and 1975); and for 2014 from the *Factsheet Rethinking Plastics* issued by the Dutch Federation of Rubber and Plastics Industry NRK (Den Haag, 2016).

[35] Lintsen et all, *The Plastics Revolution*, 157–171.

23.5.3   Understanding the Chain's Non-circularity

In contrast to the organic and mineral chains, the plastics chain has never been even close to circular. The most basic reasons are the same as those for the swift, positive reception of plastics: they are cheap; and they have a cultural resonance with 'progress' in the form of makeable materials.

Plastic packaging is emblematic. A cultural disposition towards disposability existed in Dutch society by the time plastics became an object of mass consumption. The Dutch were quite ready for "the the idea that packaging is made to be thrown away" (1957 industrial expert quote, cf. Chap. 15). A second- order cultural effect was that the modernistic ideal of the supermarket also promoted pre-packaged food products and beverages that dramatically increased the volume of plastic packaging, goaded of course by governmental food safety regulations. By the early 21st century the idea had become deeply entrenched: plastics are "an iconic application of linearity", as the WEF[35] put it.

These factors are clearly reflected in the figures. Between 1970 and 2010, the share of plastics in domestic waste increased from 5 to 17%. Tellingly, in the same period the share of glass decreased from 11 to less than 5%. By 2015, packaging constituted some two-thirds of the share of plastics in domestic waste (around 14%). This consisted mostly of plastic bags and foil (3,5%), shape fast packaging (3%) and bottles (1,9%, of which 1,2% for non-food products).[36]

The persistent non-circularity of the sector has other roots as well. As Lintsen et al.[37] note, the production of basic polymers is based on an economy of scale, their processing into end products on a logic of scope (in terms of amounts and diversity of end products). These logics have become institutionally entrenched in the sector's platform-like institutional arrangement. This has led to a virtual neglect of the waste issue in the shaping of production practices and their connections to consumption practices. This is only aggravated by the fact of the Netherlands being a global plastics player, thus further uncoupling production from the conditions of use and waste practices.

23.5.4   Energy Consumption

Together, this lack of circularity and the vast expansion of the chain explains contemporary problems with plastics waste. As a twofold sector analysis by Hoekstra et al.[38] demonstrates, increased volume is also one key factor explaining increases

---

[36] Rijkswaterstaat *Samenstelling van het huishoudelijk restafval, sorteeranalyses 2015; Gemiddelde driejaarlijkse samenstelling 2014.* (Utrecht2016), Rijkswaterstaat.

[37] H. Lintsen, M. Hollestelle and R. Hölsgens (2017), *The Plastics Revolution. How the Netherlands became a global player in plastics* Eindhoven 2017, open-access).

[38] All figures in this and the preceding paragraph are from R. Hoekstra en R. van der Helm, Paper prepared for the 18th International Input-Output Conference (Sydney 2010), June 20-25th

in the energy intensity of Dutch industry in the years 1961–1974. The abundance of natural gas since the mid-1990s increased the share of energy intensive sectors like the chemical industry, lured by golden opportunities in export markets.

The sector's sensitivity to energy costs made it especially responsive to the 1973 and 1979 oil crises. Reducing energy consumption became a guiding principle for innovation in chemical processing in general: new machinery, electronic process control and design and, more recently, 3D printing.[47] In consequence, the sector's overall energy intensity declined by 7.5% between 1961 and 2008; its growth rate for CO_2 emissions dropped from 11.4% per year (1961–1974) to zero (1975–1989), and between 1990 and 2008 emissions even dropped at a rate of 3% per year. Its share in national energy consumption became 11.6%, and in CO_2 emissions 7.2%.

Yet with an eye to the future it is important to note that the reduction in the sector's energy intensity was only half that of the Dutch economy as a whole (15%). Also, after 1990, the reduction in energy intensity was more than offset by an expansion of production, mainly driven by export markets. The changed fuel mix incorporating more natural gas was both behind relatively low emissions, and behind the increasing volume, especially through the export effect. In sum, for the decades to come, the fuel mix will be a key determinant of CO_2 emissions and exports alike.

23.6 Electricity for Households

23.6.1 Exploring the Problem

Several sectors were chiefly responsible for making the Dutch economy much more energy-intensive: air and water transportation, the chemical industry and agriculture, but certainly also the utilities. In fact, the latter accounted for one third of the total increase in CO_2 emissions between 1960 and 2008. In this sector the main driver of *changes* in energy flows was not export, but household consumption – in spite of the fact that the latter accounted for a mere 20% of all electricity consumption. Exports (by *other* Dutch industries!) take second place as drivers of change.[39] We focus here on household consumption, on practices that consume electricity and that reflect the evolution of well-being.

The sector is relevant not only for explaining key contemporary dynamics; it is also significant in absolute terms. In 2008 the sector was the single biggest consumer of energy (11,8% of total) and the single biggest producer of CO_2 emissions (29,4%). The sector is thus a major culprit in resource depletion and climate problems 'later' and 'elsewhere.'

Against this background, we will seek to understand

- the development of the dynamic relationships between household electricity consumption and overall electricity production;

---

[39] All figures in this and the next paragraph are from R. Hoekstra en R. van der Helm (2010).

- the evolution of domestic electricity demand
- the development of the relationships between electricity production, overall primary energy use and CO_2 emissions.

23.6.2 Households and Electricity Production

As in most countries, electricity in the Netherlands was initially a lighting technology, first adopted by businesses and well-to-do households, as a replacement for dangerous, smelly and noisy gas lighting. Around 1900, when electricity also became a source of motive power for trams and factories, municipalities and provinces began to establish electrical power plants (Chap. 16).

As the networks spread geographically to include additional users, marginal costs for additional connections declined: a well-known mechanism of network expansion. (Chap. 14) The resulting dynamic networks linked centralised production to decentralised consumption. That structure implied an economic incentive to further electrify households. One early example, the vacuum cleaner, was successful partly because it took over a rather laborious domestic task. Like other appliances (the electrical iron being an important second example), its use was promoted by 'trailblazers of the consumer society,' intermediaries with close ties to the product developers (Chap. 16). These intermediaries not only fostered receptivity for electrical novelties among housewives, but also transmitted housewives' experiences and needs back to R&D departments.[40]

These intermediaries were certainly driven by the economic logic of (municipal) electricity companies, but also drew on the efforts of feminists. Drawing on an international cultural ('landscape') inclination towards a modern rational lifestyle as well as on an indigenous preference for a healthy, cozy family life, contemporary feminists encouraged middle-class women to reconsider their overstuffed homes and adopt more efficient approaches to housework. The aim was also to create more free time for social engagement, especially for working-class women. In 1912, the Netherlands Association of Housewives was founded; followed in 1932 by the Women's Electricity Society. These women, informed by experience and inspired by scientific advances, undertook a variety of initiatives to rationalise the household: 'home supervisors', experimental cooking schools, conferences and exhibitions, and publications on how to manage a household in more rational and hygienic ways.[41]

[40] T. de Rijk, *Het elektrische huis: vormgeving en acceptatie van elektrische huishoudelijke apparaten in Nederland.* (Rotterdam, 1998). P. van Overbeeke & G.P.J. Verbong, 'De strijd om het huishouden' in: J.W. Schot en anderen (red.), *Techniek in Nederland in de twintigste eeuw.* (Zutphen 2000), 174–189.

[41] R. Oldenziel & Carolien Bouw (ed.). *Schoon genoeg. Huisvrouwen en huishoudtechnologie, 1898–1998.* (Nijmegen 1998). A.H. van Otterloo, 'Voeding', in J.W. Schot en anderen (eds.) *Techniek in Nederland in de twintigste eeuw* (Zutphen 2000), 140–150. L. Bervoets and R. Oldenziel (2009). 'Speaking for Consumers, Standing Up as Citizens: the Politics of Dutch Women's organizations and the Shaping of Technology, 1880–1980,' in: Adri Albert de la Bruhèze

Against this background, it is easy to see why electrification could proceed so swiftly. In Amsterdam, for instance, some 7000 homes were connected in 1913. By the end of the First World War, that number exceeded 100.000. By 1924, some 144.000 out of a total population of 165.000 were connected.[42] Sockets had become standard in new homes, as "the most effective propagators of electrification" inasmuch as they "yelled for contact." The assumption that electricity was just there, ready to hand, became deeply entrenched, helping to pave the way for a swift electrification of the household.

This central-decentral, fossil fuel-based infrastructure, as well as the cultural esteem for the 'electric home',[43] as both an expression of well-being and a symbol of faith in progress, still shape present-day electricity consumption and production. That said, privatisation of electricity production and distribution has begun to corrode electrical centralism. Also while user demand initially sprung from a desire to rationalise the household, and thus improve quality of life (Chap. 22), around 1960 the emphasis shifted towards luxury, convenience and mass consumption (see also Graph 17.1). Since the mid-1990s, ICT appliances have added another layer to this trend and produced a space for the electronic *homo ludens*.

23.6.3 The Evolution of Domestic Electricity Demand

The increase in consumptive expenditures has led to a drastic increase in domestic energy consumption, from 3,2 to 12,5 GJ/capita between 1950 and 2010. But not uniformly. Between 1982 and 1988 it decreased by 12% to remain at that level until 1993; it then increased until 2005, only to decrease again: by 2010 it was back at the 2000 level.

To understand this pattern we need more details. In 1988, an average household used 20% of its electricity for cooling, in 1993 this had fallen to 17%; in the same period lighting declined from about 25% to about 17% and the share of audio/video appliances increased from 8.5 to 12%.[44] ICT applications accounted for nearly the entire 20% growth of overall electricity consumption between 1990 and 2006. By 2008, computers, communication and auxiliary appliances accounted for 15% of the average household's electricity consumption.[45]

---

and Ruth Oldenziel (eds.), *Manufacturing Technology, manufacturing consumers. The making of Dutch consumer society.* Amsterdam: Aksant, p. 44–47.

[42] P. van Overbeeke, *Kachels, geisers en fornuizen. Keuzeprocessen en energieverbruik in Nederlandse huishoudens, 1920–1975* (Hilversum 2001: Verloren), 46.

[43] T. de Rijk, *Het elektrische huis.*

[44] K.F.B. de Paauw, J.M. Bais, *Sectorstudie huishoudens en woningen.* (Petten, 1995: E).

[45] Clevers, P. Popma, *Energiemonitor ICT 2008.* (Den Haag, 2009), Tebodin/Ministerie van Economische Zaken.

J. Ganzevles, R. van Est, *Energie 2030. Maatschappelijke keuzes van nu.* (Den Haag: Rathenau Instituut).

This evolution of domestic energy consumption is the outcome of heterogeneous developments. On the one hand, consumption increased due to the proliferation of different types of appliances and their penetration into households (e.g. dishwashers; audio/video; ICT), and more frequent use (e.g. washing machine). On the other hand, many appliances have become more energy efficient.[46] The development of more efficient light bulbs explains the decrease in the share of lighting: from 710 down to 511 kwh/yr./household between 1988 and 1993, and further down to 390 kWh in 2014.[47] For cooling, increased refrigerator efficiency was compensated by increases in volume and the rapid adoption of the freezer: total electricity consumption for cooling was 578 kWh in 1988, and 555 in 1993. Due to the increased number and variety of audio and video appliances, their power consumption increased from 247 to 391 kWh between 1988 and 1993, after which more efficient appliances began to turn the tide – down to 191 kWh by 2014.[48] In a similar pattern, the more recent ICT appliances first increased domestic consumption but with e.g. the introduction of laptops their consumption declined again.

23.6.4 The Evolution of Energy and the CO_2 Intensity of Electricity Production

The Dutch electricity market is again traditional in the sense of being a trade market: the volume of trade in electrictiy is twice what is demanded by Dutch industries and households.[49] That said, we can still say something about the energy and CO_2 intensity of electricity production. While in 1922, some 22 MJ of primary energy was needed to generate one kWh of electricity; by the 1940s, only 15 MJ was needed. Upon the introduction of natural gas (Chap. 15) and with improved efficiency, consumption per kWh rapidly declined to 10 MJ by 1964. By 2000, due to improvements in efficiency, it had further declined to around 8.4 MJ/kWh until 2003, and again from 8.1 in 2004 to 7.0 in 2011. But in 2012, the energy intensity again increased, up to 7.4 MJ/kWh by 2015, due to the replacement of gas by coal in power plants. Not surprisingly, CO_2 consumption per kWh declined by 20% between 2000 and 2011, but then increased again by 20% due to this change in the fuel mix. In

---

[46] H. Jeeninga, *Analyse energieverbruik sector huishoudens, 1982–1996. Achtergronddocument bij het rapport. 'Monitoring energieverbruik en beleid Nederland*. (Petten, 1997: ECN).

CPB, *Naar een efficiënter milieubeleid; een maatschappelijk-economische analyse van vier hardnekkige milieuproblemen*. (Den Haag, 2000): CPB/SDU.

[47] K.F.B. de Paauw, J.M. Bais, *Sectorstudie huishoudens en woningen*, 22.

[48] K.F.B. de Paauw, J.M. Bais, *Sectorstudie huishoudens en woningen*, 22. ECN, Energie-Nederland en Netbeheer Nederlan, *Energietrends, 2014* Rapport ECN-O--14-041 (z.pl 2014).

[49] ECN, Energie-Nederland en Netbeheer Nederland, *Energietrends, 2016* Rapport ECN-O--16-031 (z.pl 2016).

short, energy intensity in electricity production, just like that in plastics production, responded to changes in fuel mix related to the availability (and exhaustion) of domestic natural gas between about 1960 and 2013.

23.7 In Conclusion

In this chapter, we have discussed how currently problematic couplings between GDP growth, well-being, resource / energy use and pollution have historically emerged. Drawing upon the rich historical discussions in previous chapters, and guided by transition theory, we have shown how this may be understood on basis of the evolution of practices of production, consumption, and distribution, as well as their guiding problem definitions and value orientations and their structural and spatial embedment of these practices. We showed for three discuss supply chains (grain-for-meat; gravel-for-construction; and plastics-for-domestic-quality of life) how they became non-circular in character, and how material flows in these chains expanded so vastly, especially since 1960.

We will now turn our analytical perspective 180 degrees. Looking from the perspective of a more sustainable society, now widely desired, we will discuss strategies to make the first three value chains less massive and more circular, and to reduce the climate change effects of the fourth one, preferably down to zero. While we will, in the next chapter, contextualize this discussion also in ongoing long term trends that will shape early twenty-first century developments, we will draw on the analysis of this chapter to appreciate how historically evolved practices, structures, and spatial conditions both constrain and enable such strategies.

Literature

Beck, Ulrich, (1997): *The re-invention of politics. Rethinking Modernity in the Global Social Order* (Cambridge: Polity Press).

Bervoets, L. and R. Oldenziel (2009). 'Speaking for Consumers, Standing Up as Citizens: the Politics of Dutch Women's organizations and the Shaping of Technology, 1880–1980,' in: Adri Albert de la Bruhèze and Ruth Oldenziel (eds.), *Manufacturing Technology, manufacturing consumers. The making of Dutch consumer society.* Amsterdam: Aksant, p. 41–72.

Bijleveld, M.M. (Marijn), G.C. (Geert) Bergsma, M. (Marit) van Lieshout (2013). *Milieu-impact van betongebruik in de Nederlandse bouw Status quo en toetsing van verbeteropties.* Delft, CE Delft. Publicatienummer: 13.2828.24

Brauch, Hans Günter, Úrsula Oswald Spring, John Grin, Czeslaw Mesjasz, Patricia Kameri-Mbote, Navnita Chadha Behera, Béchir Chourou, Heinz Krummenacher (Eds., 2009): Facing Global Environmental Change: Environmental, Human, Energy, Food, Health and Water Security Concepts. Hexagon Series on Human and Environmental Security and Peace, vol. 4 (Berlin – Heidelberg – New York: Springer-Verlag).

Brauch, Hans Günter, Úrsula Oswald Spring, Czeslaw Mesjasz, John Grin, Patricia Kameri-Mbote, Béchir Chourou, Pal Dunay, Jörn Birkmann (Eds., 2011): *Coping with Global Environmental*

Change, Disasters and Security – Threats, Challenges, Vulnerabilities and Risks. Hexagon Series on Human and Environmental Security and Peace, vol. 5 (Berlin – Heidelberg – New York : Springer-Verlag).

CBS (2015). *Electriciteit in Nederland.*

Clevers, & P. Popma (2009). *Energiemonitor ICT 2008.* Den Haag: Tebodin/Ministerie van Economische Zaken.

Cordell, Dana (2010). *The Story of Phosphorus:Sustainability implications of global phosphorus scarcity for food security.* Linköping: Linköping University Press.

CPB, Naar een *efficiënter milieubeleid*; *een maatschappelijk-economische analyse van vier hardnekkige milieuproblemen. (*Den Haag, 2000): CPB/SDU.

Denkgroep Duurzame Veehouderij (2001). *Toekomst voor de veehouderij. Agenda voor een herontwerp voor de sector.* Advice to the Dutch Agricultural Minister.

de Paauw, K.F.B. & J.M. Bais (1995). *Sectorstudie huishoudens en woningen.* Petten: ECN

Ekamper, P. R. van der Erf, N. van der Gaag, K. Henkens, E. van Imhoff & F. van Poppel (2003), *Bevolkingsatlas* van *Nederland – Demografische ontwikkelingen van 1850 tot heden.* NIDI. Main results in English at https://www.nidi.knaw.nl/en/research/pd/120504

DG Environment of the EU (2011). *Science for Environment Policy | In-depth Reports | Plastic Waste: Ecological and Human Health Impacts.*

ECN, Energie-Nederland en Netbeheer Nederland, *Energietrends, 2014* Rapport ECN-O--14-041 (z.pl 2014).d

ECN, Energie-Nederland en Netbeheer Nederland, *Energietrends, 2016* Rapport ECN-O--16-031 (z.pl 2016)

J. Ganzevles, R. van Est *Energie 2030. Maatschappelijke keuzes van nu. (Den Haag: Rathenau Instituut)*

Grin, John, Henk van de Graaf & Philip Vergragt (2003). 'Een derde generatie milieubeleid: Een sociologisch perspectief en een beleidswetenschappelijk programma,' *Beleidswetenschap,* jrg. 17, nr. 1, p. 51–72.

Grin, John, 2010: "Understanding Transitions from a Governance Perspective", Part III,) in: Grin, John; Rotmans, Jan; Schot, Johan, 2010: *Transitions to Sustainable Development. New Directions in the Study of Long Term Structural Change* (New York: Routledge): 223–314.

Grin, J. (2012). Changing governments, kitchens, supermarkets, firms and farms: the governance of transitions between societal practices and supply systems, p. 35–56 in G. Spaargaren, P. Oosterveer, & A. Loeber (Eds.), *Food practices in transition: changing food consumption, retail and production in the age of reflexive modernity.* New York [etc.]: Routledge.

Grin, John, Niki Frantzeskaki, Vanessa Castan Broto and Lars Coenen (2017) 'Sustainability Transitions and the City: Linking to transition studies and looking forward ' chapter 22 (conclusions) in: Niki Frantzeskaki; Vanessa Castan---Broto; Lars Coenen (ed.s), *Urban Sustainability Transitions.*

Harmelink, M., L. Bosselaar, J. Gerdes, P. Boonekamp, R. Segers, H. Pouwelse, M. Verdonk (2012). *Berekening van de CO2-emissies, het primair fossiel energiegebruik en het rendement van elektriciteit in Nederland.* Agentschap NL/CBS/ECN/PBL.

Heckenmüller, Markus Daiju Narita, Gernot Klepper (2014). *Global Availability of Phosphorus and Its Implications for Global Food Supply: An Economic Overview.* Kiel Working Paper No. 1897. University of Kiel.

Hees, E. (2013). *Voedsel, grondstoffen en geopolitiek. Rapportage aan het Platform Landbouw, Innovatie & Samenleving.* Culemborg: CLM. P. 23–28

Hoekstra, R. en R. van der Helm (2010). Paper prepared for the 18th International Input-Output Conference. Sydney: June 20–25th.

Hesselmans, A., 1993. 'Elektriciteit', in: H.W. Lintsen en anderen (red.), Geschiedenis van de techniek in Nederland (Zutphen: Walburg Pers, deel III, 135–161.

Hodson, Mike; Marvin, Simon, 2010: "Can cities shape socio-technical transitions and how would we know if they were?", in: *Research Policy,* 39: 477–485.

Ike, P (1985) *Grind gegvalueerd : nadere beschouwing van de tot op heden opgestelde prognoses en aanbevelingen voor nieuw op te stellen prognoses.* Delft : Delftse Universitaire Pers.

B.C.E. Janssen, B.C.E. (1965). 'Het veranderend gelaat van Nederland. Grindwinning in Midden-Limburg' in: *Tijdschrift Nederlands Aardrijkskundig Genootschap,* deel 82, p. 376–380.

Jeeninga, H. (1997). Analyse energieverbruik *sector huishoudens, 1982–1996. Achtergronddocument bij het rapport. 'Monitoring* energieverbruik *en beleid Nederland'.* Petten: ECN. m

Jetten, L., B. Merkx, J. Krebbekx en G. Duivenvoorde (2011). *Onderzoek kunststof afdankstromen in Nederland,* Berenschot/Agentschap NL/DPI Value Centre/Federatie NRK.

Leslie, H.A., M.D. van der Meulen, F.M. Kleissen, A.D. Vethaak (2011). *Microplastic litter* in the Dutch marine environment: Providing facts and analysis for Dutch policymakers concerned with marine *microplastic litter.* Delft: Deltares

Lintsen, H., M. Hollestelle and R. Hölsgens (2017), *The Plastics Revolution. How the Netherlands became a global player in plastics* Eindhoven: Foundation for the History of Technology (open-access)

Luykx, Paul en Pim Slot, (ed., 1997). *Een stille revolutie? Cultuur en mentaliteit in de lange jaren vijftig.* Hilversum: Uitgeverij Verloren.

Mom, G.P.A. & R. Filarski (2008). *Van transport naar mobiliteit: De mobiliteitsexplosie (1895–2005).* Zutphen: Walburg.

Oldenziel, R. and Karin Zachman (eds.; 2008). *Cold War Kitchen. Americanization, Technology, and European Users.* Cambridge, MA: The MIT Press.

Oldenziel, R. and Albert de la Bruhèze, A. (2009) 'Theorizing the mediation junction for technology and consumption', in A. Albert de la Bruhèze and R. Oldenziel (eds) *Manufacturing technology, manufacturing consumers: The making of Dutch consumer society,* Amsterdam: Aksant.

Oldenziel, Ruth & Carolien Bouw (ed., 1998). *Schoon genoeg. Huisvrouwen en huishoudtechnologie, 1898–1998.* Nijmegen: SUN.

Oldenziel, R. & Hård, M. (2012). *Consumers, Tinkerers, Rebels: The People who Shaped Europe.* (Series: Making Europe: technology and transformations, 1850–2000). New York: Palgrave McMillan.

Otterloo, A. H. van (2000). 'Voeding', in J.W. Schot en anderen (eds.) *Techniek in Nederland in de twintigste eeuw.* Zutphen: Walburg Pers, Deel III, 235–374.

Ours, J.C. van (1985) *Gezinsconsumptie in Nederland 1951–1980.* Rotterdam: 010 Uitgeverij.

Overbeeke, P. van, G.P.J. Verbong (2000). 'De strijd om het huishouden' in: J.W. Schot en anderen (red.), *Techniek in Nederland in de twintigste eeuw.* Zutphen : Walburg Pers, deel II, 174–189.

Overbeeke, P. van (2001). *Kachels, geisers en fornuizen. Keuzeprocessen en energieverbruik in Nederlandse huishoudens, 1920–1975.* Hilversum: Verloren.

Plastics Europe (2012), *Plastics – the Facts 2012. An analysis of European plastics production, demand and waste data for 2011. Plastics Europe.* Brussels: Plastics Europe

Raven, Rob; Schot, Johan; Berkhout, Frans, 2012: "Space and scale in socio-technical transitions", in: *Environmental Innovation and Societal Transitions,* 4: 63–78.

Ridder, M. de, S. de Jong, J. Polchar and S. Lingemann *Risks and Opportunities in the Global Phosphate Rock Market.* The Haguye: The Hague Centre for Strategic Studies (HCSS). Rapport No 17 | 12 | 12.

Rijk, T. de (1998). *Het elektrische huis: vormgeving en acceptatie van elektrische huishoudelijke apparaten in Nederland. Rotterdam:* Uitgeverij 010.

Rijkswaterstaat (2016). *Samenstelling van het huishoudelijk restafval, sorteeranalyses 2015; Gemiddelde driejaarlijkse samenstelling 2014.* Utrecht : Rijkswaterstaat.

Segers, R. (2014). *Rendementen en CO_2-emissie van elektriciteitsproductie in Nederland, update 2012.* Den Haag: CBS.

Smith, Adrian; Voß, Jan-Peter; Grin, John, 2010: "Innovation studies and sustainability transitions: the allure of adopting a broad perspective, and its challenges", in: *Research Policy,* 39: 435–448.

Späth, *Philipp;* Rohracher, Harald, 2010: "'Energy regions': The transformative power of regional discourses on socio-technical futures, in: *Research Policy,* 39, 4: 449–458.

Späth, Philipp; Rohracher, Harald, 2014: "Beyond Localism: The Spatial Scale and Scaling in Energy Transitions". in: Padt,

Frans J. G.; Opdam, Paul F. M.; Polman, Nico B. P.; Termeer, Kathrien J. A. M. (Eds.): *Scale-sensitive Governance of the Environment* (Oxford: John Wiley): 106–121.

Switzer, Andrew; Bertolini, Luca; Grin, John, 2013: "Transitions of Mobility Systems in Urban Regions: A Heuristic Framework", in: *J. Environmental Policy and Planning*, 15, 2: 141–160.

Switzer, Andrew, Luca Bertolini & John Grin (2015). 'Understanding transitions in the regional transport and land-use system: Munich 1945 – 2013', *Town Planning Review*, vol. 86 no. 6. p. 699–723.

Truffer, Bernhard, Coenen, Lars, 2012: "Environmental Innovation and Sustainability Transitions in Regional Studies", in: *Regional Studies*, 46, 1: 1–21.

van der Bie, R.J. & J.P. Smits (2001). *Tweehonderd jaar. statistiek in tijdreeksen. 1800–1999*. Den Haag/Groningen: CBS/ RU Groningen.

World Economic Forum, Ellen MacArthur Foundation and McKinsey & Company, (2016). *The New Plastics Economy — Rethinking the future of plastics* (http://www.ellenmacarthurfoundation.org/publications).

Wetenschappelijke Raad voor het Regeringsbeleid (WRR) (2014). *Naar een voedselbeleid* (rapport nr. 93). Den Haag/Amsterdam: WRR/Amsterdam University Press.

Zeijts, H. van, M.M. van Eerdt, W.J. Willems, G.A. Rood, A.C. den Boer, D.S. Nijdam (2010). *Op weg naar een duurzame veehouderij. Ontwikkelingen tussen 2000 en 2010*. Den Haag/ Bilthoven : Planbureau voor de Leefomgeving (PBL).

Chapter 24
Conceivable Strategies for Sustainable Well-being

John Grin

Contents

Abstract The final chapter of this volume explores how – at this moment in time – strategies can be developed in the Netherlands to achieve sustainable well-being. In addition to key problems like the non-circularity of supply chains and the vast increase in volumes of mass flows since 1960 (cf. Chap. 22) the Netherlands is also struggling with a 'welfare paradox,' i.e. high levels of welfare accompanied by widespread alienation and cynicism. The development of a circular economy and the achievement of sustainable levels of mass-flows are the key goals. Three strategies to achieve these goals are described: regeneration, restoration, and dematerialisation. For all four cases described in Chap. 23 it is shown how each of the three strategies contributes to more circularity and more manageable mass flows. In a final section it is argued that the transition toward a circular economy harbours promises about resolving both some patent defects of the market system as well as

© The Author(s) 2018   537
H. Lintsen et al., *Well-being, Sustainability and Social Development*,
https://doi.org/10.1007/978-3-319-76696-6_24

a potential resolution of the "welfare paradox" by aiming at an economy that is integrated into society and that fosters markets and social institutions that include, rather than exclude, ordinary citizens.

Keywords Circular economy · Mass flows · Welfare paradox · Regeneration · Restoration · De-materialisation · Markets · Grain · Meat · Gravel · Plastics · Electricity

24.1 Introduction

The previous two chapters have provided a backdrop against which to develop strategies for sustainable well-being: here and now, at the end of the second decade of the twenty-first century. Chapter 22 recapitulated how, by 1960, the problems of well-being that had plagued Dutch society from the mid-nineteenth century (poverty, poor housing, poor health) had largely been resolved – accompanied by a vast expansion of resource and energy flows and associated damage to natural capital in the Netherlands and elsewhere. It also showed how, ironically, 50 years later Dutch society sees itself confronted with a welfare paradox: while the country's level of welfare is spectacularly high, certainly by international standards, there is widespread public discontent and pessimism about the future.

In the previous chapter, we analysed three exemplary value chains (grain-for-meat; gravel-for-housing; plastics-for-everyday life) and showed how two key problems emerged from evolving practices of production and consumption and their structural and spatial embedding: the non-circularity of chains and the vast increase in the volume of mass flows since around 1960. A fourth case, electricity-for-households, explored increasing energy intensity, showing how domestic electricity consumption, the efficiency of electricity generation and primary energy use and emissions evolved over time.

In this chapter we will, first, proactively evaluate, for each of the cases, basic strategies to address these key problems and then, second, discuss the relationship between these strategies and society's potential for coping with the welfare paradox. In the remainder of this introductory section, we introduce the analytical framework for the first task: elaborating three basic strategies and assessing their plausibility. This will provide a basis for the work of the next section where we will answer, for each case, three questions:

1. How, for this case, could the three strategies be elaborated?
2. How plausible is each strategy, and what could be done to promote them?
3. How (much) do these strategies and associated changes contribute to resolving the welfare paradox?

We conclude the chapter with a reflection on the mix of routes Dutch society could take towards 2050, and the ways in which and conditions under which that might contribute to dealing with the welfare paradox described in Chap. 22.

24.1.1 Basic Strategies for Resource Use

Two of our three basic strategies are related to notions that, at the time of this writing, are comprised in the notion of a circular economy. If, as the Ellen MacArthur Foundation puts it "[a] *circular economy* seeks to rebuild capital, whether this is financial, manufactured, human, social or natural" in order to ensure "flows of goods and services," then a circular economy is obviously one potential solution for the basic challenge outlined in Chap. 22, of sustaining well-being, broadly defined. It particularly takes issue with the non-linear character of supply chains, due to which key resources are wasted. To be sure, as of now few resources are actually close to global depletion. Yet in the long run, wastage is not tenable and a few resources are already suffering local depletion; in even more cases around 2050 increasing scarcity may produce political-economic and geopolitical ramifications. These were in fact the motives for the Dutch government to declare in 2016, that the Dutch economy should be circular by 2050.[1]

What exactly *is* a circular economy? Its authoritative propagator, the Ellen MacArthur Foundation characterises it as a catch-phrase for a variety of approaches: Cradle to Cradle, Performance Economy, Biomimicry, Industrial Ecology, Natural Capitalism, Blue Economy and Regenerative Design.[2] More explicitly, a circular economy is '*an economy that is restorative and regenerative by design, and which aims to keep products, components and materials at their highest utility and value at all times, distinguishing between technical and biological cycles… In a true circular economy, consumption happens only in effective biocycles; elsewh. ere use replaces consumption. Resources are generated in the biocycle or recovered and restored in the technical cycle.*'[3] Thus, there are two broad modes of circularity: regeneration of resources (the 'biological cycle') and restoration of materials and products ('technical cycle'). Ideally, resource consumption is limited to biocycles that are fully regenerative, with solar (and hence wind) energy as the sole external inputs; in all other cases resources are used rather than consumed. The proper collection of (what in a linear case would be called) waste, so as to re-introduce it into the chain, is part and parcel of both modes. All this may be summarised in three main principles.[4]

1. Preserve and enhance natural capital by controlling finite stocks and balancing renewable resource flows.
2. Optimise resource yields by circulating products, components and materials at the highest utility at all times in both technical and biological cycles.

---

[1] Grondstoffenvoorzieningszekerheid, *Tweede Kamer* 32852 no. 33; Ministerie van Infrastructuur en Milieu (2016). Rijksbrede Programma Circulair Economie. «Nederland circulair in 2050»

[2] Ellen MacArthur Foundation. *Towards a Circular Economy: Economic and business rationale for an accelerated transition* (2013), pp. 26–27.

[3] https://www.ellenmacarthurfoundation.org/circular-economy/overview/ (consulted May, 2017).

[4] Ellen MacArthur Foundation. *Towards a Circular Economy: Business rationale for an accelerated transition.* (web publication, 2015), pp. 5–8.

3. Foster system effectiveness by revealing and designing out negative externalities, including damage to other systems (food production, forests) and externalities such as land use, and air, water and noise pollution.

In the Foundation's framework, one basic strategy is *regeneration*, pertaining to the
production of renewable materials in the biological cycle. While these materials
may be consumed, the natural capital needed for production (nutrients) may, in
principle, be largely regenerated in the biological cycle. *Restoration* strategies focus
on the technical cycle. Materials, parts or derivatives from products used are fed
back into the technical cycle. In operationalising restoration as a second basic strategy, it helps to distinguish among three modalities. We can distinguish among recycling (processing materials so as to retrieve the original functionality) down-cycling
(processing for lower functionality) and upcycling (processing for higher
functionality).

The third overall strategy is de-materialisation. Here we follow the approach
taken by Potting et al.[5] We thus avoid bracketing consumption practices or treating
them simply as different modes of appropriating products emerging 'from the supply side.' Rather, we consider consumption practices as relatively autonomous, both
as parts of the problem and as an element of potential solutions.[6] Iconic examples
are wearing a sweater to maintain bodily warmth at a lower room temperature and
building a larder in homes to replace electric refrigerators[7] or vegetarianism as a
way to make more efficient use of plant protein.[8]

Refuse	Make product superfluous, by discarding its function or by fulfilling that function with a radically different artefact
Rethink	Intensify product use (e,g, through sharing it, or through multifunctional use)
Reduce	Manufacture the product more efficiently, using less resources and materials; or design it for less resource- and material-intensive use

With regard to our fourth case, finally, the three basic strategies are more efficient
appliances, more limited electrification and renewable electricity production.

---

[5] Potting, J., M. Hekkert, E. Worrell & A. Hanemaaijer (2016): Circulaire economie: Innovatie
meten in de keten, Den Haag: Planbureau voor de Leefomgeving PBL.

[6] Elisabeth Shove, Frank Trentmann, Richard Wilk. *Time, consumption and everyday life. Practice,
materiality and Culture.* (Oxford, 2007); Elisabeth Shove, Matthew Watson, Martin Hand, Jack
Ingram. *The Design of Everyday Life.* (Oxford, 2009).; Elisabeth Shove, Gordon Walker.
'CAUTION! Transitions ahead: politics, practice, and sustainable transition management' in
Environment and Planning A, 39,4: 763–770.; Hui, A. Shove, E. and Schatzki, T., *The Nexus of
Practices: Connections, constellations, practitioners*, (London, 2016); Gert Spaargaren, Peter
Oosterveer 'Citizen-consumers as agents of change in globalizing modernity: the case of sustain-
able consumption'. *Sustainability* vol. 2 no. 7 (2010), p. 1887–1908.; Gert Spaargaren, Peter
Oosterveer, Anne Loeber (Eds.), *Food practices in transition: changing food consumption, retail
and production in the age of reflexive modernity*, (New York, 2012) p. 1–31.

[7] Elisabeth Shove, Frank Trentmann, Richard Wilk. *Time, consumption and everyday life. Practice,
materiality and Culture;* Hui, A. Shove, E. and Schatzki, T., *The Nexus of Practices: Connections,
constellations, practitioners.*

[8] Spaargaren et al. *Food practices in transition.* (New York, 2012).

24.1.2 Assessing Strategies

Absent a crystal ball, any pro-active analysis of the merits and limits of these strategies requires a theory of change. In Chap. 1, we outlined the claim of transition theory that innovations will normally follow the patterns implicit in a regime: a web of incumbent practices oriented to resolving dominant social problems aligned to a particular normative orientation, and the spatial and institutional structures in which they are embedded. If dominant (perceptions of) societal problems shift, there will be pressure to change practices and their embedding. Such changes may be induced by new problem definitions and novel, often experimental, practices that address these, but also by more or less exogenous 'landscape' trends pressing on the regime.

Thus, in discussing the possibilities and opportunities for the different strategies, we will focus on the historically emerged institutional and spatial embedding of contemporary supply chains, as well as on the long-term trends that are likely to shape the further evolution of the practices and structures of production, innovation, policy-making, and consumption that constitute them.

24.2 Grain-for-Meat

The grain-for-meat case is an occasion to explore a transition in a transnational, biotic chain. In fact, two nested biotic chains may be distinguished: meat production with grain serving as a resource; and grain production with water, nutrients and arable land as resources. Thus, in an ideal-typical circular chain, natural capital would be maintained through a closed nutrient cycle between the meat production and grain production chain, and through the regeneration of soil quality (first circular principle), while minimising externalities for other eco-systems (third principle). However, the present situation is rather remote from this ideal. First, grain production is hardly regenerative and it is rife with negative externalities. Second, due to the transnational nature of the chain, nutrients are exported from countries where grain is produced, only to be, third, largely wasted during meat production and consumption – yielding, fourth, local pollution (acidification of the soil; eutrophication of surface waters) and, finally, extracting significant flows of nutrients from global stocks. The latter is especially problematic for phosphates. While the degree of depletion of mineral phosphate resources is a contested issue, the political-economic and geo-political ramifications may still be significant.

The consequences of this non-circularity are exacerbated by the vast volume of flows. As we have seen (Chap. 23), non-circularity emerged historically after nineteenth century agricultural practices sought to resolve domestic problems of food quality, availability and affordability, as well as to cull benefits from international trade. These practices were structurally and spatially embedded in a robust incumbent chain that was characterized in Chap. 23 as an analytical-chemical web, interconnecting practices from different sectors on different continents through main ports and other trade infrastructures, as well as to supportive institutional arrangements for governance and innovation. In the Netherlands, more specifically, animal and crop production became structurally differentiated, with animal production taking the lead. These same factors promoted the vast expansion of the chain, which was oriented as much to export as to domestic consumption. After about 1960, meat consumption both in the Netherlands and in export countries also boomed due to increased welfare and the emergence of mass consumption.

We will now take one specific example, soy-for-pig-breeding, and assess possible strategies for making such a massive, transnational, non-circular chain more sustainable. In doing so, we will consider the nature of the incumbent regime, as well as address long-term trends that may (be mobilized to) change it: e.g. liberalisation of global food trade and increasing recognition of food (resource) security in a post-Cold War, post-colonial multi-polar world – the latter trend lending additional salience to two other crucial trends: climate change and water scarcity.

24.2.1   Towards More Regenerative Grain Production

Two basic strategies may be distinguished to make grain production more regenerative: more sustainable production of soy elsewhere, and increased production of cattle fodder in The Netherlands. In either case, restoring phosphates in the Netherlands is an essential complement.

Land use is a significant externality of soy-for-pigs production: the largest share of land dedicated to meat consumption in Europe is used for fodder production.[9] Soy production in Latin America frequently competes with food production for the domestic population and impacts on other ecosystems there. By 2050, *ceteris paribus*, a more numerous global population will generate even greater impacts.

Soy production impinges negatively on the *regeneration of soil and water quality*; much of current soy production is a mono-culture with adverse consequences for soil structure. These problems may be partly solved by more appropriate farming methods (more nature-inclusive, more diversified, more water recycling etc.). Inasmuch as there are many Latin American soy producers all competing to sell to a limited number of Dutch customers, some arrangement for transnational collaboration between chain actors is essential. An early initiative was the Round Table on Responsible Soy, a private standard setting organization (PSO). Subsequently, in the mid-2000s, the Sustainable Trade Initiative (STI) was established, a Netherlands-based private initiative that was ultimately co-sponsored by the Dutch government.[10] Such transformational physical and informational links among chain players and between private and public parties may facilitate demand for fairly priced sustainable soy and promote an appropriate incentive structure (like legal provisions) in the

---

[9] Baltussen, W.H.M., M.A. Dolman, R. Hoste, S.R.M. Janssens, J.W. Reijs, A.B. Smit,. *Grondstofefficiëntie in de zuivel-, varkensvlees-, aardappel- en suikerketen.* (Wageningen, 2016).

[10] Eric Hees. *Voedsel, grondstoffen en geopolitiek. Rapportage aan het Platform Landbouw, Innovatie & Samenleving.* (Culemborg, 2013), pp. 27–28.

source countries.[11] Such initiatives, together with transnational certification organizations, may also re-orient the incumbent chain, and improve coordination between different market arrangements and PSOs.[12]

In principle, this strategy is congruent with the chain's traditional institutional arrangement. However, the chain's orientation to reducing costs is at odds with attempts to achieve more ecologically and socially (a fair price for primary producers) sustainable soy production. Long-term developments in the WTO framework and Europe's changing position in the global economy, not to mention the market's increasing sensitivity to societal pressure, may all help.[13] Moreover, the specific political economy of the soy market harbours paradoxical opportunities.[14] While the Netherlands is the world's second soy importer (30% of EU-27) after China, the latter's share is more than 5 times that of the EU-27 countries put together. Also, China is a major soybean crusher. Since 2010, changing food lifestyles and water scarcity have increased Chinese imports of soy beans. This (increasingly) dominant position in a market with only a few source countries (US, Brazil and Argentina account for 75% of the production) enables China to 'play' with the prices, as part of its resource security strategy.[15] Paradoxically, should European countries be prepared to pay more for sustainably produced soy, they may come to exploit this as an opportunity to include soy in a resource security strategy, as argued for by e.g. De Ridder and Hees.[16]

To be successful, such policies will also have to keep restoring nutrients high on the agenda of players outside the agri-food sector, like the Port of Amsterdam that in 2010 announced its aspiration to become a circular economy hub. While such efforts may hasten the coming of greater circularity, it is quite possible that much of the impetus toward a circular economy will come primarily from established actors in the waste and energy sectors, for whom global food issues may be less central than other problem dimensions.[17] In that case, while manure and residual soy flows may be used more (efficiently) for bio-based production, restoring nutrients may not be high on the agenda. Thus, to facilitate phosphate recovery (an objective in an emerging Dutch policy for a 50% circular economy by 2050), the national government and logistic main ports must ensure an integral approach, properly embedded in a broad societal, economic and policy context, and involving a wide range of

---

[11] https://www.idhsustainabletrade.com/sectors/soy/ Consulted May, 2017.

[12] Fransen, Luc. The Politics of Meta-Governance in Transnational Private Sustainability Governance' in *Policy Sciences*, Volume 48 (2015), No. 2, June 2015.

[13] John Grin, *Understanding Transitions from a Governance Perspective*, (Part III, 2010) in: Grin, John; Rotmans, Jan; Schot, Johan,*Transitions to Sustainable Development. New Directions in the Study of Long Term Structural Change* (New York, 2010: Routledge), pp. 223–314.

[14] Eric Hees, *Voedsel, grondstoffen en geopolitiek. Rapportage aan het Platform Landbouw, Innovatie & Samenleving.* (Culemborg, 2013), pp. 23–28.

[15] Marjolein de Ridder, Sijbren de Jong, Joshua Polchar and Stephanie Lingemann *Risks and Opportunities in the Global Phosphate Rock Market.* (The Hague, 2011).

[16] De Ridder, Marjolein et al. (2011). *Op weg naar een Grondstoffenstrategie. Quick scan ten behoeve van de grondstoffennotitie.* The Hague: The Hague Centre for Strategic Studies (HCSS); Hees, Eric (2013). *Voedsel, grondstoffen en geopolitiek.* Rapportage aan het Platform Landbouw, Innovatie & Samenleving. Culemborg: CLM.

[17] Rob Weterings, Elsbeth Roelofs, Roald Suurs, Frans van der Zee. *Tussen gouden bergen en groene business. Systeemverkenning van een biobased economy.* (Den Haag, 2011), pp. 1.

actors and stakeholders dedicated to shaping a circular economy.[18] Even then, full nutrient recovery remains a remote objective.

24.2.2 Regenerating Nutrients Regionally: Replacing Grain from Elsewhere by Dutch Inputs

Alternatively, the Netherlands may reduce pressure on natural capital elsewhere by replacing a portion of the imported soy in animal fodder by domestically cultivated soy or other foodstuffs and use that for regeneration by closing the cycle on a regional level. The main challenge to replacing soy by other foodstuffs is to find fodder material with sufficient proteins – particularly with sufficient so-called essential amino acids – and preferably less phosphates. Substances that fit the bill are fish meal and bone meal (but these are sensitive, and partly prohibited, after the BSE pandemics), peas, waste flows from diverse plant oils, protein from potatoes, milk and traditional crops like rapeseed and lupine. Economically, a mix would be most optimal.[19]

Since the mid-2010s, opportunities for such regional fodder production have been explored. The smaller scale of production and reduced opportunities for cascading with human food production (which is easy with soy and palm oil), may incur 5–10% price increases for fodder.[20] If retailers do not apply a multiplier effect on these increases, as has been the practice in similar cases, this price-hike need not be prohibitive.[21] Different strategies to mitigate these effects against the backdrop of the future global soy market (see above) include: better information flows between primary producers and actors further downstream in the chain,[22] different retail strategies[23] or alternative, more local chains.[24]

That said, regional fodder production would require a shift from the current orientation to discrete products toward more circular (collaborative) business propositions, where the costs and productivity of different products hang together.

---

[18] Rijksoverheid *Nederland circulair in 2050*. Rijksbreed programma Circulaire Economie. (Den Haag: 2016).

[19] Harry Vahl. *Alternatieven voor Zuid-Amerikaanse soja in Veevoer.* (Utrecht: 2009).

[20] Harry Vahl, *Alternatieven* (2009); Raad voor Regionaal Veevoer. *Naar 100% regionaal eiwit. Kansen en knelpunten voor eitwitrijke veevoergrondstoffen.*(Amsterdam 2016).

[21] Harry Vahl, *Alternatieven* (2009); Baltussen W.H.M. et al. *Prijsvorming van voedsel; Ontwikkelingen van prijzen in acht Nederlandse ketens van versproducten.* (Wageningen, 2014).

[22] Baltussen W.H.M et al. *Prijsvorming.*

[23] Oosterveer, Peter, 'Restructuring Food Supply: Sustainability and Supermarkets.', in Spaargaren et al. *Food practices in transition: changing food consumption, retail and production in the age of reflexive modernity.* (New York, 2012). pp. 153–176.

[24] Roep, Dirk & Han Wiskerk,'Reshaping the Foodscape: The Role of Alternative Food Networks.' 8) in Spaargaren et al. *Food practices in transition: changing food consumption, retail and production in the age of reflexive modernity.* (New York, 2012). pp. 207–22.

Interestingly, the Dutch could mobilise their own historical experience by establishing novel co-operatives of livestock and crop farms, possibly also involving non-agricultural enterprises.

But there are some fundamental problems. First, the primary sector (and associated knowledge institutes) suffer from a lack of knowledge about protein-rich cultivars like lupine, rapeseed and clover, about embedding agriculture in surrounding nature and about the use of animal manure rather than artificial fertilizers. This is rooted in the historical, institutionally and spatially embedded, shift from mixed farming to a husbandry- centered agriculture based on imports of the bulk of the animal fodder. Another fundamental problem is that, inasmuch as the livestock sector has historically far outgrown the capacity for producing animal fodder, a maximum use of regional fodder would require all the crop growing land available.[25] This strategy therefore could not do without some form of reduction of meat production.

A more recently explored possibility is the use of advanced protein sources like seaweed, algae and duck brood. Significantly, these are often phosphate rich (the latter is even used to retrieve phosphates from surface water). Their prices may be reduced thanks to cascading fodder production, bio-energy production and phosphate recovery.[26] For these advanced foodstuffs as well as the more traditional alternatives, pertinent knowledge gaps would need to be addressed while options for cascading need further exploration and development. Over time, pressures from the source countries (which will increasingly have to face the consequences of climate change for their food production and ecosystems) may stimulate such innovations, as may growing awareness about increasing resource scarcity and climate change. In fact, the first set of considerations are included in the Dutch government's support of the Sustainable Trade Initiative, and the second in its 2011 white paper ('Grondstoffennotitie') on resources policies.

24.2.3 Regenerating Phosphates

These pressures may also stimulate the development of a key complement to both strategies just discussed: recovering the phosphates wasted in meat production and consumption, through the food cycle of animals and humans, respectively. While directly closing the phosphate cycle with source countries does not seem viable, a more realistic strategy might be to feed retrieved nutrients back into fertilizer and soil improvement markets. There are basically two options. The first is to retrieve phosphates from surface water (e.g. through duck brood, see above), from waste water (e.g. through bonding P to magnesium to produce struvite; or, more recently, to iron) or from deep soil (by deeply-rooted plants). As of the mid-2010s, interesting

---

[25] Raad voor Regionaal Veevoer. *Naar 100% regionaal eiwit.* (2016).

[26] Nieuwenhuis, R. & L. Maring. *Naar nieuwe ketens voor het benutten van eendenkroos.*, (Utrecht: 2009).

novel opportunities have emerged like the 'resources factory', i.e. a wastewater treatment facility whose core business is retrieving and marketing resources from waste water rather than discarding them as waste. In the decades up to the 2050s, significant novelties may be expected, and there will especially be room for attractive, circular / cascaded business models.[27]

The second option for regeneration is collecting and re-cycling manure, or at least the nutrients it contains. As of 2016, half of the manure was processed at the farm, often by burning it (sometimes retaining the phosphate-containing ash for use as a soil improver). The rest, including a portion of the first half (sold domestically or exported) was partly used for crop production and partly for the production of bio-energy, generally without nutrient recovery. One urgent goal for regeneration is to develop cascaded forms of bio-energy production, that no longer waste nutrients. A second one is a governmental intervention in the nutrient market to make manure more competitive. The legal obligation to dispose of manure in a proper (and hence costly) manner, the historical path leading to (Sect. 23.3) the widespread availability of artificial fertilisers, as well as the ever more restrictive legal limits on fertiliser use in crop production depress manure prices. Thus by 2016, responsible manure disposal had become a serious burden for the animal sector (about € 250 million per year for the sector), while it saves the crop production sector a similar amount.[28]

24.2.4 Dematerialisation

Dematerialisation might proceed along two tracks. The first is reducing the phosphate content of animal fodder. In the early 2010s, the market undertook voluntary initiatives in this direction, partly out of shared concern but also to avoid government limits being imposed on the volume of livestock. In 2017, PBL estimated that pursuing this option might ultimately reduce phosphate emissions by 10%.[29] The other track is to reduce meat production, tackling both of the above problems by reducing the volume of the mass flows. To be sure, some reduction of meat production in the Netherlands is conceivable, if not likely, as a response to environmental problems and the risk of pandemics. For instance, in 2016, the province of Noord Brabant (home to 38% of Dutch pigs) decided to reduce the number of pig farms by about 30%, retaining only those prepared to adopt sustainable production methods.

The more radical approach is refuse: reducing meat consumption. This is at odds with deeply embedded (e.g. in nutritional education and advertising) cultural notions that meat is important for human health. This idea is rooted in earlier

[27] Bocken, Nancy M. P., Ingrid de Pauw, Conny Bakker & Bram van der Grinten (2016), 'Product design and business model strategies for a circular economy' in *Journal of Industrial and Production Engineering*. 33 (5) published online.

[28] Planbureau voor de Leefomgeving. *Evaluatie Meststoffenwet 2016-Syntheserapport*, (Den Haag, 2017), p. 14, p. 27; chapters 2, 9.

[29] Planbureau voor de Leefomgeving (2017). Evaluatie Meststoffenwet 2016-Syntheserapport, Den Haag: Planbureau voor de Leefomgeving PBL, pp. 146–151.

times, when it had some justification (although vegetarianism was always an alternative practice). Until 1960, diets were often unhealthily low on proteins. But since 2005, vegetarianism has become more popular in both domestic and European export markets. While by 2015 the number of strict vegetarians did not exceed 4% of the population, a middle-course known as 'flexitarianism' had become much more popular. In 2010 and 2015, respectively, some 43% and 55% of the population ate meat not more than 4 days a week. Motives for vegetarianism and flexitarianism vary from concerns about one's own health (antibiotics, cholesterol) to animal welfare and environmental concerns. Also, personal lifestyles are increasingly shaping food choices.[30] In this connection, meat consumption tends to be part of urban lifestyles.[31]

These developments are an instructive instance of how the shift in welfare dynamics beyond meeting primary needs may both produce and help resolve sustainability problems. Thus the future of meat consumption in northern countries like the Netherlands will be shaped by the degree to which societal concerns remain in place (partly dependent on the evolution of meat production) as well as by trade-offs among different cultural changes.

24.3   Gravel-for-Construction

As discussed in Sect. 23.4, the main problem associated with gravel use is the damage caused to nature and ecosystems by its excavation. Wartime excepted, the chain was virtually completely linear: waste was hardly ever re-used. Thus, the balance between imported gravel and domestic gravel excavation has fluctuated with the tensions between national gravel demand on the one hand, and domestic gravel excavation and processing capacity on the other.

Historically, *gravel demand* was shaped largely by the volume of construction, that in turn has been shaped by population growth and the increasing size of homes, initially to improve a primary life need (hygienic conditions) and later, especially after 1960, in response to changing lifestyles: individualisation yielded smaller households and more rooms per household. Counter-tendencies included alternative

---

[30] De Bakker, Erik en Hans Dagevos. *Vleesminnaars, vleesminderaars en vleesmijders. Duurzame eiwitconsumptie in* een *carnivore eetcultuur.* (Den Haag, 2010); Klintman, Mikael; Boström, Magnus (2012). 'Political Consumerism and the Transition Towards a More Sustainable Food Regime Looking Behind and Beyond the Organic Shelf' in: Spaargaren et al. *Food Practices in Transition: Changing Food Consumption, Retail and Production in the Age of Reflexive Modernity.* (New York, 2012), pp.114–116.; Keuchenius, Cecilia en Bram van der Lelij. *Motivaction Quickscan 2015: eetpatronen van verschillende sociale milieus, duurzaamheid en voedselverspilling.* (Den Haag, 2015), pp. 134–148.

[31] Murray, Shannon, Saman Brock & Karen S Seto. 'Urbanization, food consumption and the environment.' in Karen C. Seto et al. *The Routledge Handbook of Urbanization and Global Environmental Change.* (New York: 2015) pp. 27–40.

construction materials – especially plastics – (since 1970), and the completion of the hydraulic infrastructural 'Delta works' in 1986.

Domestic *excavation* has historically been shaped by political controversies in response to its local side-effects, as well as by strong governmental interventions. The latter was initially a response to scarcity during WW I, after 1980 to protests against landscape damage and somewhat later to a paradigmatically changed waste policy, with more emphasis on re-cycling. In spite of the latter, since 1980 the Netherlands has largely exported its gravel-related landscape damage – retaining, however, a significant domestic institutional capacity for concrete production and innovation (Chaps. 14 and 22).

The chain thus exemplifies cases in which the Netherlands exports its problems, yet retains adequate autonomous potential to make the chain more circular and less voluminous through domestic measures. These strategies will be contextualized for the present gravel regime, and in a long-term landscape. First and foremost, it is expected that population will more or less stabilise towards 2050. By 2010 natural population growth had sunk below 2 per 1000, and it is expected to become negative by 2038 for the population currently living in the Netherlands. After that date, the migration balance and annual population decline are expected to stabilise the population. Different estimates of actual migration, birth rates (life choices, developments in well-being) and death rates (developments in health care) yield a range between 16 and 20 million.[32] Two other trends, ageing (associated with declining birth and death rates) and individualisation, will yield some decline in the number of people per household.[33]

All in all, the total number of households is thus expected to increase from 7,3 million (2010), through about 8,5 million (2030) up to around 8,6 million after 2040. The demand for housing construction will decline from the 50,000 dwellings per year around which it will fluctuate until about 2025, down to 20,000 after 2040, slightly more than the demolition rate (15.000 per year after 2016).[34]

Finally, partly due to the emergence of the information society (miniaturisation and the rise of flex-working) and partly due to market failures in the real estate business, an increasing stock of unused office and industrial buildings has come into being. The associated devaluation of real estate and land that occurred by the mid-2010s adversely affected local governmental budgets and this is likely to affect future construction and land-use plans.[35]

---

[32] Stoeldraijer, Lenny, Coen van Duijn, Corina Huisman. *Kernprognose 2016–2060: 18 miljoen inwoners in 2034 voorzien.* (Den Haag: 2016).

[33] Hofstra, U, B. van Bree, J. Neele, R. de Wildt. *cenariostudie BSA-granulaten*: aanbod en afzet van 2005 tot 2025. (Delft, 2006), 29; 35; 39.

[34] Poulus, Co, Gerard van Leeuwen, David Omtzigt, Kenneth Gopal, Ruud Steijvers, Marnix Koopman. *Tussenrapportage. Prognose bevolking-, huishoudens- en woningbehoefte 2015–2050.* (Delft: 2016), pp. 21–23.

[35] Janssen-Jansen, L. & Lloyd, G. Property booms and bubbles: a demolition strategy – towards a tabula. rasa? *Journal of Surveying, Construction and Property*, 3 (2012) 2, p. 1–11; Janssen-Jansen, L., Lloyd, G., Peel, D., & van der Krabben, E. (2012). *Planning in an environment without growth: invited essay for the Raad voor de leefomgeving en infrastructuur(Rli), the Netherlands.* (The Hague: 2012).

24.3.1 Restoration: From Down-Cycling to Re-cycling

By the mid-2010s, two trends in restoration could be distinguished: a shift from down-cycling demolition waste to more valuable forms of restoration, and a shift towards designing for re-use.

Crushing concrete into concrete granulates to be used in foundations for road and hydraulic construction has long been a rather common practice, a paragon of down-cycling. Since the 2010s, recycling has been on the rise, i.e. using these granulates to replace gravel in concrete production. Established firms like Mebin now advertise with concrete in which e.g. 30, 50 or even 100% of the gravel has been replaced by granulates.[36] Simultaneously, newcomers like Paro (established in 2010 in the Port of Amsterdam) collect flows of construction waste brought to the Port by ships and sorts them, using innovative processes, into different fractions.[37] While some are down-cycled, others are deemed fit to be sold to other firms (e.g. the neighbouring *Voorbij Prefab*), that specialise in re-cycled concrete materials).[38]

These early adopters reveal that although, historically, practices in the gravel-for-construction value chain changed mostly in response to governmental pressures, by the early twenty-first century, other actors and forces have come into place. Around 2010, the Dutch concrete branch, collaborating with societal stakeholders, launched the Beton Bewust initiative that seeks to promote 100% recycling and that co-founded, in 2014, the global Concrete Sustainability Council.[39] By the mid-2010s, other actors were also bringing their influence to bear on the sector, for example bank real-estate account managers, claiming that "whoever did not yet have a sustainable business model, was actually out of the game. Circularity is the next step. The task of the bank is to promote awareness and create funding opportunities."[40] Given the historical record, however, unless re-cycling becomes clearly economically superior, some (national or EU) governmental intervention will probably be indispensable to move the sector as a whole into the era of sustainable production.

The early efforts just outlined have important technical limits. Not all waste can be recycled as yet, and not all concrete products can presently be made from 100% recycled granulates. The sector's current institutional arrangement, with close ties among concrete manufacturers, R&D infrastructure and government may nurture the necessary innovation. Also, newcomers like Paro may draw on the Dutch tradition of the *trafiek* manufacturing in main-ports. As such, it is part of a wider strategy of the Port of Amsterdam to become a hub in the global circular economy. A poten-

---

[36] ENCI, Mebin, Sagrex. Sustainability Update. *Het duurzaamheidsbeleid van de bedrijven van HeidelbergCement Benelux* (2016); www.mebin.nl (consulted June, 2017).

[37] www.paro-bv.nl (consulted June, 2017).

[38] www.voorbijprefab.nl (consulted June, 2017).

[39] www.concretesustainabilitycouncil.org/ and www.betonbewust-csc.nl/beton-verduurzamen (consulted June, 2017).

[40] Rudolf Scholten of ABN AMRO Bank, (April 2017). www.insights.abnamro.nl/2017/04/circulair-bouwen-heeft-de-toekomst/

tial risk is that this may lead to the lock-in of a sub-optimal solution: less than 100%
recycling coupled to long-distance transportation.[41]

24.3.2   Designing and Building for Re-cycling

The circularity promoted by the bank manager cited above includes designing for
re-use. "A building thus becomes a collection of materials with an eternal life," he
says – a vision summarised by his colleague as "Lego-lisation." This entails a major,
systemic innovation: a different architecture, which 'designs for deconstruction',
takes available materials as a point of departure and adopts accordingly different
aesthetics; novel business models, based on including materials and parts in a cycle
(e.g. through conferring a 'passport' with their biographies); involvement of suppliers of materials and parts in the design process art the outset; and even different
property models, in which e.g. materials and parts remain the property of their supplier, who is seen to provide a service to the owner of building.

An early adopter of such a more radical strategy is *Beton Prefab Veghel*, an innovative family firm established in 1967 as a prefab concrete floors producer. Since
2012 the firm has sought not only to maximize re-cycling in all products, but also to
design their products for re-use after 50–100 years, when their initial application is
expected to have reached the end of its lifetime. A concept for even more comprehensive steps towards re-use, *'Lekker Eigen Huis'* ('Delightful Own Home') have
been proposed by a novel consortium, TBI, a flexible network of a developer and
several construction companies (some relatively new enterprises, as well as a pre-
war family firm that transformed itself). The core asset is a smart design system that
supports a client in designing a home tailored to the clients own authentic desires,
meeting high sustainability criteria and re-using materials and parts that are available at the time.[42]

These examples show both a-priori feasibility and the need for a very significant
R&D effort, not only aimed at better recycling options, but also at novel design
methods for both materials and final construction products. While such options may
draw in part on the sector's incumbent regime, structural change may well be needed
for other aspects: new relations between customers, designers and constructors;
novelties like a material passport; and new financing schemes that adequately incorporate the value of the materials. The cultural landscape trends of increasing autonomy and individual distinctiveness, which for instance also express themselves in
the growth of DIY construction in the Netherlands, may help to overcome these
obstacles.

---

[41] Port of Amsterdam (2014). Visie 2030. Port of Amsterdam, Port of partnerships. Amsterdam:
Port of Amsterdam; Port of Amsterdam (2017). Koers naar de Amsterdam Metropolitan Port.
Strategisch plan Havenbedrijf Amsterdam, 2017 — 2021. Amsterdam: Port of Amsterdam.

[42] www.lekkereigenhuis.nl/. (consulted June 2017).

24.3.3   Dematerialisation

Dematerialisation may take four basic forms. First, a 'reduce strategy': e.g. using less gravel for the same construction performance by employing more brick (from river clay) or laminated wood and polymer materials.[43] Interestingly, the latter two may be produced through a biotic rather than a technical cycle, implying a shift from restoration to regeneration of resources (but demanding additional sustainable production of raw materials in a more bio-based economy). Second, 'reduce'may also result from a longer concrete lifetime (now often shorter than the technical lifetime of the constructions in which it is applied). Emerging options like self-healing concrete (e.g. through calcium phosphate producing bacteria) are estimated to enhance lifetime by 20%.[44]

More fundamentally, given that (see above), around 2040 home demolition and construction rates are likely to be close to each other, while business buildings will increasingly be vacated, not only 'reduce' in the sense of construction with zero gravel use becomes conceivable, but also 'refuse' in the sense of avoiding construction altogether through using superfluous business buildings for novel purposes. Janssen-Jansen propose 'planning without growth,' by adopting a more holistic view of land and property development.[45] By the mid-2010s, re-developing old industrial areas into shared spaces for communal leisure, housing, social functions and creative industry was rapidly becoming popular, and ideas for converting e.g. superfluous office space into small apartments were spreading. Janssen-Jansen argues that these options are promising in many respects, but only if property and power relations can change to suit. Smart, reflexive planning will be needed in order to identify and deal with such difficulties, drawing on the long term trends mentioned above as both the rationale for changing and as a force for change on the incumbent construction and planning regime.[46]

A final strategy for refuse is 'Design/Build for Change' (VITO, Flanders): homes and offices grow and shrink with the biography of their use, i.e. through modular design.

---

[43] Van Dam, Jan & Martien den Oever. *Catalogus Biobased bouwmaterialen. Het groen bouwen.* (Wageningen: 2012).

[44] Van Lieshout, Marit *Update Prioritering handelingsperspectieven verduurzaming betonketen 2015*, (Delft: 2015).

[45] Janssen-Jansen, L., et al. *Planning in an environment without growth* (2012).

[46] Lissandrello, Enza & John Grin (2011). 'Reflexive planning as design and work: lessons from the Port of Amsterdam,' *Planning Theory and Practice*, 12 (2), p. 223–248.

24.4   Plastics for Everyday Life

The plastics-for-everyday-life chain exemplifies the mass consumption associated with contemporary lifestyles, strong non-linearity and the global environmental consequences of local, wasteful use. The problems associated with plastics use and production are (section 23.4) that it is an iconic example of wasteful use of resources (yielding problems for 'later'); and that the waste generated threatens organisms and ecosystems as well as human health (both locally and 'elsewhere', and given the accumulative nature, especially 'later'). These problems are seriously aggravated due to the very massive nature of the chain, its non-circularity from the outset, and its semi-open-ended nature (80% of plastics produced in the Netherlands is exported).

These properties emerged in response to cultural inclinations, and structural and spatial conditions that still shape consumption and production to this day. From its inception around 1945, the use of plastics in everyday life, including the use of disposable packaging, was seen as progress through artificial materials: diffusing convenience, comfort and aesthetics to all classes. It thus exemplifies a chain associated with the mass-consumption, lifestyle-focused well-being that emerged after 1960. The same cultural trends shaped the supply side, so that supply was well-matched to consumer preferences. Supply practices were able to draw, and continue to do so, on spatiality (mainports with major fossil flows) and a helpful platform-like institutional rectangle, tying together polymer producers, polymer processors, and government sponsored knowledge institutes (typical for the era, civil society hardly played a role). Also, value creation around mainports is rooted in the Dutch trafieken tradition. Finally, the chain's non-circularity was reinforced over time by the institutional logic of the massive scale of polymer production, and the institutional logic of the enormous scope of polymer processing.

Future developments will be co-shaped by these spatial, cultural and structural forces, as well as by various long-term trends. First, the emergence of tailor-made production, symbolized by the swift emergence of 3D printing, may both increase plastics' popularity and generate new opportunities for polymer re-use in recycling.[47] Increasing world population and growing welfare in developing countries will – *ceteris paribus* – yield a growing demand for plastics. China, for example, tripled its per capita consumption between 1980 and 2010.[48] Simultaneously, the European plastics industry is under threat as China is increasing its plastic production, and India the volume of its plastics recovery and processing, while oil producing countries are seeking downstream integration and Brazil and other Latin American countries are increasing bio-based plastics production.[49] Other long terms

---

[47] Ford, Simon Mélanie Despeisse. Additive manufacturing and sustainability: an exploratory study of the advantages and challenges, *Journal of Cleaner Production*, Vol. 137, (2016) p. 1573–1587.

[48] The European House Ambrosetti. *Excellence of the Plastics Supply Chain in Relaunching Manufacturing in Italy and Europe.* (Milano, 2013), pp. 39–40.

[49] The European House Ambrosetti. *Excellence of the Plastics Supply Chain* (2013), pp. 94–96.

trends that may (be made to) affect the plastics regime are the increasing volatility and geopolitical vulnerability of fossil resources[50] the rapid expansion of use and possession of electronic and electric appliances in daily life (Sects. 23.5 and 23.6); and the politicisation of the side- effects of plastics use, which has become a concern of the plastics industry.

24.4.1 Restoration

By the mid-2010s, a range of efforts had been mobilised to make plastic chains more circular.[51] In the Netherlands, plastics collection, the most basic form, embarked upon a significant volume expansion in the mid-2010s: from 25 kton (3%) in 2009 to 162,2 (20%) kton in 2014 and 250 (30%) kton in 2016.[52] The Plastic Heroes waste collection initiative undertaken by the packaging sector in collaboration with municipalities, yielded volumes of waste plastics and triggered increasing processing capacity. While, in the mid-2010s, a significant share is still being down-cycled (making carpets from PET bottles) polymers are increasingly being broken down into monomers (mechanical recycling, already occurring with e.g. PET), or more basic molecules (chemical recycling) – but in what are as yet energy-intensive processes.[53]

On the positive side, the platform-like institutional arrangement between plastics producers and processors offers optimal conditions for connecting flows of collected waste to plastics processing enterprises. As of the mid-2010s, however, economic viability was still problematic. While a reasonable price was usually paid for PET, PE and PP, the price for other less easily processable plastic waste flows (mixed plastics, foils) tends to become negative as soon as oil prices make virgin feedstock cheaper than recyclates. Due to subsidies for domestic waste this is mostly a problem for industrial plastic waste. Thus WEF stipulate that a much better developed after-use plastics economy is crucial.[54] Ongoing innovations that make collection, cleaning and separation easier and less costly, may help here. Food safety regulations, on the other hand may yield another kind of inertia. For most types of recycled plastics there is, as of 2017, no regulation for plastic food packaging in

---

[50] World Economic Forum (WEF), *Rethinking the future of plastics,* (2016). p. 36; PlasticsEurope and European Plastics Converters Association (EuPC). *The plastics industry: a strategic partner for economic recovery and economic growth in Europe. Manifesto on the competitiveness of the plastics industry.* (Brussels: 2014), p, 11.

[51] World Economic Forum (WEF), *Rethinking the future of plastics,* (2016).

[52] KIDV & SNM *Factcheck Plastic Recycling.* (Den Haag: 2016) NRK, *Kunststof Recycling in Nederland,* (Den Haag: 2016).

[53] Rahimi, Aliurezza Jeannette M. Garcia.Chemical recycling of waste plastics for new materials production. *Nature Reviews Chemistry* 1 (2017), Article nr 0046 (online journal) doi:https://doi.org/10.1038/s41570-017-0046

[54] World Economic Forum (WEF), *Rethinking the future of plastics,* (2016), p. 33.

place (which may contain impurities).[55] The platform-like plastics institutional rectangle may help overcome these barriers, fostering the conditions listed by WEF i.e. maintaining a "cross-value chain dialogue mechanism", "matchmaking mechanisms" for recycled materials, the redesign and convergence of materials, formats, and finally, after-use systems and policy measures.[56]

A higher form of recycling is re-use of plastics for the same function. More than 20% of the total mass of plastic products could be re-used, especially short-lived plastics products, most notably packaging.[57] In the UK, a small charge for plastic bags has led to an 85% reduction of single use. Interestingly, first experiences in the Netherlands suggest that this may also make consumers bring durable shopping bags from home: a form of de-materialization.

Finally, some 30 mass% of the plastics in use as of the mid-2010s cannot be re-used or recycled absent fundamental redesign or innovation: they are too small (e.g. sweet wraps), multi-material, use uncommon materials (like pill strips) or are contaminated (coffee capsules; waste bags). While ongoing innovations remain to be implemented, dematerialization remains a competitive strategy.

24.4.2 Regeneration

A more recent and innovative kind of recycling is the recovery of plastics from waste water. In an early example, the PHARIO demonstration project, biopolymers from the polyhydroxyalkanoate (PHA) family were produced by mixed microbial cultures from surplus activated sludge, fed with volatile fatty-acid-rich liquors from a local candy industry.[58] The process is part of a cascade with biological nitrogen removal and chemical phosphorus removal. Significantly, and contrary to most biodegradable polymers, PHA polymers degrade relatively swiftly in water – reducing the plastic soup risk. The project is part of the so-called 'resources factory' concept: a waste-water-treatment-plant-turned-resource-recoverer, demonstrating that consistently high quality production is possible in an economically viable way. There are significant energy and greenhouse gas emissions benefits compared with traditional water treatment and with producing PHA bioplastics from monocultures (soy, grain, sugar cane or beet).

Residual biotic flows involving potatoes, sugar beets, grass and the like are also objects of experimentation and debate.[59] Responding to fears that using such crops

[55] KIDV & SNM, *Factcheck* (2016); Van den Oever, Martien, Karin Molenveld, Maarten van der Zee, Harriëtte Bos. *Bio-based and biodegradable plastics – Facts and Figures.* (Wageningen: 2017); Rahimi and Garcia, 2017. Chemical recycling of waste plastics for new materials production.

[56] World Economic Forum (WEF), *Rethinking the future of plastics,*(2016), p. 13.

[57] World Economic Forum (WEF), *Rethinking the future of plastics,* (2016).

[58] Bengtsson, Simon, Alan Werker, Cindy Visser, Leon Korving. *PHARIO: stepping stone to a sustainable value chain for PHA bioplastic using municipal activated sludge.* (Amersfoort: 2017).

[59] Asveld L., R. van Est & D. Stemerding. *Naar de kern van de bio-economie: De duurzame beloftes van biomassa in perspectief.* (Den Haag: 2011).

for biomass (and biofuel) production will compete with agriculture and ecosystems for land and lead to soil deterioration, sophisticated schemes have been proposed and explored that optimise the cascaded use of e.g. maize to regenerate phosphates and to extract proteins for animal fodder and carbohydrates for bio-based materials.[60] More recently, the breeding of algae and seaweed as an additional source is being explored by major companies.

Even more radical is the idea of producing bioplastics directly from carbon dioxide through artificial photosynthesis.[61] CO_2 and hydrogen may be made to react to produce methanol, which can be used as a building block for larger organic molecules. A more innovative option is bringing the CO_2 molecule as a whole into a carbon-hydrate skeleton to produce novel polymers. One promising example is polypropene carbonate, which contains 43% CO_2 and might be made from methanol and CO_2: a 100% bio-based polymer. Also higher carbohydrate-based polymers (with an even higher CO_2 content) may be made in a bio-based mode. While, by the mid-2010s interesting progress had been made, artificial synthesis remained a formidable challenge and far from trivial.[62] Also, while it may certainly capture a portion of CO_2 emissions, the overall impact on the concentration of greenhouse gases is limited, and it must not be forgotten that given the inert nature of CO_2 these are very energy-intensive processes.[63]

Overall, the use of bio-based plastics is still limited: in 2015 it was less than 1% of all plastics, though it is expected to rise to 2,5% by 2020.[64] Bioplastics tend to be more expensive than traditional ones. An analyst like Nossin is therefore pessimistic about the future role of bioplastics, although he stresses that, technically, 90% of the existing plastics could be replaced by biopolymers. At the bottom of the pyramid, biobased bulk polymers like polyethylene lose the price battle.[65] That may be different for some commodity polymers, like PET, used in bottles. For high performance polymers and specialties, price is less important than functionality, though bioplastics may fall short on the latter count as well. Others, like Harmsen & Hackmann are much less pessimistic, pointing out consumer demand, governmental pressures and the increasing scarcity of oil derivates.[66]

Long terms trends discussed in the introduction to this section – such as the increasing volatility and geopolitical vulnerability of fossil resources and the politi-

---

[60] van der Hoeven, Diederik and Paul Reinshagen. *Biomaterialen, drijfveer voor een groene economie. Strategie voor een groene samenleving.* (Den Haag: 2013).

[61] Van der Hoeven and Reinshagen. *Biomaterialen.* (2013).

[62] Purchase, R.. H. de Vriend, H. de Groot, *Kunstmatige fotosynthese. Voor de omzetting van zonlicht naar brandstof* (Leiden: 2015).

[63] Van der Hoeven and Reinshagen. *Biomaterialen.* (2013).

[64] Van den Oever, Martien, Karin Molenveld, Maarten van der Zee, Harriëtte Bos. *Bio-based and biodegradable plastics - Facts and Figures.* (Wageningen: 2017), pp. 11.

[65] Nossin, P. (2012) *Biopolymeren in breder perspectief. Nut en noodzaak*, p.21; cited by Lintsen H., Hollestelle M., Hölsgens R. *The plastics revolution. How the Netherlands became a global player in plastics.* (Eindhoven: 2017).

[66] Harmsen. P. en M. Hackmann, *Groene bouwstenen voor biobased plastics. Biobased routes en marktontwikkeling.* (Wageningen: 2012): cited by Lintsen et al. *The plastics revolution* (2017).

cisation of the side- effects of plastics use – might indeed (be mobilised to) promote bioplastics. This is especially so because, while the prices of fossil based plastics fluctuate with oil prices, biomass prices are more stable, and at larger scales (and more cascading in the use of biomass – JG) they are likely to decrease.[67] To further close the price gap, institutional innovations will be important. These will have to support circular value propositions involving the cascaded use of biomass. Such circular products and their associated markets require a different set of relationships, market rules and funding arrangements than linear chains, focused on a single end product – market combination.

24.4.3 Dematerialisation

In regard dematerialisation strategies, it should first be noted that plastic itself has led to dematerialisation through a 'refuse' strategy, partially replacing metals in e.g. home and car construction. That said, there is significant additional potential for dematerialisation through 'refuse' or 'rethink,' especially in regard to plastic's short-lived applications. In packaging, for instance, alternatives like re-usable packaging, and house-to-house delivery and simply less packaging (especially by producers and deliverers) are ready to hand.[68] The challenge is mainly cultural. As the Plastic Heroes campaign noted above has shown, it is possible to change behaviour in regard to e.g. re-usable bags. Using less plastic in e.g. food packaging however, is likely to run afoul of cherished convictions, embedded in legislation on food safety.

Other short-lived applications are endemic especially in everyday electronics and electric equipment, which is often replaced every few years. Repair is often possible, the main problem being that disproportionality between labour and materials / energy costs often make it too expensive – although niche activities began to surface in the mid-2010s like Repair Cafés or mutual services trading networks. In a further developed sharing economy, novel opportunities for dematerialisation might arise. The 'millennials' generation may be more prone to such a way of life than earlier generations, though this may extend mostly to costly items, like cars and luxury goods. Finally, by the mid-2010s combining various functions (a calculator, a web browser, a flashlight and a camera) into one artefact (a mobile phone) was widespread. It had, however, not yet led to abandoning the purchase of calculators, laptop computers, stand-alone flashlights and a proper camera. The cultural inclination to own a wide variety of objects is strong.[69]

---

[67] Van den Oever, Martien, Karin Molenveld, Maarten van der Zee, Harriëtte Bos. *Bio-based and biodegradable plastics - Facts and Figures. pp. 10.*

[68] Van Ours, J.C. *Gezinsconsumptie in Nederland 1951–1980.* (Rotterdam: 1985).

[69] Raad voor de Leefomgeving en infrastructuur. *Circulaire economie van wens naar uitvoering.* (Den Haag: 2015).

24.5   Electricity-for-Households

Obviously, the most crucial contemporary problem with electricity consumption by households is its significant contribution to climate change. Since the turn of the twentieth century, this contribution has increased due to nearly universal electrification, i.e. (1) increasing numbers and (2) more intensive use of (3) more kinds of appliances. These effects outweighed countervailing developments: (4) drastically increased efficiency of appliances, (5) a steady reduction of the energy intensity of electricity production over the twentieth century – especially since the mid-1970s, and (6) use of a different fuel mix in electricity production.

As argued in Sect. 23.6, electrification proceeded apace because from the outset electricity was presented as an asset 'waiting to be used', and user experiences were folded back in to promoting and developing new appliances. The economic logic of a network connecting centralised provision to decentralised consumption was another driver. The modern idea of progress brought with it a strong cultural inclination towards the 'electric home.' After the Second World War, domestic practices in the Netherlands became Americanised, and since the 1980s dynamic lifestyles have become unthinkable without audio, video and ICT appliances. These factors continue to drive electrification as audio and video appliances, ICT, communication and auxiliary equipment penetrate into ever expanding markets. Additional new drivers of electrification are trends like the termination of the natural gas network and increasing electric mobility.

One countervailing development, the improved efficiency of appliances, has become institutionally and discursively embedded in EU and national regulation, energy labels, parameters in comparative commodities tests et cetera. Increasing awareness of climate change and the Paris Treaty obligations provide additional momentum. Changes in the electricity generation fuel mix have been historically important as a countervailing development; the contemporary functional equivalent is the shift towards renewable resources. This development too is supported by climate change concerns and treaty obligations. The world of finance may well become an independent countervailing factor – key players, especially after 2015, have begun to abandon the fossil sector. The handwriting on the wall has included, internationally, the Rockefeller family's 2014 divestment of the fossil sector as well as Bloomberg advising investors along similar lines. Domestically, we can point to the decision by a major pension fund (ABP) to withdraw more than €5 billion from the fossil sector, as part of a systematic divestment policy.[70]

It is important to note that the change in the 'fuel mix' of centralised generation is being reinforced by a second development: a very significant increase in the share of decentralised electricity generation by households /neighbourhoods and busi-

---

[70] 'Rockefellers, Heirs to an Oil Fortune, Will Divest Charity of Fossil Fuels', *New York Times*, September 21, 2014; 'Rockefeller family charity to withdraw all investments in fossil fuel companies', *The Guardian*, March 23, 2016; 'Lage grondstofprijzen zet waarde fossiele beleggingen ABP onder druk; *Financieel Dagblad*, March 6, 2016; Corine Wortman-Kool (CEO), *Nieuw belegingsbeleid ABP is breuk met het verleden*. (www.apg.nl/pdfs/abp-pensioendoc_2016.pdf).

nesses / industrial zones as well collaborations among them. These developments have a great deal of momentum, anchored not only in increasing climate awareness and the Paris Treaty, but also in strong cultural inclinations. The latter include an appreciation of autonomy, authenticity and a new sense of inclusive entrepreneurship which integrates economic and societal objectives through novel modes of value creation. Roofs, industrial heritage and abandoned buildings become 'capital' and in the same movement a new understanding of 'the commons' is emerging. Giddens (1991) has interpreted such inclinations as typical of high modernity. They are firmly inscribed in the 'source code' of the generations born after 1980, and were reinforced in the aftermath of the financial crisis.[71] That crisis de-blocked the labour market by engendering more entrepreneurial attitudes, especially amongst starters. It also led to low interest rates and thus improved the 'competitive position' of investments in individual and collective decentralised energy generation. Moreover, it has created a sense among scholars[72] and upcoming social movements[73] that these modes of value creation are a viable and necessary alternative to the market system that evolved since 1945.

These developments have also yielded a novel type of energy actor: prosumers, i.e. citizens, firms and others who both produce and consume electricity. Together with the rather intermittent character of two key renewable electricity providers, sun and wind, this novel type of actor will profoundly change the structure and meaning of the electricity system. Against this background, we will now discuss in greater detail, first, electrification and increasing efficiency of households, and then the changing electricity networks.

---

[71] Van Steensel, Karin. *Internetgeneratie*; *de broncode ontcijferd*. (Den Haag: 2005).

[72] Kemp, R., Strasser, T., Davidson, R., Avelino, F., Pel, B., Dumitru, A., Kunze, I., Backhaus, J., O'Riordan, T., Haxeltine, A. and Weaver, P. ´The humanization of the economy through social innovation.´ *SPRU50* (Brighton: 2016); Van Bavel, Bas *The Invisible Hand? How Market Economies have Emerged and Declined Since AD 500.* (Oxford: 2017); Van der Heijden. *Na het neoliberalisme. Klimaatverandering, sociale bewegingen en politiek.* (Delft: 2017).

[73] Including the transition towns movement in especially the UK and Flanders, the Energiedörfer in Germany, de 'Nederland kantelt' movement in the Netherlands, and the transnational 'city-makers' Peer2Peer movements. Typically, these movements bring together citizens and community organizations, entrepreneurs and experts. Their structure and operations have been extensively documented in scholarly publications, see e.g.

Seyfang, G., Haxeltine, A. Growing grassroots innovations: exploring the role of community-based initiatives in governing sustainable energy transitions, *Environment and Planning C: Government and Policy*, vol. 30 (2012), p. 381–400.; Broto, Vanesa Castán, Harriet Bulkeley. 'A survey of urban climate change experiments in 100 cities', *Global Environmental Change* 23 (2013), p. 92–102.; Seyfang, G., Longhurst, N. Growing green money? Mapping community currencies for sustainable development, *Ecological Economics* vol. 86 (2013), pp. 65–77.; Loorbach, D., Avelino, F., Wittmayer, J.M., Haxeltine, A., Kemp, R. O'Riordan, T. and P. Weaver. The economic crisis as a game-changer? Exploring the role of social construction in sustainability transitions, *Ecology & Society*, 21 (2016), 4.; Scholl, Christian & René Kemp. City Labs as vehicles for innovation in urban planning processes, *Urban Planning*, 1(2016) 4: 89–10.; Longhurst, N., Avelino, F., Wittmeyer, J., Weaver, P., Dumitru, A., Heilscher, S., Cipolla, C., Afonso, R., Kunze, I. and Elle, M. (2017). *"New economic" logics and urban sustainability transitions.* (TRANSIT working paper # 8).

24.5.1 Electrification and Increasing Efficiency, 2010–2050

By the mid-2010s, at least five (relatively) novel modes of electrification may be discerned. Two are still emergent and associated with the necessary *phasing out of fossil fuels*. First, electric cars were declared the key long-term option for individual transportation in the Fuel Vision (2015) one of the elaborations of the 2014 multi-lateral Energy Agreement. The implementation of this agreement became more urgent in June, 2015, after a sustainability organization, Urgenda, won a court case mandating the government to intensify its efforts.[74] The conclusion of the Paris Treaty in November of that year added fuel to this fire. Widespread use of electric cars will both generate a significant extra demand for electricity, in part in households, as well as add considerable storage capacity to the electricity network.

The phasing out of natural gas, partly to combat climate change and partly because of public outrage about earthquakes above empty gas fields, is firmly anchored in the government's 2016 Energy Agenda, stipulating that by 2050, the natural gas grid is to be terminated. The most likely scenario will be a shift from 'heating' to 'maintaining temperature'. Homes will enjoy more rigorous thermal insulation (part of the 2014 Energy Agreement), so much so that less than 10% of current amounts of external energy supply will be needed. Temperature maintenance may rely on solar collectors, geothermal installations, heat pumps, heat nets as well as providing space in homes to heat-generating business assets, such as servers. Heat may be extracted from waste water coming from showers, washing machines et cetera. The associated apparatus, especially heat pumps and boilers for hot water supply, will imply additional electricity consumption. Tailor-made solutions will be needed, given the requirement to make regional heat plans to co-shape spatial planning and, for instance, new businesses ventures and business zoning.

A rare estimate of the increase in electricity demand due to those forms of electrification implicit in phasing out fossil fuels, mentions a 50% increase.[75] Clearly, and as we will elaborate below, achieving this transition will require novel institutional arrangements – and a significant cultural shift: a shift from the belief that 'practices for living and working may assume that energy is abundant' to the new idea that 'society will become co-responsible for energy provision, both in planning and in every-day life.' Experiences in water management, where a similar cultural shift (from controlling the water in order to support social and economic practices, to living with the water) suggest that this will be far from a smooth process. But to the extent that these changes are perceived as yielding more autonomy and space for the entrepreneurial energies of citizens and businesses, this shift may also draw on the new cultural inclinations just mentioned.

---

[74] 'Hague court orders cuts in Dutch carbon emissions', *Financial Times*; John Schwarz 'Ruling Says Netherlands Must Reduce Greenhouse Gas Emissions', *New York Times;* 'Dutch court orders state to slash carbon emissions', *Al Jazeera English Edition* (all dated June, 24, 2015).

[75] de Joode, J. 'The role of power-to-gas in the future Dutch energy system'. *Report ECN_E_14_026.* (Petten: 2014), cited in Daniëls, B. & P. Koutstaal. *De rol van de elektriciteitsvoorziening in het klimaatbeleid.* (Petten: 2016), pp. 7 8. Items included differ somewhat, but 50% is warranted as a rough, indicative estimate.

Other forms of contemporary electrification are related to the *dominant lifestyles of 21st century homo ludens*. The use of domestic ICT and mobile devices is rampant and progressing; the Internet of Things (IoT) looms on the horizon. The growth of electrification due to increasing numbers of laptops, tablets, smart mobile devices and so on per household shows no sign of abating, and novel forms of 'electricity-based leisure' may yield yet further expansion. In all likelihood, the associated lifestyles are here to stay: the desire for self-expression has become a basic need since the 1960s, and these lifestyles reflect a deep-rooted fascination with the technological culture and the cultural inclinations and communication modes driven by long terms trends like individualisation and (transnational) network society.[76] Robotisation may lead to more leisure time. Most of these factors will also promote the Internet of Things, including the robotisation of domestic practices.

History shows that improvements in efficiency may partly compensate for the additional electricity consumption resulting from electrification. Domestic electricity use associated with ICT and communication appliances increased until the early 2000s, but after that was mitigated by improved efficiency. Since 2008, ICT use has shifted from mainly desktop computers to laptops, tablets and smartphones, while new appliances like e-readers also entered the scene. As a result, there are more appliances, although each of them uses less electricity than a desktop. Auxiliary devices such as wifi routers are also swiftly becoming more efficient, partly due to EU regulation.[77]

Extrapolating from these developments, Afman & Scholten foresee that by 2020 we may expect a reduction of about 30% (compared with 2008) in ICT-related electricity consumption (from 2 down to 1,4 TWh/year for all households), and an increase of 30% (from 1 to 1,3 TWh/year) for communications.[78] For 2030, two scenarios are conceivable. First, ICT related electricity consumption will decrease by 15% (down to 1,2 TWh/year) while communication related use will slightly increase (with about 7,5% up to 1,4 TWh/year). Second, the Internet of Things may lead to a doubling of the electricity consumption associated with communication (up to 2,7 TWh/year). The authors consider both scenarios 'plausible'. While cultural analysis is missing, this study of the impact of ICT and the Internet of Things shows that associated changes in electricity consumption depend crucially on the evolution of user practices and parallel improvements in the efficiency of communications devices.

---

[76] Giddens, Anthony. *Modernity and Self-Identity. Self and Society in the Late Modern Age.* (Stanford: 1991).

[77] Afman, M.R. & T. Scholten. *Trends ICT en Energie 2013–2030.* (Delft: 2016), pp. 17–19.

[78] Afman, M.R. & T. Scholten. *Trends ICT en Energie 2013–2030. pp. 50–53.*

24.5.2   Options for the Electricity System

We may conclude that the future of the electricity system is more open than it has ever been since the 1930s. Yet, at a certain level of abstraction, we can identify three features that are likely to emerge.

First, the electricity system will be a mixed network, comprising a wide diversity of central and decentral sources and with many (potentially nearly all) former consumers becoming prosumers.[79] Energy sources will include:

- solar energy (especially decentral: roofs, walls and maybe even roads in the built environment)
- wind energy, both decentral (from farms, villages and industrial parks) and central, especially off-shore (4500 MW, or 4–5 traditional coal/gas/nuclear power plants are planned for 2023)
- hydropower (in the Netherlands, mostly pumped-hydro storage as a medium for supply-demand-match management);
- biomass, with or without carbon capture and storage (CCS). If the biomass is sustainably produced, this may lead to negative CO_2 emissions. The availability of sustainably produced biomass, that is also needed for the bio-based economy, is the most important limiting factor. Novel forms of biomass production, e.g. employing seaweed and algae, may offer partial solutions.
- coal or gas, combined with CCS;
- nuclear (in the Netherlands the most likely way would be through procurement from other EU countries – replacement of the country's sole nuclear plant is politically not salient)
- several relatively new options: osmosis, deep geothermal electricity generation, tidal energy and wave energy (some of these may account for some share by 2050).

Solar and wind will probably form the spine of the future system, but given their intermittent character, several of the other options will be necessary as well. Intermittency may also be dealt with by buffering electricity (e.g. in charged cars, batteries integrated in underground constructions of e.g. a parking garage, and pumped-hydro storage) and by electrifying functions that may be organized in a flexible way, such as producing organic fuels and other chemicals from CO_2 and hydrogen, or producing heat. Realising such provisions will add to energy costs.

The relative contributions that these various sources might make to the electricity mix in 2050 are still unclear. Many analyses are based on a comparison of the costs per ton CO_2 reduced.[80] An alternative logic would consider the costs as investments essential to attaining a competitive position in a new, sustainable economy with

---

[79] Daniëls, B. & P. Koutstaal 'De rol van de elektriciteitsvoorziening in het klimaatbeleid', (Petten: 2016); Blanford, G.J., Aalbers, R.F.T., J.C. Bollen, K. Folmer. 'Technological Uncertainty in Meeting Europe's Decarbonisation Goals', *CPB Discussion Paper 301*, (Den Haag: 2015).

[80] Blanford et al., (2015) 'Technological Uncertainty in Meeting Europe's Decarbonisation Goals'.

moderate sensitivity to international developments. In addition, in the new economy, business models (e.g. for bio-ethanol produced in a cascade) will often be circular rather than linear as assumed in these analyses. Also, they typically assume that investments will be forthcoming from major parties in order to fund the centralised production of electricity. Meanwhile, however, decentralised solar power is mushrooming, thanks to the cultural inclinations and financial conditions discussed above. The only certainty seems that the future system will be much more heterogeneous and flexible than our present one.

Second, citizens and businesses will no longer be able to assume that energy is simply abundant, but will increasingly have to cut their coats according to their cloth. First, the network structure and its management will increasingly depend on regional constraints *and* the other way around. What has said above about heating plans may apply in an analogous way to regional opportunities for solar, wind and other forms of electricity generation. Second, the intermittent and variable nature of sun and wind and the changing demand over the course of days and seasons, make matching supply and demand much more of a challenge.

Finally, at least over the medium-term, there is a tension between further electrification and meeting Paris Treaty objectives. Regarding security and reliability of supply in a mixed network, a lot of questions remain to be answered. The same goes for household demand, the combined effect of increased efficiency and continued electrification. Also, little is certain about the rate of electrification of car transport and heating[81] Other scenarios and visions suggest how society could – and should – deal with its electricity system to facilitate continuing electrification, assuming only 'moderate' use of ICT[82] All scenarios share one salient point: recognition of the tensions between meeting the requirements of the Paris Treaty and ensuring supply security and reliability.

In sum, although major changes are afoot, there is still ample reason to support Ganzevles and Van Est's[83] recommendation that *reducing* electricity consumption should become a national objective, canonizing and reinforcing a development already going on since 2004.[84] In Daniëls & Koutstaal's words: "the development of electricity demand is no longer an exogenous given to be taken into account in electricity generation, but also a consequence of choices on the ways to reduce

[81] Blanford et al. (2015); Schoots, K. and P. Hammingh. *Nationale Energieverkenning 2015.* (Petten: 2015); Daniëls and Koutstaal. 'De rol van de elektriciteitsvoorziening in het klimaatbeleid' (Petten: 2016); Ros, J et al. *Opties voor energie- en klimaatbeleid,* (Den Haag: 2016); Enexis. *Eindeloze energie. Inspiratieboek.* (Den Bosch: 2016); International Energy Agency *World Energy Outlook 2015.* (Paris: 2016).

[82] Greenpeace International, Global Wind Energy Council and Solar Power Europe. *Energy Revolution. A sustainable world Energy Outlook 2015. 100% renewable energy for all.*(Amsterdam: 2015), Urgenda *Nederland 100% duurzame energie in 2030. Het kan als je het wilt! www.urgenda. nl/visie/actieplan-2030/* (Amsterdam: 2016)

[83] Ganzevles, Jurgen en Rinie van Est. *Energie 2030. Maatschappelijke keuzes van nu.* (Den Haag: 2011).

[84] ECN, PBL, CBS and RvO. *Energieverkenning Nederland, 2015.* (Petten: 2015). pp. 74–79.

emissions."[85] Crucially, this also pertains to the various options for circularity discussed in previous sections.

The network's heterogeneity and flexibility, and the need for continuous tuning of demand and supply, require novel modes of network coordination and management. Similarly, business models fitting such a system also need to be developed. Given the importance of place and differences in conceivable economic logics, what eventually may result is a system of different, regionally oriented, systems.

Let us consider one perspective, consisting of four scenarios put forth by Enexis, an incumbent network manager.[86] The scenarios may co-exist, and even (more or less lightly and visibly) overlap. In two scenarios, large-scale, central generation dominates. In new housing estates, storage capacity, heat pumps etc. are installed in the neighbourhoods; the second scenario is appropriate to established, perhaps historic urban quarters, depending almost exclusively on external, centralised resources. Either way, network managers control the system, drawing on a flexible configuration of sources, and performing demand management through smart meters and, sometimes, enforcing limited use. Here, the business model resembles the current one: network managers continuously buy and sell electricity from a variety of sources. These options may draw on incumbent institutional arrangements, but presuppose significant institutional trust, and stable relations between citizens and large players.

In two other scenarios, a regional population takes responsibility to ensure an energy-efficient and comfortable way of life, with an emphasis on regional electricity generation. Homes generate solar energy; heat is provided from biogas produced by local farmers from manure; other, tailor-made innovative solutions are installed; there is ICT-based demand-supply management and a regionally oriented, sharing economy. The region may or may not be coupled to the national grid. In the former case, the community may frequently sell excess electricity at profitable prices to other regions; in the latter case, inhabitants choose a form of autonomy or even autarky. Either way, ICT will be used to optimize the business model and realize a fair distribution of costs and benefits. More radically, distributed management is also very conceivable, based, for instance, on the block chain system. Local communities could thus arrange intermittency management through supply-demand matching and storage among their number, with similar communities elsewhere and even with central providers. Both scenarios may be likely outcomes in regions where inclinations for autonomy are strong and trust in big players in industry and government relatively weak, and where there is some minimal level of social capital.

National and EU political choices will also set important boundary conditions. Innovation policy to promote the development of central and decentral modes of electricity generation; physical planning to enable regionally tailored solutions; and legislation to provide an institutional framework. As of the mid-2010s, government seemed to be favouring 'both' options, both in terms of physical planning (working

---

[85] Daniëls and Koutstaal (2016). pp. 6.
[86] Enexis. *Eindeloze energie. Inspiratieboek.* (Den Bosch: 2016).

both on large scale off-shore wind parks and decentralizing much of the other measures from the Energy Agreement); as well as in terms of legislation (adopting a new framework law on spatial planning, enabling local initiatives by municipalities, citizens and others parties by simplifying the planning process). Innovation policy was still underdeveloped, de facto leaving a lot to the discretion of main market players; while more specific legislation to regulate the electricity market was also still in statu nascendi. Absent more governmental effort, the bets are on the side of the established main players. That said, that prospect may generate a sense of discomfort amongst those with limited trust in main players, and may even foment rebellion among those who are taking energy matters into their own hands.

24.6   Beyond the Welfare Paradox – Towards a New Economy and Society?

The above discussion makes it clear that strategies to resolve contemporary problems of well-being imply vast changes, not only in technologies, but also in social and economic practices of production and consumption. Pursuing them will obviously require a major effort, which will be met with inertia, contestation and resistance. We have weighed the chances of these strategies by considering them in the context of their historically grown structural and spatial embedding, as well as of their long-term sectoral trends. In this concluding section we will add one more generic trend, and discuss two important promises of the transition we have been discussing that, in the most favourable of circumstances, might help generate support for the strategies proposed.

The generic trend is the transition from what Carlota Perez has called the fourth techno-economic paradigm, based on oil, physical mobility on the basis of the combustion engine and mass production and consumption, toward a new paradigm, based on information, network collaboration and tailor- made production as its key characteristics.[87] Clearly, a transition to circular production and smart grids, and to more ongoing alignment between demand and supply, is congruent with this shift, as further argued in Perez.[88] An important qualification she makes is that any acceleration and amplification of this trends depends utterly on governmental agency to effect the necessary institutional transformation.

The first promise associated with the transition discussed is that the idea of 'making ends meet', through circularity and alignment between demand and supply of electricity, may mitigate the source of what economic historian Bas van Bavel has identified as the key source of the repeated failures of free markets – the fact that the

---

[87] Perez, Carlota, *Technological Revolutions and Financial Capital: The Dynamics of Bubbles and Golden Ages*, (Cheltenham: 2002).

[88] Perez, "The financial crisis and the future of innovation: A view of technical change with the aid of history", *Working Papers in Technology Governance and Economic Dynamics*, No. 28, (Tallinn: 2002).

market mechanism focuses on transactions of economic commodities, but hardly on the inputs into the economy: land, financial capital and natural capital.[89]

A second promise, potentially, would be that these strategies might help mitigate the welfare paradox identified in Chap. 22: income and fulfilment of basic needs have significantly increased; yet there are widespread sentiments of dissatisfaction and institutional distrust. It is important to appreciate what goes on beneath the surface of widespread satisfaction. Sociologist Arlie Hochschild, in an important ethnographic study on dissatisfied citizens in Louisiana, finds that these people feel marginalized, a moral minority, 'strangers in their own land.'[90] Their wages remain stable or decline; they experience swift demographic change and they detest a cosmopolitan dominant culture of 'white wine and cappuccino.' Though they suffer from pollution and unemployment due to corporations' behavior, they are neglected by the government, which has failed them and which they perceive to be run by a cosmopolitan elite that sees them as bad and backward. Similarly, Belgian cultural-historian David Van Reybrouck, during a long stay in Germany, witnessed similar feelings among discontented people there – alienation prevailing most of all in the former East.[91] He stresses the fact that these people may vote, but lack opportunities to voice their concerns. Anthropologist and investigative journalist Joris Luyendijk undertook a series of interviews with discontented citizens in the Netherlands, and came to similar findings; he sees a connection with the fact that people are increasingly aware that not only banks (as he had found in an earlier book), but also other corporations, hospitals, and so on have become the private playground of investors, to whom they are mere objects of profit making.[92] Such analyses also echo Van Bavel's findings that the contemporary market system is in deep crisis, because crucial inputs into the economy have been monopolized by an elite. As a consequence, there is barely anything resembling a free market, and society is less and less able to make autonomous choices.[93]

If these are the underlying problems, then Kemp et al.'s proposal to re-embed the economy in society and nature points towards an interesting route for solutions.[94] Building on Polanyi, they observe a triple movement in society: marketisation, the spread of market thinking and coordination throughout society, state-based social protection and (here they extend Polanyi's analysis) a humanisation of the economy, re-embedding it into society and nature.[95] With the latter movement, based on a major European research project on social innovation, they refer to "less hierarchi-

---

[89] Van Bavel, Bas *The Invisible Hand? How Market Economies have Emerged and Declined Since AD 500.* (Oxford: 2017).

[90] Hochschild, Arlie Russell. *Strangers in Their Own Land: Anger and Mourning on the American Right.* (New York: 2016).

[91] Interview with Sander van Walsum in *De Morgen*, October 10, 2017.

[92] Interviews of Luyendijk can be found at: www.kunnenwepraten.nl

[93] Van Bavel, Bas (2017). The Invisible Hand? How Market Economies have Emerged and Declined Since AD 500. (352 p.). Oxford: Oxford University Press.

[94] Kemp, R., Strasser, T., Davidson, R., Avelino, F., Pel, B., Dumitru, A., Kunze, I., Backhaus, J., O'Riordan, T., Haxeltine, A. and Weaver, P. 'The humanization of the economy through social innovation'. Paper presented at *SPRU50*, (Brighton, United Kingdom, 2016).

[95] Karl; Polanyi, *The Great Transformation. The Political and Economic Origins of Our Time.* (Boston: 1944).

cal forms of decision-making in business and public organisations, new and non-profit ownership models, greater self-determination in work, ethics-based forms of consumption (…) commons-based peer production, social enterprises, alternative currencies, communal ways of living and working, the sharing of urban spaces and participatory budgeting (…) in all sectors." (p. 8) It "involves people from different walks of life" and is partly a reaction to marketisation and associated phenomena, and partly an expression of people's need for autonomy, social bonds and meaningful relationships with others, "and desires to engage in meaningful activity (…) and contribute to a better world that is more equal, fair and respectful of people and nature." (p. 11).[96]

It is not difficult to see how elements of the strategies discussed above fit into this picture: re-connecting different agricultural sectors to each other and to ecosystems, re-developing real-estate into shared spaces with communal functions, setting up mixed central-decentral energy systems et cetera. Importantly, Kemp et al. point out that such systems may also fulfill the need for self-determination (understood as autonomy, competence and relatedness) that has been identified as a primary need in a cross-cultural study by Chen et al.[97] Thus, such changes may contribute to reducing feelings of alienation and lack of self-determination that figured so prominently in the above studies into public discontent.

That is the optimist account. In a more pessimist scenario, a darker movement, dominated by radical leaders and subversive forces, wielding internet-based disinformation, may win the battle for people's minds and hearts. As Cox (put it in a remarkably insightful analysis, "Civil society has become the crucial battleground for recovering citizen control of public life."[98] One does not need to be a neo-marxist to share his point that the outcome will depend not only on technical advances but also, and especially, on "resurrecting a spirit of association in civil society together with a continuing effort by the organic intellectuals (…) to think through and act towards an alternative social order at local, regional and global levels."

Literature

Afman, M.R. & T. Scholten (2016). *Trends ICT en Energie 2013–2030*. CE Publication nr. 16.3F48.05 Delft: CE Delft.
Asveld L., R. van Est & D. Stemerding (red., 2011*), Naar de kern van de bio-economie: De duurzame beloftes van biomassa in perspectief*. Den Haag: Rathenau Instituut

---

[96] Kemp, et al. (2016) 'The humanization of the economy through social innovation'.

[97] Kemp, et al. (2016) 'The humanization of the economy through social innovation'; Chen, B., Vansteenkiste, M., Beyers, W., Soenens, B., & Van Petegem, S.. Autonomy in family decision-making among Chinese adolescents: Disentangling the dual meaning of autonomy, *Journal of Cross-Cultural Psychology*. 44 (2013) 7, pp. 1184–1209.

[98] Cox, R. 'Civil Society at the Turn of the Millennium: Prospects for an Alternative World Order', *Review of International Studies* 25 (1999) 1, p. 3–28.

Auld, G. (2015). 'Certification as Governance' in M. Dubnick and D. Bearfield (eds.) *Encyclopedia of Public Administration and Public Policy*, Abingdon: Taylor and Francis.

Auld, G. (2010). 'Assessing certification as governance: effects and broader consequences for coffee', *Journal of Environment & Development*, 19(2), 215–241.

Baltussen W.H.M., M. Kornelis, M.A. van Galen, K. Logatcheva, P.L.M. van Horne, A.B. Smit, S.R.M. Janssens, A. de Smet, N.F. van Zelst, V.M. Immink, E.B. Oosterkamp, A. Gerbrandy, W.B. van Bockel en T.M.L. Pham, (2014). *Prijsvorming van voedsel; Ontwikkelingen van prijzen in acht Nederlandse ketens van versproducten.* LEI Nota14–112. Wageningen: LEI Wageningen UR (University & Research centre),

Baltussen, W.H.M., M.A. Dolman, R. Hoste, S.R.M. Janssens, J.W. Reijs, A.B. Smit, (2016). *Grondstofefficiëntie in de zuivel-, varkensvlees-, aardappel- en suikerketen.* Wageningen, LEI Wageningen UR (University & Research centre), LEI Nota 2016–013.

Beck, Ulrich, (1997): *The re-invention of politics. Rethinking Modernity in the Global Social Order.* Cambridge: Polity Press.

Bengtsson, Simon, Alan Werker, Cindy Visser, Leon Korving (2017). *PHARIO: stepping stone to a sustainable value chain for PHA bioplastic using municipal activated sludge.* Report 2017–5. Amersfoort: STOWA.

Bijleveld, M.M. (Marijn), G.C. (Geert) Bergsma, M. (Marit) van Lieshout (2013). *Milieu-impact van betongebruik in de Nederlandse bouw Status quo en toetsing van verbeteropties.* Publication nr: 13.2828.24. Delft, CE Delft.

Blanford, G.J., Aalbers, R.F.T., J.C. Bollen, K. Folmer. (2015) 'Technological Uncertainty in Meeting Europe's Decarbonisation Goals', *CPB Discussion Paper 301*, Den Haag: CPB

Bocken, Nancy M. P., Ingrid de Pauw, Conny Bakker & Bram van der Grinten (2016), ´Product design and business model strategies for a circular economy´ in *Journal of Industrial and Production Engineering*. 33 (5) published online.

Brauch, Hans Günter, Úrsula Oswald Spring, John Grin, Czeslaw Mesjasz, Patricia Kameri-Mbote, Navnita Chadha Behera, Béchir Chourou, Heinz Krummenacher (Eds., 2009). *Facing Global Environmental Change: Environmental, Human, Energy, Food, Health and Water Security Concepts.* Hexagon Series on Human and Environmental Security and Peace, vol. 4. Berlin: Springer.

Brauch, Hans Günter, Úrsula Oswald Spring, Czeslaw Mesjasz, John Grin, Patricia Kameri-Mbote, Béchir Chourou, Pal Dunay, Jörn Birkmann (Eds., 2011): *Coping with Global Environmental Change, Disasters and Security – Threats, Challenges, Vulnerabilities and Risks.* Hexagon Series on Human and Environmental Security and Peace, vol. 5. Berlin: Springer.

Broto, Vanesa Castán, Harriet Bulkeley (2013). 'A survey of urban climate change experiments in 100 cities', *Global Environmental Change* 23 (1), p. 92–102.

Bulkeley, H, V Castán Broto (2013).'Government by experiment? Global cities and the governing of climate change.' *Transactions of the Institute of British Geographers* vol. 38 (3), 361–375

CBS (2015). *Electriciteit in Nederland.* Den Haag / Voorburg: CBS.

Chen, B., Vansteenkiste, M., Beyers, W., Soenens, B., & Van Petegem, S. (2013). Autonomy in family decision-making among Chinese adolescents: Disentangling the dual meaning of autonomy, *Journal of Cross-Cultural Psychology.* 44 (7), pp. 1184–1209

Clevers, & P. Popma (2009). *Energiemonitor ICT 2008.* Den Haag: Tebodin/Ministerie van Economische Zaken.

Cordell, Dana (2010). *The Story of Phosphorus: Sustainability implications of global phosphorus scarcity for food security.* Linköping: Linköping University Press.

Cox, R. (1999), 'Civil Society at the Turn of the Millennium: Prospects for an Alternative World Order', *Review of International Studies* 25 (1), p. 3–28

CPB (2000). Naar een *efficiënter milieubeleid; een maatschappelijk-economische analyse van vier hardnekkige milieuproblemen.* Den Haag: CPB/SDU.

Daniëls, B. & P. Koutstaal (2016). 'De rol van de elektriciteitsvoorziening in het klimaatbeleid'. ECN-E-16-058bran. Petten: ECN.

Denkgroep Duurzame Veehouderij (2001). *Toekomst voor de veehouderij. Agenda voor een herontwerp voor de sector.* Advice to the Dutch Agricultural Minister.

De Bakker, Erik en Hans Dagevos (2010). *Vleesminnaars, vleesminderaars en vleesmijders. Duurzame eiwitconsumptie in een carnivore eetcultuur.* LEI-rapport 2010–003. Den Haag: LEI.

De Joode, J. (2014). The role of power-to-gas in the future Dutch energy system. Report ECN_E_14_026. Petten: ECN.

De Paauw, K.F.B. & J.M. Bais (1995). *Sectorstudie huishoudens en woningen.* Petten: ECN

De Ridder, Marjolein, Sijbren de Jong, Joshua Polchar and Stephanie Lingemann *Risks and Opportunities in the Global Phosphate Rock Market.* Report No 17-12-12. The Hague: The Hague Centre for Strategic Studies (HCSS).

De Ridder, Marjolein et al. (20 co-authors from HCSS, CE Delft and TN) (2011). *Op weg naar een Grondstoffenstrategie. Quick scan ten behoeve van de grondstoffennotitie.* The Hague: The Hague Centre for Strategic Studies (HCSS).

De Rijk, Timo (1998). Het *elektrische huis: vormgeving en acceptatie van elektrische huishoudelijke apparaten in Nederland.* Rotterdam: Uitgeverij 010

DG Environment of the EU (2011). *Plastic Waste: Ecological and Human Health Impacts.* Science for Environment Policy, In-depth Reports. (web publication)

Ekamper, P. R. van der Erf, N. van der Gaag, K. Henkens, E. van Imhoff & F. van Poppel (2003), *Bevolkingsatlas* van *Nederland – Demografische ontwikkelingen van 1850 tot heden.* NIDI. Main results in English at https://www.nidi.knaw.nl/en/research/pd/120504

Ellen MacArthur Foundation (2013). *Towards a Circular Economy: Economic and business rationale for an accelerated transition,* Report 2013–1, (online publication)

Ellen MacArthur Foundation (2015). *Towards a Circular Economy: Business rationale for an accelerated transition.* (online publication)

ECN, PBL, CBS and RvO (2015). *Energieverkenning Nederland, 2015.* Petten: ECN.

ECN, Energie-Nederland en NetbeheerNederland (2016). *Energietrends 2016.* Petten: ECN.

ENCI Mebin, Sagrex (2016). Sustainability Update. *Het duurzaamheidsbeleid van de bedrijven van HeidelbergCement Benelux.* Maastricht: ENCI.

Enexis (2016). *Eindeloze energie. Inspiratieboek.* Den Bosch: Enexis.

Ford, Simon Mélanie Despeisse (2016), Additive manufacturing and sustainability: an exploratory study of the advantages and challenges, *Journal of Cleaner Production,* Vol. 137, p. 1573–1587

Fransen, Luc (2015). ´The Politics of Meta-Governance in Transnational Private Sustainability Governance´ in *Policy Sciences,* Volume 48, No. 2, June 2015.

Ganzevles, Jurgen en Rinie van Est (ed.) (2011). *Energie 2030. Maatschappelijke keuzes van nu.* Den Haag: Rathenau Instituut.

Giddens, Anthony (1991). *Modernity and Self-Identity. Self and Society in the Late Modern Age.* Stanford, CA: Stanford University Press.

Greenpeace International, global wind energy Council and solar Power Europe (2015). *Energy Revolution. A sustainable world Energy Outloook 2015. 100% renewable energy for all.* Amsterdam: Greenpeace.

Grin, John, Henk van de Graaf & Philip Vergragt (2003). 'Een derde generatie milieubeleid: Een sociologisch perspectief en een beleidswetenschappelijk programma,' *Beleidswetenschap,* jrg. 17, nr. 1, p. 51–72.

Grin, John, (2010). "Understanding Transitions from a Governance Perspective", Part III,) in: Grin, John; Rotmans, Jan; Schot, Johan, 2010: *Transitions to Sustainable Development. New Directions in the Study of Long Term Structural Change.* New York: Routledge. pp. 223–314.

Grin, John. (2012). Changing governments, kitchens, supermarkets, firms and farms: the governance of transitions between societal practices and supply systems, p. 35–56 in Spaargaren et.al., *Food practices in transition: changing food consumption, retail and production in the age of reflexive modernity.* New York: Routledge.

Grin, John, Niki Frantzeskaki, Vanessa Castan Broto and Lars Coenen (2017) 'Sustainability Transitions and the City: Linking to transition studies and looking forward ' chapter 22 (conclusions) in: Niki Frantzeskaki; Vanessa Castan-Broto; Lars Coenen and Derk Loorbach (ed.s), *Urban Sustainability Transitions.* New York: Routledge.

Harmelink, Lex Bosselaar, Joost Gerdes, Piet Boonekamp, Reinoud Segers, Hans Pouwelse, Martijn Verdonk (2012). *Berekening van de CO2-emissies, het primair fossiel energiegebruik en het rendement van elektriciteit in Nederland.* Den Haag: Agentschap NL/CBS/ECN/PBL.

Harmsen, P. en M. Hackmann (2012), *Groene bouwstenen voor biobased plastics. Biobased routes en marktontwikkeling.* Wageningen: Wageningen UR.

Heckenmüller, Markus Daiju Narita, Gernot Klepper (2014). *Global Availability of Phosphorus and Its Implications for Global Food Supply: An Economic Overview.* Kiel Working Paper No. 1897. Kiel: University of Kiel.

Hees, Eric (2013). *Voedsel, grondstoffen en geopolitiek. Rapportage aan het Platform Landbouw, Innovatie & Samenleving.* Culemborg: CLM.

Hochschild, Arlie Russell (2016). Strangers in Their Own Land: Anger and Mourning on the American Right. New York: The New Press.

Hodson, Mike; Marvin, Simon, (2010): "Can cities shape socio-technical transitions and how would we know if they were?" In: *Research Policy*, 39: 477–485.

Hofstra, U, B. van Bree, J. Neele, R. de Wildt (2006). *cenariostudie BSA-granulaten*: aanbod en afzet van 2005 tot 2025. Delft: Rijkswaterstaat

Hui, A. Shove, E. and Schatzki, T. (Eds) (2016) *The Nexus of Practices: Connections, constellations, practitioners,* London: Routledge.

Ike, P (1985) *Grind gegvalueerd : nadere beschouwing van de tot op heden opgestelde prognoses en aanbevelingen voor nieuw op te stellen prognoses.* Delft : Delftse Universitaire Pers.

International Energy Agency (IEA, 2016). *World Energy Outlook 2015.* Paris: IEA.

Janssen-Jansen, L., & Lloyd, G. (2012). Property booms and bubbles: a demolition strategy – towards a tabula. rasa? *Journal of Surveying, Construction and Property*, 3(2), p. 1–11.

Janssen-Jansen, L., Lloyd, G., Peel, D., & van der Krabben, E. (2012). *Planning in an environment without growth: invited essay for the Raad voor de leefomgeving en infrastructuur(Rli), the Netherlands.* The Hague: Raad voor de leefomgeving en infrastructuur.

Jeeninga, H. (1997). *Analyse energieverbruik sector huishoudens, 1982–1996. Achtergronddocument bij het rapport. 'Monitoring energieverbruik en beleid Nederland'.* Petten: ECN.

Jetten, L., B. Merkx, J. Krebbekx en G. Duivenvoorde (2011), *Onderzoek kunststof afdankstromen in Nederland,* Den Haag: Berenschot/Agentschap NL/DPI Value Centre/Federatie NRK.

Kemp, R., Strasser, T., Davidson, R., Avelino, F., Pel, B., Dumitru, A., Kunze, I., Backhaus, J., O'Riordan, T., Haxeltine, A. and Weaver, P. (2016). The humanization of the economy through social innovation. Paper presented at *SPRU50 anniversary conference.* Brighton: SPRU.

Keuchenius, Cecilia en Bram van der Lelij (2015). *Motivaction Quickscan 2015: eetpatronen van verschillende sociale milieus, duurzaamheid en voedselverspilling.* Den Haag: Voedingscentrum.

KIDV (Kennisinstituut Duurzame Verpakkingen), (2015). *Minder tasjes, duurzaam (her)gebruik en goed gedrag. Project 'Verminderen milieudruk (plastic) draagtassen in het winkelkanaal'.* Den Haag: KIDV.

KIDV & SNM (Kennisinstituut Duurzame Verpakkingen.& Stichting Natuur & Milieu; with support from research and consultancy agency CE Delft) (2016). *Factcheck Plastic Recycling.* Den Haag: Kennisinstituut Duurzame Verpakkingen.

Klintman, Mikael; Boström, Magnus (2012). 'Political Consumerism and the Transition Towards a More Sustainable Food Regime Looking Behind and Beyond the Organic Shelf' in: Spaargaren, Gert; Loeber, Anne; Oosterveer, Peter (Eds.). *Food Practices in Transition: Changing Food Consumption, Retail and Production in the Age of Reflexive Modernity*: New York: Routledge.

Lintsen, H., Hollestelle M., Hölsgens R. (2017) *The plastics revolution. How the Netherlands became a global player in plastics.* Eindhoven: Stichting Historie der Techniek.

Lissandrello, Enza & John Grin (2011). Reflexive planning as design and work: lessons from the Port of Amsterdam, *Planning Theory and Practice*, 12 (2), p. 223–248.

Longhurst, N., Avelino, F., Wittmeyer, J., Weaver, P., Dumitru, A., Heilscher, S., Cipolla, C., Afonso, R., Kunze, I. and Elle, M. (2017). *"New economic" logics and urban sustainability transitions.* TRANSIT working paper # 8. (online publication).

Loorbach, D., Avelino, F., Wittmayer, J.M., Haxeltine, A., Kemp, R. O'Riordan, T. and P. Weaver (2016). The economic crisis as a game-changer? Exploring the role of social construction in sustainability transitions, *Ecology & Society*, 21(4).

Luykx, Paul en Pim Slot, (ed.) (1997). *Een stille revolutie? Cultuur en mentaliteit in de lange jaren vijftig*. Hilversum: Uitgeverij Verloren.

Mom, G.P.A. & R. Filarski (2008). *Van transport naar mobiliteit: De mobiliteitsexplosie (1895–2005)*. Zutphen: Walburg pers.

Murray, Shannon, Saman Brock & Karen S Seto (2015). 'Urbanization, food consumption and the environment.' Chapter 2 (p. 27–40) in Karen C. Seto, William D. Solecki, Corrie A Griffith (eds.). *The Routledge Handbook of Urbanization and Global Environmental Change*. New York: Taylor & Francis.

Nieuwenhuis, R. & L. Maring (2009). *Naar nieuwe ketens voor het benutten van eendenkroos.*, Utrecht: InnovatieNetwork.

Nossin, P. (2012) *Biopolymeren in breder perspectief. Nut en noodzaak* (no place).

NRK (2016), *Kunststof Recycling in Nederland*, NRK: Den Haag

Oldenziel, R. and Karin Zachman (eds.) (2008). *Cold War Kitchen. Americanization, Technology, and European Users*. Cambridge, MA: The MIT Press.

Oldenziel, R. and Albert de la Bruhèze, A. (2009) 'Theorizing the mediation junction for technology and consumption', in A. Albert de la Bruhèze and R. Oldenziel (eds) *Manufacturing technology, manufacturing consumers: The making of Dutch consumer society*, Amsterdam: Aksant.

Oldenziel, R. & Hård, M. (2012). *Consumers, Tinkerers, Rebels: The People who Shaped Europe*. (Series: Making Europe: technology and transformations, 1850–2000). New York: Palgrave McMillan.

Oosterveer, Peter (2012). 'Restructuring Food Supply: Sustainability and Supermarkets.', ch 7 (p. 153–176) in in G. Spaargaren, P. Oosterveer, & A. Loeber (Eds.), *Food practices in transition: changing food consumption, retail and production in the age of reflexive modernity*. New York: Routledge.

Parfitt, Julian, Marc Barthel, Sarah Macnaughton (2010). 'Food waste within food supply chains: quantification and potential for change to 2050', *Philosophical Transactions of the Royal Society B, Biological sciences,* Volume 365, issue 1554.

Perez, Carlota (2002a). *Technological Revolutions and Financial Capital: The Dynamics of Bubbles and Golden Ages,* Cheltenham: Edward Elgar.

Perez, Carlota (2002b). "The financial crisis and the future of innovation: A view of technical change with the aid of history", *Working Papers in Technology Governance and Economic Dynamics*, No. 28, Tallinn: The Other Canon Foundation/Tallinn University of Technology.

Planbureau voor de Leefomgeving (2017). *Evaluatie Meststoffenwet 2016-Syntheserapport*, Den Haag: Planbureau voor de Leefomgeving PBL, pp. 146–151.

PlasticsEurope (2013). *Plastics – The Wonder Material*. (online publication).

PlasticsEurope and European Plastics Converters Association (EuPC) (2014). *The plastics industry: a strategic partner for economic recovery and economic growth in Europe. Manifesto on the competitiveness of the plastics industry*. Brussels: EuPC

Polanyi, Karl. (1944) *The Great Transformation. The Political and Economic Origins of Our Time.* Boston: Beacon Press.

Port of Amsterdam (2014). *Visie 2030. Port of Amsterdam, Port of partnerships*. Amsterdam: Port of Amsterdam.

Port of Amsterdam (2017). Koers naar de Amsterdam Metropolitan Port. Strategisch plan Havenbedrijf Amsterdam, 2017 — 2021. Amsterdam: Port of Amsterdam.

Potting, J., M. Hekkert, E. Worrell & A. Hanemaaijer (2016): *Circulaire economie: Innovatie meten in de keten*, Den Haag: Planbureau voor de Leefomgeving PBL.

Poulus, Co, Gerard van Leeuwen, David Omtzigt, Kenneth Gopal, Ruud Steijvers, Marnix Koopman (2016). *Tussenrapportage. Prognose bevolking-, huishoudens- en woningbehoefte 2015-2050*. Delft: ABF Research.

Purchase, R. H. de Vriend, H. de Groot (2015), *Kunstmatige fotosynthese. Voor de omzetting van zonlicht naar brandstof.* Leiden: Wageningen UR

Raad voor de leefomgeving en infrastructuur (2015). *Circulaire economie van wens naar uitvoering.* Den Haag: Raad voor de leefomgeving en infrastructuur.

Raad voor Regionaal Veevoer (2016). *Naar 100% regionaal eiwit. Kansen en knelpunten voor eitwitrijke veevoergrondstoffen.* Amsterdam: Milieudefensie.

Rahimi, Aliurezza Jeannette M. Garcia (2017), Chemical recycling of waste plastics for new materials production. *Nature Reviews Chemistry* 1, Article nr 0046 (online journal) doi:https://doi.org/10.1038/s41570-017-0046

Raven, Rob; Schot, Johan; Berkhout, Frans (2012): "Space and scale in socio-technical transitions", in: *Environmental Innovation and Societal Transitions,* 4: 63–78.

Rijksoverheid (2016). Nederland circulair in 2050. Rijksbreed programma Circulaire Economie (www.circulaireeconomienederland.nl). Den Haag: Rijksoverheid.

Rijkswaterstaat (2016). *Samenstelling van het huishoudelijk restafval, sorteeranalyses 2015 ; Gemiddelde driejaarlijkse samenstelling 2014.* Utrecht : Rijkswaterstaat.

Roep, Dirk & Han Wiskerke (2012).,'Reshaping the Foodscape: The Role of Alternative Food Networks.' Ch 9 (p. 207–228) in G. Spaargaren, P. Oosterveer, & A. Loeber (Eds.), *Food practices in transition: changing food consumption, retail and production in the age of reflexive modernity.* New York: Routledge.

Ros, J et al. (2016). *Opties voor energie- en klimaatbeleid,* Den Haag: PBL.

Scholl, Christian & René Kemp (2016), City Labs as vehicles for innovation in urban planning processes, *Urban Planning,* 1(4): 89–102

Schoots, K. & P. Hammingh (2015). *Nationale Energieverkenning 2015.* Petten: ECN.

Segers, R. (2014). *Rendementen en CO_2-emissie van elektriciteitsproductie in Nederland, update 2012.* Den Haag: CBS.

Seyfang, G., Haxeltine, A. (2012). Growing grassroots innovations: exploring the role of community-based initiatives in governing sustainable energy transitions, *Environment and Planning C: Government and Policy,* vol. 30, p. 381–400.

Seyfang, G., Longhurst, N. (2013). Growing green money? Mapping community currencies for sustainable development, *Ecological Economics* vol. 86, pp. 65–77.

Shove, Elisabeth, Frank Trentmann, Richard Wilk (eds.) (2009). *Time, consumption and everyday life. Practice, materiality and Culture.* Oxford: Berg Publishers

Shove, Elisabeth, Matthew Watson, Martin Hand, Jack Ingram (2007). *The Design of Everyday Life.* Oxford: Berg Publishers

Shove, Elisabeth; Walker, Gordon, (2007): 'CAUTION! Transitions ahead: politics, practice, and sustainable transition management' in *Environment and Planning A,* 39,4: 763–770.

Smith, Adrian; Voß, Jan-Peter; Grin, John, (2010): "Innovation studies and sustainability transitions: the allure of adopting a broad perspective, and its challenges", in: *Research Policy,* 39: 435–448.

Spaargaren, Gert & Peter Oosterveer (2010). 'Citizen-consumers as agents of change in globalizing modernity: the case of sustainable consumption'. *Sustainability* vol. 2 no. 7, p. 1887–1908.

Spaargaren, Gert, Peter Oosterveer, & Anne Loeber (Eds.) (2012). *Food practices in transition: changing food consumption, retail and production in the age of reflexive modernity,* New York: Routledge.

Späth, Philipp; Rohracher, Harald, (2010): "'Energy regions': The transformative power of regional discourses on socio-technical futures, in: *Research Policy,* 39,4: 449–458.

Späth, Philipp; Rohracher, Harald, (2014): "Beyond Localism: The Spatial Scale and Scaling in Energy Transitions". in: Padt, Frans J. G.; Opdam, Paul F. M.; Polman, Nico B. P.; Termeer, Kathrien J. A. M. (Eds.): *Scale-sensitive Governance of the Environment.* Oxford: John Wiley. p. 106–121.

Stoeldraijer, Lenny, Coen van Duijn, Corina Huisman (2016). *Kernprognose 2016–2060: 18 miljoen inwoners in 2034 voorzien.* Den Haag: CBS.

Switzer, Andrew; Bertolini, Luca; Grin, John, (2013): "Transitions of Mobility Systems in Urban Regions: A Heuristic Framework", in: *J. Environmental Policy and Planning*, 15,2: 141–160.

Switzer, Andrew, Luca Bertolini & John Grin (2015). 'Understanding transitions in the regional transport and land-use system: Munich 1945–2013', *Town Planning Review*, vol. 86 no. 6. p. 699–723.

The European House Ambrosetti (2013). *Excellence* of the *Plastics Supply Chain in Relaunching Manufacturing in Italy and Europe.* Milano: Ambrosetti.

Truffer, Bernhard, Coenen, Lars, (2012): "Environmental Innovation and Sustainability Transitions in Regional Studies", in: *Regional Studies*, 46,1: 1–21.

Urgenda (2017). *Nederland 100% duurzame energie in 2030. Het kan als je het wilt!* www.urgenda.nl/visie/actieplan-2030/ (online publication)

Vahl, Harry (2009). *Alternatieven voor Zuid-Amerikaanse soja in Veevoer. Natuur en Milieu.* Utrecht: Natuur en Milieu.

Van Bavel, Bas (2017). *The Invisible Hand? How Market Economies have Emerged and Declined Since AD 500.* (352 p.). Oxford: Oxford University Press.

Van Dam, Jan & Martien den Oever (2012). *Catalogus Biobased bouwmaterialen. Het groen bouwen.* Wageningen: Wageningen UR.

Van den Oever, Martien, Karin Molenveld, Maarten van der Zee, Harriëtte Bos (2017). *Bio-based and biodegradable plastics – Facts and Figures.* report number 1722. Wageningen: Wageningen Food & Biobased Research,.

Van der Bie, R.J. & J.P. Smits (2001). *Tweehonderd jaar. statistiek in tijdreeksen. 1800–1999.* Den Haag/Groningen: CBS/RU Groningen.

Van der Heijden (2017). *Na het neoliberalisme. Klimaatverandering, sociale bewegingen en politiek.* Delft: Eburon.

Van der Hoeven, Diederik en Paul Reinshagen (2013). *Biomaterialen, drijfveer voor een groene economie. Strategie voor een groene samenleving.* Den Haag: Wetenschappelijke en Technologische Commissie voor de biobased economy.

Van Lieshout, Marit (2015) Update *Prioritering handelingsperspectieven verduurzaming betonketen 2015*, nr: 15.2A59.22. Delft: CE Delft

Van Ours, J.C. (1985) *Gezinsconsumptie in Nederland 1951–1980.* Rotterdam: 010 Uitgeverij

Van Steensel, Karin (2005). *Internetgeneratie; de broncode ontcijferd.* Den Haag: SMO.

Van Zeijts, H, M.M. van Eerdt, W.J. Willems, G.A. Rood, A.C. den Boer, D.S. Nijdam (2010). *Op weg naar een duurzame veehouderij. Ontwikkelingen tussen 2000 en 2010.* Den Haag/ Bilthoven: Planbureau voor de Leefomgeving (PBL).

Weterings, Rob, Elsbeth Roelofs, Roald Suurs, Frans van der Zee (2011). *Tussen gouden bergen en groene business. Systeemverkenning van een biobased economy.* Den Haag: TNO / Hague Centre for Strategic Studies

World Economic Forum (wef), Ellen MacArthur Foundation and McKinsey & Company, (2016). *The New Plastics Economy — Rethinking the future of plastics* (http://www.ellenmacarthur-foundation.org/publications). (online publication)